High-Speed Penetration Dynamics

Engineering Models and Methods

High-Speed Penetration Dynamics

Engineering Models and Methods

Gabi Ben-Dor
Anatoly Dubinsky
Tov Elperin
Ben-Gurion University of the Negev, Israel

World Scientific
NEW JERSEY • LONDON • SINGAPORE • BEIJING • SHANGHAI • HONG KONG • TAIPEI • CHENNAI

Published by

World Scientific Publishing Co. Pte. Ltd.
5 Toh Tuck Link, Singapore 596224
USA office: 27 Warren Street, Suite 401-402, Hackensack, NJ 07601
UK office: 57 Shelton Street, Covent Garden, London WC2H 9HE

British Library Cataloguing-in-Publication Data
A catalogue record for this book is available from the British Library.

HIGH-SPEED PENETRATION DYNAMICS
Engineering Models and Methods

ISBN 978-981-4439-04-6

Printed in Singapore by World Scientific Printers.

Preface

The subject of this book is related to the previous monograph by the authors (Ben-Dor et al., 2006a). However, the contents of the book and the amount of included material are completely different.

In this book we consider normal high-velocity penetration (with zero angle of attack) of rigid (non-deformable) projectiles into a shield. High-speed penetration is accompanied mainly by local interaction of a striker with a shield and corresponds to the range of impact velocities from several hundreds up to 1,500 m/s. The case of relatively small impact velocities when a striker interacts with a whole plate and hypervelocity penetration that can be described by hydrodynamic models, are not considered in the monograph.

In writing the book the authors pursued two goals: (i) to give a comprehensive description of the engineering models for calculating high-speed penetration of rigid projectiles into various media (concrete, metals, geological shields) and (ii) to demonstrate the capabilities and efficiency of using approximate models for investigating conventional and non-standard problems of penetration mechanics.

The first goal is accomplished by surveys which describe empirical (phenomenological) and analytical models of penetration into concrete, metals and geological shields. This part of the book (more than a quarter of the book) is in fact a handbook since it contains a very detailed summary of the models that allows their use for practical calculations.

The second goal is achieved by presenting the results of the investigations which the authors conducted during recent years. Presentation includes comprehensive surveys of the state-of-the art in relevant research fields. The problems considered can be separated into two large classes. The first class includes the traditional problems where a considerable progress has been achieved during the last years: analysis of the effect of spacing and layering on protective properties of plates and optimization of the multilayered shields; optimization of the shape of the projectiles penetrating into various shields; optimization of two-component ceramic-metal armor. The second class includes nontraditional problems and approaches: modeling of penetration and optimal control of penetrators equipped with jet thruster; investigation of the efficiency and optimization of segmented impactors; new approaches for developing penetration models (inverse problems); estimating ballistic properties of penetrators under the conditions of incomplete information.

Each chapter is accompanied by a list of the notations used repeatedly in this chapter; in addition, generally all notations are explained in the text of the chapter when they are used for the first time. Consequently, the presentation of the material in the book is convenient and allows reading different chapters independently. However, even a cursory examination of Chapters 1 and 2, which present the most widely used universal approaches, is advised for better understanding of some sections.

Unless otherwise indicated SI units are used.

To master the material presented in the monograph, it is sufficient for the reader to be familiar with basic engineering courses in mathematics and mechanics. Nevertheless, if the reader is not interested in the mathematical proofs in the chapters dealing with investigation of applied problems, these proofs can be skipped without impairing an understanding of the obtained results.

The authors believe that this monograph will become a reliable and indispensable reference guide for anyone interested in using engineering models in high-speed penetration mechanics.

The authors are indebted to Mrs.V. Orlov for her help in technical editing of the manuscript and preparation of the camera-ready version.

G. Ben-Dor
A. Dubinsky
T. Elperin

Contents

Part 1

Some Conventional Approaches to Penetration Modeling

Within the broad class of approximate engineering models we distinguish between two sub-classes: empirical (semi-empirical, phenomenological) models and analytical models. Separate chapters are devoted to applications of the models belonging to each of these two subclasses to penetration into concrete, metal and geological shields.

Hereafter we use the terms "empirical model" for the relations (for instance, between "integral characteristics" of penetration, namely, the impact velocity and the depth of penetration for a semi-infinite shield and the ballistic limit velocity and the thickness of the plate for a shield of a finite thickness) that have been obtained by statistical analysis of the experimental results and are not based on the physical laws.

Three remarks are in place in this regard.

Firstly, the analysis of the accuracy of the model, which involves the same data that have been used for model derivation, is not completely correct.

Secondly, while the model with multiple parameters, which is based on a few experimental points, can be quite accurate in these points, the good predictive properties of the model cannot be guaranteed. Consider the limiting case of a model where the number of parameters is equal to the number of the experimental points. Clearly, the errors of the approximation in these experimental points are equal zero, while the accuracy of the model in other points can be very low.

Thirdly, it is not worthwhile to overvalue the statistical estimates of the reliability of the model since many of these estimates are based on a number of assumptions of questionable validity.

Analytical models can be physically substantiated although usually their justification requires a large number of assumptions. We consider mainly the relatively simple engineering models which are characterized by the following features: either they determine the relations between the "integral characteristics" of penetration in the explicit form (in algebraic form or including quadratures) or they describe local interaction between the shield and the penetrator at the points of the penetrator-shield contact surface that yields such relations.

This Part 1 includes two chapters describing general approaches to penetration modeling.

Chapter 1 is dedicated to the localized interaction approach which allows us to derive formulas for the "integral characteristics" (the depth of penetration, the ballistic limit velocity, the residual velocity) applying the most widely used models which describe local interaction between the shield and penetrator at the points of the penetrator-shield contact surface ("local models"). Hence we avoid the necessity of presenting repeatedly formulas for the "integral characteristics" applied for physically different penetration conditions.

Chapter 2 is dedicated to the cavity expansion approach. This approach provides the most widely used and universal method for deriving the "local models" and determining the dependencies of the parameters of the models vs. mechanical properties of the shield material. In this chapter we describe general concepts of derivation of the localized interaction model using the cavity expansion model. Some particular cavity expansion models are described in Chapters 4, 6 and 8.

Chapter 1

Localized Interaction Models (LIMs)

Nomenclature

a_0, a_1, a_2	parameters determining the model;
A_0, A_1, A_2	parameters, Table 1.6.1, Eqs. (1.7.3)–(1.7.5);
b	thickness of the shield;
B_0, B_1, B_2	parameters, Eqs. (1.6.5), (1.6.11), (1.6.16), (1.7.4);
BLV	ballistic limit velocity;
D	drag force acting at the projectile;
D_0	function, Eqs. (1.6.2) and (1.7.2);
D_{flat}	resistance of the flat part of the nose of impactor;
D_{av}, $\tilde{D}_{av}$	average drag force;
DOP	depth of penetration;
e_0, e_2	parameters, Eqs. (1.4.22)–(1.423);
f_0, f_2	functions, Eqs. (1.4.5)–(1.4.8) and (1.4.19);
h	instantaneous depth of penetration, Fig. 1.1.2;
H	depth of penetration (DOP);
k_{cone}	tangent of semi-angle of a conical-shaped nose of projectile;
K	function, Eq. (1.7.10);
K_{CRH}	caliber radius head of a ogive-nosed projectile, Eq. (1.6.28);
L	length of the nose of the impactor, Fig. 1.1.2;
$\overline{L}$	$=L/R$;
L_0	length of a cylindrical part of the impactor, Fig. 1.1.2;
LIM	localized interaction model;

m	mass of projectile;
$\vec{n}$	inner normal vector at the surface of projectile;
Q	function, Eq. (1.4.12);
r	radius of the flat bluntness of the nose of impactor;
$\bar{r}$	$= r/R$;
R	radius of the shank of impactor;
t	time;
T	$= Q(b+L)$, Eq. (1.4.16);
u_0, u_1	function, Eq. (1.3.2);
u	$= -\vec{v}^0 \cdot \vec{n}^0 = \cos\upsilon$, also Eq. (1.3.1), (1.3.24);
U	function, Eqs. (1.3.7) and (1.3.24);
v_{imp}	impact (initial) velocity of projectile;
v	instantaneous velocity of impactor (normal impact);
$\vec{v}$	vector of the surface element velocity of the projectile;
v_{bl}	ballistic limit velocity (BLV) of projectile;
v_{res}	residual velocity of projectile;
V	function describing the dependence v vs. h and v_{imp}, Eq. (1.3.14);
x, ρ, ϕ	cylindrical coordinates associated with the impactor, Fig. 1.1.2;
δ	function in the expression for the drag force acting at the nose of impactor, see Eqs. (1.3.8)–(1.3.9);
θ, Θ	functions determining the impactor-shield interaction surface, Figs. 1.2.1, 1.2.2, 1.2.3, 1.2.4 and 1.2.5, Eqs. (1.2.2) and (1.2.4);
ϑ	semi-vertex angle of cone-nosed impactor;
κ	parameter, Eq. (1.4.22)–(1.4.23);
μ_{fr}	friction coefficient;
$\Xi^{(\Theta)}$	function, Eq. (1.4.29);
$\Xi^{(\theta)}$	function, Eq. (1.4.34);
ρ	see x, ρ, ϕ;
ρ_{og}	radius of the circle of ogive, Fig. 1.6.1;
$\vec{\tau}$	tangent vector at the projectile surface;

υ	angle between the vector $\vec{n}$ and the vector $(-\vec{v})$, Fig. 1.1.1;
ϕ	see x, ρ, ϕ;
φ	function, Eq. (1.4.21);
Φ	function determining the shape of the impactor, Fig. 1.1.2;
Φ_x	$= \partial\Phi/\partial x$ or $d\Phi/dx$;
Φ_ϕ	$= \partial\Phi/\partial\phi$;
Ψ_1	parameters, Eqs. (1.7.6) and (1.7.8);
Ψ_2	parameters, Eqs. (1.6.36), (1.6.39) and (1.6.41);
Ω_n, Ω_τ	functions determining the LIM;
Ω_0	function, Eq. (1.1.6);
Superscripts	
0	unit vector.

1.1 Basics of the Localized Interaction Theory

Many engineering models for penetration modeling belong to the category of the so-called localized interaction models (LIMs), in which the integral effect of the interaction between a shield and a penetrating projectile is described as a superposition of the independent local interactions of the projectile surface elements with the shield. Every local interaction is primarily determined by the local velocity of the surface element and the angle between the local surface velocity vector and the local normal vector to the projectile surface as well as by some global parameters that take into account the integral characteristics of the shield (e.g., hardness, density, etc.).

The unified description of different versions of LIM which are widely used in the penetration mechanics can be presented in the following form:

$$d\vec{F} = \begin{cases} [\Omega_n(u,v)\vec{n}^0 + \Omega_\tau(u,v)\vec{\tau}^0]ds & if \quad u_* < u < 1 \\ \Omega_n(1,v)\vec{n}^0 ds & if \quad u = 1 \\ 0 & if \quad u \le u_* \end{cases} \quad (1.1.1)$$

$$\vec{\tau}^0 = -(\vec{v}^0 + u \cdot \vec{n}^0)/\sqrt{1-u^2}\,, \quad (1.1.2)$$

$$u = -\vec{v}^0 \cdot \vec{n}^0 = \cos\upsilon\,, \quad (1.1.3)$$

where (see Fig. 1.1.1) $d\vec{F}$ is the force acting at the surface element dS of the projectile that is in contact with the shield, $\vec{n}^0$ and $\vec{\tau}^0$ are the inner normal and tangent unit vectors at a given location on the projectile surface, respectively, $\vec{v}^0$ is an unit vector of the surface element velocity of the projectile, $\vec{v}$, υ is the angle between the vector $\vec{n}^0$ and the vector $(-\vec{v}^0)$. The non-negative functions $\Omega_n(u,v)$ and $\Omega_\tau(u,v)$, the normal stress and the tangential stress, respectively, determine the model of the projectile-shield interaction and depend also on the parameters that characterize, primarily, the properties of the shield. The unit tangent vector $\vec{\tau}^0$ lies in the plane of the vectors $\vec{v}^0$ and $\vec{n}^0$ and is normal to the vector $\vec{n}^0$; its direction is chosen such that $\vec{v}^0 \cdot \vec{\tau}^0 < 0$, i.e., the friction force is directed in the positive direction of the vector $\vec{\tau}^0$.

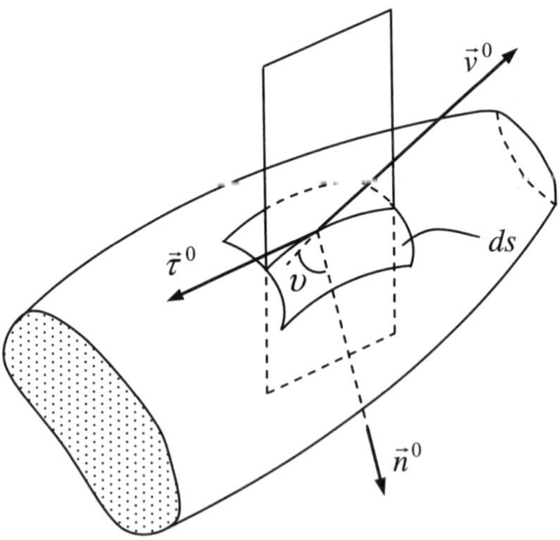

Fig. 1.1.1. Description of the LIM.

Parameter u_* ($0 \le u_* < 1$) determines the maximum magnitude of the angle υ, $\upsilon_* = \cos^{-1} u_*$, whereby the impactor still interacts with the shield. It is assumed that for $\upsilon = \upsilon_*$ the contact between the lateral surface of the impactor and the shield is disrupted. Since the magnitude of υ_* is not known, it is commonly assumed that $u_* = 0$. Taking into account the possibility of "cavitation" (loss of contact between a portion of the lateral surface of the penetrator and the shield) during penetration into solid media is one of the research directions to improve LIMs (Bazhenov and Kotov, 2010, 2011).

The first and the second formulas in Eqs. (1.1.1) describe interaction between the impactor and the shield upon their contact while the third equation determines the condition when there is no contact. The case with $u = 1$ is described separately by the second equation because the choice of the direction of the tangent vector $\vec{\tau}^0$ in this situation is undetermined.

The resultant force acting on the projectile at each instant of time is determined by integrating $d\vec{F}$ over the surface of the projectile-shield contact at the same instant, S. Let us now consider a normal impact (impact velocity is normal to the impacted plate) of a rigid "symmetric body", whereby projectile executes a translational motion under the effect of the drag force D, which can be written in the general case as follows:

$$D = (-\vec{v}^0) \cdot \iint_S d\vec{F} = \iint_{S_{perp}} (-\vec{v}^0) \cdot d\vec{F} + \iint_{S_{lat}} (-\vec{v}^0) \cdot d\vec{F} \,, \qquad (1.1.4)$$

where S_{perp} is a part of the contact surface S that is normal to the impactor velocity (as a rule, the flat bluntness) and S_{lat} is the lateral surface of the impactor. Substituting $d\vec{F}$ from Eq. (1.1.1) into Eq. (1.1.4), we obtain:

$$D = \Omega_n(1, v) \iint_{S_{perp}} dS + \iint_{S_{lat}} \Omega_0(u, v) dS \,, \qquad (1.1.5)$$

where

$$\Omega_0(u,v) = u\Omega_n(u,v) + \sqrt{1-u^2}\,\Omega_\tau(u,v). \qquad (1.1.6)$$

Hereafter we consider normal penetration and use the following notations (Fig. 1.1.2). The coordinate h, the instantaneous depth of penetration, is defined as the distance between the nose of the impactor and the front surface of the shield, and L is the length of the nose of the impactor. The cylindrical coordinates x, ρ, ϕ are associated with the impactor, and the equation $\rho = \Phi(x,\phi)$, where Φ is some function, determines the shape of the impactor.

Generally, we consider impactors with flat bluntness and a cylindrical part of the length L_0, and assume that this cylindrical part does not interact with the shield. In other words, all the above formulas refer to the nose of the impactor which is located between the cross-sections $x = 0$ and $x = L$.

In spite of the presence of the cylindrical part sometimes for brevity we will use the term "conical impactor" or "cone" instead of "conical-nosed impactor".

History of development of LIT is surveyed in our earlier monograph (Ben-Dor *et al.*, 2006a). Emergence and development of LIT in the field of gas dynamics is comprehensively described in the monographs by Alekseeva and Barantsev, 1976; Bunimovich and Dubinsky, 1995 and

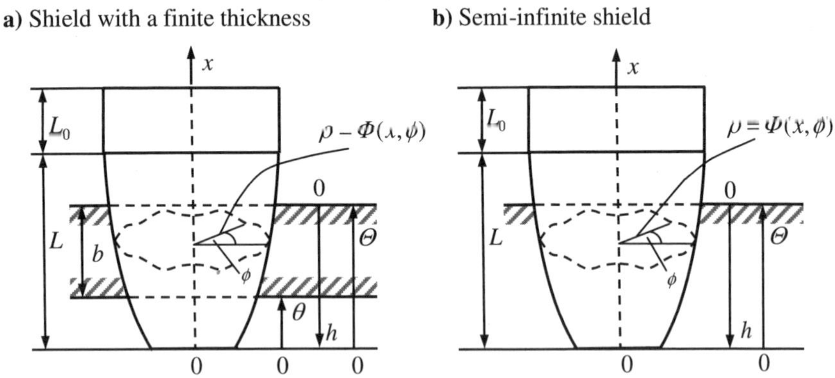

Fig. 1.1.2. The notations.

Miroshin and Khalidov, 1991, 2002, as well as in the reviews of Bunimovich and Dubinsky, 1996, 1997. Progress achieved by LIT in gas dynamics stimulated attempts to extend this approach also to penetration mechanics. Notably, a number of monographs published in the last decade of the 20th century (e.g. Bunimovich and Dubinsky, 1995; Vedernikov and Shchepanovsky, 1995; Vedernikov *et al.*, 1995; Ostapenko, 1997) have separate sections dealing with application of the approximate models in both fields, gas dynamics and penetration mechanics.

A similar to LIT concept forms the basis for the approach that is known in the penetration mechanics as the differential area force law (DAFL). This model was proposed by the AVCO Corporation in the early 1970s (Hadala, 1975; Heuzé, 1990, Bernard and Creighton, 1979). In this method the projectile is divided into a number of small sub-regions, and the total force acting on the impactor is obtained by summing (integrating) the forces acting at the sub-regions.

Regarding the particular realization of this procedure for penetration into soils (actually it was developed for this purpose) it must be noted that DAFL differs significantly from LIT: (i) expression for the local force explicitly depends on the instantaneous depth of penetration and (ii) analytical expression for the local force is quite involved. This is the reason that DAFL does not allow deriving explicit formulas for the penetration depth, and is mainly applied in numerical simulations of penetration for the non-normal impact (Heuzé, 1990).

A similar to LIT approach can be also found in the study by Recht, 1990.

Hereafter we consider comprehensively the two-term and three-term models which describe penetration of rigid strikers having the shape of bodies of revolution. Additional formulas for the design of strikers having different shapes and descriptions of different classes of models can be found in our earlier monograph (Ben-Dor *et al.*, 2006a).

1.2 Impactor-Shield Interaction Surface

1.2.1 *Semi-infinite shield*

The formalism of the description of the impactor-shield interaction surface in the case of a semi-infinite shield is illustrated in Fig. 1.2.1. Generally, two stages of penetration can be considered. The first stage, entry into the shield, occurs when $0 \le h \le L$ (Fig. 1.2.1a). In this case, the flat bluntness of the impactor (if any) and the part of its lateral surface between the cross-sections $x = 0$ and $x = h$ interact with the shield. The second stage (Fig. 1.2.1b), i.e., motion inside the shield, is characterized by full immersion of the bluntness and of the lateral surface of the impactor into the shield and occurs when $x \ge L$. Therefore, the moving area of the impactor-shield interaction can be described as follows (see Fig. 1.2.2):

$$0 \le x \le \Theta(h), \tag{1.2.1}$$

where

$$\Theta(h) = \begin{cases} h & if\ 0 \le h \le L \\ L & if\ h \ge L \end{cases} \tag{1.2.2}$$

In some instances, it is convenient to define function $\Theta(h)$ also for the negative values of the variable h, assuming that $\Theta(h) = 0$ for $h < 0$.

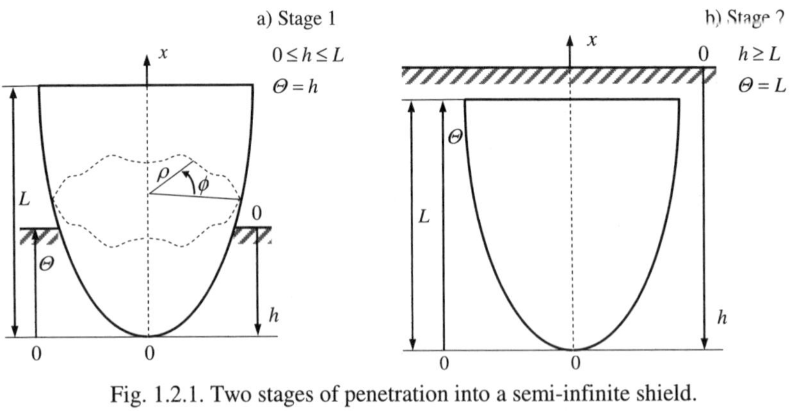

Fig. 1.2.1. Two stages of penetration into a semi-infinite shield.

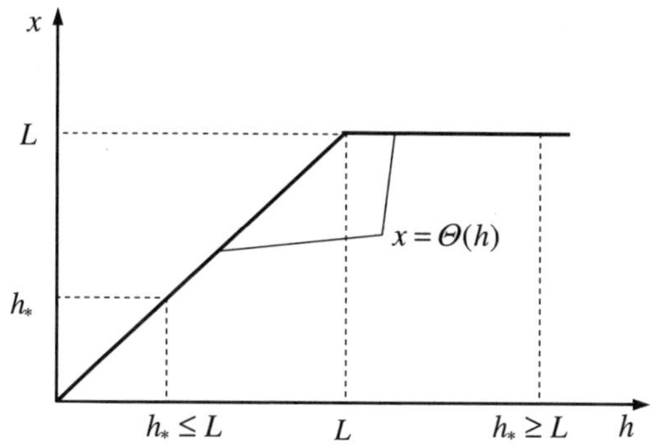

Fig. 1.2.2. Penetration into a semi-infinite shield. Description of the area of impactor shield interaction.

1.2.2 *Shield having a finite thickness*

Consider a shield with a finite thickness, b. First, let $L \leq b$. In this case perforation can be considered as a three-stage process shown in Fig. 1.2.3. At the first stage (entry of the impactor into the shield, $0 \leq h \leq L$), the flat bluntness of the impactor (if any) and the part of the lateral surface between the cross-sections $x = 0$ and $x = h$ interact with the shield. At the second stage (full immersion of the impactor into the shield, $L \leq h \leq b$), the flat bluntness of the impactor (if any) and the entire lateral surface ($0 \leq x \leq L$) interact with the shield. At the third stage (emergence of the impactor from the shield, $b \leq h \leq b + L$), the flat bluntness of the impactor (if any) does not interact with the shield, while part of the lateral surface of the impactor between the cross-sections $x = h - b$ and $x = L$ interacts with the shield. The case $L \geq b$ is illustrated in Fig. 1.2.4 and can be analyzed in a similar manner to the case when $L \leq b$. In both cases, the moving area of the impactor-shield interaction can be described as follows (see Fig. 1.2.5):

$$\theta(h) \leq x \leq \Theta(h), \qquad (1.2.3)$$

where function $\Theta(h)$ is defined by Eq. (1.2.2) and

$$\theta(h)=\begin{cases}0 & \text{if} \quad 0\le h\le b\\ h-b & \text{if} \quad b\le h\le b+L\end{cases}. \tag{1.2.4}$$

In some instances, it is convenient to define functions $\theta(h)$ and $\Theta(h)$ for a shield having a finite thickness as follows: $\theta(h)=\Theta(h)=0$ for $h<0$ and $\theta(h)=L$ for $h>b+L$. The latter definition implies that $\theta(h)=\Theta(h)=L$.

Therefore Eq. (1.2.3) can be used as a unified description of the area of the impactor-shield interaction, taking into account that $h\ge 0$ and $\theta=0$ for a semi-infinite shield and $0\le h\le b+L$ and $\theta(h)$ is defined by Eq. (1.2.4) for a shield having a finite thickness.

The model can be simplified if we do not take into account the stage where penetrator is only partially immersed in the shield. Since such simplification is used frequently, in Sections 1.5-1.7 we consider this subclass of the models.

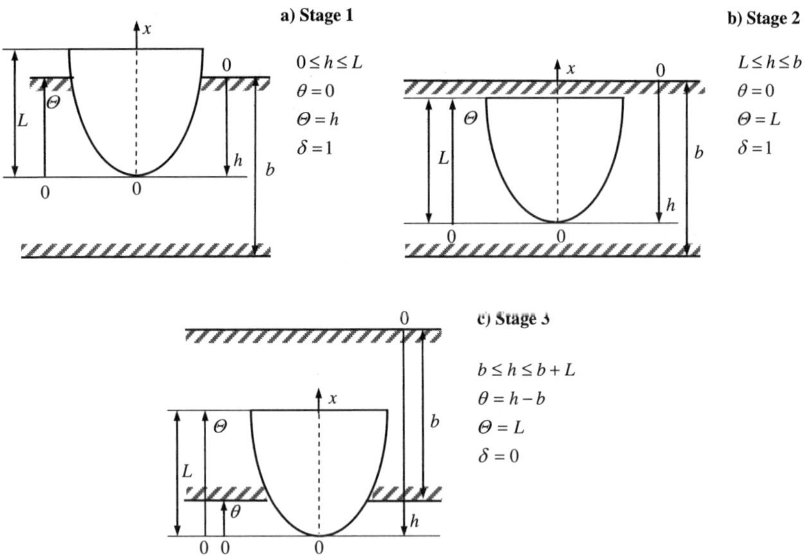

Fig. 1.2.3. Three stages of penetration into a shield having a finite thickness. The case $L\le b$.

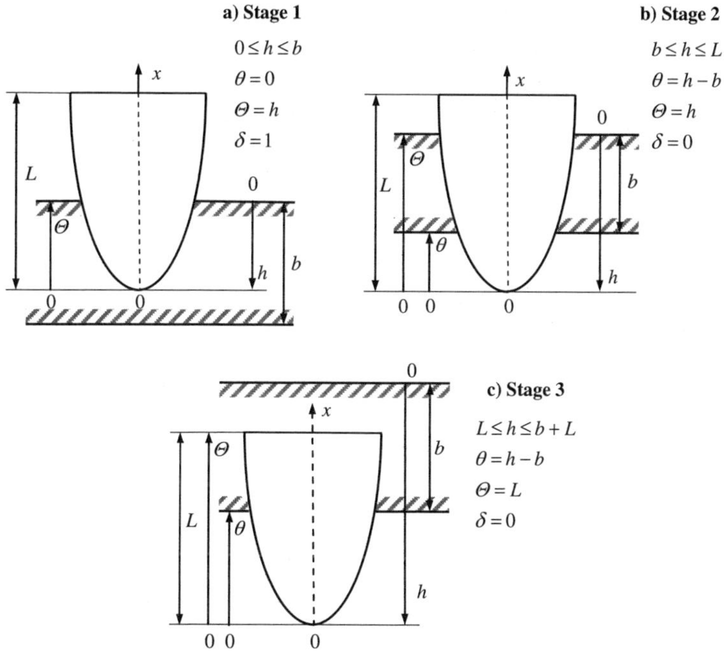

Fig. 1.2.4. Three stages of penetration into a shield having a finite thickness. The case $L \ge b$.

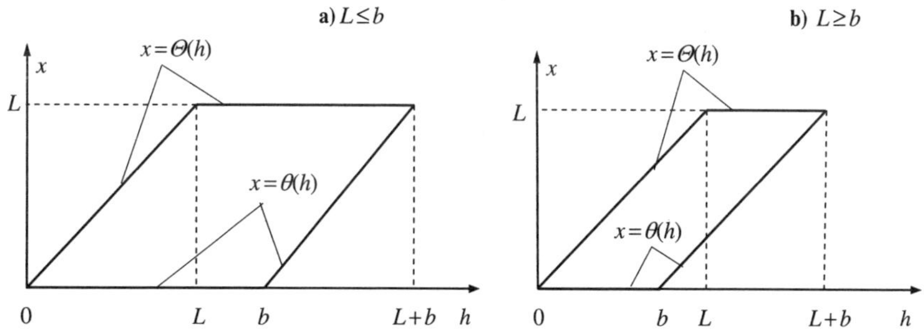

Fig. 1.2.5. Penetration into a shield having a finite thickness. Description of the area of the impactor-shield interaction.

1.3 General Relationships for 3-D Impactors

1.3.1 *Drag force. Equation of motion*

Using the adopted system of coordinates and notations allows us to derive an expression for the drag force. Since

$$u(x,\phi) = \vec{x}^0 \cdot \vec{n}^0 = \frac{u_1(x,\phi)}{u_0(x,\phi)}, \quad dS = u_0(x,\phi)dxd\phi, \tag{1.3.1}$$

$$u_0(x,\phi) = \sqrt{\Phi^2(\Phi_x^2+1)+\Phi_\phi}, \quad u_1(x,\phi) = \Phi\Phi_x, \tag{1.3.2}$$

Eq. (1.1.5) can be transformed into the following form:

$$D(h,v) = \Omega_n(1,v)\sigma_{nose}\,\delta(h) + \int_{\theta(h)}^{\Theta(h)} dx \int_0^{2\pi} \Omega_0\big(u(x,\phi),v\big)u_0(x,\phi)d\phi, \tag{1.3.3}$$

where σ_{nose} is the nose area of the impactor,

$$\sigma_{nose} = \frac{1}{2}\int_0^{2\pi} \Phi^2(0,\phi)d\phi\,. \tag{1.3.4}$$

Friction between the impactor and the shield is taken into account as follows:

$$\Omega_\tau = \mu_{fr}\Omega_n\,, \tag{1.3.5}$$

where μ_{fr} is a friction coefficient. Taking into account Eqs. (1.1.6), (1.3.1)–(1.3.2) and (1.3.5) we can rewrite Eq. (1.3.3) as follows:

$$D(h,v) = \Omega_n(1,v)\sigma_{nose}\,\delta(h) + \int_{\theta(h)}^{\Theta(h)} dx \int_0^{2\pi} \Omega_n\big(u(x,\phi),v\big)U(x,\phi)d\phi, \tag{1.3.6}$$

where

$$U(x,\phi) = [u + \mu_{fr}\sqrt{1-u^2}\,]u_0 = \Phi\Phi_x + \mu_{fr}\sqrt{\Phi^2+\Phi_\phi^2}\,. \tag{1.3.7}$$

Function $\delta(h)$ in Eq. (1.3.6) for the total drag force describes the resistance force acting at the nose of the impactor:

$$\delta(h) = 1 \tag{1.3.8}$$

in the case of penetration into a semi-infinite shield, and

$$\delta(h) = \begin{cases} 0 & if \quad h \le 0 \\ 1 & if \quad 0 \le h \le b, \\ 0 & if \quad h > b \end{cases} \tag{1.3.9}$$

when an impactor penetrates into a shield having a finite thickness. Clearly, we do not consider here penetration phenomena that are accompanied by plug formation.

Equation of motion of the impactor having mass m,

$$m\ddot{h} = -D(h, \dot{h}), \tag{1.3.10}$$

with the initial conditions,

$$h(0) = 0, \quad \dot{h}(0) = v(0) = v_{imp}, \tag{1.3.11}$$

allows us to determine the depth of penetration h and the velocity of the impactor v as a function of time t and impact velocity v_{imp}.

Since the right-hand side of Eq. (1.3.10) does not depend on time in the explicit form, the order of this differential equation can be decreased. Considering v as a function h, $v = v(h)$, and taking into account that

$$\ddot{h} = \frac{d^2h}{dt^2} = \frac{dv}{dt} = \frac{dv}{dh}\frac{dh}{dt} = v'(h)v(h) \tag{1.3.12}$$

we can rewrite Eq. (1.3.10) as follows:

$$mv\frac{dv}{dh} + D(h, v) = 0. \tag{1.3.13}$$

Let

$$v = V(h; v_{imp}) \tag{1.3.14}$$

be the solution of Eq. (1.3.13) with the initial condition

$$v(0) = v_{imp} . \tag{1.3.15}$$

The depth of penetration (DOP) into a semi-infinite shield for a given impact velocity, H , that is defined as the depth at which the impactor slows down to zero velocity, can be found from the following equation:

$$0 = V(H; v_{imp}) . \tag{1.3.16}$$

Once the function $V(h; v_{imp})$ is determined, the law of motion of the impactor can be found as the solution of the differential equation for h with the appropriate initial condition:

$$\frac{dh(t)}{dt} = V(h; v_{imp}), \quad h(0) = 0 . \tag{1.3.17}$$

The latter equation can be written as follows:

$$t(h, v_{imp}) = \int_0^h \frac{dz}{V(z; v_{imp})} . \tag{1.3.18}$$

Equation (1.3.18) allows us to deduce the inverse dependence, h vs. t, if required. Therefore, Eqs. (1.3.14) and (1.3.18) determine the function $v(t)$ in a parametric form.

1.3.2 *Residual and ballistic limit velocities. Depth of penetration*

In the case of a shield having a finite thickness, the ballistic limit velocity (BLV), v_{bl}, is usually defined in analytical models as the initial velocity of the impactor that is required for the impactor to emerge from the shield with zero velocity. Therefore, v_{bl} is determined from the following equation:

$$V(b+L;v_{bl})=0\,. \tag{1.3.19}$$

The expression for the residual velocity, the velocity of the impactor with which the impactor emerges from the shield, reads:

$$v_{res}=V(b+L;v_{imp}),\ v_{imp}\geq v_{bl}\,. \tag{1.3.20}$$

In the case of a semi-infinite shield the depth of penetration (DOP), H, for the known impact velocity, v_{imp}, is found from the equation:

$$V(H;v_{imp})=0\,. \tag{1.3.21}$$

Therefore, general characteristics of penetration can be obtained through solving a first order ordinary differential equation.

1.3.3 *Impactor Having a Shape of Body of Revolution*

1.3.3.1 *General formulas*

If the impactor is a body of revolution then

$$\Phi=\Phi(x)\,. \tag{1.3.22}$$

Consequently,

$$\sigma_{nose}=\pi r^2\,, \tag{1.3.23}$$

where r is radius of the flat bluntness of the impactor nose, and Eqs. (1.3.1) and (1.3.7) imply that

$$u=u(x)=\Phi_x\Big/\sqrt{\Phi_x^2+1}\,,\quad U=U(x)=\Phi(\Phi_x+\mu_{fr})\,. \tag{1.3.24}$$

Using Eqs. (1.3.22)–(1.3.24) the expression for the drag force, Eq.(1.3.6), can be rewritten as follows:

$$\frac{D(h,v)}{\pi}=\Omega_n(1,v)r^2\delta(h)+2\int_{\theta(h)}^{\Theta(h)}\Omega_n\left(\frac{\Phi_x}{\sqrt{\Phi_x^2+1}},v\right)(\Phi_x+\mu_{fr})\Phi dx\,. \tag{1.3.25}$$

1.3.3.2 *Cone-nosed impactor. Finite thickness shield*

If the nose of the impactor has the shape of a sharp cone with semi-angle ϑ then

$$\Phi(x) = \tan\vartheta\, x \tag{1.3.26}$$

and Eq. (1.3.25) can be written as follows:

$$D(h,v) = 2\pi\Omega_n(\sin\vartheta, v)(\tan\vartheta + \mu_{fr})\tan\vartheta \int_{\theta(h)}^{\Theta(h)} x dx\,. \tag{1.3.27}$$

For a finite thickness shield, Eq. (1.3.13) with a drag force D given by Eq. (1.3.27) yields:

$$\int_{v_{res}}^{v} \frac{\tilde{v} d\tilde{v}}{\Omega_n(\sin\vartheta, \tilde{v})} = \frac{2\pi(\tan\vartheta + \mu_{fr})\tan\vartheta}{m}\zeta(h), \tag{1.3.28}$$

where

$$\zeta(h) = \int_{h}^{b+L}\left(\int_{\theta(\tilde{h})}^{\Theta(\tilde{h})} x dx\right) d\tilde{h}\,. \tag{1.3.29}$$

Substituting $v = v_{imp}$ and $h = 0$ into Eq. (1.3.28) we obtain the relationship between v_{imp} and v_{res}:

$$\int_{v_{res}}^{v_{imp}} \frac{v dv}{\Omega_n(\sin\vartheta, v)} = \frac{2\pi(\tan\vartheta + \mu_{fr})\tan\vartheta}{m}\zeta(0), \tag{1.3.30}$$

where

$$\zeta(0) = \int_{0}^{b+L}\left(\int_{\theta(h)}^{\Theta(h)} x dx\right) dh = \int_{0}^{L}\left(\int_{h=x}^{h=x+b} dh\right) x dx = \frac{bL^2}{2}\,. \tag{1.3.31}$$

We changed the order of integration (see Fig. 1.2.5) when $\zeta(0)$ was calculated.

Equation (1.3.30) implies a formula for v_{bl} when $v_{res}=0$ and $v_{imp}=v_{bl}$.

1.4 Projectiles Having a Shape of Bodies of Revolution. Two-Term Models

1.4.1 *Arbitrary body of revolution*

In this section we consider the following class of models:

$$\Omega_n(u,v)=\omega(u)v^2+a_0, \quad \Omega_\tau(u,v)=\mu_{fr}\Omega_n(u,v), \tag{1.4.1}$$

where function $\omega(u)$ and parameter a_0 determine the model.

Along with this general model we also consider two particular models:

$$\omega(u)=a_2u^2 \Rightarrow \Omega_n(u,v)=a_2u^2v^2+a_0 \tag{1.4.2}$$

and

$$\omega(u)=\frac{a_2u^2}{1-u^2} \Rightarrow \Omega_n(u,v)=\frac{a_2u^2}{1-u^2}v^2+a_0, \tag{1.4.3}$$

where two parameters, a_0 and a_2, depend on the mechanical properties of the shield material. These two models are often used in penetration mechanics. In particular, Eq. (1.4.2) and Eq. (1.4.3) are related with spherical and cylindrical two-term cavity expansion models, respectively (see Chapter 2).

Substituting Ω_n from Eq . (1.4.1) into Eq. (1.3.25) we obtain:

$$D(h,v)=f_2(h)v^2+f_0(h), \tag{1.4.4}$$

where in a general case

$$f_0(h)=\pi a_0\left[r^2\delta(h)+2\int_{\theta(h)}^{\Theta(h)}(\Phi_x+\mu_{fr})\Phi dx\right], \tag{1.4.5}$$

$$f_2(h) = \pi \left[\omega(1)r^2\delta(h) + 2\int\limits_{\theta(h)}^{\Theta(h)} \omega\left(\frac{\Phi_x}{\sqrt{\Phi_x^2+1}}\right)(\Phi_x + \mu_{fr})\Phi dx \right]. \quad (1.4.6)$$

In the particular case of the model given by Eq. (1.4.2)

$$f_2(h) = \pi a_2 \left[r^2\delta(h) + 2\int\limits_{\theta(h)}^{\Theta(h)} \frac{\Phi_x^2(\Phi_x + \mu_{fr})\Phi}{\Phi_x^2+1} dx \right], \quad (1.4.7)$$

while in the case of the model given by Eq. (1.4.3) for sharp impactors without flat bluntness

$$f_2(h) = 2\pi a_2 \int\limits_{\theta(h)}^{\Theta(h)} (\Phi_x + \mu_{fr})\Phi_x^2 \Phi dx . \quad (1.4.8)$$

Equation of motion of the impactor, Eq. (1.3.13), with $D(h,v)$ given by Eq. (1.4.4), is a linear ordinary differential equation with respect to v^2:

$$\frac{m}{2}\frac{dv^2}{dh} + f_2(h)v^2 + f_0(h) = 0 . \quad (1.4.9)$$

The solution of this equation with the initial condition

$$v(0) = v_{imp} \quad (1.4.10)$$

reads:

$$v^2(h) = \frac{1}{Q(h)}\left[v_{imp}^2 - \frac{2}{m}\int\limits_0^h f_0(\tilde{h})Q(\tilde{h})\, d\tilde{h} \right], \quad (1.4.11)$$

where

$$Q(h) = exp\left(\frac{2}{m}\int\limits_0^h f_2(\zeta)d\zeta \right). \quad (1.4.12)$$

In the case of a semi-infinite shield, Eq. (1.4.11) yields an equation for the DOP, H, after substituting $v(H) = 0$:

$$\int_0^H f_0(h)Q(h)\,dh = \frac{m}{2}v_{imp}^2, \tag{1.4.13}$$

where Eqs. (1.4.5) and (1.4.7) are used with $\delta(h)=1$.

In the case of a shield having a finite thickness, the BLV, v_{bl}, can be calculated from Eq. (1.4.11) by substituting $h=b+L, v(h)=0$ and $v_{imp}=v_{bl}$:

$$v_{bl}^2 = \frac{2}{m}\int_0^{b+L} f_0(h)Q(h)\,dh\,. \tag{1.4.14}$$

Assuming that $v_{imp} \geq v_{bl}$, we can write Eq. (1.4.11) for the residual velocity, $v_{res} = v(b+L)$ as follows:

$$v_{res}^2 = \frac{1}{T}\left[v_{imp}^2 - \frac{2}{m}\int_0^{b+L} f_0(h)Q(h)\,dh\right], \tag{1.4.15}$$

where

$$T = Q(b+L)\,. \tag{1.4.16}$$

It is assumed that Eq. (1.3.9) is used for calculation of $\delta(h)$ in the case of shield of a finite thickness.

Eqs. (1.4.14) and (1.4.15) imply the following simple relationship between the impact velocity, the residual velocity and the BLV:

$$\left(\frac{v_{imp}}{v_{bl}}\right)^2 - T\left(\frac{v_{res}}{v_{bl}}\right)^2 = 1, \quad v_{imp} \geq v_{bl}\,. \tag{1.4.17}$$

1.4.2 *Sharp conical-shaped impactor*

1.4.2.1 *General equations*

If a nose of projectile has a shape of a conical body of revolution with semi-angle $\vartheta = \tan^{-1} k_{cone}$, then

$$\Phi(x) = k_{cone}x \tag{1.4.18}$$

and formulas for functions $f_0(h)$ and $f_2(h)$ in Eqs. (1.4.5) and (1.4.6) can be written as follows:

$$f_0(h) = e_0\varphi(h), \quad f_2(h) = e_2\varphi(h)\,, \tag{1.4.19}$$

where in general case:

$$\begin{gathered} e_0 = \kappa e_2, \quad e_2 = 2\pi\omega(z)(k_{cone} + \mu_{fr})k_{cone}, \\ \kappa = \frac{a_0}{\omega(z)}, \quad z = \frac{k_{cone}}{\sqrt{k_{cone}^2 + 1}}, \end{gathered} \tag{1.4.20}$$

$$\varphi(h) = \int\limits_{\theta(h)}^{\Theta(h)} x dx = 0.5[\Theta^2(h) - \theta^2(h)]\,. \tag{1.4.21}$$

In the particular case of the model given by Eq. (1.4.2)

$$e_0 = \kappa e_2, \quad e_2 = \frac{2\pi a_2 k_{cone}^3 (k_{cone} + \mu_{fr})}{k_{cone}^2 + 1}, \quad \kappa = \frac{a_0}{a_2}\frac{k_{cone}^2 + 1}{k_{cone}^2}, \tag{1.4.22}$$

while in the case of the model given by Eq. (1.4.3)

$$e_0 = \kappa e_2, \quad e_2 = 2\pi a_2 (k_{cone} + \mu_{fr}) k_{cone}^3, \; \kappa = \frac{a_0}{a_2} k_{cone}^2\,. \tag{1.4.23}$$

Taking into account Eq. (1.4.19) the derivative of the function $Q(h)$ in Eq. (1.4.12) can be written as follows:

$$\begin{aligned} \frac{dQ}{dh} &= \frac{2}{m} f_2(h) \exp\left(\frac{2}{m} \int\limits_0^h f_2(\zeta)\, d\zeta \right) \\ &= \frac{2}{m} f_2(h) Q(h) = \frac{2e_2}{me_0} f_0(h) Q(h) \end{aligned} \tag{1.4.24}$$

and, consequently,

$$\int\limits_0^h f_0(\tilde{h}) Q(\tilde{h})\, d\tilde{h} = \frac{m\kappa}{2} \int\limits_0^h \frac{dQ(\tilde{h})}{d\tilde{h}} d\tilde{h} = \frac{m\kappa}{2} [Q(h) - 1]\,. \tag{1.4.25}$$

Eq. (1.4.11), that determines the dependence v vs. h, reads:

$$v^2(h) = \frac{v_{imp}^2 - \kappa[Q(h)-1]}{Q(h)}, \tag{1.4.26}$$

where

$$\begin{aligned} Q(h) &= \exp\left(\frac{2e_2}{m}\int_0^h d\hat{h} \int_{\theta(\hat{h})}^{\Theta(\hat{h})} x dx\right) \\ &= \exp\left(\frac{e_2}{m}\int_0^h [\Theta^2(\hat{h}) - \theta^2(\hat{h})] d\hat{h}\right) \end{aligned}. \tag{1.4.27}$$

1.4.2.2 *Semi-infinite shield*

In the case of a semi-infinite shield, $\theta(h)=0$, $\Theta(h)$ is determined by Eq. (1.2.2) and the expression for $Q(h)$ can be written as follows:

$$Q(h) = \exp\left(\frac{e_2}{m}\Xi^{(\Theta)}(h)\right), \tag{1.4.28}$$

where

$$\Xi^{(\Theta)}(h) = \int_0^h \Theta^2(\hat{h}) d\hat{h} = \begin{cases} \int_0^h \hat{h}^2 d\hat{h} & \text{if} \quad 0 \le h \le L \\ \int_0^L \hat{h}^2 d\hat{h} + \int_L^h L^2 d\hat{h} & \text{if} \quad h > L \end{cases} \tag{1.4.29}$$

$$= \begin{cases} h^3/3 & \text{if} \quad 0 \le h \le L \\ L^2(3h-2L)/3 & \text{if} \quad h > L \end{cases}.$$

Substituting $v(h)=0$ and $h=H$ into Eq. (1.4.26) we obtain the equation for the depth of penetration, H:

$$Q(H) = 1 + v_{imp}^2 / \kappa, \tag{1.4.30}$$

which has the following solution:

$$H = \begin{cases} \sqrt[3]{\dfrac{3m}{e_2}\ln(1+v_{imp}^2/\kappa)} & if\ 0 \le v_{imp} \le \hat{v}_{imp} \\ \dfrac{m}{e_2 L^2}\ln(1+v_{imp}^2/\kappa)+\dfrac{2}{3}L & if\ v_{imp} > \hat{v}_{imp} \end{cases}, \quad (1.4.31)$$

where

$$\hat{v}_{imp} = \sqrt{\kappa\left[\exp\left(\frac{e_2 L^3}{3m}\right)-1\right]}. \quad (1.4.32)$$

Eq. (1.4.26) describes the dependence v vs. h, where $Q(h)$ is given by Eq. (1.4.28).

1.4.2.3 *Shield having a finite thickness*

In the case of a shield having a finite thickness, the functions $\theta(h)$ and $\Theta(h)$ are determined by Eqs. (1.2.4) and (1.2.2), respectively; the expression for $Q(h)$, Eq. (1.4.27), can be written as follows:

$$Q(h) = \exp\left(\frac{e_2}{m}[\Xi^{(\Theta)}(h) - \Xi^{(\theta)}(h)\right), \quad 0 \le h \le b+L, \quad (1.4.33)$$

where $\Xi^{(\Theta)}(h)$ is determined by Eq. (1.4.29) while

$$\Xi^{(\theta)}(h) = \int_0^h \theta^2(\hat{h})d\hat{h} = \begin{cases} 0 & if\quad 0 \le h \le b \\ \int_b^h (\hat{h}-b)^2 d\hat{h} & if\quad b < h \le b+L \end{cases}$$
$$= \begin{cases} 0 & if\quad 0 \le h \le b \\ (h-b)^3/3 & if\quad b < h \le b+L \end{cases}. \quad (1.4.34)$$

The dependence v vs. h can be obtained by substituting $Q(h)$ from Eq. (1.4.33) into Eq. (1.4.26) and taking into account Eqs. (1.4.29) and (1.4.34).

Since

$$\Xi^{(\theta)}(b+L)=L^3/3,\quad \Xi^{(\Theta)}(b+L)=L^2(3b+L)/3, \tag{1.4.35}$$

then Eq. (1.4.33) implies that

$$Q(b+L)=T,\quad T=\exp(e_2bL^2/m) \tag{1.4.36}$$

and Eq. (1.4.26) allows to determine the expression for the residual velocity:

$$v_{res}^2=v^2(b+L)=[v_{imp}^2-\kappa(T-1)]/T,\quad v_{imp}\geq v_{bl}, \tag{1.4.37}$$

where

$$v_{bl}=\sqrt{\kappa(T-1)} \tag{1.4.38}$$

is the BLV because $v_{res}=0$ when $v_{imp}=v_{bl}$.

1.5 Averaged LIMs. General Approach

1.5.1 *Introduction*

The drag force depends upon the instantaneous penetration depth because the projectile-shield contact area varies at the stage of the incomplete immersion of the projectile in the shield. This dependence renders the model quite involved. The latter shortcoming is usually eliminated by replacing the variable integration limits by constant integration limits in the expression for the drag force, $\theta(h)=0$ and $\Theta(h)=L$, although such simplification can impair the accuracy of the model (Li *et al.*, 2004). Ostapenko *et al.*, 1994 validated the following approximate estimations: the stage of incomplete immersion of the projectile in the semi-infinite shield can be disregarded if $v_{imp}>10(L/R)$ for penetration into soil and $v_{imp}>130(L/R)$ for penetration into metal shields.

In Section 1.5 we show (for strikers having a shape of bodies of revolution) that such simplification can be interpreted as a result of application of a certain averaging procedure. In Section 1.6 we present a comprehensive description of two-term models which are used hereafter. In some cases we use formulas pertinent to the three-term model for penetration into semi-infinite shields; these formulas can be found in Section 1.7.

We start from the Eq. (1.3.25) for the drag force D where the function Ω_n is arbitrary and

$$\Omega_\tau(u,v) = \mu_{fr}\Omega_n(u,v), \quad u = \Phi_x \Big/ \sqrt{\Phi_x^2 + 1}. \tag{1.5.1}$$

Eq. (1.3.25), can be rewritten as follows:

$$D(h,v) = \pi\Omega_n(1,v)\, r^2 \delta(h) + \int_{\theta(h)}^{\Theta(h)} \psi(h,v)dx, \tag{1.5.2}$$

where

$$\psi(x,v) = 2\pi\Omega_n\left(\frac{\Phi_x}{\sqrt{\Phi_x^2+1}}, v\right)(\Phi_x + \mu_{fr})\Phi. \tag{1.5.3}$$

1.5.2 *Shield having a finite thickness*

Assume that the impactor perforates the shield. Let us calculate the spatial average over h $(0 \le h \le b+L)$ of the drag force $D(h,v)$ given by Eq. (1.5.2). The expression for the average drag force, D_{av}, reads:

$$D_{av}(v) = \frac{1}{b+L}\int_0^{b+L} D(h,v)dh = \frac{\pi\Omega_n(1,v)\, r^2}{b+L}\int_0^{b+L}\delta(h)dh$$
$$+ \frac{1}{b+L}\int_0^{b+L} dh \int_{\theta(h)}^{\Theta(h)} \psi(x,v)dx \tag{1.5.4}$$

Taking into account the definition of the function $\delta(h)$ given by Eq. (1.3.9) we obtain:

$$\int_0^{b+L} \delta(h)dh = \int_0^b dh = b\,. \tag{1.5.5}$$

The double integral in right side of Eq. (1.5.4) can be transformed by changing the order of integration (see Fig. 1.2.5):

$$\int_0^{b+L} dh \int_{\theta(h)}^{\Theta(h)} \psi(x,v)dx = \int_0^L \psi(x,v)dx \int_{h=x}^{h=x+b} dh = b\int_0^L \psi(x,v)dx\,. \tag{1.5.6}$$

Using Eqs. (1.5.5)–(1.5.6) we can rewrite Eq. (1.5.4) as follows:

$$D_{av}(v) = \frac{b}{b+L} D_0(v)\,, \tag{1.5.7}$$

where

$$D_0(v) = \pi\Omega_n(1,v)\,r^2 + \int_0^L \psi(x,v)dx\,. \tag{1.5.8}$$

The equation of motion of impactor, $mvdv/dh + D_{av} = 0$, can be written as the following differential equation with separable variables:

$$mv\frac{dv}{dh} + \frac{b}{b+L} D_0(v) = 0\,. \tag{1.5.9}$$

The solution of Eq. (1.5.9) with the initial condition $v(0) = v_{imp}$ reads:

$$\int_v^{v_{imp}} \frac{\tilde{v}\,d\tilde{v}}{D_0(\tilde{v})} = \frac{b}{(b+L)m} h\,. \tag{1.5.10}$$

Substituting $h = b + L,\ v = 0$ and $v_{imp} = v_{bl}$ in Eq. (1.5.10), we obtain the expression for the BLV:

$$\int_0^{v_{bl}} \frac{\tilde{v}\,d\tilde{v}}{D_0(\tilde{v})} = \frac{b}{m}\,. \tag{1.5.11}$$

In order to obtain the expression for the residual velocity let us substitute $h = b + L$ and $v = v_{res}$ in Eq. (1.5.10):

$$\int_{v_{res}}^{v_{imp}} \frac{\tilde{v}d\tilde{v}}{D_0(\tilde{v})} = \frac{b}{m}. \tag{1.5.12}$$

It can be easily shown that formulas for the BLV and for the residual velocity coincide for the "exact" and averaged models if the impactor has the shape of a sharp cone.

1.5.3 *Semi-infinite shield*

Averaging $D(h,v)$ given by Eq. (1.5.2) over h ($0 \le h \le H$), where H, as before, is the DOP and $\delta(h) = 1$, we obtain the expression for the average drag force, $\tilde{D}_{av}$:

$$\begin{aligned}\tilde{D}_{av}(v) &= \frac{1}{H}\int_0^H D(h,v)dh \\ &= \pi\Omega_n(1,v)r^2 + \frac{1}{H}\int_0^H dh \int_0^{\Theta(h)} \psi(x,v)dx.\end{aligned} \tag{1.5.13}$$

Assuming that $H >> L$ and changing the order of integration in the integral in Eq. (1.5.13), we arrive at the following relation (see Fig. 1.2.2):

$$\begin{aligned}\frac{1}{H}\int_0^H dh \int_0^{\Theta(h)} \psi(x,v)dx &= \frac{1}{H}\int_0^L \psi(x,v)dx \int_x^H dh \\ &= \int_0^L \left(1 - \frac{x}{H}\right)\psi(x,v)dx \approx \int_0^L \psi(x,v)dx.\end{aligned} \tag{1.5.14}$$

Therefore we proved that

$$\tilde{D}_{av}(v) = D_0(v) \tag{1.5.15}$$

and the equation of impactor motion can be written as follows:

$$mv\frac{dv}{dh}+D_0(v)=0\,. \qquad (1.5.16)$$

The solution of Eq. (1.5.16) with the initial condition $v(0)=v_{imp}$,

$$\int_{v}^{v_{imp}}\frac{\tilde{v}d\tilde{v}}{D_0(\tilde{v})}=\frac{h}{m}, \qquad (1.5.17)$$

allows us to determine the expression for the DOP setting $v=0$ and $h=H$:

$$\int_{0}^{v_{imp}}\frac{\tilde{v}d\tilde{v}}{D_0(\tilde{v})}=\frac{H}{m}. \qquad (1.5.18)$$

1.6 Averaged Two-Term Models

1.6.1 *General two-term model*

The equation of the general two-term model reads:

$$\Omega_n(u,v)=\omega(u)v^2+\omega_0(u),\quad \Omega_\tau(u,v)=\mu_{fr}\Omega_n(u,v), \qquad (1.6.1)$$

where $\omega(u)$ and $\omega_0(u)$ are the known functions which determine the model.

After substituting Ω_n from Eq. (1.6.1), expression for D_0 from Eq. (1.5.8) can be rewritten as follows:

$$D_0(v)=A_2v^2+A_0, \qquad (1.6.2)$$

where

$$A_0=\pi\left[r^2\omega_0(1)+2\int_0^L\omega_0\left(\frac{\Phi_x}{\sqrt{\Phi_x^2+1}}\right)(\Phi_x+\mu_{fr})\Phi dx\right], \qquad (1.6.3)$$

$$A_2 = \pi\left[\omega(1)r^2 + 2\int_0^L \omega\left(\frac{\Phi_x}{\sqrt{\Phi_x^2+1}}\right)(\Phi_x + \mu_{fr})\Phi dx\right]. \tag{1.6.4}$$

The contribution of the flat part of the nose of impactor, D_{flat}, in the total drag force is as follows ($0 < \omega(1) < \infty$):

$$D_{flat} = B_2 v^2 + B_0,\ B_0 = \pi r^2 \omega_0(1),\ B_2 = \pi r^2 \omega(1), \tag{1.6.5}$$

In Sections 1.6.2-1.6.4 we derive formulas, based on the general two-term model and its sub-models, for the finite width and semi-infinite shields.

1.6.2 *Shield having a finite thickness*

1.6.2.1 *General two-term model*

Eqs. (1.5.10)–(1.5.12) yield the following relations when $D_0(v)$ is given by Eq. (1.6.2):

$$v^2(h) = \left(v_{imp}^2 + \frac{A_0}{A_2}\right)\exp\left(-\frac{2bA_2}{m(b+L)}h\right) - \frac{A_0}{A_2},\quad 0 \le h \le b+L, \tag{1.6.6}$$

$$v_{bl}^2 = \frac{A_0}{A_2}\left[\exp\left(\frac{2bA_2}{m}\right) - 1\right], \tag{1.6.7}$$

$$\begin{aligned} v_{res}^2 &= \left(v_{imp}^2 + \frac{A_0}{A_2}\right)\exp\left(-\frac{2bA_2}{m}\right) - \frac{A_0}{A_2} \\ &= (v_{imp}^2 - v_{bl}^2)\exp\left(-\frac{2bA_2}{m}\right),\quad v_{imp} \le v_{bl} \end{aligned} \tag{1.6.8}$$

where A_0 and A_2 are determined in Eqs. (1.63) and (1.6.4), correspondingly.

1.6.2.2 *Sub-model* $\Omega_n(u,v) = \omega(u)v^2 + a_0$

In the case when $\omega_0(u) = a_0$ (a_0 is a constant),

$$\Omega_n(u,v) = \omega(u)v^2 + a_0, \quad \Omega_\tau(u,v) = \mu_{fr}\Omega_n(u,v)\,, \tag{1.6.9}$$

Eqs. (1.6.4) and (1.6.6)–(1.6.8) do not change, while Eqs. (1.6.3) and (1.6.5) can be written as follows:

$$A_0 = \pi a_0\left[r^2 + 2\int_0^L (\Phi_x + \mu_{fr})\Phi dx\right], \tag{1.6.10}$$

$$D_{flat} = B_2 v^2 + B_0,\, B_0 = \pi r^2 a_0,\, B_2 = \pi r^2 \omega(1)\,. \tag{1.6.11}$$

If friction is not taken into account ($\mu_{fr} = 0$) Eq. (1.6.10) can be written in the following form:

$$A_0 = \pi a_0 R^2\,, \tag{1.6.12}$$

taking into account that

$$\int_0^L \Phi_x \Phi dx = \frac{1}{2}\int_0^L \frac{d\Phi^2}{dx}dx = \frac{1}{2}(R^2 - r^2), \quad R = \Phi(L)\,. \tag{1.6.13}$$

1.6.2.3 *Sub-model* $\Omega_n(u,v) = a_2 u^2 v^2 + a_0$

In the case of the sub-model

$$\Omega_n(u,v) = a_2 u^2 v^2 + a_0\,, \tag{1.6.14}$$

Eqs. (1.6.10) and (1.6.6)–(1.6.8) do not change while

$$A_2 = \pi a_2\left[r^2 + 2\int_0^L \frac{\Phi_x^2(\Phi_x + \mu_{fr})\Phi}{\Phi_x^2 + 1}dx\right], \tag{1.6.15}$$

$$D_{flat} = B_2 v^2 + B_0,\, B_0 = \pi r^2 a_0,\, B_2 = \pi r^2 a_2\,. \tag{1.6.16}$$

If $\mu_{fr} = 0$, then taking into account Eq. (1.6.13) we can rewrite Eq. (1.6.15) in the following form:

$$A_2 = \pi a_2 \left[r^2 + 2\int_0^L \frac{\Phi_x^3 \Phi}{\Phi_x^2 + 1} dx \right] = \pi a_2 \left[R^2 - 2\int_0^L \frac{\Phi_x \Phi}{\Phi_x^2 + 1} dx \right], \quad (1.6.17)$$

while Eq. (1.6.12) remains valid for A_0.

1.6.2.4 *Sub-model* $\Omega_n(u,v) = a_2[u^2/(1-u^2)]v^2 + a_0$

In the case of the sub-model given by Eq. (1.4.3), when

$$\Omega_n(u,v) = a_2[u^2/(1-u^2)]v^2 + a_0, \quad (1.6.18)$$

the formulas in Eqs. (1.6.6)–(1.6.8),) remain valid with

$$A_0 = 2\pi a_0 \int_0^L (\Phi_x + \mu_{fr})\Phi dx, \quad (1.6.19)$$

$$A_2 = 2\pi a_2 \int_0^L (\Phi_x + \mu_{fr})\Phi_x^2 \Phi dx. \quad (1.6.20)$$

If $\mu_{fr} = 0$ then Eq. (1.6.12) for A_0 remains valid.

Since $\omega(1) = \infty$, the model is applicable only for sharp impactors when $r = 0$ or for calculation of the drag force acting the lateral surface of the impactor.

1.6.3 *Semi-infinite shield*

Eqs. (1.5.17) and (1.5.18) yield the following formulas for the general two-term model determined by Eq. (1.6.1):

$$v^2(h) = \left(v_{imp}^2 + \frac{A_0}{A_2} \right) \exp\left(-\frac{2A_2}{m} h \right) - \frac{A_0}{A_2}, \quad (1.6.21)$$

$$H = \frac{m}{2A_2} \ln\left(1 + \frac{A_2}{A_0} v_{imp}^2\right), \qquad (1.6.22)$$

where A_0 and A_2 are determined by Eqs. (1.6.3). and (1.6.4).

In the case of the sub-models considered above, A_0 and A_2 are calculated using the formulas in the corresponding chapters.

1.6.4 *Ogive-shaped impactors*

1.6.4.1 *Description of the ogive shape*

Particular attention in penetration mechanics has been given to ogive-nosed impactors. The generatrix of the nose of ogive-nosed projectiles is an arc of the circle (see Fig. 1.6.1) which is described in the general case of truncated ogive-nosed impactor by the following equation:

$$\Phi(x) = \sqrt{\rho_{og}^2 - (x - L)^2} + R - \rho_{og}, \quad 0 \le x \le L, \qquad (1.6.23)$$

where ρ_{og} is the radius of the circle with the center at the line $x = L$. Eq. (1.6.23) describes the increased convex upwards arc of the circle. Since this arc must intersect the semi-axis of positive ordinate, this implies the following constraint:

$$\rho_{og} \ge L, \qquad (1.6.24)$$

$$r = \Phi(0) = \sqrt{\rho_{og}^2 - L^2} + R - \rho_{og} \ge 0. \qquad (1.6.25)$$

Depending on the relationship between the parameters, ρ_{og}, L and R, there are several possible versions of the nose shape.

If Eq. (1.6.24) is satisfied and $\rho_{og} < R$ then the constraint given by Eq. (1.6.25) is satisfied and $r > 0$, i.e., the impactor has a flat bluntness (truncated ogive) and has the form shown in Fig. 1.6.1a.

If $\rho_{og} \ge R$ then the solution of the inequality in Eq. (1.6.25) reads:

$$\rho_{og} \geq \rho_{og}^{*}, \quad \rho_{og}^{*} = \frac{R^2 + L^2}{2R}. \tag{1.6.26}$$

Since

$$\rho_{og} - L \geq \frac{(R-L)^2}{2R} \geq 0, \tag{1.6.27}$$

the constraint given by Eq. (1.6.24) is satisfied in this case. The shape of the generatrix is shown in Fig. 1.6.1c. In particular, $r = 0$, i.e. projectile has a sharp ogive shape (see Figs. 1.6.1b,c.) when $\rho_{og} = \rho_{og}^{*}$. The projectiles having the shape that corresponds to the case when $\rho_{og} \geq R$ (in particular, for $r = 0$) are most often found in the literature on penetration mechanics.

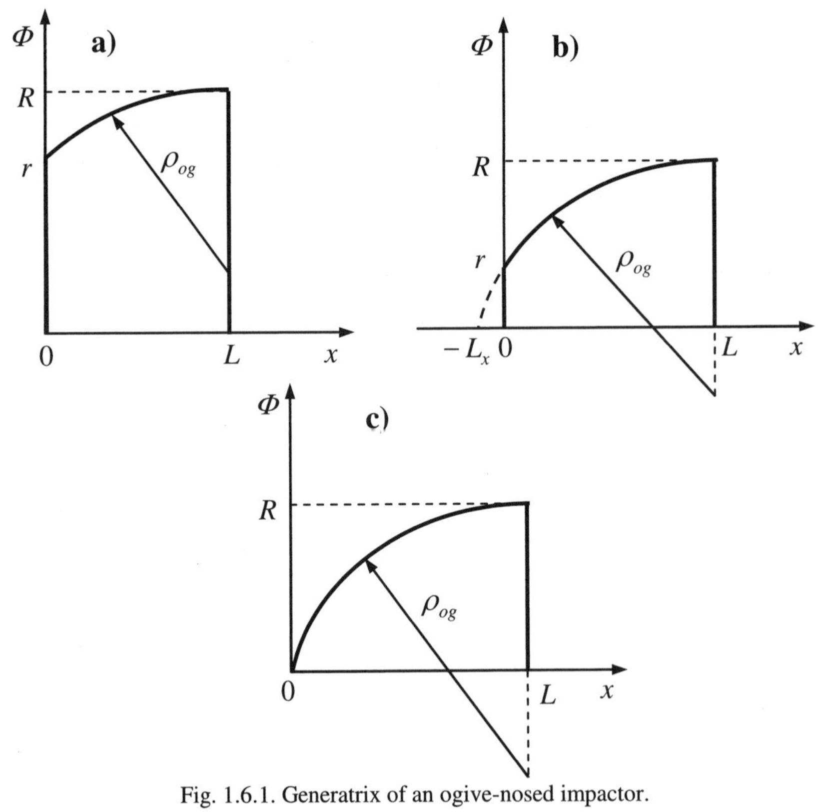

Fig. 1.6.1. Generatrix of an ogive-nosed impactor.

Using the dimensionless parameters

$$K_{CRH}=\frac{\rho_{og}}{2R},\quad \overline{L}=\frac{L}{R},\quad \overline{r}=\frac{r}{R},\tag{1.6.28}$$

where K_{CRH} is the so called caliber radius head, the above analysis shows that the projectile has a shape shown in Fig. 1.6.1a if

$$0.5\overline{L}\le K_{CRH}\le 0.5,\tag{1.6.29}$$

the shape shown in Fig. 1.6.1b if

$$\begin{aligned}&K_{CRH}>\max[0.5,0.25(\overline{L}^2+1)],\\&or\ K_{CRH}>0.25(\overline{L}^2+1)\ while\ \overline{L}\ge 1\end{aligned}\tag{1.6.30}$$

and the shape shown in Fig. 1.6.1c if

$$K_{CRH}=0.25(\overline{L}^2+1)\ while\ \overline{L}\ge 1.\tag{1.6.31}$$

The values, $K_{CRH}=0.5$ and $\overline{L}=1$, correspond to the hemispherical nose.

Therefore the admissible values of K_{CRH} and L must satisfy the following constrains:

$$\begin{cases}0.5\overline{L}\le K_{CRH}\le 0.5 & if\quad 0\le\overline{L}\le 1\\ K_{CRH}\ge 0.25(\overline{L}^2+1) & if\quad \overline{L}>1\end{cases}.\tag{1.6.32}$$

The domain that is determined by these constrains is shown in Fig. 1.6.2.

If the inequalities in Eq. (1.6.32) are satisfied then Eq. (1.6.25) yields the following expression for $\overline{r}$:

$$\overline{r}=1-2K_{CRH}+\sqrt{4K_{CRH}^2-\overline{L}^2}.\tag{1.6.33}$$

1.6.4.2 *Sub-model* $\Omega_n(u,v)=a_2u^2v^2+a_0,\ \mu_{fr}=0$

For the sub-model

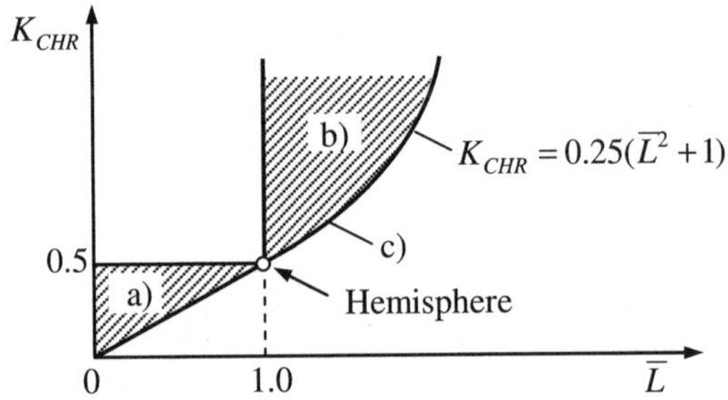

Fig. 1.6.2. Domain on the plane ($K_{CRH}, \bar{L}$) corresponding to the shapes a)-c) of the nose of impactor which are shown in Fig. 1.6.1.

$$\Omega_n(u,v) = a_2 u^2 v^2 + a_0, \mu_{fr} = 0, \tag{1.6.34}$$

in the general case of truncated impactor, integral in Eq. (1.6.17) can be calculated and the expression for A_2 can be written as follows:

$$A_2 = \pi a_2 R^2 \Psi_2, \tag{1.6.35}$$

where

$$\Psi_2 == \frac{(1-\bar{r})^2[3\bar{r}^2 + (2\bar{r}+1)(8K_{CRH} - 1)]}{24K_{CRH}^2}. \tag{1.6.36}$$

For a non-truncated ogive-shaped nose when

$$r = 0, \quad K_{CRH} = 0.25(\bar{L}^2 + 1), \tag{1.6.37}$$

the constrains in Eq. (1.6.32) are satisfied for

$$\bar{L} \geq 1, \text{ or } K_{CRH} \geq 0.5 \tag{1.6.38}$$

and Eq. (1.6.36) yields

$$\Psi_2 = \frac{8K_{CRH} - 1}{24K_{CRH}^2}. \tag{1.6.39}$$

1.6.4.3 *Sub-model* $\Omega_n(u,v) = a_2[u^2/(1-u^2)]v^2 + a_0$, $\mu_{fr} = 0$

For the sub-model

$$\Omega_n(u,v) = a_2[u^2/(1-u^2)]v^2 + a_0,\ \mu_{fr} = 0, \tag{1.6.40}$$

in the case of a non-truncated ogive-shaped nose when the constrains given by Eqs. (1.6.37) and (1.6.38) are satisfied, the integral in Eq. (1.6.20) can be calculated and the result is given by Eq. (1.6.35), where

$$\Psi_2 = \frac{1}{2}(4K_{CRH} - 1)(8K_{CRH} - 3) - 4K_{CRH}^2 \ln\left(\frac{1}{2K_{CRH}} - 1\right), \tag{1.6.41}$$

and $0 < K_{CRH} < 0.5$.

1.6.5 *Summary of two-term models*

Table 1.6.1 Summary of two-term models (BR – arbitrary body of revolution, ON - ogive-nosed body of revolution).

Model number	Ω_p	Impactor's shape	Friction	A_0	A_2
1	$\omega(u)v^2 + \omega_0(u)$	BR		Eq. (1.6.3)	Eq. (1.6.4)
2	$\omega(u)v^2 + a_0$	BR		Eq. (1.6.10)	Eq. (1.6.4)
		BR	$\mu_{fr} = 0$	Eq. (1.6.12)	Eq. (1.6.4)
3	$a_2 u^2 v^2 + a_0$	BR		Eq. (1.6.10)	Eq. (1.6.15)
		BR	$\mu_{fr} = 0$	Eq. (1.6.12)	Eq. (1.6.17)
		ON	$\mu_{fr} = 0$	Eq. (1.6.10)	Eq. (1.6.35), Eq. (1.6.36)
		ON, $r = 0$	$\mu_{fr} = 0$	Eq. (1.6.10)	Eq .(1.6.35), Eq. (1.6.39)
4	$a_2 \frac{u^2 v^2}{1-u^2} + a_0$	BR		Eq. (1.6.19)	Eq. (1.6.20)
		BR	$\mu_{fr} = 0$	Eq. (1.6.12)	Eq. (1.6.20)
		ON, $r = 0$	$\mu_{fr} = 0$	Eq. (1.6.19)	Eq. (1.6.35), Eq. (1.6.40)

For convenience the main formulas for A_0 and A_2 in Sections 1.6.1-1.6.4 are summarized in Table 1.6.1. Characteristics v_{bl} and v_{res} for a finite width shield and H for a semi-infinite shield are calculated in all cases using Eqs. (1.6.7), (1.6.8) and (1.6.22), correspondingly. Blank in the column "Friction" indicates that the model is presented for a general case, $\mu_{fr} \neq 0$.

1.7 Averaged Three-Term Model

Let us consider an averaged three-term model without friction,

$$\Omega_n(u,v) = a_2 u^2 v^2 + a_1 uv + a_0, \quad \mu_{fr} = 0. \tag{1.7.1}$$

For semi-infinite shield, the relationship

$$D_0(v) = A_2 v^2 + A_1 v + A_0 \tag{1.7.2}$$

is valid instead Eq. (1.6.2), where A_0 and A_2 are determined by Eqs. (1.6.12) and (1.6.17) while the formula for A_1 reads:

$$A_1 = \pi a_1 \left[r^2 + 2\int_0^L \frac{\Phi\Phi_x(\Phi_x + \mu_{fr})}{\sqrt{\Phi_x^2 + 1}} dx \right]. \tag{1.7.3}$$

The averaged resistance force acting at the flat part of the nose of a projectile, D_{flat}, is as follows:

$$D_{flat} = B_2 v^2 + B_1 v + B_0, \quad B_i = \pi r^2 a_i, \quad i = 0,1,2. \tag{1.7.4}$$

In the case of truncated ogive-shaped impactors, integral in Eq. (1.7.3) can be calculated ($\mu_{fr} = 0$) and the expression for A_1 can be written in the following form:

$$A_1 = \pi a_2 R^2 \Psi_1, \tag{1.7.5}$$

where

$$\Psi_1 == \bar{r}^2 + 2K_{CRH}(1-2K_{CRH})\sin^{-1}\left(\frac{\bar{L}}{2K_{CRH}}\right)$$
$$+\frac{\bar{L}[8K_{CRH}^2 - (2K_{CRH} + \bar{r} - 1)(2\bar{r} - 2K_{CRH} + 1)}{6K_{CRH}} \tag{1.7.6}$$

$\bar{r}$ is determined by Eq. (1.6.33) and inequalities in Eq. (1.6.32) are satisfied.

For a non-truncated ogive-shaped nose when

$$r = 0, \quad \bar{L} = \sqrt{4K_{CRH} - 1}, \quad K_{CRH} \geq 0.5, \tag{1.7.7}$$

Eq. (1.7.6) yields:

$$\Psi_1 = 2K_{CRH}(1-2K_{CRH})\sin^{-1}\left(\frac{\sqrt{4K_{CRH}-1}}{2K_{CRH}}\right)$$
$$+\frac{\sqrt{4K_{CRH}-1}}{6K_{CRH}}(12K_{CRH}^2 - 4K_{CRH} + 1) \tag{1.7.8}$$

Equation of motion of the impactor implies the following formula for the DOP:

$$H = mK(v_{imp}, A_0, A_1, A_2), \tag{1.7.9}$$

where

$$K(v, g_0, g_1, g_2) = \int_0^v \frac{\zeta d\zeta}{g_2\zeta^2 + g_1\zeta + g_0}. \tag{1.7.10}$$

Integral in Eq. (1.7.10) can be found analytically (see also Zook, 1977):

$$K(v, g_0, g_1, g_2) = \begin{cases} \dfrac{1}{2g_2}\left[\ln\left(\dfrac{\chi}{g_0}\right) - g_1\omega\right] & if \quad \Delta \neq 0 \\ \dfrac{1}{g_2}\left[\ln\left(\dfrac{v+\varepsilon}{\varepsilon}\right) - \dfrac{\varepsilon}{v+\varepsilon}\right] & if \quad \Delta = 0 \end{cases} \tag{1.7.11}$$

where

$$\omega = \begin{cases} \dfrac{1}{\sqrt{-\Delta}} \ln\left[\dfrac{(\eta - \sqrt{-\Delta})(g_1 + \sqrt{-\Delta})2}{(\eta + \sqrt{-\Delta})(g_1 - \sqrt{-\Delta})}\right] & if \quad \Delta < 0 \\ \dfrac{2}{\sqrt{\Delta}}\left[\tan^{-1}\left(\dfrac{\eta}{\sqrt{\Delta}}\right) - \tan^{-1}\left(\dfrac{g_1}{\sqrt{\Delta}}\right)\right] & if \quad \Delta > 0 \end{cases} \tag{1.7.12}$$

$$\chi = g_2 v^2 + g_1 v + g_0, \eta = 2g_2 v + g_1, \varepsilon = \sqrt{g_0 / g_2}, \Delta = 4g_0 g_2 - g_1^2. \tag{1.7.13}$$

1.8 Oversimplified Models

LIMs yield the dependencies for the instantaneous drag, D, vs. the instantaneous velocity, v, (and in the case of the non-averaged models, vs. the instantaneous penetration depth, h). Coefficients in these relations depend upon the shape of a penetrator and can be calculated using the known formulas. It is assumed that the model which determines the local interaction between a shield and a penetrator is known.

Along with the described above approach, a different simplified approach is used. In this approach the local interaction between a shield and a striker is not modeled, and a particular dependence, that includes as a rule empirical coefficients, of the drag force acting at the striker vs. v and (less common) vs. h, is assumed. Such approach is useful for determining experimental dependencies and analyzing general laws governing penetration.

Heimdahl and Schulz, 1986 studied the motion of an impactor for an arbitrary function $D(v)$. A number of investigators have proposed power law dependences for different media (Mileiko and Sarkisyan, 1981; Mileiko *et al.*, 1994; Forrestal *et al.*, 1984; Forrestal *et al.*, 1986). Various approaches to determine the drag force acting on the body as a function of its velocity or velocity and penetration depth were considered by Stone, 1994; Zook, 1977; Beth, 1946; Allen *et al.*, 1957a; Bernard, 1978; Dehn, 1979, 1986, 1987. A brief analysis of such models and references to the early studies can be found in Goldsmith, 1960 and Backman and Goldsmith, 1978.

Chapter 2

Cavity Expansion Approximations

Nomenclature

a_0, a_1, a_2	parameters of LIMs in Eq. (2.4.3), (2.4.4), (2.4.6);
BLV	ballistic limit velocity;
CCE	cylindrical cavity expansion;
DOP	depth of penetration;
h	instantaneous depth of penetration, Fig. 2.3.1;
LIM	localized interaction model;
SCE	spherical cavity expansion;
u	$=\cos \upsilon$;
v	instantaneous velocity of impactor;
v_n	local normal velocity at the surface of projectile;
V	velocity of the surface of the hole;
x	coordinate associated with the impactor, Fig. 2.3.1;
y	radius of the hole;
α_c	coefficient of a CCE model;
β_c	coefficient of a CCE model;
α_s	coefficient of a SCE model;
β_s	coefficient of a SCE model;
γ_s	coefficient of a SCE model;
ξ	coordinate, Fig. 2.3.1;
ρ	coordinate associated with the impactor, Fig. 2.3.1;
μ_{fr}	friction coefficient;
σ_r	normal stress at the boundary of the cavity;
υ	angle between the direction of the projectile motion and the local external normal to the surface;

Φ function determining the shape of the impactor, Fig. 2.3.1;

Ω_n, Ω_τ functions determining the LIM.

2.1 Introduction

Cavity expansion approach is a fairly universal approximate method that allows devising analytical models of penetration mechanics describing the local interaction between the shield and the penetrator at the points of the penetrator-shield contact surface. These models include explicitly parameters which determine mechanical properties of the material of a shield. Some models based on the cavity expansion approach allow us to determine the instantaneous local interaction force between the shield and the striker during its motion inside the shield. This is achieved by postulating a certain relation between penetration and expansion of a cavity in the material of the shield.

Application of cavity expansion approach requires solving two fundamental problems.

The first problem is to determine a law of expansion of a cavity inside a material (generally from the zero initial radius). This problem is usually considered for the cases with axial or spherical symmetry. Multiplicity of formulations of this problem is related with numerous models of the material of the shield (compressible/incompressible, elastic-plastic with different strain-stress laws, etc.).

It is a common practice to distinguish between three classes of models depending on the level of description of the cavity dynamics. Static (or occasionally called quasi-static) models describe stresses at the surface of a cavity in a static state. Quasi-dynamic models are the basis of cavity expansion approach in penetration mechanics. These models determine stresses at the cavity surface as a function of the constant rate of increase of the radius of the cavity. Dynamic models take into account acceleration of the cavity surface. Some particular models are considered in the chapters on penetration into concrete, metal and geological shields.

The second problem is justifying a particular way of using a solution of cavity expansion problem in penetration mechanics. If the solution of the axially symmetric problem is used, then the method is called a cylindrical cavity expansion (CCE) approach/approximation while application of spherically symmetric solution is called a spherical cavity expansion (SCE). This problem is considered in the present chapter in relation to application of the quasi-static models for determining the local interaction forces between the striker and the shield during motion of the striker inside the shield.

Bishop *et al.*, 1945 pioneered the application of cavity expansion models in penetration mechanics. They obtained solutions describing the static expansion of cylindrical and spherical cavities in an infinite medium from the zero initial radius, and used these solutions to determine the forces acting at a conical impactor. Goodier, 1965 applied a dynamic cavity expansion solution to the problem of high speed penetration of a spherical-nosed projectile. A survey of the state-of-the-art up to the late 1950s concerning the problems of expansion of cavities in solids was written by Hopkins, 1960. Useful information on this topic is summarized in the monograph of Yu, 2000. Application of cavity expansion models in penetration mechanics have been described and analyzed by Teland, 1999 and Satapathy, 1997.

Studies directly associated with the application of cavity expansion methods in mathematical modeling of ballistic impact are discussed further in the sections devoted to penetration into concrete, metal and geological shields.

The most extensive research on cavity expansion models was conducted at Moscow State University (the results were published mainly in Russian) and Sandia Research Laboratories. Isbell *et al.*, 1992 noted that some of the results were obtained independently in the former Soviet Union and in the West. During recent years cavity expansion approach is vigorously pursued in Technion (Israel).

Investigations that were performed at Moscow State University in the 1950s and at the beginning of 1960s in the field of soil dynamics were summarized in the monograph by Rahmatulin *et al.*, 1964 with thorough coverage of penetration modeling. The authors described dynamic solutions for the expansion of cylindrical and spherical cavities in soil. Penetration of a striker into a soil was analyzed by applying the suggested cylindrical cavity expansion models. The authors determined the DOP taking into account the incomplete immersion of the impactor in the shield at the initial stage of penetration. The results of subsequent investigations, mainly associated with the development and application of the CCE approach, were summarized in the monographs of Sagomonyan, 1974, 1988. A dynamic CCE approach was used to model perforation of a metal plate, brittle plate and geological media by sharp and blunt impactors.

Extensive studies of cavity expansion approach in penetration dynamics were performed at the Sandia Research Laboratories by Forrestal and his colleagues. They proposed a large variety of spherical and cylindrical cavity expansion models for materials with different mechanical properties. Simple quasi-dynamic two-term or three-term models for engineering applications were developed by using theoretical studies of cavity expansion problems, numerical simulations, and experimental investigations. A limited number of impactor nose shapes (cone, sphere and ogive) were considered comprehensively, although the approach is applicable to a wide class of bodies of revolution. Forrestal and coauthors did not consider separately the stage of penetration with incomplete immersion of the impactor when normal penetration was modeled.

2.2 Spherical Cavity Expansion Approximation

The SCE approximation is widely used in impactor-shield interaction models in a quasi-dynamic version when expansion of a spherically symmetrical cavity from a zero initial radius at a constant velocity is described by the following formula:

$$\sigma_r = \sigma_r(V), \tag{2.2.1}$$

where σ_r is the normal stress at the boundary of the cavity (this function is assumed to be known from the solution of the cavity expansion problem), V is the velocity of the boundary of the hole.

Calculating the interaction force between the shield and the projectile is accomplished as follows. Consider some point at the surface of the impactor moving with the instantaneous velocity v inside the shield. The normal velocity of the point at this location is $v_n = v\cos\upsilon = uv$, where $u = \cos\upsilon$, υ is the angle between the direction of the projectile motion and the local external normal to the surface. It is assumed that the normal stress produced in the shield at this location is equal to the stress at the boundary of the cavity that expands with a constant velocity, $V = v_n$. Therefore the normal stress at the surface of the impactor is described by a LIM with

$$\Omega_n(u,v) = \sigma_r(uv)\,. \tag{2.2.2}$$

Friction between a projectile and a shield can be taken into account by using the relationship $\Omega_\tau(u,v) = \mu_{fr}\Omega_n(u,v)$, where μ_{fr} is the friction coefficient.

Application of the SCE model to calculating the force acting at a striker at some location at the surface of a striker can be illustrated by simple geometric considerations. Let us consider a sphere which has the same tangential plane as the striker at this location. This sphere can be viewed as a surface of the expanding cavity. Then the normal stress at the surface of the impactor equals to the normal stress at the cavity surface. Since σ_r does not depend upon the radius of the cavity, the ambiguity in selecting the radius of the sphere does not affect the result. Clearly, the spheres are different at different locations at the surface of the impactor. However, since the surface of impactors having an ogive-shaped nose can be obtained by rotating an arc of a circle, the external spherical surface of a cavity is the same for all locations at the impactor. Consequently, this geometrical interpretation is particularly descriptive for ogive-nosed impactors.

Applications of the dynamic SCE models in penetration mechanics, to impactors having several typical shapes are described in Bernard and Hanagud, 1975. Calculating the force acting at a striker at some location at the surface of a striker using the dynamic SCE model is ambiguous and does not allow a universal geometric interpretation. This is the reason why the dynamic SCE models are not widely used in penetration mechanics.

2.3 Cylindrical Cavity Expansion Approximation

Another widely used in penetration mechanics approach is known as the cylindrical cavity expansion (CCE) approximation (model, method, etc.). Sometimes other names are used, e.g., the method of plane sections (Sagomonyan, 1960; Rahmatulin *et al.*, 1964) and the disks model (Yankelevsky and Adin, 1980).

CCE method can be justified more readily than a SCE method. In CCE approach, normal penetration of a slender body of revolution is usually considered, and it is assumed that particles of the material of the shield move in a radial direction during penetration by the impactor. The shield can be viewed as consisting of infinitely thin layers, and in each layer a cavity its expansion caused by the moving impactor is modeled. This approach facilitates calculating the stress at the boundary of the hole in each layer and, consequently, the force acting at the penetrator at each location on the lateral surface of a projectile.

The CCE approach can be described in a general case of a dynamic model of hole expansion for each layer as follows:

$$\sigma_r = \sigma_r(y, \dot{y}, \ddot{y}), \quad \dot{y} = V = dy/dt, \quad \ddot{y} = d^2 y/dt^2, \qquad (2.3.1)$$

where y is the radius of the hole.

Let $\rho = \Phi(x)$ be the equation of the surface of the projectile. Then, for the infinitesimal layer with the coordinate ξ (see Fig. 2.3.1), the conditions that the surface of the hole coincides with the surface of the impactor and that the velocity and acceleration of the hole surface are equal to the radial components of the same kinematical characteristics of the impactor yield (Rahmatulin *et al.*, 1964):

$$y = \Phi(h - \xi) \qquad (2.3.2)$$

$$\dot{y} = \Phi'(x)\dot{h}, \ddot{y} = \Phi''(x)\dot{h}^2 + \Phi'(x)\ddot{h}, \quad x = h - \xi. \qquad (2.3.3)$$

Hereafter the dot over a symbol denotes time derivative. The expression for the normal stress at the surface of the impactor, p, p, is obtained by substituting $y, \dot{y}$ and $\ddot{y}$ from Eqs. (2.3.2) and (2.3.3) into Eq. (2.3.1):

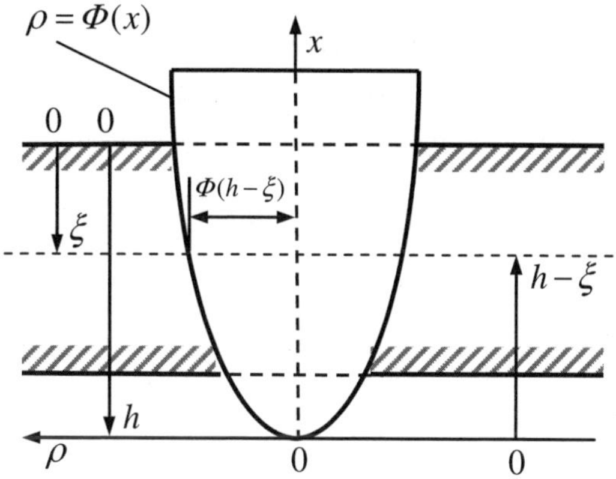

Fig. 2.3.1 The cylindrical cavity expansion approximation.

$$p = \sigma_r(\Phi, \Phi'\dot{h}, \Phi''\dot{h}^2 + \Phi'\ddot{h}), \quad \Phi = \Phi(x). \tag{2.3.4}$$

Note that Eq. (2.3.4) implies equation of motion of impactor that is similar to Eq. (1.3.10).

The case of a quasi-dynamic model is similar to the case described by Eq. (2.2.1) for CCE model when

$$\sigma_r = \sigma_r(\dot{y}) = \sigma_r(V), \tag{2.3.5}$$

and Eq. (2.3.4) reads:

$$p = \sigma_r(\Phi' v), \quad \dot{h} = v. \tag{2.3.6}$$

After substituting

$$\Phi' = \cot \nu = u / \sqrt{1-u^2}, \tag{2.3.7}$$

into Eq. (2.3.6) one can conclude that a quasi-dynamic CCE model is reduced in this case to a LIM with

$$\Omega_n(u, v) = \sigma_r\left(\frac{uv}{\sqrt{1-u^2}}\right). \tag{2.3.8}$$

2.4 Cavity Expansion Approximations and LIMs

We demonstrated that quasi-dynamic cavity expansion models belong to a class of LIMs. Two-term cavity expansion models are the most widely used models in penetration mechanics:

$$\sigma_r(V) = \alpha_s + \beta_s V^2, \tag{2.4.1}$$

$$\sigma_r(V) = \alpha_c + \beta_c V^2, \tag{2.4.2}$$

where subscripts "s" and "c" refer to the SCE model and the CCE model, correspondingly, $\alpha_s, \beta_s, \alpha_c, \beta_c$ are coefficients of the models. The above theory implies that Eqs. (2.4.1) and (2.4.2) determine the following two LIMs, respectively:

$$\Omega_n(u,v) = a_0 + a_2 u^2 v^2, \quad a_0 = \alpha_s, a_2 = \beta_s. \tag{2.4.3}$$

$$\Omega_n(u,v) = a_0 + a_2 \frac{u^2 v^2}{1-u^2} \quad a_0 = \alpha_c, a_2 = \beta_c. \tag{2.4.4}$$

Sometimes the following three-term CCE model with coefficients $\alpha_c, \beta_c, \gamma_c$ is used:

$$\sigma_r(V) = \alpha_s + \gamma_s V + \beta_s V^2, \tag{2.4.5}$$

which corresponds to the following LIM:

$$\Omega_n(u,v) = a_0 + a_1 uv + a_2 u^2 v^2, \quad a_0 = \alpha_s, a_1 = \gamma_s, a_2 = \beta_s. \tag{2.4.6}$$

Friction between impactor and a shield can be taken into account by using the relationship $\Omega_\tau(u,v) = \mu_{fr} \Omega_n(u,v)$.

The LIMs given by Eqs. (2.4.3), (2.4.4) and (2.4.6) are considered in details in Chapter 1 where formulas for the resistance force, the DOP, the BLV, etc. are presented. Formulas for LIMs of other types can be obtained using general equations that are also given in this chapter. Some additional information that can facilitate practical application of LIMs can be found in our earlier monograph (Ben-Dor *et al.*, 2006a).

In conclusion we would like to make the following remarks. Formulas for the resistance force, the DOP, the BLV, etc. which appear in numerous papers on quasi-dynamic cavity expansion models are derived using these particular models. Clearly, many formulas which are derived in Chapter 1 on the basis of the general approach, coincide with formulas published in earlier papers on penetration mechanics. Although these formulas are presented in Chapter 1 without referring to the earlier studies, this does not imply that the authors of this monograph claim credit for these formulas. These formulas are derived using standard mathematical manipulations and establishing priority in this case is quite difficult and hardly appropriate.

Part 2

Penetration into Concrete Shields

Engineering models for calculating penetration into concrete shields are described in the dedicated surveys by Kennedy, 1976; Adeli and Amin, 1985; Williams, 1994; Teland, 1998; Li *et al.*, 2005; Murthy *et al.*, 2010; Rahman *et al.*, 2010b; Zaidi *et al.*, 2010b. Overviews on this topic can be also found in the papers and reviews by Linderman *et al.*, 1974; Walter and Wolde-Tinsae, 1984; Brown, 1986; Corbett *et al.*, 1996; Dancygier and Yankelevsky, 1996; Yankelevsky, 1997; Dancygier, 2000; Ben-Dor *et al.*, 2005b; Vossoughi *et al.*, 2007; Guirgis and Guirguis, 2009; Daudeville and Malécot, 2011, monographs by Bulson, 1997; Bangash and Bangash, 2006; Carlucci and Jacobson, 2008; Bangash, 2009; Szuladziński, 2010, and also in the report published by the United States Department of Energy DOE, 2006. A few studies evaluate performance of some models by comparing their predictions with experimental results.

This Part 2 of the book includes two chapters.

Chapter 3 is dedicated to the empirical models. We have tried to include most of the suggested models (including those that did not appear in the previous reviews), to present them in the unified form and supplement the presentation with some additional analyses when they are warranted. Using experimental results we evaluated the performance of the models and compared their accuracy.

Chapter 4 is devoted to the analytical models. A major emphasis in this chapter is given to the two-stage models which are widely used for describing penetration into concrete shields. We present a classification of these models, and using a unified approach derive most of the formulas for calculating characteristics of penetration.

Bibliography to Part 2 includes more than 130 items.

Chapter 3

Empirical Models

Nomenclature

b	thickness of the shield;
b_{perf}	perforation thickness;
b_{scab}	scabbing thickness;
BLV	ballistic limit velocity;
c	maximum aggregate size of concrete;
$c_{\times}$	half the concrete aggregate size;
CRH	caliber radius head, Eq. (1.6.28);
d	maximum diameter of impactor, reference size;
DOP	depth of penetration;
$\overline{e}$	parameter defined in Eq. (3.9.1);
E	Young modulus of material of projectile;
E_{steel}	Young modulus of steel;
f_c'	unconfined compressive strength (Pa);
H	depth of penetration;
$\overline{H}_{\min}^{(i)}, \overline{H}_{\max}^{(i)}$	parameters in functions $\Psi_{perf}(\overline{H})$ and $\Psi_{scab}(\overline{H})$, Eq. (3.2.9);
$\overline{H}_*^{(i)}$	parameter defined by Eq. (3.2.13);
$H_{\exp}$	DOP obtained in experiment;
I	parameter, Eq. (3.16.3);
I_*	parameter, Eq. (3.19.5);
$I_{\min}^{(i)}, I_{\max}^{(i)}$	parameters of penetration model in Eq. (3.16.1);
J	parameter defined in Eq. (3.18.2);
K_{CRH}	caliber radius head defined in Eq. (1.6.28);
K_{shape}	nose shape coefficient of impactor in Young model;

$K_{shape}^{(1)}$	nose shape coefficient defined in Eq. (3.7.2);
$K_{shape}^{(2)}$	nose shape coefficient defined in Eq. (3.9.2);
$K_{shape}^{(3)}$	nose shape coefficient defined in Eq. (3.22.3);
$K_{shape}^{(4)}$	nose shape coefficient defined in Eq. (3.18.3);
m	mass of impactor;
v_{imp}	impact velocity;
v_{res}	residual velocity;
v_0	reference velocity, $1000\,m/s$;
v_{sl}	scabbing limit velocity;
v_{bl}	ballistic limit velocity;
$\alpha_{pen}, \beta_{pen}, \gamma_{pen}$	parameters of penetration model in Eq. (3.19.1);
$\alpha_{perf}, \beta_{perf}, \gamma_{perf}$	parameters of perforation model in Eq. (3.19.2);
$\alpha_{scab}, \beta_{scab}, \gamma_{scab}$	parameters of scabbing model in Eq. (3.19.3);
$\alpha^{(i)}, \beta^{(i)}, \gamma^{(i)}$	parameters in functions $\Psi_{perf}(\bar{H})$ and $\Psi_{scab}(\bar{H})$, Eq. (3.2.9);
$\tilde{\alpha}_{pen}^{(i)}, \tilde{\beta}_{pen}^{(i)}$	parameters of penetration model in Eq. (3.16.1);
μ	parameter of the model (can be different in different models);
$\tilde{\mu}$	parameter defined in Eq. (3.16.4);
Ξ_{perf}	function that defines dependence $\bar{b}_{perf}$ vs. $\bar{v}_{imp}$;
Ξ_{scab}	function that defines dependence $\bar{b}_{scab}$ vs. $\bar{v}_{imp}$;
ρ_{sh}	density of concrete;
ϕ	function that defines dependence $\bar{H}$ vs. $\bar{v}_{imp}$;
Ψ_{perf}	function that defines dependence $\bar{b}_{perf}$ vs. $\bar{H}$;
Ψ_{scab}	function that defines dependence $\bar{b}_{scab}$ vs. $\bar{H}$.

Note. Bar over parameters indicates dimensionless parameters that are normalized either by a diameter of a striker d when a parameter has a dimension of length or by a characteristic velocity $v_0 = 1000\,m/s$ when a parameter has a dimension of velocity.

3.1 Introduction

The local response of the shield is initiated with spalling (Fig. 3.1.1a) and subsequently can result in penetration, scabbing of the shield material from the back face of the shield (Fig. 3.1.1b), and eventual perforation of the shield (Fig. 3.1.1c) when the missile penetrates the shield (DOE, 2006).

Throughout the book we consider normal penetration (with zero angle of attack) of rigid (non-deformable) projectiles into a shield. If otherwise is not indicated, we consider a lightly reinforced concrete shield and a flat-nosed projectile. Formulas suggested for the projectiles with a circular cross-section may be applied for the projectiles having a more complicated shape by replacing the diameter by the equivalent diameter which is based on the perimeter (Barr, 1990; Walter and Wolde-Tinsae, 1984).

Definitions of key terms are presented in the following (DOE, 2006; Kennedy, 1976).

The depth of penetration (DOP), H , is defined as the depth to which a projectile penetrates into a massive (semi-infinite) concrete shield, for a given impact velocity.

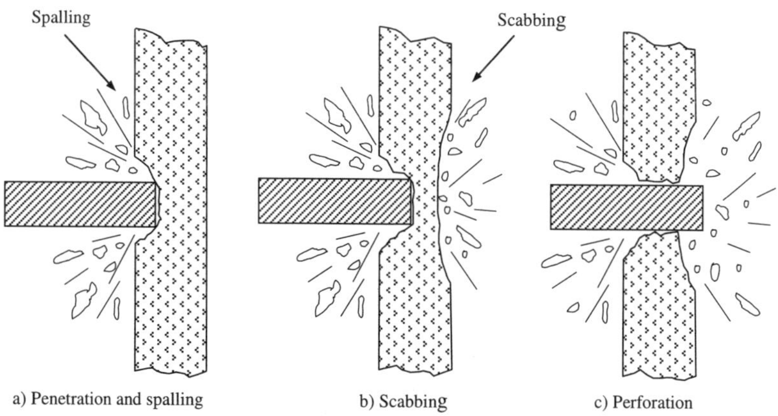

Fig. 3.1.1. Local response of the shield.

The scabbing thickness, b_{scab}, for a given impact velocity, is defined as the shield thickness that is just large enough to prevent the peeling off of the back face of the panel opposite to the face of impact. In other words, scabbing thickness is the minimum thickness of the shield which is required to prevent scabbing.

The perforation thickness, b_{perf}, for a given impact velocity, is defined as the shield thickness that is just large enough to allow a missile to emerge from the back face of the shield with zero exit velocity. In other words, the perforation thickness is the minimum thickness of the shield required to prevent perforation.

Ballistic limit velocity (BLV), v_{bl}, is defined, as before, as the minimum impact (initial) velocity required for perforating the shield with a given thickness.

Similarly to the BLV, we can introduce *the scabbing limit velocity*, v_{sl}, as the minimum impact (initial) velocity that is required for scabbing the shield with a given thickness.

Note that DOE, 2006 recommends using in practical calculations of the shield thicknesses that prevent from scabbing and perforation, the values of b_{scab} and b_{perf} obtained from the empirical formulas, and enlarged by 10% and 20%, correspondingly.

We assume that $b_{scab} < b_{perf}$ in the range of validity of the models.

3.2 Unified Approach

Empirical model is determined by the triad of basic equations of the following form:

$$\bar{H} = \phi(\bar{v}_{imp}), \tag{3.2.1}$$

$$\bar{b}_{perf} = \Psi_{perf}(\bar{H}), \tag{3.2.2}$$

$$\bar{b}_{scab} = \Psi_{scab}(\bar{H}), \tag{3.2.3}$$

where ϕ, Ψ_{perf}, Ψ_{scab} are the known functions. Consequently, the analysis of penetration into the finite thickness shield involves the depth of penetration (DOP) into a semi-infinite shield which is not directly related with the problem. Therefore, it is convenient to present the model for a finite thickness shield by a pair of equations that are obtained after eliminating $\overline{H}$ from Eqs. (3.2.1)–(3.2.3):

$$\overline{b}_{perf} = \Xi_{perf}(\overline{v}_{imp}), \tag{3.2.4}$$

$$\overline{b}_{scab} = \Xi_{scab}(\overline{v}_{imp}), \tag{3.2.5}$$

where

$$\Xi_{perf}(z) = \Psi_{perf}\left(\phi(z)\right), \quad \Xi_{scab}(z) = \Psi_{scab}\left(\phi(z)\right). \tag{3.2.6}$$

Taking into account that Ξ_{perf} and Ξ_{scab} are increasing functions, Eqs. (3.2.4)–(3.2.6) yield the following formulas for $\overline{v}_{bl}$ and $\overline{v}_{sl}$:

$$\overline{v}_{bl} = \Xi_{perf}^{-1}(\overline{b}) = \phi^{-1}\left(\Psi_{perf}^{-1}(\overline{b})\right), \tag{3.2.7}$$

$$\overline{v}_{sl} = \Xi_{scab}^{-1}(\overline{b}) = \phi^{-1}\left(\Psi_{scab}^{-1}(\overline{b})\right), \tag{3.2.8}$$

Fig. 3.2.1 illustrates derivation of Eq. (3.2.7) . The curve described by the equation $\overline{b} = \Xi_{perf}(\overline{v}_{imp})$ separates the domain of parameters $\overline{v}_{imp}$ and $\overline{b}$ into two sub-domains. The sub-domain under the curve corresponds to the perforation of the shield while the sub-domain above the curve corresponds to non-perforation. Let $\overline{b}_0$ be an arbitrary thickness of the shield. Inspection of Fig. 3.2.1 shows that the minimum value of $\overline{v}_{imp}$ for which perforation occurs (by definition this value equals the BLV) is $\Xi_{perf}^{-1}(\overline{b}_0)$, and consequently, Eq. (3.2.7) is valid. The validity of Eq. (3.2.8) can be proved similarly.

Consider some typical classes of the penetration models. This will allow us in describing the particular models to present only the values of the coefficients without rewriting the bulky formulas.

Most types of the dependences between $\overline{b}_{perf}$ and $\overline{H}$ and between $\overline{b}_{scab}$ and $\overline{H}$ can be described as follows:

$$\bar{b}_{p/s} = \Psi_{p/s}(\bar{H}) = \begin{cases} \beta^{(1)} + \alpha^{(1)}\bar{H} - \gamma^{(1)}\bar{H}^2 & if \quad \bar{H}_{\min}^{(1)} \le \bar{H} \le \bar{H}_{\max}^{(1)} \\ \cdots & \cdots \quad \cdots \\ \beta^{(i)} + \alpha^{(i)}\bar{H} - \gamma^{(i)}\bar{H}^2 & if \quad \bar{H}_{\min}^{(i)} \le \bar{H} \le \bar{H}_{\max}^{(i)} \\ \cdots & \cdots \quad \cdots \\ \beta^{(n)} + \alpha^{(n)}\bar{H} - \gamma^{(n)}\bar{H}^2 & if \quad \bar{H}_{\min}^{(n)} \le \bar{H} \le \bar{H}_{\max}^{(n)} \end{cases} \tag{3.2.9}$$

where it is assumed that all parameters $\alpha^{(i)}$, $\beta^{(i)}$ and $\gamma^{(i)}$ are non negative and functions $\Psi_{p/s}$ are continuous at the joint points of the adjacent segments:

$$\bar{H}_{\max}^{(i)} = \bar{H}_{\min}^{(i+1)}, \quad i = 1, 2, ..., n-1. \tag{3.2.10}$$

$$\begin{aligned} &\beta^{(i)} + \alpha^{(i)}\bar{H}_{\max}^{(i)} - \gamma^{(i)}[\bar{H}_{\max}^{(i)}]^2 \\ &= \beta^{(i)} + \alpha^{(i)}\bar{H}_{\min}^{(i+1)} - \gamma^{(i)}[\bar{H}_{\min}^{(i+1)}]^2, \quad i = 1, 2, ..., n-1 \end{aligned} \tag{3.2.11}$$

The subscript p/s denotes "perforation" or "scabbing". Clearly, the parameters $\alpha^{(i)}, \beta^{(i)}, \gamma^{(i)}, \bar{H}_{\min}^{(i)}, \bar{H}_{\max}^{(i)}$ should have the same additional subscripts which are omitted for simplicity.

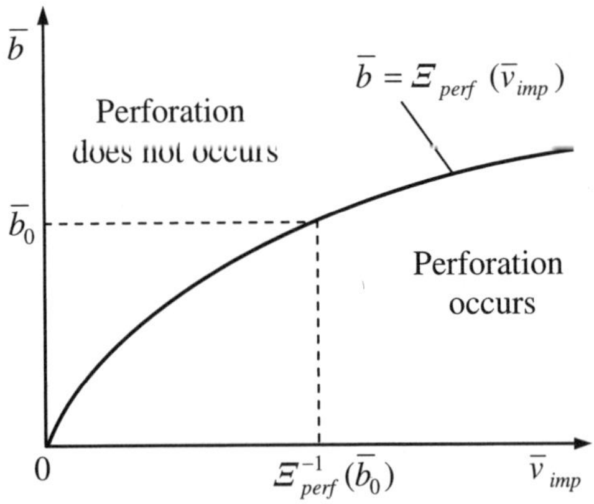

Fig. 3.2.1 Ballistic limit velocity and perforation thickness

If $\gamma^{(i)} = 0$, then $\Psi_{p/s}(\overline{H})$ is an increasing function at the i th interval ($i = 1, 2, ..., n$). If $\gamma^{(i)} > 0$, $\Psi_{p/s}(\overline{H})$ is an increasing function when

$$\overline{H}^{(i)}_{\max} < \overline{H}^{(i)}_{*}, \quad i = 1, 2, ..., n, \tag{3.2.12}$$

where

$$\overline{H}^{(i)}_{*} = 0.5\alpha^{(i)} / \gamma^{(i)}, \quad i = 1, 2, ..., n. \tag{3.2.13}$$

The constraints given by Eq. (3.2.12) must be taken into account when admissible ranges of the parameters are indicated.

Taking into account Eq. (3.2.9), Eq. (3.2.4) and Eq. (3.2.7) can be rewritten as follows:

$$\overline{b}_{p/s} = \Xi_{p/s}(\overline{v}_{imp})$$
$$= \begin{cases} \beta^{(1)} + \alpha^{(1)}\phi(\overline{v}_{imp}) - \gamma^{(1)}[\phi(\overline{v}_{imp})]^2 & if \quad \phi^{-1}(\overline{H}^{(1)}_{\min}) \le \overline{v}_{imp} \le \phi^{-1}(\overline{H}^{(1)}_{\max}) \\ \dots & \dots \quad \dots \\ \beta^{(i)} + \alpha^{(i)}\phi(\overline{v}_{imp}) - \gamma^{(i)}[\phi(\overline{v}_{imp})]^2 & if \quad \phi^{-1}(\overline{H}^{(i)}_{\min}) \le \overline{v}_{imp} \le \phi^{-1}(\overline{H}^{(i)}_{\max}) \\ \dots & \dots \quad \dots \\ \beta^{(n)} + \alpha^{(n)}\phi(\overline{v}_{imp}) - \gamma^{(n)}[\phi(\overline{v}_{imp})]^2 & if \quad \phi^{-1}(\overline{H}^{(n)}_{\min}) \le \overline{v}_{imp} \le \phi^{-1}(\overline{H}^{(n)}_{\max}) \end{cases} \tag{3.2.14}$$

$$\overline{v}_{bl/sl} = \Xi^{-1}_{perf}(\overline{b})$$
$$= \begin{cases} \phi^{-1}(\overline{H}^{(1)}) & if \quad \beta^{(1)} + \alpha^{(1)}\overline{H}^{(1)}_{\min} - \gamma^{(1)}[\overline{H}^{(1)}_{\min}]^2 \le \overline{b} \le \beta^{(1)} + \alpha^{(1)}\overline{H}^{(1)}_{\max} - \gamma^{(1)}[\overline{H}^{(1)}_{\max}]^2 \\ \dots & \dots \quad \dots \\ \phi^{-1}(\overline{H}^{(i)}) & if \quad \beta^{(i)} + \alpha^{(i)}\overline{H}^{(i)}_{\min} - \gamma^{(i)}[\overline{H}^{(i)}_{\min}]^2 \le \overline{b} \le \beta^{(i)} + \alpha^{(i)}\overline{H}^{(i)}_{\max} - \gamma^{(i}[\overline{H}^{(i)}_{\max}]^2 \\ \dots & \dots \quad \dots \\ \phi^{-1}(\overline{H}^{(n)}) & if \quad \beta^{(n)} + \alpha^{(n)}\overline{H}^{(n)}_{\min} - \gamma^{(1)}[\overline{H}^{(n)}_{\min}]^2 \le \overline{b} \le \beta^{(n)} + \alpha^{(n)}\overline{H}^{(n)}_{\max} - \gamma^{(n)}[\overline{H}^{(n)}_{\max}]^2 \end{cases} \tag{3.2.15}$$

where the subscript bl/sl denotes the BLV and scabbing limit velocity, and $\overline{H}^{(i)}$ is a root of the equation

$$\beta^{(i)} + \alpha^{(i)}\overline{H}^{(i)} - \gamma^{(i)}[\overline{H}^{(i)}]^2 = \overline{b}, \tag{3.2.16}$$

that is given by the following formula:

$$\bar{H}^{(i)} = \sqrt{\left[\frac{\alpha^{(i)}}{2\gamma^{(i)}}\right]^2 + \frac{\bar{b} - \beta^{(i)}}{\gamma^{(i)}}} - \frac{\alpha^{(i)}}{2\gamma^{(i)}} \quad if \quad \gamma^{(i)} > 0, \qquad (3.2.17)$$

$$\bar{H}^{(i)} = \frac{\bar{b} - \beta^{(i)}}{\alpha^{(i)}} \quad if \quad \gamma^{(i)} = 0. \qquad (3.2.18)$$

3.3 Modified Petry Formulas

Petry formula was originally developed in 1910 (Petry, 1910). Its modified version (Kennedy, 1976) can be written similarly to Li *et al.*, 2005 as follows:

$$\bar{H} = \phi(\bar{v}_{imp}) = \frac{0.0795 K_p m}{d^3} \log_{10}(1 + 50\bar{v}_{imp}^2), \qquad (3.3.1)$$

where K_p is a coefficient depending on the type of the concrete. The coefficients in Eq. (3.3.1) are selected so that the dimensional parameters m and d are measured in SI units while British units (ft^3/lb) are retained for K_p in order to use generally adopted values of this parameter.

Amirikian, 1950 with reference to Samuely and Hamann, 1939 prescribes the following values of the parameter K_p for typical types of concretes: 0.00799 for a massive concrete, 0.00426 for a normal reinforced concrete, and 0.00284 for a specially reinforced concrete (this version of the model Kennedy, 1976 specifies as the modified Petry I model) and suggests to use additional dependence between K_p and f_c' for specially reinforced concrete represented in graphical form (specified by Kennedy, 1976 as the modified Petry II model). In a more convenient, analytical form, this dependence is presented in Walter and Wolde-Tinsae, 1984:

$$K_p = 6.34 \cdot 10^{-3} \exp(-0.2937 \cdot 10^{-7} f_c'). \qquad (3.3.2)$$

Amirikian, 1950 suggests taking into account the thickness of the shield when the DOP is calculated. He recommends to use Eq. (3.3.1) if $\bar{b} \geq 3\bar{H}$, assumes that perforation begins when $\bar{b} = 2\bar{H}$, i.e.,

$$\bar{b}_{perf} = \Psi_{perf}(\bar{H}) = 2\bar{H}, \tag{3.3.3}$$

and proposes, instead of Eq. (3.3.1), using the following expression for the DOP, $\bar{H}_{\times}$, in the intermediate domain, $2\bar{H} \leq \bar{b} \leq 3\bar{H}$:

$$\bar{H}_{\times} = \Lambda(\bar{H}) = \left[1 + \exp\left(4(\bar{b}/\bar{H} - 2)\right)\right]\bar{H}. \tag{3.3.4}$$

Kennedy, 1976 specifies as the modified Petry formulas for the scabbing thickness the following relation:

$$\bar{b}_{scab} = \Psi_{scab}(\bar{H}) = 2.2\bar{H}. \tag{3.3.5}$$

Eqs. (3.3.3) and (3.3.5) can be written in the form given by Eq. (3.2.14) with the coefficients shown in Table 3.3.1.

Taking into account the modifications proposed by Amirikian, 1950 the DOP for the shield with the thickness $\bar{b}$ can be written as follows:

$$\bar{H}_{\times} = \begin{cases} \phi(\bar{v}_{imp}) & if \quad \bar{v}_{imp} \leq \phi^{-1}(\bar{b}/3) \\ \Lambda\left(\phi(\bar{v}_{imp})\right) & if \quad \phi^{-1}(\bar{b}/3) < \bar{v}_{imp} < \phi^{-1}(\bar{b}/2) \\ perforation & if \quad \bar{v}_{imp} \geq \phi^{-1}(\bar{b}/2) \end{cases}, \tag{3.3.6}$$

where

$$\phi^{-1}(z) = \sqrt{0.02\left[10^{d^3 z/(0.795 K_p m)} - 1\right]}. \tag{3.3.7}$$

Formulas for other basic ballistic characteristics are given by Eqs. (3.2.14) , (3.2.15) and (3.2.18) taking into account the expression for ϕ^{-1} from Eq. (3.3.7).

Amde *et al.*, 1997 gives the following Petry formula for the residual velocity referring to Gilbert Associates (1974):

$$v_{res} = v_{imp}\sqrt{1 - (0.5b/H)} \tag{3.3.8}$$

Table 3.3.1. Petry model for perforation and scabbing. Coefficients in Eq. (3.2.9). $n = 1$.

	$\alpha^{(1)}$	$\beta^{(1)}$	$\gamma^{(1)}$	$\overline{H}^{(1)}_{\min}$	$\overline{H}^{(1)}_{\max}$
Perforation	0	2.0	0	-	-
Scabbing	0	2.2	0	-	-

It should be noted that classifying Petry formulas, Eq. (3.3.1), as empirical models is done here partly in order to follow the tradition. Indeed, their form implies that these formulas were derived by integrating the equation of motion of the penetrator under the assumption of the two-term quadratic dependence (without the linear term) of the drag force vs. the velocity.

3.4 Ballistic Research Laboratory (BRL) Formulas

The BRL formula did not take into account the influence of the unconfined compressive strength on the protective properties of the shield, and the value $f_c' = 20.7\,MPa$ was assumed (Kennedy, 1976). In order to remedy this shortcoming the modified BRL formula was suggested. As Kennedy, 1976 emphasizes, this modified formula, in contrast to the common approach, directly predicts the perforation thickness:

$$\overline{b}_{perf} = \frac{13m}{d^{2.8}\sqrt{f_c'}}\overline{v}_{imp}^{1.33}, \tag{3.4.1}$$

while the scabbing thickness can be estimated as

$$\overline{b}_{scab} = 2\overline{b}_{perf}\,. \tag{3.4.2}$$

Then

$$\overline{v}_{bl} = \phi^{-1}(\overline{b}), \quad \overline{v}_{sl} = \phi^{-1}(\overline{b}/2), \tag{3.4.3}$$

where

$$\phi^{-1}(z) = d^{2.1}(f_c')^{0.375}\left(\frac{z}{13m}\right)^{0.75}. \tag{3.4.4}$$

Note that Chelapati *et al.*, 1972 obtained similar result starting from the expression for the DOP. They introduced the DOP, $\bar{H}$, using the analogue of Eq. (3.4.1) with the right-hand side divided by 1.3, that yields $\bar{b}_{perf} = 1.3\bar{H}$.

3.5 Whiffen Formula

Bulson, 1997 with reference to Whiffen, 1943 presents the following model:

$$\bar{H} = \phi(\bar{v}_{imp}) = \frac{2.6m}{d^3\bar{c}^{0.1}\sqrt{f_c'}}(1.87\bar{v}_{imp})^{\tilde{n}}, \quad \tilde{n} = \frac{97.5}{\sqrt[4]{f_c'}}, \quad \bar{c} = \frac{c}{d}, \tag{3.5.1}$$

where c is the maximum aggregate size of the concrete.

The latter equation is based on the experiments with the ogival-nosed projectiles conducted for the following ranges of parameters: $0.8 \le K_{CRH} \le 3.5$, $5\,MPa < f_c' < 70\,MPa$, $12mm < d < 965mm$, $0.02 < \bar{c} < 2$, $v_{imp} < 1130\,m/s$, $136g < m < 10{,}000\,kg$. This formula fits the experimental data within a scatter band of the order of $\pm 15\%$.

Note that Teland, 1998 refers to this model as "TBAA formula".

3.6 Army Corporations of Engineers (ACE) Formula

The rewritten using variables $\bar{H}$ and $\bar{v}_{imp}$ and SI units ACE formula (ACE, 1946; Gwaltney, 1968) reads:

$$\bar{H} = \phi(\bar{v}_{imp}) = \frac{11.1\,m}{d^{2.785}\sqrt{f_c'}}\bar{v}_{imp}^{1.5} + 0.5. \tag{3.6.1}$$

There are two versions of the dependences between the perforation/scabbing thickness and the DOP which differ very slightly (Gwaltney, 1968; Chelapati *et al.*, 1972; Li *et al.*, 2005). Following Kennedy,

Table 3.6.1. ACE model for perforation and scabbing. Coefficients in Eq. (3.2.9), $n = 1$.

	$\alpha^{(1)}$	$\beta^{(1)}$	$\gamma^{(1)}$	$\bar{H}_{\min}^{(1)}$	$\bar{H}_{\max}^{(1)}$
Perforation	1.24	1.32	0	1.35	13.5
Scabbing	1.36	2.12	0	0.65	11.75

1976 we have selected the version of this formula with the coefficients shown in Table 3.6.1 which is associated with the model given by Eq. (3.2.9).

Formulas for other basic ballistic characteristics are given by Eqs. (3.2.14)–(3.2.15) and (3.2.18) taking into account the following expression for ϕ^{-1} from Eq. (3.6.1):

$$\phi^{-1}(z) = \left[\frac{d^{2.785}\sqrt{f_c'}}{11.1m}(z - 0.5) \right]^{0.67}. \tag{3.6.2}$$

3.7 Ammann and Whitney Formula

The Ammann and Whitney formula has been suggested for predicting the perforation of small explosively generated fragments with the impact velocity larger than $300\,\text{m/s}$ (Kennedy, 1976):

$$\bar{H} = \phi(\bar{v}_{imp}) = \frac{15K_{shape}^{(1)}m}{d^{2.8}\sqrt{f_c'}}\bar{v}_{imp}^{1.8}, \tag{3.7.1}$$

where

$$K_{shape}^{(1)} = \begin{cases} 0.72 & \textit{for flat nose} \\ 0.84 & \textit{for blunt nose} \\ 1.00 & \textit{for average nose (spherical end)} \\ 1.14 & \textit{for very sharp nose} \end{cases} \tag{3.7.2}$$

3.8 Modified National Defense Research Committee (NDRC) Formula

The modified NDRC formula can be written in a commonly accepted form as the implicit function with respect to the DOP (NDRC, 1946; Kennedy, 1976). Using the dimensionless variables $\bar{H}$ and $\bar{v}_{imp}$ and SI units for the dimensional parameters this formula reads:

$$G(\bar{H}) = \mu \bar{v}_{imp}^{1.8}, \tag{3.8.1}$$

where

$$\mu = 9.55 \frac{K_{shape}^{(1)} m}{d^{2.8} \sqrt{f_c'}}, \tag{3.8.2}$$

$$G(H) = \begin{cases} 0.25\bar{H}^2 & if \quad \bar{H} \le 2 \\ \bar{H} - 1 & if \quad \bar{H} > 2 \end{cases}, \tag{3.8.3}$$

and the effect of the unconfined compressive strength is included in the model following the suggestion of Kennedy, 1976.

In the unified form given by Eq. (3.2.1), this model can be rewritten as follows:

$$\bar{H} = \phi(\bar{v}_{imp}) = \begin{cases} 2\sqrt{\mu} \bar{v}_{imp}^{0.9} & if \quad \bar{v}_{imp} \le 1/\mu^{5/9} \\ \mu \bar{v}_{imp}^{1.8} + 1 & if \quad \bar{v}_{imp} > 1/\mu^{5/9} \end{cases}. \tag{3.8.4}$$

The modified NDRC model for the perforation and scabbing (Kennedy, 1976) can be written in the form of Eq. (3.2.9) with the coefficients presented in Table 3.8.1. The dependences $\bar{b}_{perf}$ and $\bar{b}_{scab}$ vs. $\bar{v}_{imp}$ as well as $\bar{v}_{bl}$ and $\bar{v}_{sl}$ vs. $\bar{b}$ are determined by Eqs. (3.2.14)–(3.2.18), where

$$\phi^{-1}(z) = \begin{cases} (0.5z/\sqrt{\mu})^{10/9} & if \quad z \le 2 \\ [(z-1)/\mu]^{5/9} & if \quad z \ge 2 \end{cases}. \tag{3.8.5}$$

Table 3.8.1. Modified NDRC model for perforation and scabbing. Coefficients in Eq. (3.2.9), $n = 2$.

	i	$\beta^{(i)}$	$\alpha^{(i)}$	$\gamma^{(i)}$	$\bar{H}_{\min}^{(i)}$	$\bar{H}_{\max}^{(i)}$	$\bar{H}_{*}^{(i)}$
Perforation	1	0	3.19	0.718	-	1.35	2.22
	2	1.32	1.24	0	1.35	13.5	-
Scabbing	1	0	7.91	5.06	-	0.65	0.78
	2	2.12	1.36	0	0.65	11.75	-

Inspection of Table 3.8.1 shows that inequalities given by Eq. (3.2.12) are satisfied.

According to DOE, 2006, NDRC formula was derived for the following ranges of the impact parameters: $v_{imp} > 152 m/s$, $2.54 cm < b > 40.6 cm$ and $b/d \geq 3$. However, following the suggestion by Kennedy, 1976 this formula can be extrapolated beyond these ranges.

3.9 Kar Formula

Kar formula (Kar, 1978, 1978a, 1979; Bangash and Bangash, 2006) is an improved modified NDRC formula that takes into account the size of the aggregates of concrete and type of the projectile material.

In the unified form given by Eq. (3.2.1), Kar formula is given by Eq. (3.8.4) where

$$\mu = 9.55 \frac{K_{shape}^{(2)} m \bar{e}^{-6.25}}{d^{2.8} \sqrt{f_c'}}, \quad \bar{e} = \left(\frac{E}{E_{steel}} \right)^{0.2}, \tag{3.9.1}$$

E and E_{steel} are the Young's moduli of the material of the projectile and of the steel, respectively. The nose shape parameter, $K_{shape}^{(2)}$, is determined by the following formula:

$$K_{shape}^{(2)} = \begin{cases} 0.72 & for\ flat-nosed\ projectile \\ \min(\hat{K}_{nose}, 1.17) & for\ ogive-nosed\ projectile \end{cases}, \tag{3.9.2}$$

Table 3.9.1. Kar model for perforation and scabbing. Coefficients in Eq. (3.2.9), $n = 2$.

	i	$\beta^{(i)}$	$\alpha^{(i)}$	$\gamma^{(i)}$	$\overline{H}_{\min}^{(i)}$	$\overline{H}_{\max}^{(i)}$	$\overline{H}_{*}^{(i)}$
Perforation	1	$\overline{c}_\times$	3.19	0.718	-	1.35	2.22
	2	$\overline{c}_\times + 1.32$	1.24	0	1.35	13.5	-
Scabbing	1	$\overline{c}_\times$	$7.91\overline{e}$	$5.06\overline{e}$	-	0.65	0.71
	2	$\overline{c}_\times + 2.12\overline{e}$	$1.36\overline{e}$	0	0.65	11.75	-

where

$$\hat{K}_{nose} = 0.72 + 0.25\sqrt{K_{CRH} - 0.25}\,. \tag{3.9.3}$$

Kar model for perforation and scabbing can be written in the form of Eq. (3.2.9) where the coefficients are presented in Table 3.9.1 and $\overline{c}_\times$ is the half-size of the concrete aggregate.

The dependencies $\overline{b}_{perf}$ and $\overline{b}_{scab}$ vs. $\overline{v}_{imp}$ as well as $\overline{v}_{bl}$ and $\overline{v}_{sl}$ vs. $\overline{b}$ are given by Eqs. (3.2.14)–(3.2.18) where $\phi^{-1}(z)$ is determined by Eq. (3.8.5).

3.10 Healey-Weissman Formula

The Healey-Weissman penetration formula (Healey and Weissman, 1974; Li *et al.*, 2005) can be considered as a version of the Kar formula where Eq. (3.9.1) that determines μ is replaced by the following expression:

$$\mu = 10.95\left(\frac{E}{E_{steel}}\right)\frac{K_{shape}^{(2)} m}{d^{2.8}\sqrt{f_c'}}\,. \tag{3.10.1}$$

The formula for the residual velocity reads (Healey *et al.*, 1975; Kar, 1979):

$$\frac{v_{res}}{v_{imp}} = \begin{cases} [1 - (b/b_{perf})^2]^{0.555} & if \quad b \le 2d \\ [1 - (b/b_{perf})]^{0.555} & if \quad b > 2d \end{cases}. \tag{3.10.2}$$

3.11 Bechtel Formula

The Bechtel Corporation proposed the following formula for scabbing thickness that is valid for hard cylindrical projectiles (Rotz, 1975, 1976; Bangash and Bangash, 2006):

$$\bar{b}_{scab} = \Xi_{scab}(\bar{v}_{imp}) = \frac{1.23 \cdot 10^3 m^{0.4}}{d^{1.2}\sqrt{f_c'}} \bar{v}_{imp}^{0.5} . \quad (3.11.1)$$

The latter equation is based on 12 tests with solid missiles and 9 tests with half-pipe missiles which were conducted for the following ranges of the parameters (Teland, 1998): $37\,m/s < v_{imp} < 144\,m/s$, $20.3\,cm < d < 21.8\,cm$, $30.5\,cm < b < 61\,cm$ and $7.6\,cm < b < 22.9\,cm$, $30\,MPa < f_c' < 40\,MPa$, $3.6\,kg < m < 97.1\,kg$.

3.12 Stone and Webster Formula

The following formula was proposed for calculating the scabbing thickness (Jankov *et al.*, 1976i; Bangash and Bangash, 2006, Li *et al.*, 2005):

$$\bar{b}_{scab} = \Xi_{scab}(\bar{v}_{imp}) = \frac{100}{d}\left(\frac{m}{K_{\bar{b}}}\right)^{1/3} \bar{v}_{imp}^{2/3} , \quad (3.12.1)$$

where $K_{\bar{b}}$ is the dimensional coefficient. Li *et al.*, 2005 suggested the following formula for $K_{\bar{b}}$:

$$K_{\bar{b}} = 0.013\bar{b} + 0.33, \quad 1.5 \le \bar{b} \le 3 . \quad (3.12.2)$$

The Stone and Webster formula is based on 7 tests with solid missiles and 21 tests with half-pipe missiles which were conducted for the following ranges of the parameters (Teland, 1998): $27\,m/s < v_{imp} < 157\,m/s$, $4.1\,cm < d < 8.9\,cm$, $11.4\,cm < b < 15.2\,cm$, $22MPa < f_c' < 30\,MPa$, $1.9\,kg < m < 12.8\,kg$.

3.13 CEA-EDF Formula

The CEA-EDF (Commisariat àl'Enerfie Atomoque–Electricité de France) formula reads (Berriaud *et al.*, 1978):

$$\bar{b}_{perf} = \Xi_{perf}(\bar{v}_{imp}) = \frac{146m^{0.5}}{d^{1.5}(f_c')^{0.375}\rho_c^{0.125}}\bar{v}_{imp}^{0.75}, \qquad (3.13.1)$$

where ρ_{sh} is the density of the concrete. Formula (3.13.1) implies that

$$\bar{v}_{bl} = \Xi_{perf}^{-1}(\bar{b}) = \frac{1.3\cdot 10^{-3} d^2 \rho_{sh}^{1/6}\sqrt{f_c'}}{m^{2/3}}\bar{b}^{4/3}. \qquad (3.13.2)$$

This model is valid in the following range of parameters: $v_{imp} < 200\,m/s$, $150\,kg/m^3 < \rho_{sh} < 300\,kg/m^3$, $20\,kg < m < 300\,kg$, $0.35 < b/d < 4.17$, $23\,MPa < f_c' < 46\,MPa$.

3.14 Degen Formula

Based on the available experimental data, Degen, 1980 suggested formula for the perforation thickness that can be written in the form given by Eq. (3.2.9) with the coefficients shown in Table 3.14.1, where $\bar{H}$ is determined from the modified NDRC model, Eq. (3.8.4). The dependence $\bar{v}_{sl}$ vs. $\bar{b}$ is given by Eq. (3.2.15) where $\phi^{-1}(z)$ is determined by Eq. (3.8.5).

The experiments were conducted in the following ranges of the parameters: $25\,m/s < v_{imp} < 310\,m/s$, $15\,kg < m < 134\,kg$, $28\,MPa < f_c' < 43\,MPa$, $10\,cm < d < 31\,cm$, $15\,cm < b < 60\,cm$.

Table 3.14.1. Degen model for perforation. Coefficients in Eq. (3.2.9), $n = 2$.

	i	$\beta^{(i)}$	$\alpha^{(i)}$	$\gamma^{(i)}$	$\bar{H}_{\min}^{(i)}$	$\bar{H}_{\max}^{(i)}$	$\bar{H}_{*}^{(i)}$
Perforation	1	0	2.2	0.3	-	1.52	3.67
	2	0.69	1.29	0	1.52	13.42	-

The concrete reinforcement varied in the range between $160\,kg/m^3$ and $350\,kg/m^3$, and penetrators having the flat, conical and hemispherical nose shapes were used.

3.15 Chang Formula

Chang formulas (Chang, 1981) for the flat-nosed projectiles penetrating into a reinforced concrete shield read:

$$\bar{b}_{perf} = \Xi_{perf}(\bar{v}_{imp}) = \frac{497}{d^{1.5}}\sqrt{\frac{m}{f_c'}}\,\bar{v}_{imp}^{0.75}, \qquad (3.15.1)$$

$$\bar{b}_{scab} = \Xi_{scab}(\bar{v}_{imp}) = \frac{321}{d^{1.2}}\left(\frac{m}{f_c'}\right)^{0.4}\bar{v}_{imp}^{2/3}. \qquad (3.15.2)$$

These formulas are based on the experiments conducted in the following ranges of the parameters:

$17\,m/s < v_{imp} < 312\,m/s$, $110\,g < m < 344\,kg$,

$23\,MPa < f_c' < 46\,MPa$, $5.1\,cm < b < 61\,cm$,

$2\,cm < d < 30.5\,cm$.

Eqs. (3.15.1) and (3.15.2) imply that

$$\bar{v}_{bl} = \Xi_{perf}^{-1}(\bar{b}) = 2.54\cdot 10^{-4} d^2 \left(\frac{f_c'}{m}\right)^{2/3}\bar{b}^{4/3}, \qquad (3.15.3)$$

$$\bar{v}_{sl} = \Xi_{scab}^{-1}(\bar{b}) = 1.8\cdot 10^{-4} d^{1.8} \left(\frac{f_c'}{m}\right)^{0.6}\bar{b}^{1.5}. \qquad (3.15.4)$$

3.16 Haldar–Miller Formula

Consider the following class of models:

$$\bar{H} = \begin{cases} \tilde{\alpha}_{pen}^{(1)} I + \tilde{\beta}_{pen}^{(1)} & if \quad I_{\min}^{(1)} \le I \le I_{\max}^{(1)} \\ \tilde{\alpha}_{pen}^{(2)} I + \tilde{\beta}_{pen}^{(2)} & if \quad I_{\min}^{(2)} \le I \le I_{\max}^{(2)} \\ \tilde{\alpha}_{pen}^{(3)} I + \tilde{\beta}_{pen}^{(3)} & if \quad I_{\min}^{(3)} \le I \le I_{\max}^{(3)} \end{cases} \tag{3.16.1}$$

where

$$I_{\max}^{(i)} = I_{\min}^{(i+1)}, \quad \tilde{\alpha}_{pen}^{(i)} I_{\max}^{(i)} + \tilde{\beta}_{pen}^{(i)} = \tilde{\alpha}_{pen}^{(i+1)} I_{\min}^{(i+1)} + \tilde{\beta}_{pen}^{(i+1)}, \quad i = 1, 2, \tag{3.16.2}$$

$$I = \tilde{\mu} \bar{v}_{imp}^{2}, \tag{3.16.3}$$

$$\tilde{\mu} = 10^{6} \frac{K_{shape}^{(1)} m}{d^{3} f_{c}'}. \tag{3.16.4}$$

These models can be written in the unified form as follows:

$$\bar{H} = \phi(\bar{v}_{imp}) = \begin{cases} \tilde{\alpha}_{pen}^{(1)} \tilde{\mu} \bar{v}_{imp}^{2} + \tilde{\beta}_{pen}^{(1)} & if \quad \sqrt{I_{\min}^{(1)} / \tilde{\mu}} \le \bar{v}_{imp} \le \sqrt{I_{\max}^{(1)} / \tilde{\mu}} \\ \tilde{\alpha}_{pen}^{(2)} \tilde{\mu} \bar{v}_{imp}^{2} + \tilde{\beta}_{pen}^{(2)} & if \quad \sqrt{I_{\min}^{(2)} / \tilde{\mu}} \le \bar{v}_{imp} \le \sqrt{I_{\max}^{(2)} / \tilde{\mu}} \\ \tilde{\alpha}_{pen}^{(3)} \tilde{\mu} \bar{v}_{imp}^{2} + \tilde{\beta}_{pen}^{(3)} & if \quad \sqrt{I_{\min}^{(3)} / \tilde{\mu}} \le \bar{v}_{imp} \le \sqrt{I_{\max}^{(3)} / \tilde{\mu}} \end{cases} \tag{3.16.5}$$

The Haldar–Miller model for penetration (Haldar and Miller, 1982) is determined by Eqs. (3.16.1)–(3.16.5) with coefficients presented in Table 3.16.1 below.

Haldar and Miller considered their model as an improved modified NDRC model.

Table 3.16.1. Haldar–Miller model for penetration. Coefficients in Eq. (3.16.1).

i	$\tilde{\alpha}_{pen}^{(i)}$	$\tilde{\beta}_{pen}^{(i)}$	$I_{\min}^{(i)}$	$I_{\max}^{(i)}$
1	0.22024	-0.02725	0.3	2.5
2	0.446	-0.592	2.5	3.0
3	0.06892	0.53886	3.0	21.0

3.17 Haldar–Hamieh–Miller Formula

This penetration model (Haldar *et al.*, 1983; Haldar and Hamieh, 1984) is described by Eqs. (3.16.1)–(3.16.5) with the coefficients shown in Table 3.17.1.

The Haldar–Hamieh–Miller formula yields the following expression for the scabbing thickness after substituting I from Eq. (3.16.3):

$$\bar{b}_{scab} = 0.0342\tilde{\mu}\bar{v}_{imp}^2 + 3.3437, \quad 4.58/\sqrt{\tilde{\mu}} \le \bar{v}_{imp} \le 19.6/\sqrt{\tilde{\mu}}\,, \tag{3.17.1}$$

and the following formula for the scabbing limit velocity:

$$\bar{v}_{sl} = 5.41\sqrt{(\bar{b} - 3.3437)/\tilde{\mu}}, \quad 4.1 \le \bar{b} \le 16.5\,. \tag{3.17.2}$$

Table 3.17.1. Haldar–Hamieh–Miller model for penetration. Coefficients in Eq. (3.16.1).

i	$\tilde{\alpha}_{pen}^{(i)}$	$\tilde{\beta}_{pen}^{(i)}$	$I_{\min}^{(i)}$	$I_{\max}^{(i)}$
1	0.2251	-0.0308	0.3	4.0
2	0.0567	0.6740	4.0	21.0
3	0.0299	1.1875	21.0	455

3.18 Hughes' Formula

The Hughes' formula for penetration reads (Hughes, 1984):

$$\bar{H} = 0.19K_{shape}^{(4)} \frac{J}{1 + 12.3\ln(1 + 0.03J)}, \tag{3.18.1}$$

where

$$J = 10^6 \frac{m\bar{v}_{imp}^2}{d^3 f_t}, \tag{3.18.2}$$

$$K_{shape}^{(4)} = \begin{cases} 1.00 & \textit{for flat nose} \\ 1.12 & \textit{for a blunt nose} \\ 1.26 & \textit{for spherical nose} \\ 1.39 & \textit{for very sharp nose} \end{cases} \tag{3.18.3}$$

and f_t is the tensile strength of concrete. With the reference to ACI, 1978, Hughes, 1984 recommends using the following relationship between f_t and f_c':

$$f_t = 630\sqrt{f_c'}\,. \tag{3.18.4}$$

The dependencies $\overline{b}_{perf}$ vs. $\overline{H}$ and $\overline{b}_{scab}$ vs. $\overline{H}$ are the same as given by Eq. (3.2.9) ($n = 2$) with the coefficients shown in Table 3.18.1.

Table 3.18.1. Hughes' model for perforation and scabbing. Coefficients in Eq. (3.2.9), $n = 2$.

	i	$\beta^{(i)}$	$\alpha^{(i)}$	$\gamma^{(i)}$	$\overline{H}_{\min}^{(i)}$	$\overline{H}_{\max}^{(i)}$
Perforation	1	0	3.6	0	-	0.7
	2	1.4	1.58	0	0.7	-
Scabbing	1	0	5.0	0		0.7
	2	2.3	1.74	0	0.7	-

Since the dependence $\overline{H} = \phi(\overline{v}_{imp})$ does not allow to express $\overline{v}_{imp}$ as a function of $\overline{H}$ in the explicit form, Eqs. (3.2.14) and (3.2.15), which include the inverse function, ϕ^{-1}, also cannot be written in the explicit form.

Hughes, 1984 notes that his formulas are valid in the range $J < 3500$ but they will be conservative for $J < 40$ and $\overline{b} < 3.5$.

3.19 Adeli–Amin Formula

Using a large set of experimental data (Sliter, 1980), Adeli and Amin, 1985 proposed a model that, similarly to the Haldar–Miller and Haldar–Hamieh–Miller models, is based on the parameter I given by Eqs. (3.16.3) and (3.16.4). This model can be written as follows:

$$\bar{H} = \psi(\alpha_{pen}, \beta_{pen}, \gamma_{pen}; I), \tag{3.19.1}$$

$$\bar{b}_{perf} = \psi(\alpha_{perf}, \beta_{perf}, \gamma_{perf}; I), \tag{3.19.2}$$

$$\bar{b}_{scab} = \psi(\alpha_{scab}, \beta_{scab}, \gamma_{scab}; I), \tag{3.19.3}$$

where

$$\psi(\alpha, \beta, \gamma; I) = \beta + \alpha I - \gamma I^2 \tag{3.19.4}$$

and the parameters α, β, γ with the corresponding subscripts are given in Table 3.19.1. The parameter

$$I_* = 0.5\alpha/\gamma \tag{3.19.5}$$

is shown in the last column of Table 3.19.1. This parameter is the upper limit of the increasing function ψ vs. I. The requirement similar to the inequality in Eq. (3.2.12) reads:

$$I < I_*. \tag{3.19.6}$$

After substituting $I = \tilde{\mu}\bar{v}_{imp}^2$ into Eqs. (3.19.1)–(3.19.3), this model can be written in the unified form.

Eqs. (3.19.2)–(3.19.4) and (3.16.3) yield the following relationships for calculating the BLV and scabbing limit velocity:

$$\bar{v}_{bl} = \sqrt{\psi^{-1}(\alpha_{perf}, \beta_{perf}, \gamma_{perf}; \bar{b})/\tilde{\mu}}, \tag{3.19.7}$$

Table 3.19.1. Adeli and Amin model. Coefficients in Eqs. (3.19.1)–(3.19.3).

	β	α	γ	I_*
Penetration	0.0416	0.1698	0.0045	18.8
Perforation	0.906	0.3214	0.0106	15.1
Scabbing	1.8685	0.4035	0.0114	17.7

$$\bar{v}_{sl} = \sqrt{\psi^{-1}(\alpha_{scab}, \beta_{scab}, \gamma_{scab}; \bar{b}) / \tilde{\mu}}, \qquad (3.19.8)$$

where

$$\psi^{-1}(\alpha, \beta, \gamma; z) = 0.5[\alpha - \sqrt{\alpha^2 - 4\gamma(z - \beta)}] / \gamma. \qquad (3.19.9)$$

Instead of Eq. (3.19.1), Adeli and Amin allow using the following formula:

$$\bar{H} = 0.0123 + 0.196I - 0.008I^2 + 0.0001I^3, \qquad (3.19.10)$$

where function $\bar{H}(I)$ increases when $I < I_* = 19.1$.

Adeli–Amin formulas are valid in the following ranges of parameters (Adeli and Amin, 1985): $27\,m/s < v_{imp} < 312\,m/s$, $110\,g < m < 344\,kg$, $0.7 < b/d < 18$, $d < 30\,cm$, $\bar{H} < 2.0$, $0.3 < I < 21$. The range for the parameter I should be decreased taking into account Eq. (3.19.6) and values of I_* in Table 3.19.1.

3.20 CRIEPI Formula

The model that is known as the Central Research Institute of Electric Power Industry (CRIEPI) of Japan formula gives the following relationship for the DOP (Ohnuma *et al.*, 1985; Li *et al.*, 2005):

$$\bar{H} = \phi(\bar{v}_{imp}) = \frac{26.1m}{d^{1.8}}\left(\frac{16.7 \cdot 10^4}{(f_c')^{2/3}} - 1\right)\left[\frac{d + 0.25}{(1.25\bar{b} + 1)\bar{b}}\right]\bar{v}_{imp}^2. \qquad (3.20.1)$$

Ohnuma *et al.*,1985 recommend to use Eq. (3.20.1) for $v_{imp} \le 50\,m/s$. Clearly, the constraints $f_c' < 68.2$ and $\bar{H} \le \bar{b}$ must be valid.

For calculating the perforation and scabbing thicknesses the following formulas are proposed (Ohnuma *et al.*,1985):

$$\bar{b}_{perf} = \Xi_{perf}(\bar{v}_{imp}) = \frac{447}{d^{1.5}}\sqrt{\frac{m}{f_c'}}\,\bar{v}_{imp}^{0.75}, \qquad (3.20.2)$$

$$\bar{b}_{scab} = \Xi_{scab}(\bar{v}_{imp}) = \frac{306}{d^{1.2}}\left(\frac{m}{f_c'}\right)^{0.4} \bar{v}_{imp}^{2/3}. \quad (3.20.3)$$

These formulas differ by the values of the coefficients from the Chang model (Eqs. (3.15.1) and (3.15.2)).

3.21 Vretblad (British) Formula

Teland, 1998 with the reference to Vretblad, 1988 presents the following penetration model which he labels as the "British formula":

$$\bar{H} = \phi(\bar{v}_{imp}) = \frac{0.76 \cdot 10^{-3}(1 - 0.6 \cdot 10^{-8} f_c')m}{d^3 \bar{c}_{\times}} \bar{v}_{imp}^{1.5}. \quad (3.21.1)$$

3.22 UKAEA–CEBG–NNC Formulas

This Section is mainly based on the "Guidelines for the Design and Assessment of Concrete Structures Subjected to Impact" (Barr, 1990). Some additional information on this subject can be also found in Fullard and Barr, 1989, Fullard *et al.*, 1991.

Barr, who compiled this document, notes that the major contributions have been made by UK Atomic Energy Authority (UAEA), Central Electrical Generating Board (CEGB) and National Nuclear Corporation. Therefore we use the names of these organizations in the name of the model.

For calculation of the DOP of a solid missile penetrating into a reinforced concrete barrier having a sufficient thickness to suffer no scabbing, the following model is recommended:

$$g(\bar{H}) = \mu \bar{v}_{imp}^{1.8}, \quad (3.22.1)$$

where

$$\mu = 9.55 \frac{K_{shape}^{(3)} m}{d^{2.8}\sqrt{f_c'}}, \tag{3.22.2}$$

$$K_{shape}^{(3)} = \begin{cases} 0.72 & \text{for flat nose} \\ 0.84 & \text{for spherical nose} \\ 1.00 & \text{for a blunt (conic frustum or ogive) nose} \\ 1.14 & \text{for sharp (conical) nose} \end{cases} \tag{3.22.3}$$

$$g(\bar{H}) = \begin{cases} 0.55\bar{H} - \bar{H}^2 & \text{if} \quad \bar{H} < 0.22 \\ 0.25\bar{H}^2 + 0.0605 & \text{if} \quad 0.22 \le \bar{H} \le 2 \\ \bar{H} - 0.9395 & \text{if} \quad \bar{H} > 2 \end{cases}. \tag{3.22.4}$$

In the unified form given by Eq. (3.2.1), the UKAEA–CEBG–NNC model can be rewritten as follows:

$$\bar{H} = \phi(\bar{v}_{imp}) = \begin{cases} 0.275 - \sqrt{0.0756 - \mu\bar{v}_{imp}^{1.8}} & \text{if} \quad \bar{v}_{imp} < 0.233/\mu^{5/9} \\ 2\sqrt{\mu\bar{v}_{imp}^{1.8} - 0.0605} & \text{if} \quad 0.233/\mu^{5/9} \le v_{imp} \le 1.033/\mu^{5/9} \\ \mu\bar{v}_{imp}^{1.8} + 0.9395 & \text{if} \quad \bar{v}_{imp} > 1.033/\mu^{5/9} \end{cases} \tag{3.22.5}$$

The ranges of the parameters in this formula are as follows: +20% to -20% for $\bar{H} > 0.75$ and +100% to -50% for $\bar{H}$ <0.75; 25*m/s* < v_{imp}<300*m/s*; $5 \cdot 10^3\, kg/m^3 < m/d^3 < 2 \cdot 10^5\, kg/m^3$ and $22\, MPa < f_c' < 44\, MPa$.

For prediction of the scabbing thickness the following formula is suggested:

$$\bar{b}_{scab} = \Xi_{scab}(\bar{v}_{imp}) = 5.3\mu^{1/3}\bar{v}_{imp}^{-0.6}. \tag{3.22.6}$$

The accuracy of this formula is ±40% for $2 < \bar{b}_{scab} < 5.6$ within the following ranges of the parameters: $29\, m/s < v_{imp} < 238\, m/s$, $26\, MPa < f_c' < 44\, MPa$ and $1.5 \cdot 10^3\, kg/m^3 < m/(d^2 b_{scab}) < 4 \cdot 10^4\, kg/m^3$.

Eq. (3.22.6) implies that

$$\bar{v}_{scab} = \Xi_{scab}^{-1}(\bar{b}) = \frac{0.062}{\mu^{5/9}}\bar{b}^{5/3}. \tag{3.22.7}$$

Formula for the BLV of a flat nose missile with a circular or non circular cross-section against reinforced concrete shield reads:

$$\bar{v}_{bl} = \begin{cases} \bar{v}_{bl}^{*} & if \quad \bar{v}_{bl}^{*} \le 0.07 \\ \bar{v}_{bl}^{*}(4\cdot \bar{v}_{bl}^{*2}+1) & if \quad \bar{v}_{bl}^{*} > 0.07 \end{cases}, \tag{3.22.8}$$

where

$$\bar{v}_{bl}^{*} = \frac{1.3\cdot 10^{-3} k_R k_p^{2/3} \rho_{sh}^{1/6} d^2 \sqrt{f_c'^{*}}}{m^{2/3}} \sqrt{r+0.3}\,\bar{b}^{4/3}, \tag{3.22.9}$$

$$f_c'^{*} = \begin{cases} f_c' & if \quad f_c' \le 37\,MPa \\ 37\,MPa & if \quad f_c' > 37\,MPa \end{cases} \tag{3.22.10}$$

$$k_R = \begin{cases} 1.2 - 0.6 c_R/b & if \quad 0.12 \le c_R/b \le 0.49 \\ 1 & if \quad c_R/b < 0.12 \; or \; c_R/b > 0.49 \end{cases} \tag{3.22.11}$$

$$r = \frac{100}{b}\left(\frac{a_F}{3c_F} + \frac{2a_R}{3c_R}\right), \tag{3.22.12}$$

$$k_p = p/(\pi d), \tag{3.22.13}$$

a_F (a_R) is the cross section area of a single front (rear) rebar, c_F (c_R) is a front (rear) rebar spacing (see Fig. 3.22.1a), p is the perimeter of the impactor, f_c' is the characteristic compressive strength of concrete measured for 150 mm diameter, 300 mm long cylinders; $k_p = 1$ for a missiles having a circular cross-section.

This model is valid in the following ranges of the parameters: $11\,m/s < v_{bl} < 300\,m/s$, $0 < r < 0.75$, $0.2 < p/(\pi b) < 3$, $150\,kg/m^3 < m/(p^2 b)$ $< 10^4\,kg/m^3$, $f_c' > 15\,MPa$.

In the case when a steel plate is installed at the distal face of the shield (see Fig. 3.22.1b) for improving its protective effectiveness, the following formula is recommended instead of Eq. (3.22.9):

$$v_{bl}^{*} = \frac{1.3\rho_{sh}^{1/6}\sqrt{f_c'^{*}}}{m^{2/3}}\left(\frac{p}{\pi}\right)^{2/3}\sqrt{B+0.3}\,b^{4/3}, \tag{3.22.14}$$

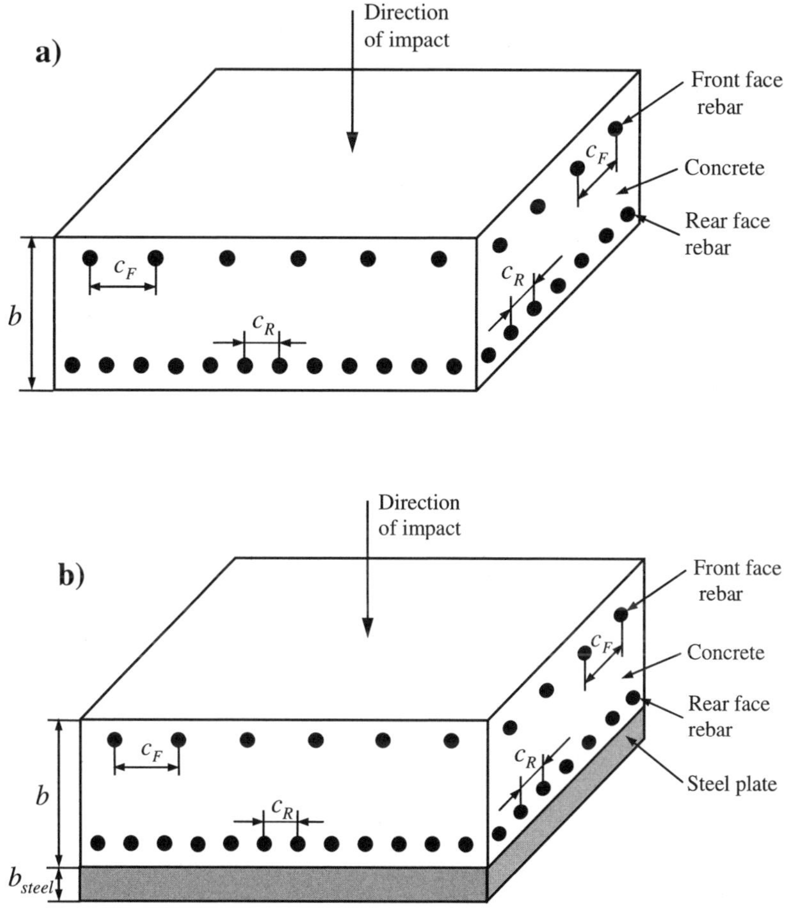

Fig. 3.22.1. Reinforced concrete barrier without liner (a) and with liner (b). Notations.

where

$$B = r + 100 b_{steel} / b \tag{3.22.15}$$

and b_{steel} is the steel plate thickness.

This formula is valid in the following ranges of the parameters: $45\, m/s < v_{bl} < 300\, m/s$, $0 < r < 0.75$, $150\, kg/m^3 < m/(p^2 b) < 10^4\, kg/m^3$, $1.2 < B < 4.3$, $0.2 < p/(\pi b) < 3$, $f_c' > 15\, MPa$.

3.23 Young Formula

3.23.1 *Original model*

After some algebra, the Young equation (Young, 1997) can be written as follows:

$$\bar{H} = \phi(\bar{v}_{imp}) = \frac{K_S K_{shape} K_m}{d^{2.4}} P(\bar{v}_{imp}), \tag{3.23.1}$$

where the coefficients K_S and K_{shape} are defined below,

$$P(\bar{v}_{imp}) = \begin{cases} P_a(\bar{v}_{imp}) & if \quad \bar{v}_{imp} < \bar{v}_* \\ P_b(\bar{v}_{imp}) & if \quad \bar{v}_{imp} \geq \bar{v}_* \end{cases} \tag{3.23.2}$$

$$P_a(\bar{v}_{imp}) = \alpha_1 \ln(1 + \alpha_2 \bar{v}_{imp}^2), \quad \alpha_1 = 9.48 \cdot 10^{-4}, \alpha_2 = 215, \tag{3.23.3}$$

$$P_b(\bar{v}_{imp}) = k(\bar{v}_{imp} - \bar{v}_0), \quad k - 0.0213, \quad \bar{v}_0 - 0.0305, \tag{3.23.4}$$

$$\bar{v}_* = 0.061. \tag{3.23.5}$$

$$K_m = \begin{cases} 0.46 m^{0.85} & if \quad m < 182 \\ m^{0.7} & if \quad m \geq 182 \end{cases} \tag{3.23.6}$$

The coefficient K_{shape} depends on the shape of the impactor, and it is determined by the following formula:

$$K_{shape}=\begin{cases}0.18\overline{L}_{nose}-0.09\Delta\overline{L}_{nose}+0.56 & for\ ogive\ nose\\ 0.25\overline{L}_{nose}-0.125\Delta\overline{L}_{nose}+0.56 & for\ conic\ nose\end{cases}, \tag{3.23.7}$$

where $\overline{L}_{nose}$ and $\Delta\overline{L}_{nose}$ are the dimensionless (measured in impactor's diameter units) length of the nose and the reduction of this length because of the bluntness (if any), correspondingly.

For sharp ogive nosed shapes, the following equation can also be used:

$$K_{shape}=0.56+0.18\sqrt{K_{CRH}-0.25}\ . \tag{3.23.8}$$

If the nose of the impactor is neither ogive nor cone, Young, 1997 recommends approximating the actual nose shape by the ogive or conic shapes. If the bluntness is less than 10% of the penetrator's diameter, it can be ignored.

The coefficient K_S depends on the properties of the concrete and it is determined by the following expression:

$$K_S=\frac{0.247K_e(11-K_D)}{(t_c'\overline{b}')^{0.06}(f_c')^{0.3}}, \tag{3.23.9}$$

where

$$t_c'=\min(t_c,1), \tag{3.23.10}$$

$$\overline{b}'=\min(\overline{b},6), \tag{3.23.11}$$

$$K_e=\max\left[\left(\frac{K_F F}{\overline{W}}\right)^{0.3},1\right], \tag{3.23.12}$$

$$F=\begin{cases}20 & \text{for reinforced concrete}\\ 30 & \text{for nonreinforced concrete}\end{cases}, \tag{3.23.13}$$

$$K_F=\begin{cases}0.5 & if\quad \overline{b}\le 2\\ 1 & if\quad \overline{b}>2\end{cases}, \tag{3.23.14}$$

K_D is the volumetric rebar content, t_c is cure time in days, $\overline{W}$ is the shield width. Bars over variables denote the dimensionless parameters which are normalized as indicated in the beginning of this section.

If there is no data for calculating the coefficient K_S, the default value of $K_S = 0.9$ can be used.

This model is recommended for $7\,MPa < f_c' < 124\,MPa$, $m > 5\,kg$, $\overline{H} \geq 3$, $v_{imp} < 1220\,m/s$.

3.23.2 *Modifications of the models*

Ben-Dor *et al.*, 2008a, 2008c proposed two modification of the model that affords continuity of $P(\overline{v}_{imp})$ and its first derivative which are discontinuous at the point $\overline{v}_{imp} = \overline{v}_*$. This discontinuity has no physical meaning and can cause problems when applying these relations.

3.23.2.1 *First modification*

Ben-Dor *et al.*, 2008c modified the model for the relatively small $\overline{v}_{imp}$ in order to construct a smooth approximation for the function $P(\overline{v}_{imp})$. The problem is to determine the parameters, $\alpha_1 > 0$, $\alpha_2 > 0$ and $\overline{v}_*$ such that

$$P_a(\overline{v}_*) = P_b(\overline{v}_*), \quad P_a'(\overline{v}_*) = P_b'(\overline{v}_*). \tag{3.23.15}$$

Since the goal is to preserve the Young approximation for large $\overline{v}_{imp}$, and since the function $\overline{P}_b(\overline{v}_{imp})$ is defined only for $\overline{v}_{imp} \geq 2\overline{v}_0$, one can impose the constraint that $\overline{v}_* \geq 2\overline{v}_0$.

Substituting P_a and P_b from Eqs. (3.23.3) and (3.23.4) into Eq. (3.23.15) yields the following relations:

$$\alpha_1 \ln(1+\alpha_2 v_*^2) = k(v_* - v_0), \quad 2\alpha_1\alpha_2 v_* = k(1+\alpha_2 v_*^2). \tag{3.23.16}$$

After the change of variables we find that

$$\alpha_1 = \frac{k_1(\overline{v}_* - \overline{v}_0)}{\ln x}, \quad \alpha_2 = \frac{x-1}{\overline{v}_*^2}. \tag{3.23.17}$$

It can be easily seen that the first in Eqs. (3.23.16) is satisfied identically while the second yields the equation, $f(x) = 0$, where

$$f(x) = x\ln x - 2(x-1)(1-\overline{v}_0/\overline{v}_*), \quad f'(x) = \ln x - 1 + 2\overline{v}_0/\overline{v}_* . \qquad (3.23.18)$$

Inspection of Eq. (3.23.17) shows that we must consider the independent variable, x, in the range $x > 1$.

In the case of Young approximation ($\overline{v}_* = 2\overline{v}_0$), $f'(x) > 0$ for $x > 1$ and, consequently, $f(x)$ increases as x increases. Since $f(1) = 0$, the equation $f(x) = 0$ does not have roots in the range $x > 1$. Consequently, Young approximation can not be improved by keeping $\overline{v}_* = 2\overline{v}_0$ and changing only α_1 and α_2.

If $\overline{v}_* > 2\overline{v}_0$, then $f(x)$ decreases from $f(1) = 0$ to $f(x_0) < 0$ when x varies from 1 to $x_0 = \exp(1 - 2\overline{v}_0/\overline{v}_*)$ and tends to ∞ as x tends to infinity. Consequently, a single root of equation $f(x) = 0$, $x_* > x_0$, always exists for every $\overline{v}_* > 2\overline{v}_0$. After selecting $\overline{v}_*$ and determining x_* as a solution of the equation $f(x) = 0$, the parameters α_1 and α_2 are determined from Eq. (3.23.18) for $x = x_*$.

For

$$\overline{v}_* = 0.065 \qquad (3.23.19)$$

we obtain $x_* = 1.1282$ and

$$\alpha_1 = 6.092 \cdot 10^{-3}, \ \alpha_2 = 30.34 . \qquad (3.23.20)$$

3.23.2.2 *Second modification*

Ben-Dor *et al.*, 2008a proposed a smooth approximation of the function $P(\overline{v}_{imp})$. This approximation is continuous and has continuous first and second order derivatives for all $\overline{v}_{imp} > 0$. Toward this end, they modified the model for the relatively small $\overline{v}_{imp}$ keeping the Young approximation for large $\overline{v}_{imp}$. In this case the function $P_b(\overline{v}_{imp})$ is given by Eq. (3.23.4) while the following approximation of the function $P_a(\overline{v}_{imp})$ is sought for:

$$P_a(\overline{v}_{imp}) = (\beta_2 \overline{v}_{imp}^2 + \beta_1 \overline{v}_{imp} + \beta_0)\overline{v}_{imp}^2,\ \overline{v}_{imp} \le \overline{v}_*,\ \overline{v}_* = s\overline{v}_0,\ s \ge 2. \qquad (3.23.21)$$

Now the problem is reduced to finding such $\beta_0, \beta_1, \beta_2$ and s that

$$P_a(\overline{v}_*) = P_b(\overline{v}_*),\ P_a'(\overline{v}_*) = P_b'(\overline{v}_*),\ P_a''(\overline{v}_*) = P_b''(\overline{v}_*),\ \overline{v}_* = s\overline{v}_0, \qquad (3.23.22)$$

where functions P_a and P_b are determined by Eqs. (3.23.21) and (3.23.4), correspondingly,

$$\begin{aligned} &P_a'(\overline{v}_{imp}) = 4\beta_2 \overline{v}_{imp}^3 + 3\beta_1 \overline{v}_{imp}^2 + 2\beta_0 \overline{v}_{imp}, \quad P_b'(\overline{v}_{imp}) = k \\ &P_a''(\overline{v}_{imp}) = 12\beta_2 \overline{v}_{imp}^2 + 6\beta_1 \overline{v}_{imp} + 2\beta_0, \quad P_b''(\overline{v}_{imp}) = 0 \end{aligned}. \qquad (3.23.23)$$

In addition, simple physical reasoning reveals that P_a must be an increasing function.

Eq. (3.23.22) yields a system of linear equations with respect to β_0, β_1 and β_2 which has the following solution:

$$\beta_0 = \frac{3k(s-2)}{s^2 \overline{v}_0},\quad \beta_1 = \frac{k(8-3s)}{s^3 \overline{v}_0^2},\quad \beta_2 = \frac{k(s-3)}{s^4 \overline{v}_0^3}. \qquad (3.23.24)$$

Let us prove that for any $s \ge 2$ the function $P_a(\overline{v}_{imp})$ increases when $0 < \overline{v}_{imp} \le s\overline{v}_0$. This occurs if $P_a'(\overline{v}_{imp}) > 0$ in the same range of $\overline{v}_{imp}$. After substituting β_0, β_1 and β_2 given by formulas (3.23.24) into Eq. (3.23.23) for $P_a'(\overline{v}_{imp})$ and changing variables,

$$V = \overline{v}_{imp} / (s\overline{v}_0), \qquad (3.23.25)$$

it can be easily seen that the condition $P_a'(\overline{v}_{imp}) > 0$ is equivalent to the condition that $\omega(V) > 0$ for $0 < V \le 1$ where

$$\omega(V) = \frac{1}{k\overline{v}_0 V} P_a'(s\overline{v}_0 V) = 4(s-3)V^2 + 3(8-3s)V + 6(s-2). \qquad (3.23.26)$$

Let $2 \le s < 3$. Since $\omega(0) \ge 0$ and $\omega(1) = s > 0$, the convex upward function $\omega(V) > 0$ for $0 < V \le 1$.

The discriminant of the quadratic equation $\omega(V)=0$ is $\Delta(s)=3s(16-5s)$. Since $\Delta<0$ and $s-3>0$ when $s>16/5$, then $\omega(V)>0$ for $s>16/5=3.2$.

Let $3\le s<3.2$. Then there are two different roots of the equation $\omega(V)=0$, V_1 and V_2. The Viet formula implies that both roots are positive. The smaller root, V_1, is $V_1(s)=[3(3s-8)-\sqrt{\Delta(s)}]/[8(s-3)]$. It is easy to prove that $V_1>1$ if $3\le s<3.2$. Since $\omega(0)>0$ and all the roots are larger than 1, $\omega(V)>0$ for $0<V\le 1$.

In order to complete the analysis, let us consider $\omega(V)$ for two remaining values of s. If $s=3$ then $\omega(V)=3(2-V)>0$, and if $s=3.2$ then $\omega(V)=0.8(V-3)^2>0$, for $0<V\le 1$.

Therefore we proved that $P_a(\overline{v}_{imp})$ is an increasing function in the interval $0<\overline{v}_{imp}\le s\overline{v}_0$ when $s\ge 2$. In order to obtain a more convenient approximation let us set $s=8/3$.
then

$$\beta_0=0.196,\quad \beta_1=0,\quad \beta_2=-4.95,\quad \overline{v}_*=s\overline{v}_0=0.0813 \tag{3.23.27}$$

in Eq. (3.23.21).
Young approximation and the suggested modifications are illustrated in Fig. 3.23.1.

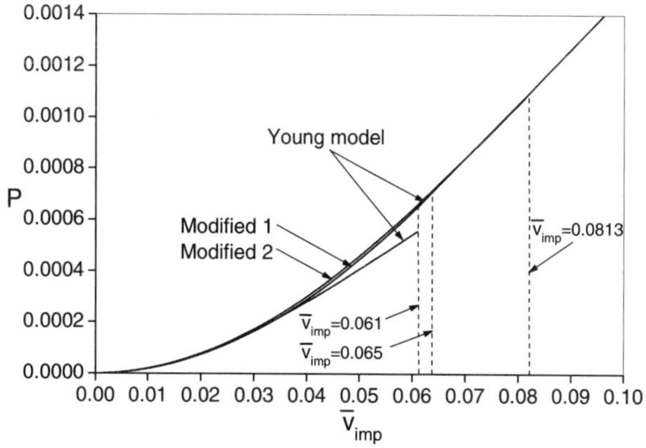

Fig. 3.23.1. Young model and its modifications.

3.24 UMIST Formulas

This section is based on the studies by Li *et al.*, 2005 and Li *et al.*, 2006 which refer to the approach suggested in the investigations conducted in the UMIST-University of Manchester Institute of Science and Technology (Reid and Wen, 2001; BNFL, 2003).

3.24.1 *Penetration model*

The formula for the DOP reads (Li *et al.*, 2005):

$$\bar{H} = \phi(\bar{v}_{imp}) = \frac{0.88 \cdot 10^6 K_{shape}^{(3)} m}{\sigma_t(\bar{v}_{imp}) d^3} \bar{v}_{imp}^2, \tag{3.24.1}$$

where σ_t is the rate-dependent characteristic strength of the concrete,

$$\sigma_t(\bar{v}) = \zeta_0(f_c') + \zeta_1(f_c')\bar{v}, \tag{3.24.2}$$

$$\zeta_0(f_c') = 4.2 f_c' + 1.35 \cdot 10^8, \quad \zeta_1(f_c') = 14 f_c' + 0.45 \cdot 10^8. \tag{3.24.3}$$

Eq.(3.24.1) has been validated for $\bar{H} < 2.5$, $3m/s < v_{imp} < 66m/s$, $35kg < m < 2500kg$, $5cm < d < 60cm$.

3.24.2 *Perforation and scabbing model and its analysis*

On the basis of formulas presented in Li *et al.*, 2005 (Li *et al.*, 2006 described a similar model for the flat-nosed projectiles when the concrete reinforcement is ignored), the relations for the BLV and the scabbing limit velocity can be written as follows:

$$\frac{\bar{v}_{bl}^2}{\kappa \sigma_t(\bar{v}_{bl})} = \begin{cases} \chi_{perf}^{(1)}(\bar{b}) & if \quad 0.5 < \bar{b} \le 1 \\ \chi_{perf}^{(2)}(\bar{b}) & if \quad 1 \le \bar{b} < 5 \\ \chi_{perf}^{(3)}(\bar{b}) & if \quad \bar{b} \ge 5 \end{cases} \tag{3.24.4}$$

$$\frac{\overline{v}_{sl}^2}{\kappa\sigma_t(\overline{v}_{sl})} = \begin{cases} \chi_{scab}^{(1)}(\overline{b}) & if \quad 0.5 < \overline{b} < 5 \\ \chi_{scab}^{(2)}(\overline{b}) & if \quad \overline{b} \geq 5 \end{cases}, \tag{3.24.5}$$

where

$$\kappa = 2 \cdot 10^{-8} d^3 / m, \tag{3.24.6}$$

$$\begin{aligned} \chi_{perf}^{(1)}(\overline{b}) &= \eta(1.506\overline{b}^2 - 0.506\overline{b}), \\ \chi_{perf}^{(2)}(\overline{b}) &= \eta(2\overline{b}^3 - \overline{b}), \\ \chi_{perf}^{(3)}(\overline{b}) &= 78.5\overline{b} - 235.5, \end{aligned} \tag{3.24.7}$$

$$\begin{aligned} \chi_{scab}^{(1)}(\overline{b}) &= \eta(1.386\overline{b}^2 - 0.5441\overline{b}) / K_{shape}^{(3)} \\ \chi_{scab}^{(2)}(\overline{b}) &= (56.52\overline{b} - 24.3) / K_{shape}^{(3)} \end{aligned} . \tag{3.24.8}$$

Parameter η is associated with the reinforcement of the shield:

$$\eta = \begin{cases} 1.5r(d/c_{FR}) + 0.5 & if \quad d/c_{FR} < \sqrt{d/d_{FR}} \\ 1.5r(d/d_{FR}) + 0.5 & if \quad d/c_{FR} \geq \sqrt{d/d_{FR}} \end{cases}, \tag{3.24.9}$$

where $r = 100 a_{FR}/(c_{FR} b)$ is determined by Eq. (3.22.12) with $a_F = a_R = a_{FR} = \pi d_{FR}^2/4$ and $c_F = c_R = c_{FR}$; a_{FR} and d_{FR} are the cross section area and the diameter of the rebar, correspondingly, c_{FR} is the rebar spacing (see Fig. 3.22.1a).

These scabbing and perforation models are valid (Li *et al.*, 2005) for $2.2\,cm < d < 60\,cm$, $1\,kg < m < 2622\,kg$, $20\,MPa < f_c' < 79\,MPa$, $0 < r < 4$, $5.1\,cm < b < 64\,cm$, $0 < v_{imp} < 427\,m/s$.

Since σ_t is a linear function of $\overline{v}$, determining $\overline{v} = \overline{v}_{bl}$ or $\overline{v} = \overline{v}_{sl}$ in terms of $\overline{b}$ requires solving the quadratic equation $f(\overline{v}) = 0$, where

$$f(\overline{v}) = \overline{v}^2 - (\kappa\zeta_1\Theta)\overline{v} - \kappa\zeta_0\Theta, \tag{3.24.10}$$

Θ is the right side of Eq. (3.24.4) or Eq. (3.24.5). It can be easily shown that the equation $f(\overline{v}) = 0$ has a single positive root:

$$\overline{v} = \omega(\Theta), \quad \omega(\Theta) = 0.5\kappa\zeta_1\Theta + \sqrt{\kappa\Theta(0.25\kappa\zeta_1\Theta + \zeta_0)} \,. \tag{3.24.11}$$

Taking into account Eq. (3.24.11), the expressions for $\overline{v}_{bl}$ and $\overline{v}_{sl}$ as functions of $\overline{b}$ can be written as follows:

$$\overline{v}_{bl} = \begin{cases} \omega\left(\chi^{(1)}_{perf}(\overline{b})\right) & if \quad 0.5 < \overline{b} \le 1 \\ \omega\left(\chi^{(2)}_{perf}(\overline{b})\right) & if \quad 1 \le \overline{b} < 5 \\ \omega\left(\chi^{(3)}_{perf}(\overline{b})\right) & if \quad \overline{b} \ge 5 \end{cases}, \tag{3.24.12}$$

$$\overline{v}_{sl} = \begin{cases} \omega\left(\chi^{(1)}_{scab}(\overline{b})\right) & if \quad 0.5 < \overline{b} < 5 \\ \omega\left(\chi^{(2)}_{scab}(\overline{b})\right) & if \quad \overline{b} \ge 5 \end{cases}. \tag{3.24.13}$$

In order to show that $\overline{v}_{bl}$ and $\overline{v}_{sl}$ are increasing function of $\overline{b}$ let us calculate the first order derivatives:

$$\frac{d\overline{v}_{bl}}{d\overline{b}} = \begin{cases} \omega'\left(\chi^{(1)}_{perf}(\overline{b})\right)\eta(3.012\overline{b} - 0.506) & if \quad 0.5 < \overline{b} \le 1 \\ \omega'\left(\chi^{(2)}_{perf}(\overline{b})\right)\eta(6\overline{b}^2 - 1) & if \quad 1 \le \overline{b} < 5 \\ \omega'\left(\chi^{(3)}_{perf}(\overline{b})\right)78.5 & if \quad \overline{b} \ge 5 \end{cases}, \tag{3.24.14}$$

$$\frac{d\overline{v}_{sl}}{d\overline{b}} = \begin{cases} \omega'\left(\chi^{(1)}_{scab}(\overline{b})\right)\eta(2.772\overline{b} - 0.5441)/N^* & if \quad 0.5 < \overline{b} < 5 \\ \omega'\left(\chi^{(2)}_{scab}(\overline{b})\right) 56.52/K^{(3)}_{shape} & if \quad \overline{b} \ge 5 \end{cases}, \tag{3.24.15}$$

where

$$\omega'(\Theta) = 0.5\kappa\left[\zeta_1 + \frac{0.5\kappa\zeta_1\Theta + \zeta_0}{\sqrt{\kappa\Theta(0.25\kappa\zeta_1\Theta + \zeta_0)}}\right]. \tag{3.24.16}$$

Since $\omega'(\Theta) > 0$ and the derivatives, $d\chi^{(i)}_{perf}(\overline{b})/d\overline{b}$ ($i = 1,2,3$) and $d\chi^{(j)}_{scab}(\overline{b})/d\overline{b}$ ($j = 1,2$), given by Eq. (3.24.14) and Eq. (3.24.15) are positive, in the corresponding ranges, it can be concluded that $\overline{v}_{bl}(\overline{b})$ and $\overline{v}_{sl}(\overline{b})$ are the increasing function. Consequently, (i) the suggested model adequately describes physics of the penetration phenomenon, and

(ii) there exist the single-valued inverse functions $\bar{b}_{perf}(\bar{v}_{imp})$ and $\bar{b}_{scab}(\bar{v}_{imp})$.

For the function $\bar{v}_{bl} = \omega\left(\chi_{perf}^{(i)}(\bar{b})\right)$ ($i = 1,2,3$) defined on the interval, $\bar{b}_{\min}^{(i)} \le \bar{b} \le \bar{b}_{\max}^{(i)}$, determining the function $\bar{b}_{perf}(\bar{v}_{imp})$ reduces to solving the following equation with respect to $\bar{b}_{perf}$:

$$\frac{\bar{v}_{imp}^2}{\kappa\sigma_t(\bar{v}_{imp})} = \chi_{perf}^{(i)}(\bar{b}_{perf}). \tag{3.24.17}$$

The latter equation is valid for $\bar{v}_{imp}$ between boundary values: $(\bar{v}_{imp})_{\min}^{(i)} = \omega\left(\chi_{perf}^{(i)}(\bar{b}_{\min}^{(i)})\right)$ and $(\bar{v}_{imp})_{\max}^{(i)} = \omega\left(\chi_{perf}^{(i)}(\bar{b}_{\max}^{(i)})\right)$. Similarly, the dependence $\bar{b}_{scab}$ vs. $\bar{v}_{imp}$ can be determined.

When the right-hand side of Eq. (3.24.17) is a linear or quadratic function, the equation can be easy solved. For $i = 2$, the cubic algebraic equation is obtained:

$$\bar{b}_{perf}^3 - p\bar{b}_{perf} - q = 0, \tag{3.24.18}$$

where

$$p = 0.5, \quad q = \frac{\bar{v}_{imp}^2}{2\eta\kappa\sigma_t(\bar{v}_{imp})} > 0. \tag{3.24.19}$$

For arbitrary positive p and q, the unique positive root of this equation is given by the following formulas:

$$\bar{b}_{perf} = \begin{cases} 2\sqrt{p/3}\cos(\lambda/3) & if \quad D \le 0 \\ 2\sqrt{p/3}ch(\hat{\lambda}/3) & if \quad D > 0 \end{cases}, \tag{3.24.20}$$

where

$$D = \left(\frac{q}{2}\right)^2 - \left(\frac{p}{3}\right)^3, \quad s = \frac{3\sqrt{3}}{2}\frac{q}{p^{3/2}}, \quad \lambda = \cos^{-1}s, \quad \hat{\lambda} = ch^{-1}s. \tag{3.24.21}$$

Let p and q be given by Eq. (3.24.19). Since $\overline{b}_{perf}$ varies from 1 to 5 on the considered interval (see Eq. (3.24.4)), then $\overline{b}_{perf}^3 - 0.5\overline{b}_{perf}$ and, consequently, q, increases from 0.5 to 122.5. Therefore $D \geq 0.25^2 - (1/6)^3 \approx 0.058 > 0$ and the solution of Eq. (3.24.18) reads:

$$\overline{b}_{perf} = (2/\sqrt{6})ch(\hat{\lambda}/3), \quad \hat{\lambda} = ch^{-1}(3\sqrt{6}\,q). \qquad (3.24.22)$$

It is important to keep in mind that the functions defined by Eq. (3.24.4) and Eq. (3.24.5) are discontinuous in the point $\overline{b} = 5$. The latter implies that the function $\overline{v}_{bl} = \overline{v}_{bl}(\overline{b})$ assumes different values, $\overline{v}'_{bl}$ and $\overline{v}''_{bl}$, to the left and to the right of this point. Therefore the inverse function, $\overline{b}_{perf} = \Xi_{perf}(\overline{v}_{imp})$, can not be determined for $\overline{v}'_{bl} < \overline{v}_{imp} < \overline{v}''_{bl}$. The discontinuity of the function $\overline{v}_{sl} = \overline{v}_{sl}(\overline{b})$ causes similar problems.

3.25 Malaysia-UTHM Models

Zaidi *et al.*, 2010a from University Tun Hussein Onn Malaysia (UTHM) proposed a linear model for ogive-nosed impactors which describes the dependence between the DOP, $\overline{H}$, and the parameter $mv_{imp}^2/(f'_c d^3)$, with coefficients depending on the CRH of the impactor, K_{CRH}:

$$\overline{H} = \frac{0.5 \cdot 10^6 m\overline{v}_{imp}^2}{f'_c d^3} q_1 + q_0, \qquad (3.25.1)$$

where

$$\begin{aligned} q_0 &= -0.6256K_{CRH}^3 + 7.9656K_{CRH}^2 - 32.892K_{CRH} + 48.008 \\ q_1 &= 0.014K_{CRH}^3 - 0.1585K_{CRH}^2 + 0.5606K_{CRH} - 0.5405 \end{aligned}. \qquad (3.25.2)$$

This formula can be applied in the following ranges of parameters: $139\,m/s < v_{imp} < 1225\,m/s$, $13.5\,MPa \leq f'_c \leq 108\,MPa$, $64\,g < m < 13.2\,kg$, $2 \leq K_{CRH} \leq 6$, $13\,mm < d < 76.2\,mm$.

Rahman *et al.*, 2010a and Latif *et al.*, 2011 considered perforation by flat nosed (cylindrical) impactor.

Rahman *et al.*, 2010a proposed a formula for the scabbing limit velocity that can be written as follows:

$$\bar{v}_{sl} = 10^{-3} d \sqrt{\frac{2 f_c' d}{m} \psi_1(\bar{b})}, \qquad (3.25.3)$$

where

$$\psi_1(\bar{b}) = \begin{cases} 0.87\bar{b} - 0.29 & \text{if} \quad 0.69 \le \bar{b} \le 3.0 \\ 3.31\bar{b} - 7.58 & \text{if} \quad 3.0 < \bar{b} \le 6.0 \\ 4279.181\bar{b} - 25662.82 & \text{if} \quad 6.0 < \bar{b} \le 14.86 \end{cases}. \qquad (3.25.4)$$

The latter formula is valid in the following ranges of parameters: $0.69 < b/d < 14.86$, $24.15\,MPa \le f_c' \le 50.2\,MPa$, $17.5\,mm < d < 305.0\,mm$, $0.92\,kg < m < 309\,kg$, $29\,m/s < v_{sl} < 427\,m/s$, $50.8\,mm < b < 609.6\,mm$.

Latif *et al.*, 2011 suggested a formula for the BLV velocity that can be written as follows:

$$\bar{v}_{bl} = 10^{-3} d \sqrt{\frac{2 f_c' d}{m} \psi_2(\bar{b})}, \qquad (3.25.5)$$

where

$$\psi_2(\bar{b}) = 0.174\bar{b}^3 + 0.169\bar{b}^2 + 0.0577\bar{b} + 0.2969 \qquad (3.25.6)$$

for

$0.92\,kg < m < 309\,kg$, $\quad 29\,m/s < v_{bl} < 427\,m/s$,
$0.69 < b/d < 14.86$, $\quad 24.15\,MPa \le f_c' \le 50.2\,MPa$,
$17.5\,mm < d < 305.0\,mm$.

3.26 TM 5-855-1 Formulas

TM 5-855-1, 1986 formulas are proposed for describing penetration by the standard fragment (see Fig. 3.26.1).

The mass of the fragment that is used in formulas is calculated as follows:

$$m = (5\pi/24)\,\rho_{imp} d^3 = 0.654\rho_{imp} d^3 . \tag{3.26.1}$$

TM 5-1300, 1990 provides the following formula for the fragment's mass (after conversion to SI units): $m = 5.149 \cdot 10^3 d^3$, i.e., the density is assumed to be equal $\rho_{imp} = 7873 kg/m^3$.

After some transformations, formulas from TM 5-855-1, 1986 can be written in the following form:

$$H = \begin{cases} \dfrac{13.46 m^{0.37} \overline{v}_{imp}^{0.9}}{(f_c')^{0.25}} & if \quad \overline{v}_{imp} \le \overline{v}_{imp}^* \\ \dfrac{1.09 \cdot 10^3 m^{0.4} \overline{v}_{imp}^{1.8}}{(f_c')^{0.5}} + 0.395 m^{0.33} & if \quad \overline{v}_{imp} > \overline{v}_{imp}^* \end{cases} \tag{3.26.2}$$

$$b_{perf} = 0.0311 H m^{0.033} + 2.95 m^{0.33} , \tag{3.26.3}$$

$$b_{scab} = 0.0334\, H m^{0.033} + 4.465 m^{0.33} , \tag{3.26.4}$$

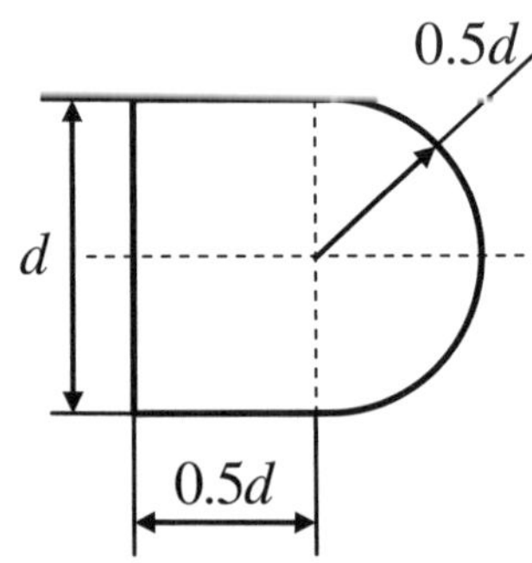

Fig. 3.26.1. Standard fragment shape.

$$\overline{v}_{res} = \begin{cases} \overline{v}_{imp}[1-(b/b_{perf})^2]^{0.555} & if \quad \overline{v}_{imp} \le \overline{v}_{imp}^* \\ \overline{v}_{imp}[1-(b/b_{perf})]^{0.555} & if \quad \overline{v}_{imp} > \overline{v}_{imp}^* \end{cases}, \tag{3.26.5}$$

where

$$\overline{v}_{imp}^* = 5.13 \cdot 10^{-3} (f_c')^{0.278} / m^{0.044} . \tag{3.26.6}$$

3.27 Folsom Model for Penetration into a Shield with a Predrilled Hole

Folsom, 1987 proposed the formula for the DOP of ogive nosed projectile into a predrilled shield that can be written in the interpretation of Teland, 2001 in the following form:

$$\overline{H} = \frac{17.3 m \overline{v}_{imp}^{1.5}}{f_c' d^{2.785}} \cdot \frac{1-038\eta^2}{1-\eta^2} - \frac{4\Omega(K_{CRH}, \eta)}{1-\eta^2} + \sqrt{K_{CRH} - 0.25} , \tag{3.27.1}$$

where d_0 is the diameter of the hole,

$$\begin{aligned} \Omega(z,\eta) &= \left(\frac{1-\eta^2}{4} - z + 2z^2 \right) g \\ &+ \frac{1-2z}{2} \left[g\sqrt{z^2 - g^2} + z^2 \sin^{-1}(g/z) \right] - \frac{g^3}{3}, \\ g(z,\eta) &= \sqrt{(1-\eta)[z - 0.25(1-\eta)]}, \quad \eta = d_0/d . \end{aligned} \tag{3.27.2}$$

3.28 Some Other Models and Related Problems

Walter and Wolde-Tinsae, 1984 proposed modified versions of several empirical formulas that allow taking into account the presence of a steel plate at the distal face of the shield by introducing the artificial thickness of the shield. Riera, 1989 presented the relationship between H and v_{imp} as an implicit function and suggested and discussed equations for determining v_{bl} and v_{sl}. Al-Hachamee and Azeez, 2010 proposed formulas for H, b_{perf} and b_{scab} on the basis of the results of 20 experiments from the literature. Dancygier, 1997 proposed a method that allowed including the reinforcement ratio as a parameter in the existing semi-empirical perforation formulas.

Me-Bar, 1997 proposed a method for scaling ballistic penetration phenomena for concrete shields. The author suggested separating the energy absorbed by the shield during penetration into the energy expended for surface effects and the energy expended for volume effects. Then using energy balance one can derive the expressions that account for the irregularity in scaling. Dancygier, 2000 investigated the effects of impact by similarly (not necessarily proportional) shaped impactors on reinforced concrete barriers. Based on the widely known empirical formulae, the expressions for the velocities ratios between the simulator (reference impactor) and the simulated impactors that are required in order to yield the same DOPs and the same perforation limit thicknesses, were derived.

3.29 Comparison Between the Models and Their Experimental Validation

3.29.1 *Brief review*

Plots which allow comparing the dependencies obtained using various phenomenological models (functions in Eqs. (3.2.1) – (3.2.8)), can be found in the studies by Kennedy, 1976; Yankelevsky, 1997; Teland, 1999; Li and Tong, 2003, etc.

The analysis of the models based on comparing their predictions with experimental results is complicated by that (i) the number of the experiments is small, (ii) they are usually conducted in a narrow range of the parameters, and (iii) information concerning the properties of concrete in the shield (reinforcement, aggregate size, etc.) is insufficient. It must also be noted that in some cases the evaluation of the accuracy of the model is performed using the same experimental data as those used for elaborating the model.

Kennedy, 1976, examined ACE, NDRC, BBL, Amman –Whitney and Perty models using experimental data and recommended the Modified

NDRC formula for calculating b_{scab} and b_{perf} in the following range of the parameters: $d < 41\,sm$, $0.02\,kg/m^3 < m/d^3 < 0.54\,kg/m^3$, $30\,m/s < v_{imp} < 900\,m/s$. He concluded that this formula generally agrees with the test results with the accuracy $\pm 20\%$ in this range of the parameters.

Sliter, 1980 collected experimental data on penetration of cylindrical projectiles into finite width shields and calculated the DOP using the NDRC model, the scabbing thickness using the NDRC, Bechtel and Stone & Webster models, and the perforation thickness using the NDRC and the CEA-EDF models. He found that the values of the DOP given by the NDRC formula are within experimental scatter $\pm 25\%$ for $v_{imp} > 152\,m/s$ while the agreement between the predicted and the observed values of the DOP is unsatisfactory for the relatively low impact velocities. As far as scabbing prediction concerns, Sliter, 1980i noted that the NDRC, Bechtel and Stone & Webster models may be used equally well for the missiles with relatively small diameters. Regarding perforation, he indicated that damage is predicted better by the CEA-EDF formula than by the NDRC model.

On the basis of the experimental data from Sliter, 1980, Adeli and Amin, 1985 examined the Modified NDRC, Haldar & Miller, Hughes and Adeli–Amin models for penetration, Petry I, Petry II, ACE, Modified NDRC, BRL, Bechtel, Chang, Hughes and Adeli–Amin models for scabbing, and Petry I, Petry II, ACE, the Modified NDRC, BRL, CEA-EDF, Degen, Chang, Hughes and Adeli–Amin models for perforation. Their conclusions are as follows.

1) *Penetration modeling*. For $\bar{H}_{\exp} \geq 0.6$, where

$$\bar{H}_{\exp} = H_{\exp}/d\,, \tag{3.29.1}$$

and $H_{\exp}$ is the experimentally observed DOP, the Modified NDRC, Haldar & Miller, Hughes and Adeli–Amin models "tend to agree with the experimental results within $\pm 25\%$". For $\bar{H}_{\exp} < 0.6$, Petry II, Haldar & Miller and the cubic Adeli–Amin models perform better than other models. The ACE and Petry I models considerably overestimated the DOP. Consequently, for calculating the DOP Adeli and Amin,1985 recommended applying the quadratic Adeli–Amin formula for $v_{imp} < 145\,m/s$, the quadratic Adeli–Amin model or the Modified NDRC formula for $145\,m/s \le v_{imp} < 305\,m/s$.

2) *Scabbing modeling.* Adeli–Amin, Bechtel and Chang models perform better and and, generally, are the least conservative among all the models. The Hughes model is the most conservative while Petry I, Petry II and BRL equations are the poorest in predicting the scabbing thickness. Consequently, for scabbing modelling the quadratic Adeli–Amin, Bechtel or Chang formula for $v_{imp} < 310\,m/s$ are recommended.

3) *Perforation modeling.* The CEA-EDF, Degen, Chang and Adeli–Amin models perform better than other models although the predictions of the Modified NDRC and Petry I formulas also agree with the experimental data. The predictions of the Petry II, BRL and ACE formulas do not show good agreement with the results of experiments. Consequently, the CEA-EDF, Degen, Chang and Adeli–Amin models for $v_{imp} < 310\,m/s$ are recommended for perforation modeling.

Walter and Wolde-Tinsae, 1984 used the results of 45 experiments for comparing the performance of the basic empirical models. They concluded that among 9 empirical models the best predictions of the perforation/non-perforation events are provided by Petry I, Petry II Degen, BRL and CEA-EDF formulas whereby the percentage of the correct prediction is in the range between 71 and 75% while the percentage of the correct prediction is minimal for the NDRC and Kar formulas (53%).

DOE, 2006 recommends the Chang formula for calculating b_{scab} and b_{perf}.

3.29.2 *Evaluation of the performance of models. Finite width shield*

Evaluation of the performance of different empirical models in this section is based on the results of experimental data collected by Sliter, 1980. The input data for the evaluated models are adopted from Sliter, 1980, and only those experiments that satisfy the range of validity of the model are used for evaluating each of the models. The UMIST model is not included to the list of the analyzed models because most of the experiments from Sliter, 1980 are outside the range of the validity of this model. The comparison of the UMIST model and the NDRC model with other set of experimental data can be found in Li *et al.*, 2006.

Detailed results of the calculations together with the experimental data are shown in Table 3.29.1. Parameters b_{perf} and b_{scab} in the average model are determined as arithmetic means of the values of the corresponding parameter in the models where these parameters were calculated. It is assumed that penetration (event "N"), scabbing (event "S") or perforation (event "P") occurs if $b < b_{scab}$, $b_{scab} \le b < b_{perf}$ or $b \ge b_{perf}$, correspondingly.

The denominator in the fraction in the column "Penetration" ("Scabbing", "Perforation") in Table 3.29.2 denotes the number of experiments where penetration (scabbing, perforation) was observed, while the numerator is the number of calculations which predicted the corresponding phenomenon. The last column contains similar characteristics for all types of damage (penetration, scabbing, perforation) so that the numerator of the fraction equals the sum of the numerators of the fractions in the preceding columns and the dominator is the sum of the dominators of the fractions in the preceding columns. Clearly only the experiments that were calculated using the empirical models are taken into account.

In Table 3.29.3 and Table 3.29.4 the damage predicted by empirical models is classified into two types: perforation and non-perforation, wherein the latter type includes penetration and scabbing. Comments regarding Analytical Model will be presented in Section 4.2.2; here this model is not analyzed.

In Table 3.29.3: «blank» has the same meaning as in Table 3.29.1, and «+» denotes agreement between calculations and experiments (perforation or non-perforation in both cases).

The meaning of the symbols, «-» and «×», requires a more detailed explanation. In the cases when there is a disagreement between the experimental and theoretical results, the following parameter is introdued:

$$\varepsilon = \frac{b - b_{perf}}{b_{perf}}, \tag{3.29.2}$$

Table 3.29.1. Comparison of the empirical models using the type of shield damage with experimental data from Sliter, 1980 ("P" – perforation, "S" –scabbing, "N" – penetration, "blank" – experimental data are outside the range of validity of the model).

Exp #	m kg	d mm	$f_c^{'}$ MPa	b mm	v_{imp} m/s	Experiment	Model							
							Petry II	BRL	ACE	NDRC	Chang	Adeli–Amin	CRIERI	Average
1	342.468	110	37	300	28	P	N	P	P	S	P	S	S	S
2	339.56	155	43	175	28	P	N	S		P	P	P	P	P
3	339.56	155	40	301	27	P	N	S		S	S		S	S
4	339.56	155	40	350	29	S	N	S		S	S	S	S	N
5	226.5	305	40	400	110	P	N	S		P	P		S	S
6	226.5	305	40	400	90	S	N	N		P	S		S	S
7	226.5	305	40	400	106	P	N	N		P	S		S	S
8	159.456	305	40	400	110	S	N	N		P	S		S	S
9	226.5	305	40	400	97	S	N	N		P	S		S	S
10	159.456	305	40	400	108	S	N	N		P	S		S	S
11	159.456	305	34	400	133	S	N	S		P	P		S	S
12	190.713	305	34	400	94	S	N	N		P	S		S	S
13	159.456	305	34	400	110	S	N	N		P	S		S	S
14	191.619	305	34	400	123	P	N	S		P	P		S	S
15	191.619	305	34	400	110	S	N	N		P	P		S	S
16	226.5	305	34	500	117	S	N	N		P	S		S	S
17	226.5	305	34	500	107	S	N	N		P	S		S	S
18	159.456	305	34	500	147	S	N	N		P	S		S	S
19	159.456	305	34	500	77	S	N	N		S	S		S	S
20	191.619	305	34	500	134	S	N	N		P	S		S	S
21	226.5	305	38	500	127	S	N	S		P	S		S	S
22	159.456	305	38	500	160	S	N	N		P	S		S	S
23	191.619	305	38	500	144	P	N	S		P	S		S	S
24	191.619	305	38	500	79	S	N	N		S	S		S	S
25	191.619	305	38	500	120	S	N	N		P	S		S	S
26	294.903	305	36	601	143	P	N	S		P	S		S	S
27	159.456	305	36	601	173	S	N	N		P	S		S	S
28	239.637	200	36	600	73	S	N	N		S	S	S	S	N
29	299.433	305	36	601	100	S	N	N		P	S		S	S
30	159.456	305	36	601	143	S	N	N		P	S		S	S
31	239.637	200	36	398	86	P	N	S	P	P	P	P	P	S
32	239.637	200	36	398	72	S	N	S		P	P	S	S	S
33	299.433	305	36	400	89	P	N	S		P	P		S	S
34	159.456	305	36	400	86	S	N	N		P	S		S	S
35	299.433	305	36	400	65	S	N	N		P	S		S	S

Exp #	m kg	d mm	$f_c^{'}$ MPa	b mm	v_{imp} m/s	Experiment	Model							
							Petry II	BRL	ACE	NDRC	Chang	Adeli–Amin	CRIERI	Average
36	29.943	200	41	260	113	S	N	N		P	S	S	S	S
37	49.875	200	41	260	112	S	N	N		P	S	P	S	S
38	29.943	250	44	260	150	S	N	N		P	S	P	S	S
39	45.345	200	44	260	138	P	N	N		P	S	P	S	S
40	29.943	200	36	260	150	S	N	N		P	S	P	S	S
41	29.943	250	34	260	187	S	N	N		P	P	P	S	S
42	33.93	278	39	209	148	S	N	N		P	P		S	P
43	33.93	278	38	209	102	S	N	N		P	S		S	S
44	51.868	278	40	209	99	S	N	N		P	S		S	S
45	55.946	278	42	209	102	P	N	N		P	S		S	P
46	32.707	300	44	207	153	S	N	N		P	S		S	P
47	17.441	200	38	260	238	S	N	N		P	S	P	S	S
48	17.441	200	42	260	269	S	N	N		P	P	P	S	S
49	25.051	250	36	260	252	S	N	N		P	P	P	P	P
50	25.051	250	40	260	309	S	N	S		P	P	P	P	P
51	34.926	300	40	258	220	S	N	N		P	P	P	S	P
52	96.942	203	40	305	37	S	N	N		S	N		N	N
53	96.942	203	31	305	65	S	N	N		P	S	S	S	S
54	96.489	203	31	305	104	P	N	S		P	P	P	P	S
55	96.489	203	31	457	49	N	N	N		S	N		N	N
56	96.489	203	34	457	103	S	N	N		S	S	S	S	N
57	96.489	203	33	609	90	N	N	N		S	N	N	N	N
58	97.395	203	31	609	115	S	N	N		S	S	S	S	N
59	5.662	76	22	114	27	N	N	N		S	N		N	N
60	5.753	76	26	152	31	N	N	N		S	N		N	N
61	5.753	76	26	152	41	N	N	N		S	N		N	N
62	5.708	76	30	152	56	N	N	N		S	S	S	S	N
63	5.753	76	26	152	62	S	N	N		S	S	S	S	N
64	10.646	76	26	152	29	N	N	N		S	N		N	N
65	10.646	76	26	152	39	S	N	N		S	S	S	S	N
66	3.642	25	41	75	66	P	P	P	P	P	P		P	P
67	3.642	25	33	150	46	N	N	S	N	N	N	N	N	N
68	3.642	25	41	150	65	N	S	S	S	N	S		S	N
69	3.642	25	41	150	67	N	P	P	S	N	S		S	S
70	3.642	25	40	225	62	N	N	S	N	N	N		N	N
71	3.624	25	25	450	92	N	S	S	N	N	N		N	N
72	3.624	25	24	300	133	N	P	P		N	S		S	P
73	1.11	45	33	231	166	N	N	N		N	N	N	N	N
74	1.11	45	34	231	164	N	N	N		N	N	N	N	N

Exp #	m kg	d mm	f_c' MPa	b mm	v_{imp} m/s	Experiment	Model							
							Petry II	BRL	ACE	NDRC	Chang	Adeli–Amin	CRIERI	Average
75	1.11	45	36	231	151	N	N	N		N	N	N	N	N
76	1.11	45	40	231	169	N	N	N		N	N	N	N	N
77	1.11	45	33	231	204	N	N	N	N	N	N	N	N	N
78	1.11	45	34	231	206	N	N	N	N	N	N	N	N	N
79	1.11	45	36	231	217	N	N	N	N	N	N	N	N	N
80	1.11	45	40	231	206	N	N	N	N	N	N	N	N	N
81	2.247	45	33	231	155	N	N	N	N	N	S	S	N	N
82	2.247	45	34	231	155	N	N	N	N	N	N	N	N	N
83	2.247	45	36	231	157	N	N	N	N	N	N	N	N	N
84	2.247	45	41	231	157	N	N	N	N	N	N	N	N	N
85	2.247	45	33	231	213	S	S	S	S	N	S		S	N
86	2.247	45	34	231	213	S	S	S	S	N	S		S	N
87	2.247	45	36	231	203	S	N	S	S	N	S		S	N
88	2.247	45	40	231	203	S	N	S	S	N	S		S	N
89	2.247	45	41	231	300	S	P	P	P	S	S		S	S
90	2.247	45	41	231	312	P	P	P	P	S	S		S	S
91	2.247	45	46	231	312	S	S	P	P	S	S		S	S
92	2.247	45	49	231	308	P	N	P	P	S	S		S	S
93	0.892	40	42	350	279	N	N	N	N	N	N		N	N
94	0.892	40	38	400	226	N	N	N	N	N	N	N	N	N
95	0.892	40	35	400	146	N	N	N		N	N	N	N	N
96	0.458	40	42	400	234	N	N	N		N	N	N	N	N
97	0.458	40	38	350	146	N	N	N		N	N	N	N	N
98	0.458	40	40	400	164	N	N	N		N	N	N	N	N
99	0.458	40	40	400	291	N	N	N	N	N	N	N	N	N
100	0.109	20	42	250	254	N	N	N	N	N	N		N	N
101	0.118	20	42	300	190	N	N	N	N	N	N	N	N	N
102	0.118	20	39	300	277	N	N	N	N	N	N		N	N

where b_{perf} is the perforation thickness calculated using the empirical model. Eq. (3.29.2) implies that $b=(1+\varepsilon)b_{perf}$. Therefore, increasing or decreasing (depending on the sign of ε) b_{perf} by a factor $(1+\varepsilon)b_{perf}$, it is possible to equalize experimental with theoretical results. Symbol «×» denotes that there exists such value of ε, with the absolute value not larger than $\varepsilon_{\max}=6\%$, that replacing the theoretically obtained b_{perf} by

Table 3.29.2. Comparison of empirical models using the type of shield damage with experimental data from Sliter, 1980. Summary.

Model	Penetration	Scabbing	Perforation	Penetration, scabbing, perforation
Petry II	$\frac{31}{35}=89\%$	$\frac{3}{51}=6\%$	$\frac{2}{16}=13\%$	$\frac{36}{102}=35\%$
BRL	$\frac{29}{35}=83\%$	$\frac{9}{51}=18\%$	$\frac{4}{16}=25\%$	$\frac{42}{102}=41\%$
ACE	$\frac{17}{19}=89\%$	$\frac{4}{6}=67\%$	$\frac{5}{5}=100\%$	$\frac{26}{30}=87\%$
NDRC	$\frac{28}{35}=80\%$	$\frac{11}{51}=22\%$	$\frac{12}{16}=75\%$	$\frac{51}{102}=50\%$
Chang	$\frac{30}{35}=86\%$	$\frac{41}{51}=80\%$	$\frac{8}{16}=50\%$	$\frac{79}{102}=77\%$
Adeli-Amin	$\frac{20}{22}=91\%$	$\frac{9}{18}=50\%$	$\frac{4}{5}=80\%$	$\frac{33}{45}=73\%$
CRIERI	$\frac{31}{35}=89\%$	$\frac{48}{51}=94\%$	$\frac{4}{16}=25\%$	$\frac{83}{102}=81\%$
Average	$\frac{34}{35}=97\%$	$\frac{32}{51}=63\%$	$\frac{6}{16}=31\%$	$\frac{71}{102}=70\%$

$(1+\varepsilon)b_{perf}$ equalizes theoretical with experimental results. Symbol «-» denotes that such $\varepsilon \le \varepsilon_{\max}$ does not exist .

Denominator in the fraction in the column "Perforation" ("Absence of perforation") of Table 3.29.4 denotes the number of experiments where perforation (absence of perforation, i.e., penetration or scabbing) was observed, while the numerator is the number of calculations in which the corresponding phenomenon was predicted. The next column («perforation and not perforation" for $\varepsilon_{\max}=0$) contains similar indexes but for both outcomes. The numerator of the fraction is equal to the number of correct predictions of perforation and non-perforation (the sum of the numerators of the fractions in the preceding columns) while the dominator equals to the number of experiments (the sum of the dominators of the fractions in the preceding columns). The last two columns contain similar indexes calculated using the corrected b_{perf} for two values of $\varepsilon_{\max}$. Clearly, only the experiments that were analyzed using the empirical models are taken into account.

Table 3.29.3. Comparison of the empirical models with the experimental data from Sliter, 1980 using the perforation/non-perforation index (explanations is given in the main text).

Exp #	m kg	d mm	f_c' MPa	b mm	v_{imp} m/s	Experiment	Model: Petry II	BRL	ACE	NDRC	Chang	Adeli–Amin	CRIERI	Average	Analytical
1	342.468	110	37	300	28	P	+	+	+	×	+	-	-	×	-
2	339.56	155	43	175	28	P	+	×	+		+	+	+	+	+
3	339.56	155	40	301	27	P	+	-	+		-	-	-	-	-
4	339.56	155	40	350	29	S	-	+	-		+	+	+	-	+
5	226.5	305	40	400	110	P	+	-	+		+	+	-	-	×
6	226.5	305	40	400	90	S	-	-	-		+	+	+	+	+
7	226.5	305	40	400	106	P	+	+	+		×	+	-	-	×
8	159.456	305	40	400	110	S	-	-	-		+	+	+	+	+
9	226.5	305	40	400	97	S	-	-	-		+	-	+	+	+
10	159.456	305	40	400	108	S	-	-	-		+	+	+	+	+
11	159.456	305	34	400	133	S	-	+	-		-	-	+	+	×
12	190.713	305	34	400	94	S	-	-	-		+	+	+	+	×
13	159.456	305	34	400	110	S	-	-	-		+	-	+	+	×
14	191.619	305	34	400	123	P	+	-	+		+	+	×	-	+
15	191.619	305	34	400	110	S	-	-	-		×	-	+	+	×
16	226.5	305	34	500	117	S	-	-	-		+	+	+	+	+
17	226.5	305	34	500	107	S	-	-		-	+		+	+	+
18	159.456	305	34	500	147	S	-	-		-	+		+	+	+
19	159.456	305	34	500	77	S	-	-		+	+		+	-	+
20	191.619	305	34	500	134	S	-	-		-	+		+	+	+
21	226.5	305	38	500	127	S	-	+		-	+		+	+	+
22	159.456	305	38	500	160	S	-	-		-	+		+	+	+
23	191.619	305	38	500	144	P	+	-		+	×		-	-	-
24	191.619	305	38	500	79	S	-	-		+	+		+	-	+
25	191.619	305	38	500	120	S	-	-		-	+		+	+	+
26	294.903	305	36	601	143	P	+	-		+	×		-	-	-
27	159.456	305	36	601	173	S	-	-		-	+		+	+	+
28	239.637	200	36	600	73	S	-	-		+	+	+	+	-	+
29	299.433	305	36	601	100	S	-	-		×	+		+	-	+
30	159.456	305	36	601	143	S	-	-		×	+		+	-	+
31	239.637	200	36	398	86	P	+	-		+	+	+	+	+	-
32	239.637	200	36	398	72	S	-	+		-	×	+	+	+	+
33	299.433	305	36	400	89	P	+	-		+	+		×	-	+
34	159.456	305	36	400	86	S	-	-		-	+		+	+	+
35	299.433	305	36	400	65	S	-	-		-	+		+	+	+

Exp #	m kg	d mm	$f_c^{'}$ MPa	b mm	v_{imp} m/s	Experiment	Model: Petry II	BRL	ACE	NDRC	Chang	Adeli–Amin	CRIERI	Average	Analytical
36	29.943	200	41	260	113	S	-	-		×	+	+	+	+	+
37	49.875	200	41	260	112	S	-	-		-	+	×	+	+	+
38	29.943	250	44	260	150	S	-	-		-	+	-	+	+	-
39	45.345	200	44	260	138	P	+	+		+	×	+	-	-	×
40	29.943	200	36	260	150	S	-	-		-	+	-	+	+	×
41	29.943	250	34	260	187	S	-	-		-	×	-	+	+	-
42	33.93	278	39	209	148	S	-	-		-	×		+	+	-
43	33.93	278	38	209	102	S	-	-		-	+		+	+	-
44	51.868	278	40	209	99	S	-	-		-	+		+	+	-
45	55.946	278	42	209	102	P	+	+		+	-		-	-	+
46	32.707	300	44	207	153	S	-	-		-	+		+	+	-
47	17.441	200	38	260	238	S	-	-		-	+	-	+	+	-
48	17.441	200	42	260	269	S	-	-		-	×	-	+	+	-
49	25.051	250	36	260	252	S	-	-		-	-	-	×	×	-
50	25.051	250	40	260	309	S	-	+		-	-	-	-	-	-
51	34.926	300	40	258	220	S	-	-		-	×	-	+	+	-
52	96.942	203	40	305	37	S	-	-		+	-		-	-	+
53	96.942	203	31	305	65	S	-	-		×	+	+	+	+	+
54	96.489	203	31	305	104	P	+	-		+	+	+	+	+	×
55	96.489	203	31	457	49	N	+	+		+	+	+	+	+	+
56	96.489	203	34	457	103	S	-	-		+	+	+	+	-	+
57	96.489	203	33	609	90	N	+	+		+	+	+	+	+	+
58	97.395	203	31	609	115	S	-	-		+	+	+	+	-	+
59	5.662	76	22	114	27	N	+	+		+	+		+	+	+
60	5.753	76	26	152	31	N	+	+		+	+		+	+	+
61	5.753	76	26	152	41	N	+	+		+	+		+	+	+
62	5.708	76	30	152	56	N	+	+		+	+	+	+	+	+
63	5.753	76	26	152	62	S	-	-		+	+	+	+	-	+
64	10.646	76	26	152	29	N	+	+		+	+		+	+	+
65	10.646	76	26	152	39	S	-	-		+	+	+	+	-	+
66	3.642	25	41	75	66	P	+	+	+	+	+		+	+	-
67	3.642	25	33	150	46	N	+	+	+	+	+	+	+	+	+
68	3.642	25	41	150	65	N	+	+	+	+	+		+	+	+
69	3.642	25	41	150	67	N	+	+	+	+	+		+	+	+
70	3.642	25	40	225	62	N	+	+	+	+	+		+	+	+
71	3.624	25	25	450	92	N	+	+	+	+	+		+	+	+
72	3.624	25	24	300	133	N	+	+		+	+		+	+	+
73	1.11	45	33	231	166	N	+	+		+	+	+	+	+	+
74	1.11	45	34	231	164	N	+	+		+	+	+	+	+	+

Exp #	m kg	d mm	$f_c^{'}$ MPa	b mm	v_{imp} m/s	Experiment	Model								
							Petry II	BRL	ACE	NDRC	Chang	Adeli–Amin	CRIERI	Average	Analytical
75	1.11	45	36	231	151	N	+	+		+	+	+	+	+	+
76	1.11	45	40	231	169	N	+	+		+	+	+	+	+	+
77	1.11	45	33	231	204	N	+	+	+	+	+	+	+	+	+
78	1.11	45	34	231	206	N	+	+	+	+	+	+	+	+	+
79	1.11	45	36	231	217	N	+	+	+	+	+	+	+	+	+
80	1.11	45	40	231	206	N	+	+	+	+	+	+	+	+	+
81	2.247	45	33	231	155	N	+	+	+	+	+	+	+	+	+
82	2.247	45	34	231	155	N	+	+	+	+	+	+	+	+	+
83	2.247	45	36	231	157	N	+	+	+	+	+	+	+	+	+
84	2.247	45	41	231	157	N	+	+	+	+	+	+	+	+	+
85	2.247	45	33	231	213	S	+	+	+	-	+		+	-	+
86	2.247	45	34	231	213	S	+	+	+	-	+		+	-	+
87	2.247	45	36	231	203	S	-	+	+	-	+		+	-	+
88	2.247	45	40	231	203	S	-	+	+	-	+		+	-	+
89	2.247	45	41	231	300	S	-	×	-	+	+		+	-	+
90	2.247	45	41	231	312	P	+	+	+	-	×		-	+	-
91	2.247	45	46	231	312	S	+	×	-	+	+		+	-	+
92	2.247	45	49	231	308	P	+	+	+	-	-		-	+	-
93	0.892	40	42	350	279	N	+	+	+	+	+		+	+	+
94	0.892	40	38	400	226	N	+	+	+	+	+	+	+	+	+
95	0.892	40	35	400	146	N	+	+		+	+	+	+	+	+
96	0.458	40	42	400	234	N	+	+		+	+	+	+	+	+
97	0.458	40	38	350	146	N	+	+		+	+	+	+	+	+
98	0.458	40	40	400	164	N	+	+		+	+	+	+	+	+
99	0.458	40	40	400	291	N	+	+	+	+	+	+	+	+	+
100	0.109	20	42	250	254	N	+	+	+	+	+		+	+	+
101	0.118	20	42	300	190	N	+	+	+	+	+	+	+	+	+
102	0.118	20	39	300	277	N	+	+	+	+	+		+	+	+

Analysis of the data presented in Table 3.29.2 and Table 3.29.4 conducted taking into account the number of the experiments included in the evaluation of each model, reveals the advantage of the Chang model and the CRIERI model over other models.

Table 3.29.4. Comparison of the empirical models with the experimental data from Sliter, 1980 using the perforation/non-perforation index. Summary (explanation is given in the main text).

Model	Perforation	Absence of perforation	Perforation and not perforation		
			$\varepsilon_{max}=0$	$\varepsilon_{max}=6\%$	$\varepsilon_{max}=15\%$
Petry II	$\frac{2}{16}=13\%$	$\frac{83}{86}=97\%$	$\frac{85}{102}=83\%$	$\frac{86}{102}=84\%$	$\frac{88}{102}=86\%$
BRL	$\frac{4}{16}=25\%$	$\frac{82}{86}=95\%$	$\frac{86}{102}=84\%$	$\frac{90}{102}=88\%$	$\frac{91}{102}=89\%$
ACE	$\frac{5}{5}=100\%$	$\frac{23}{25}=92\%$	$\frac{28}{30}=93\%$	$\frac{28}{30}=93\%$	$\frac{28}{30}=93\%$
NDRC	$\frac{12}{16}=75\%$	$\frac{50}{86}=58\%$	$\frac{62}{102}=61\%$	$\frac{67}{102}=66\%$	$\frac{71}{102}=71\%$
Chang	$\frac{8}{16}=50\%$	$\frac{77}{86}=90\%$	$\frac{85}{102}=83\%$	$\frac{96}{102}=94\%$	$\frac{100}{102}=98\%$
Adeli-Amir	$\frac{4}{5}=80\%$	$\frac{31}{40}=78\%$	$\frac{35}{45}=78\%$	$\frac{36}{45}=80\%$	$\frac{38}{45}=84\%$
CRIERI	$\frac{4}{16}=25\%$	$\frac{84}{86}=98\%$	$\frac{82}{102}=86\%$	$\frac{91}{102}=89\%$	$\frac{98}{102}=96\%$
Average	$\frac{5}{16}=31\%$	$\frac{83}{86}=97\%$	$\frac{88}{102}=86\%$	$\frac{90}{102}=88\%$	$\frac{95}{102}=93\%$
Analytical model	$\frac{4}{16}=25\%$	$\frac{70}{86}=81\%$	$\frac{74}{102}=73\%$	$\frac{83}{102}=81\%$	$\frac{87}{102}=85\%$

3.29.3 *Evaluation of the performance of models. Semi-infinite shield*

Evaluation of the performance of different models for semi-infinite shields is based on the experimental data on ogive-nosed impactors collected by Hansson, 2003. These data are presented in Table 3.29.5, where H_{exp} denotes the DOP observed in the experiment. The values of H_{exp} in Table 3.29.5 are adopted from Magnusson *et al.*, 2001 (experiments 1 to 10), Buzaud *et al.*, 1999a, Buzaud *et al.*, 1999b (experiments 11 and 12), Gran and Frew, 1997 (experiments 13 to 15), Forrestal *et al.*, 2003 (experiments 16 to 30), Riedel *et al.*, 1999 (experiments 31 to 34), Hiermaier and Thoma, 1997 (experiment 35), Gran *et al.*, 1999 (experiments 36 to 38), Baty *et al.*, 2003 (experiment 39), Frew *et al.*, 1998 (experiments 40 to 53), Forrestal *et al.*, 1994 (experiments 54 to 70) and Forrestal *et al.*, 1996 (experiments 71 to 94). These experiments were conducted in the following ranges of parameters (Hansson, 2003.):

$13\,mm \le d \le 365\,mm$, $132\,m/s \le v_{imp} \le 1050\,m/s$,

$64\,g \le m \le 485\,kg$, $21.6\,MPa \le f_c' \le 140\,MPa$,

$55\,mm \le H_{\exp} \le 1.96\,m$, $1.5 \le K_{CRH} \le 6$.

In the calculations for comparing the performance of different empirical models we included only those models which are valid for the ogive-shaped impactors: Ammann-Whitney model, Modified NDRC model, Degen model, Kar model, Healey-Weissman model, Haldar-Hamieh-Miller model, Hughes' model, UKAEA–CEBG–NNC model, Young model, Whiffen and UTHM formulas. The UMIST, Haldar–Miller and Adeli–Amin formulas are not included to the list of analyzed models because most or all experiments are outside the range of validity of these models. The value of the parameter $\overline{c}=1$ is used in Whiffen model.

Detailed results of the calculations are shown in Figs. 3.29.1a-k where circles indicate experiments for which the constraints for application of the model (if any) are satisfied while triangles denote the experiments that fall outside the range of validity of the particular model. The Average model is devised using the values of the DOP which are equal to the arithmetic mean of the values of the DOPs for all models where the model's constraints are satisfied.

The error, $\tilde{\varepsilon}$, is calculated for each experiment as follows:

$$\tilde{\varepsilon} = \frac{H - H_{\exp}}{H_{\exp}} 100\%, \qquad (3.29.3)$$

where $H_{\exp}$ is the DOP observed in the experiment. Horizontal lines denote the values of $\tilde{\varepsilon}$ which are considered as the minimum $\tilde{\varepsilon}_{\min}$ and the maximum $\tilde{\varepsilon}_{\max}$ values of $\tilde{\varepsilon}$; in some cases we excluded outliers with anomalously high deviations from the mean.

Table 3.29.5. Experimental data on ogive-nosed impactors collected by Hansson, 2003.

Exp. #	m kg	d mm	f_c' MPa	v_{imp} m/s	K_{CRH}	$H_{\exp}$ mm
1	6.28	75.0	35.0	484	1.69	680
2	6.28	75.0	35.0	483	1.69	660
3	6.28	75.0	35.0	482	1.69	660
4	6.28	75.0	35.0	647	1.69	990
5	6.28	75.0	95.0	653	1.69	560
6	6.28	75.0	130.0	647	1.69	440
7	6.28	75.0	75.0	571	1.69	410
8	6.28	75.0	140.0	620	1.69	450
9	6.28	75.0	140.0	612	1.69	540
10	6.28	75.0	140.0	619	1.69	510
11	0.5	25.0	38.4	292	3.00	123
12	0.5	25.0	38.4	132	3.00	55
13	2.3	50.8	43.0	320	3.00	185
14	2.3	50.8	43.0	310	3.00	175
15	2.3	50.8	43.0	316	3.00	157
16	13	76.2	23.0	139	3.00	240
17	13	76.2	23.0	200	3.00	420
18	13.1	76.2	23.0	250	3.00	620
19	13.2	76.2	23.0	284	3.00	760
20	13.1	76.2	23.0	337	3.00	930
21	13.1	76.2	23.0	379	3.00	1180
22	13.1	76.2	23.0	238	6.00	580
23	13.1	76.2	23.0	379	6.00	1250
24	12.9	76.2	39.0	238	3.00	300
25	12.9	76.2	39.0	276	3.00	380
26	12.9	76.2	39.0	314	3.00	450
27	12.9	76.2	39.0	370	3.00	530
28	13	76.2	39.0	456	3.00	940
29	12.9	76.2	39.0	313	6.00	610
30	12.9	76.2	39.0	449	6.00	990
31	485	363.0	37.0	259	2.07	1350
32	433	363.0	39.0	261	2.07	960
33	6.63	90.8	40.5	238	2.03	185
34	6.64	90.8	42.4	254	2.03	210
35	0.595	25.0	50.0	389	4.00	215
36	4.37	50.8	40.5	516	3.00	694

Exp. #	m kg	d mm	f_c' MPa	v_{imp} m/s	K_{CRH}	$H_{\exp}$ mm
37	4.38	50.8	41.6	510	3.00	666
38	4.38	50.8	42.4	505	3.00	689
39	30.4	105.0	34.5	273	3.00	463
40	0.478	20.3	58.4	442	3.00	287
41	0.478	20.3	58.4	610	3.00	491
42	0.478	20.3	58.4	805	3.00	840
43	0.478	20.3	58.4	1009	3.00	1300
44	0.478	20.3	58.4	791	3.00	730
45	0.478	20.3	58.4	994	3.00	1160
46	0.606	20.3	58.4	797	3.00	1010
47	0.734	20.3	58.4	803	3.00	1226
48	1.62	30.5	58.4	445	3.00	460
49	1.62	30.5	58.4	584	3.00	790
50	1.62	30.5	58.4	796	3.00	1230
51	1.62	30.5	58.4	980	3.00	1950
52	1.62	30.5	58.4	992	3.00	1960
53	1.62	30.5	58.4	972	3.00	1960
54	0.906	26.9	35.2	277	2.00	173
55	0.91	26.9	37.8	410	2.00	310
56	0.907	26.9	38.1	431	2.00	411
57	0.912	26.9	33.5	499	2.00	480
58	0.91	26.9	38.4	567	2.00	525
59	0.905	26.9	36.9	590	2.00	729
60	0.901	26.9	40.1	591	2.00	513
61	0.903	26.9	35.4	631	2.00	607
62	0.905	26.9	34.7	642	2.00	620
63	0.901	26.9	36.0	773	2.00	866
64	0.904	26.9	32.4	800	2.00	958
65	0.907	26.9	90.5	561	2.00	353
66	0.898	26.9	91.0	584	2.00	384
67	0.908	26.9	95.0	608	2.00	422
68	0.905	26.9	101.0	622	2.00	437
69	0.907	26.9	94.0	750	2.00	630
70	0.9	26.9	108.0	793	2.00	605
71	0.064	12.9	21.6	492	3.00	170
72	0.064	12.9	21.6	618	3.00	250

Exp. #	m kg	d mm	$f_c^{'}$ MPa	v_{imp} m/s	K_{CRH}	$H_{\exp}$ mm
73	0.064	12.9	21.6	788	3.00	450
74	0.064	12.9	21.6	910	3.00	550
75	0.064	12.9	21.6	1029	3.00	750
76	0.064	12.9	21.6	473	4.25	170
77	0.064	12.9	21.6	660	4.25	270
78	0.064	12.9	21.6	775	4.25	410
79	0.064	12.9	21.6	921	4.25	570
80	0.064	12.9	21.6	1050	4.25	760
81	0.478	20.3	62.8	450	3.00	300
82	0.478	20.3	62.8	612	3.00	480
83	0.478	20.3	62.8	821	3.00	760
84	0.478	20.3	62.8	926	3.00	950
85	0.478	20.3	62.8	987	3.00	920
86	0.478	20.3	62.8	1024	3.00	940
87	1.61	30.5	51.0	405	3.00	370
88	1.61	30.5	51.0	446	3.00	420
89	1.61	30.5	51.0	545	3.00	560
90	1.61	30.5	51.0	651	3.00	780
91	1.61	30.5	51.0	804	3.00	1050
92	1.61	30.5	51.0	821	3.00	1230
93	1.61	30.5	51.0	900	3.00	1410
94	1.61	30.5	51.0	1009	3.00	1750

Hereafter in calculating $\tilde{\varepsilon}_{\min}$ and $\tilde{\varepsilon}_{\max}$ (and in other calculations) for each model we took into account only those experiments that correspond to the range of the validity of the model (denoted by circles in Fig. 3.29.1).

The results of calculations are summarized in Table 3.29.6 where the name of the model is indicated in the first column. In some cases the whole range of variation of $\overline{H}_{\exp}$ is divided into two sub-ranges. The number of experiments in each sub-range or the whole range, where the value of the parameter $\tilde{\varepsilon}$ falls in the interval between $\tilde{\varepsilon}_{\min}$ and $\tilde{\varepsilon}_{\max}$, is indicated in column 2 (index N), the end-points of the range (or sub-range) are presented in column 3, $\tilde{\varepsilon}_{\min}$ and $\tilde{\varepsilon}_{\max}$ - in columns 4 and 5, respectively. The indexes in the rest of the columns are described below.

Inspection of Table 3.29.6 and Fig. 3.29.1 shows that for the Modified NDRC, Degen, Kar, Haldar-Hamieh-Miller, UKAEA–CEBG–NNC, Hughes and the Average models, $\tilde{\varepsilon}_{max} < 0$ or slightly larger than 0. Consequently, practically all predicted values of the DOP are smaller than experimental values, and the predictions of the models are expected to be biased towards smaller magnitudes of the DOP.

Since also in other models the interval of variation of $\tilde{\varepsilon}$ is strongly asymmetric with respect to $\tilde{\varepsilon} = 0$ ($\tilde{\varepsilon}_{max}$ strongly deviates from $-\tilde{\varepsilon}_{min}$), the predictions of the models are biased in these cases. Therefore it is not recommended using directly in practice the predictions of one of these models for the indicated range of penetration conditions.

Let us introduce for each model the correction factor, $\hat{k} > 0$, which allows reducing or increasing the predicted value of the DOP, i.e. replace the calculated values, H, by $\hat{k}H$. Then, taking into account Eq. (3.29.3), the «new» errors of model predictions, $\hat{\varepsilon}$, can be expressed through the «old», $\tilde{\varepsilon}$, as follows:

$$\begin{aligned} \hat{\varepsilon} &= \frac{\hat{k}H - H_{exp}}{H_{exp}} \cdot 100\% \\ &= \left[\hat{k}\left(\frac{H}{H_{exp}}\right) - 1\right] \cdot 100\% = [\hat{k}(0.01\tilde{\varepsilon} + 1) - 1] \cdot 100\% \end{aligned} \tag{3.29.4}$$

The value of $\hat{k}$ is selected by applying the following requirement: all values of $\hat{\varepsilon}$ must fall in the interval between the minimum, $-\hat{\varepsilon}_*$, and the maximum, $+\hat{\varepsilon}_*$, the values of the errors of the corrected model. Eq. (3.29.4) implies that $\hat{\varepsilon}$ increases when $\tilde{\varepsilon}$ is increased.

Therefore, the above requirement yields the following system of equations:

$$\hat{k}(0.01\tilde{\varepsilon}_{min} + 1) - 1 = -0.01\hat{\varepsilon}_*, \quad \hat{k}(0.01\tilde{\varepsilon}_{max} + 1) - 1 = 0.01\hat{\varepsilon}_*, \tag{3.29.5}$$

which has the following solution:

$$\hat{k} = \frac{2}{2 + 0.01(\tilde{\varepsilon}_{\min} + \tilde{\varepsilon}_{\max})}, \quad \hat{\varepsilon}_* = \frac{\tilde{\varepsilon}_{\max} - \tilde{\varepsilon}_{\min}}{2 + 0.01(\tilde{\varepsilon}_{\min} + \tilde{\varepsilon}_{\max})}. \qquad (3.29.6)$$

Values of the parameters $\hat{k}$ and $\hat{\varepsilon}_*$ for all models are presented in Table 3.29.6. Now the performances of the corrected models can be evaluated using only one fundamental parameter, $\tilde{\varepsilon}$, which is related with their errors. It must be noted that in the analysis of the performance of a certain empirical model which is based on the same experimental data, Hansson, 2003 also suggested introducing a correction factor although he did not devise the unambiguous procedure for calculating this parameter.

In selecting the best model, we will take into account the following considerations: the number of sub-ranges; the error interval of the corrected model, $2\hat{\varepsilon}_*$; proximity of the absolute value of the coefficient $\hat{k}$ to 1; the number of points, N. Inspection of Table 3.29.6 shows that at the first stage it is appropriate to select for further consideration those models which span the whole range $2 \le \bar{H}_{\exp} \le 65$ without dividing it into sub-ranges and having the parameter $\hat{\varepsilon}_*$ that only slightly differs from the minimal (among all models) value. These models include the Modified NDRC, Degen, Kar, Healey-Weissman and The Average models where parameter $\hat{\varepsilon}_*$ halls in the interval, $26\% \le \hat{\varepsilon}_* \le 27\%$. Among the remaining five models the performance of the Modified NDRC, Degen and Kar models is inferior to the rest since they are characterized by a comparatively large values of parameter $\hat{k}$.

Therefore, the most appropriate models are the Healey-Weissman and the Average models, which are corrected using the correction factors 1.26 и 1.25, respectively, and which are expected to have the error in the range $\pm 27\%$. Fig. 3.29.2 illustrates the distribution of errors for these corrected models.

Along with the suggested above method let us consider two more traditional methods for determining the correction coefficient for the model.

The first method of least squares for determining the correction coefficient, $\overset{\frown}{k}$, is based on minimizing the sum of the squared residuals of the model:

$$\sum (\overset{\frown}{k}\bar{H} - \bar{H}_{\exp})^2 \Rightarrow \min, \qquad (3.29.7)$$

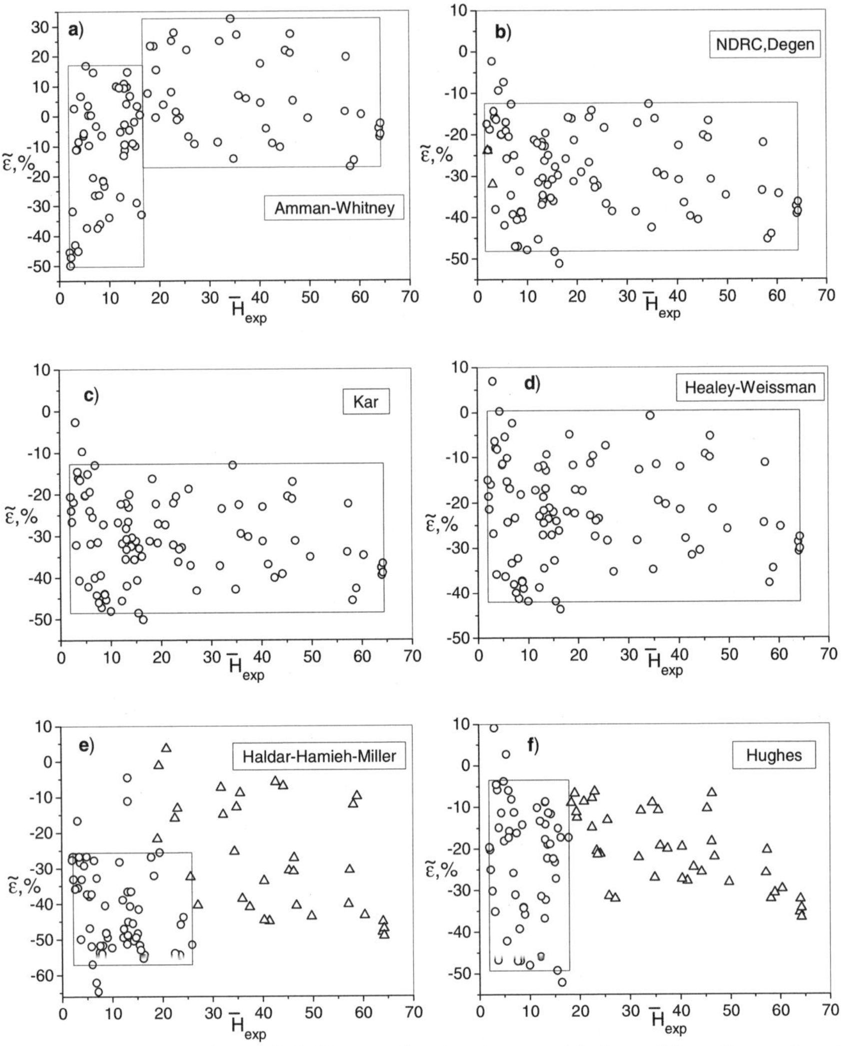

Fig. 3.29.1. Comparison of DOPs predicted using empirical model with results of experiments, taking (circles) and not taking (triangles) into account limitations of models.

where the summation is performed over those «experimental points» which satisfy the requirements of the applicability of the model and where $\tilde{\varepsilon}_{\min} \le \tilde{\varepsilon} \le \tilde{\varepsilon}_{\max}$.

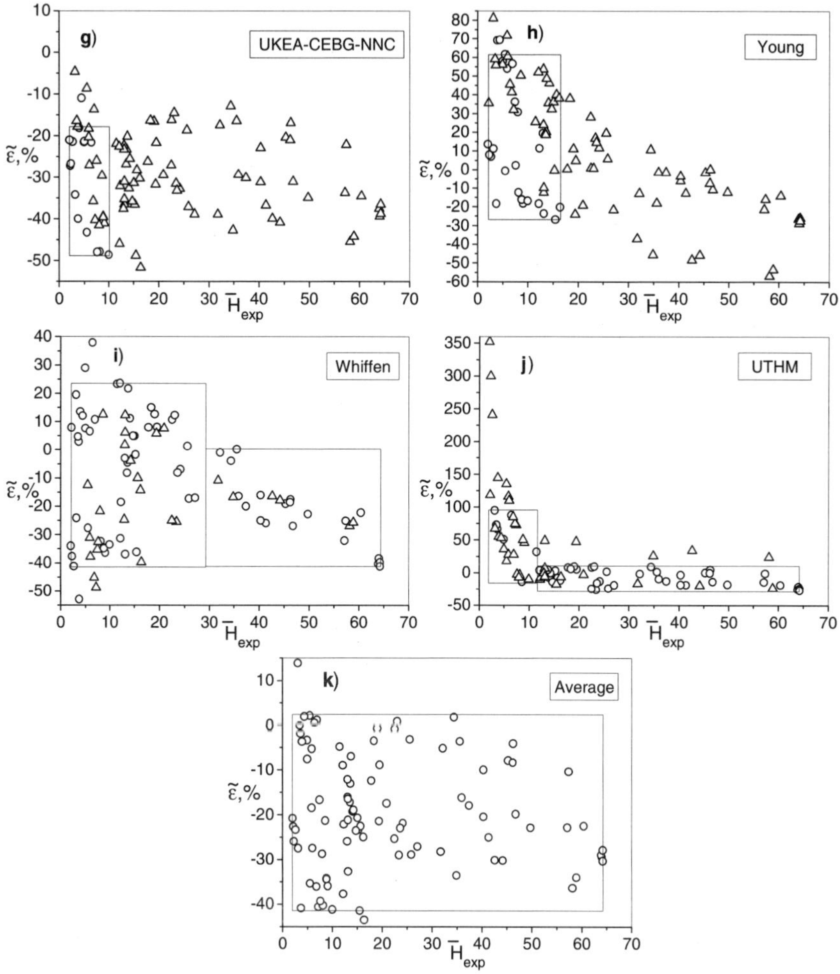

Fig. 3.29.1. Continued

The latter condition implies possible elimination of several outliers when setting the bounds $\tilde{\varepsilon}_{\min}$ and $\tilde{\varepsilon}_{\max}$. Eq. (3.29.6) yields the following expression for $\widehat{k}$:

$$\widehat{k} = \sum \bar{H}\bar{H}_{\exp} \Big/ \sum \bar{H}^2 . \tag{3.29.8}$$

Table 3.29.6. Summary of data for evaluating performance of penetration models on the basis of the indexes $\hat{k}$ and $\hat{\varepsilon}_*$.

Model	N	$\overline{H}_{\text{exp}}$	$\tilde{\varepsilon}_{\text{min}}$	$\tilde{\varepsilon}_{\text{max}}$	$\hat{k}$	$\hat{\varepsilon}_*$
Amman-Whitney	53	2 ÷ 18	-50 %	+17 %	1.20	40 %
	41	18 ÷ 65	-17 %	+33 %	0.93	23 %
NDRC,Degen	88	2 ÷ 65	-49 %	-12 %	1.44	26 %
Kar	91	2 ÷ 65	-49 %	-13 %	1.44	26 %
Healey-Weissman	92	2 ÷ 65	-42 %	0 %	1.26	27 %
Haldar-Hamieh-Miller	56	2 ÷ 26	-57 %	-25 %	1.70	27 %
Hughes	51	2 ÷ 18	-50 %	-3 %	1.36	31 %
UKEA-CEBG-NNC	14	2 ÷ 10	-50 %	-18 %	1.50	23 %
Young	24	2 ÷ 17	-30 %	+62 %	0.85	38 %
Whiffen	44	2 ÷ 28	-41 %	+24 %	1.10	35 %
	20	28 ÷ 65	-41 %	0 %	1.26	26 %
UTHM	8	2 ÷ 12	-14 %	+95 %	0.71	39 %
	43	12 ÷ 65	-27 %	+9 %	1.10	20 %
Average	92	2 ÷ 65	-42 %	+2 %	1.25	27 %

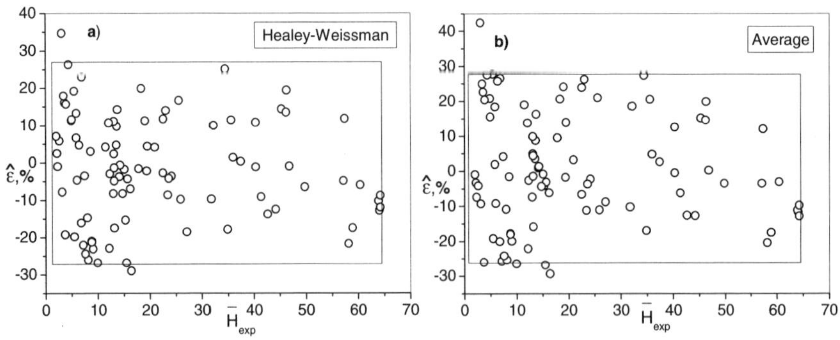

Fig. 3.29.2. Comparison of the DOPs predicted using the recommended corrected empirical model with the results of experiments.

Table 3.29.7 Summary of data for evaluation of the penetration models on the basis of indexes $\widehat{k}$ and $\Delta\widehat{\varepsilon}$.

Model	N	$\overline{H}_{\exp}$	$\hat{k}$	$\hat{\varepsilon}_*$	$\widehat{k}$	$\widehat{\varepsilon}_{\min}$	$\widehat{\varepsilon}_{\max}$	$\frac{\Delta\widehat{\varepsilon}}{2}$
Amman-Whitney	53	2 ÷ 18	1.20	40%	1.05	-48%	23%	35%
	41	18 ÷ 65	0.93	23%	0.96	-20%	27%	24%
NDRC,Degen	88	2 ÷ 65	1.44	26%	1.46	-25%	27%	26%
Kar	91	2 ÷ 65	1.44	26%	1.47	-24%	28%	26%
Healey-Weissman	92	2 ÷ 65	1.26	27%	1.29	-25%	29%	27%
Haldar-Hamieh-Miller	56	2 ÷ 26	1.70	27%	1.81	-22%	35%	28%
Hughes	51	2 ÷ 18	1.36	31%	1.27	-36%	22%	29%
UKEA-CEBG-NNC	14	2 ÷ 10	1.50	23%	1.60	-18%	31%	24%
Young	24	2 ÷ 17	0.85	38%	0.95	-30%	54%	42%
Whiffen	44	2 ÷ 28	1.10	35%	1.01	-41%	25%	33%
	20	28 ÷ 65	1.26	26%	1.36	-20%	36%	28%
UTHM	8	2 ÷ 12	0.71	39%	0.72	-38%	40%	39%
	43	12 ÷ 65	1.10	20%	1.14	-17%	24%	20%
Average	92	2 ÷ 65	1.25	27%	1.26	-26%	29%	27%

The results of calculations are shown in the four last columns in Table 3.29.7, where $\widehat{\varepsilon}_{\min}$ and $\widehat{\varepsilon}_{\max}$ are the bounds for the errors (taking into account the deleted outliers):

$$\begin{aligned}\widehat{\varepsilon}_{\min} &= [\widehat{k}(0.01\tilde{\varepsilon}_{\min}+1)-1]\cdot 100\%,\\ \widehat{\varepsilon}_{\max} &= [\widehat{k}(0.01\tilde{\varepsilon}_{\max}+1)-1]\cdot 100\%\end{aligned} \tag{3.29.9}$$

and

$$\Delta\widehat{\varepsilon} = \widehat{\varepsilon}_{\max} - \widehat{\varepsilon}_{\min}\,. \tag{3.29.10}$$

The second method for determining the correction factor, $\breve{k}$, is based on minimizing the sum of the squared relative residuals of the model (squared relative errors):

$$\sum[(\hat{k}\bar{H}-\bar{H}_{\exp})/\bar{H}_{\exp}]^2 \Rightarrow \min, \tag{3.29.11}$$

that yields the following formula for $\breve{k}$:

$$\breve{k}=\frac{\sum \bar{H}/\bar{H}_{\exp}}{\sum[\bar{H}/\bar{H}_{\exp}]^2}. \tag{3.29.12}$$

The results of calculations are shown in the four last columns in Table 3.29.8, where $\breve{\varepsilon}_{\min}$ and $\breve{\varepsilon}_{\max}$ are the bounds for the errors (taking into account the deleted outliers).

Table 3.29.8. Summary of data for evaluation of the penetration models on the basis of indexes $\breve{k}$ and $\Delta\breve{\varepsilon}$.

Model	N	$\bar{H}_{\exp}$	$\hat{k}$	$\hat{\varepsilon}_*$	$\breve{k}$	$\breve{\varepsilon}_{\min}$	$\breve{\varepsilon}_{\max}$	$\frac{\Delta\breve{\varepsilon}}{2}$
Amman-Whitney	53	2 ÷ 18	1.20	40%	1.10	-45%	28%	37%
	41	18 ÷ 65	0.93	23%	0.93	-23%	23%	23%
NDRC,Degen	88	2 ÷ 65	1.44	26%	1.40	-28%	22%	25%
Kar	91	2 ÷ 65	1.44	26%	1.42	-27%	24%	25%
Healey-Weissman	92	2 ÷ 65	1.26	27%	1.26	-27%	26%	27%
Haldar Hamieh Miller	56	2 ÷ 26	1.70	27%	1.70	-27%	26%	27%
Hughes	51	2 ÷ 18	1.36	31%	1.27	-36%	22%	29%
UKEA-CEBG-NNC	14	2 ÷ 10	1.50	23%	1.44	-26%	18%	22%
Young	24	2 ÷ 17	0.85	38%	0.86	-37%	39%	38%
Whiffen	44	2 ÷ 28	1.10	35%	1.03	-39%	27%	33%
	20	28 ÷ 65	1.26	26%	1.29	-24%	29%	27%
UTHM	8	2 ÷ 12	0.71	39%	0.61	-47%	19%	33%
	43	12 ÷ 65	1.10	20%	1.08	-21%	18%	19%
Average	92	2 ÷ 65	1.25	27%	1.22	-29%	24%	27%

$$\breve{\varepsilon}_{\min} = [\overset{\frown}{k}(0.01\tilde{\varepsilon}_{\min} + 1) - 1] \cdot 100\%,$$
$$\breve{\varepsilon}_{\max} = [\overset{\frown}{k}(0.01\tilde{\varepsilon}_{\max} + 1) - 1] \cdot 100\% \qquad (3.29.13)$$

and

$$\Delta\breve{\varepsilon} = \breve{\varepsilon}_{\max} - \breve{\varepsilon}_{\min}\,. \qquad (3.29.14)$$

Comparing the magnitudes of $0.5\Delta\overset{\frown}{\varepsilon}$, $0.5\Delta\breve{\varepsilon}$ and $\hat{\varepsilon}_*$ in Table 3.29.7 and Table 3.29.8 reveals that the method used for determining the correction coefficient only weakly affects the range of variation of the relative error and model selection, provided that the selection procedure employed $0.5\Delta\overset{\frown}{\varepsilon}$ or $0.5\Delta\breve{\varepsilon}$ instead of $\hat{\varepsilon}_*$. However, using the correction factors $\overset{\frown}{k}$ and $\breve{k}$, rather than $\hat{k}$, yields different absolute values for the bounds of the relative errors of the models. Consequently, these correction factors are inconvenient in practical applications since traditionally the error is characterized by one parameter that describes maximum two-sided error bounds of the model.

We suggested a comparatively simple approach for comparison and evaluation of the performance of different penetration models. It is feasible to employ other approaches that are more heavily based on the methods of mathematical statistics. In particular, different models can be interpreted as regression equations, and correction factor for each model can be interpreted as a regression coefficient. Clearly, models with several regression coefficients can be also considered. However, also in this case, the unambiguous procedures for solving the problem cannot be devised (Draper and Smith, 1998).

Chapter 4

Analytical Models

Nomenclature

a_0, a_1, a_2	parameters of the model, Eq. (4.1.5);
A_0, A_1, A_2	parameters, Eqs. (1.6.10), (1.6.15), (1.7.3);
b	thickness of the shield;
b_{perf}	perforation thickness;
B_0, B_1, B_2	parameters, Eq. (1.7.4);
c	parameter of the model, Eq. (4.1.2);
C	parameter of the model, Eq. (4.1.73);
C'	parameter of the model, Eq. (4.1.75);
d	$=2R$, shank diameter of impactor;
DOP	depth of penetration;
D	drag force;
$D^{(1)}$	drag force at the first stage of penetration;
$D^{(2)}$	drag force at the second stage of penetration;
$D^{(2)}_{\times}$	resistance force acting on the auxillary sharp ogive-nosed impactor;
D_{flat}	resistance of the flat part of the impactor's nose
D_*	$=D_{flat}(v_{imp})$;
f_c'	unconfined compressive strength (Pa);
G	function, Eq. (4.2.6);
h	instantaneous depth of the penetration, Fig. 4.1.1;
h_0	depth at which occurs transition from the cratering stage to the tunneling stage;
H	depth of penetration;
I	parameter, Eq. (4.1.43);
L	length of the nose of the impactor, Fig. 4.1.1;

$L_\times$	see Fig. 1.6.1b;
m	mass of the impactor;
$\vec{n}^{(0)}$	unit inner normal vector to the projectile's surface, Fig. 4.1.1;
N	parameter, Eq. (4.1.43);
r	radius of the impactor's nose flat bluntness;
R	radius of the impactor's shank;
s	parameter of the model, Eqs. (4.1.32)–(4.1.34);
u	$= -\vec{v}^{\,0} \cdot \vec{n}^{\,0}$;
v	instantaneous velocity of impactor;
v_0	velocity of impactor at $h = h_0$;
v_{imp}	impact velocity;
$\vec{v}^{\,0}$	unit vector of the velocity of the projectile, Fig. 4.1.1;
x	coordinate associated with the impactor, Fig. 4.1.1;
y	coordinate associated with the impactor, Fig. 4.1.1;
α	cone angle of the cone-shaped plug, Fig. 4.2.1;
β_0, β_1	parameters, Eq. (4.2.7);
β_2	parameters, Eq. (4.2.12);
η	parameter, Eq. (4.1.58);
κ_0, κ_1	parameters, Eqs. (4.1.75), (4.1.76);
λ	parameter, Eq. (4.1.15);
μ_{fr}	friction coefficient;
ρ_{sh}	density of concrete;
τ_f	shear strength of shield material (concrete);
Φ	function determining the shape of the impactor, Fig. 4.1.1;
Ψ_2	function, Eq. (1.6.39);
ψ	function, Eq. (4.1.11);
Ω_n, Ω_τ	functions determining the model, Eq. (4.1.5).

4.1 Semi-Infinite Shields

4.1.1 *Systematization of models*

In this section we use the conventional notations shown in Fig. 4.1.1.

Two-stage models are the most often used models to describe penetration into semi-infinite concrete shields whereby the first stage (cratering) occurs for $0 \le h \le h_0$ while the second stage (tunneling) comprises the subsequent motion of the impactor ($h > h_0$), where h_0 depends on the length of the impactor's nose.

In order to systematize the models describing penetration into semi-finite concrete shields, it is convenient to denote them as $p\,\&\,q$ models, where p and q are associated with the sub-models used at the first stage of penetration and at the second stage of penetration, correspondingly (Ben-Dor *et al.*, 2009h). In the existing models, it is assumed that at the first stage the resistance force is a function of the instantaneous distance between the nose of the impactor and the front surface of the shield, h. Hereafter the following notations are used: $p=1$ if the resistance force is proportional to h and $p=2$ if the dependence between the resistance force and h is linear with non-zero constant term. At the second stage, it is assumed that the resistance force is, generally, a quadratic function of the instantaneous velocity of impactor. In the case when all three terms in this quadratic polynomial are present, $q=3$, and $q=2$ when the linear term is missing; if the resistance force is independent of the velocity of impactor then $q=1$.

The main developments in the field of two-stage models are summarized in the following (more detailed description of these models and their analysis are given in Section 4.1.2). Originally, Forrestal *et al.*, 1994 suggested a $1\,\&\,2$ model that includes the "resistance parameter" that is determined from experiments while Frew *et al.*, 1998 suggested to calculate this parameter using analytical formula for this parameter as a function of the unconfined compressive strength of the concrete. Later, Forrestal *et al.*, 2003 returned to their original method for determining this parameter.

Whereas these models were developed for the ogive-shaped impactors, Chen and Li, 2002 and Ben-Dor *et al.*, 2003b used generalized

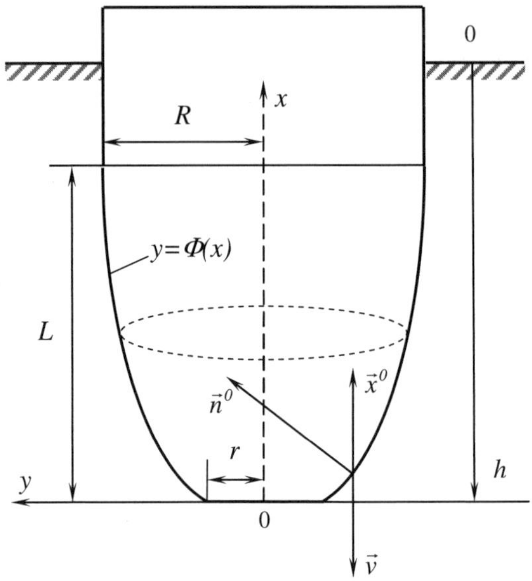

Fig. 4.1.1. Notations.

formulas for arbitrary bodies of revolution, including the projectiles with flat bluntness when the resistance force at the second stage of penetration is calculated. Forrestal and Tzou, 1997 proposed using the $1\&3$ model instead of $1\&2$ model that allows, by appropriate choice of the coefficients, to take into account the compressibility and/or cracking of the material of the shield. Ben-Dor *et al.*, 2006c suggested a similar generalization of this model.

However, these refinements were performed only for the sub-model describing the second stage of penetration, while the one-term sub-model assuming sharp impactors was used at the first stage. In order to overcome this shortcoming Lixin *et al.*, 2000 and Teland and Sjøl, 2004 suggested $2\&2$ models taking into account flat bluntness of impactors at both stages of penetration. Finally, Ben-Dor *et al.*, 2009h suggested a generalized $2\&3$ model which is a combination and refinement of the $1\&3$ model by Forrestal and Tzou, 1997 and $2\&2$ model by Teland and Sjøl, 2004. Since this model is the most general among the $p\&q$ models, it is convenient to use this model as a basic model in describing the class of such models, and consider other models as particular cases of this general model.

The bibliography of the one-stage models is less abundant, and they are analyzed in Section 4.1.3. Hereafter these models are denoted as $0\&q$ models where the parameter q has the same meaning as for the two-term models.

On the basis of the model proposed by Forrestal *et al.*, 1994 and Frew *et al.*, 1998, Sjøl and Teland, 2001, 2003 suggested a three-term model (see also Gebbeken *et al.*, 2009).

Some models for semi-infinite shields and shields having a finite thickness were suggested by Shiqiao *et al.*, 2004, Shiqiao *et al.*, 2006, Shiqiao *et al.*, 2009; Seifoori and Liaghat, 2011.

Hereafter the description of the models is based on the "from general-to-specific" concept. Reference to a particular publication does not imply that the authors of this publication formulated the model exactly in the same form as stated in this book. Very often in the original study the model is presented in a simplified form (e.g. for ogival nose projectiles or neglecting friction) while generalization of the model is straightforward.

4.1.2 *Two-stage models*

4.1.2.1 *General 2&3 model*

Let us consider the model that employs the following expression for the instantaneous resistance force, D (Ben-Dor *et al.*, 2009h):

$$D = \begin{cases} D^{(1)}(h) & if \quad 0 \le h \le h_0 \\ D^{(2)}(v) & if \quad h \ge h_0 \end{cases}, \tag{4.1.1}$$

where the superscript indicates the number of the stage of penetration,

$$D^{(1)}(h) = c\,h + D_*, \tag{4.1.2}$$

$$D_* = D_{flat}(v_{imp}) = B_2 v_{imp}^2 + B_1 v_{imp} + B_0, \tag{4.1.3}$$

$$D^{(2)}(v) = A_2 v^2 + A_1 v + A_0\,, \tag{4.1.4}$$

c is a constant that is determined below. At the second stage of penetration the model is described by Eq. (1.7.1),

$$\Omega_n(u,v) = a_2 u^2 v^2 + a_1 uv + a_0, \quad \Omega_\tau(u,v) = \mu_{fr}\Omega_n(u,v), \tag{4.1.5}$$

parameters A_i and B_i ($i = 0,1,2$) and function $D_{flat}(v)$ are determined by Eqs. (1.6.10), (1.6.15), (1.7.3) and (1.7.4). Hence the three-term model is used at the second stage of penetration. At the first stage, the linear dependence between the resistance force $D^{(1)}(h)$ and the instantaneous penetration depth is assumed, whereby $D^{(1)}(0) = D_*$ is equal to the resistance of the impactor's flat bluntness determined from the model at the second stage of penetration, as proposed Teland and Sjøl, 2004.

The second Newton law and the initial condition for the first stage of penetration can be written as follows:

$$mv(dv/dh) = -D^{(1)}(h), \quad v(0) = v_{imp}\,, \tag{4.1.6}$$

where v_{imp} is the impact velocity and m is mass of the impactor. The motion of the impactor at this stage of penetration is described by the following equation:

$$m\int_{v_{imp}}^{v} VdV = -\int_{0}^{h}(c\tilde{h} + D_*)d\tilde{h}\,. \tag{4.1.7}$$

Equation (4.1.7) implies that

$$m(v_{imp}^2 - v^2) = ch^2 + 2D_* h, \quad 0 \le h \le h_0\,. \tag{4.1.8}$$

Since the second stage of penetration begins when $h = h_0$, the motion of impactor at this stage is described by the following equation:

$$mv(dv/dh) = -(A_2 v^2 + A_1 v + A_0) \tag{4.1.9}$$

with the initial condition $v(h_0) = v_0$, where the value of v_0 is determined below. The solution of Eq. (4.1.9) reads:

$$h = h_0 + m[\psi(v_0) - \psi(v)], \quad v \le v_0, \tag{4.1.10}$$

where for brevity the notation

$$\psi(v) = K(v, A_0, A_1, A_2), \tag{4.1.11}$$

is used for the function $K(v, A_0, A_1, A_2)$ (see Eq. (1.7.10)).

Following Forrestal *et al.*, 1994 we use the conditions of continuity of the resistance force and velocity of impactor at $h = h_0$:

$$ch_0 + D_* = A_2 v_0^2 + A_1 v_0 + A_0, \quad m(v_{imp}^2 - v_0^2) = ch_0^2 + 2D_* h_0. \tag{4.1.12}$$

Equations (4.1.12) allow us to determine c and v_0:

$$c = c(v_{imp}) = \frac{m(v_{imp}^2 - v_0^2) - 2D_* h_0}{h_0^2}, \tag{4.1.13}$$

$$v_0 = v_0(v_{imp}) = \frac{\sqrt{A_1^2 + 4(A_2 + \lambda)(\lambda v_{imp}^2 - A_0 - D_*)} - A_1}{2(A_2 + \lambda)}, \tag{4.1.14}$$

where

$$\lambda = m/h_0. \tag{4.1.15}$$

$$v_{imp} \ge \sqrt{(A_0 + D_*)/\lambda}. \tag{4.1.16}$$

For $v = 0$, Eq. (4.1.10) yields the following expression for the DOP, H:

$$H = h_0 + m\psi(v_0). \tag{4.1.17}$$

In the special case of *2&3* model proposed by Ben-Dor *et al.*, 2009h, $h_0 = 4R$ and the magnitudes of the coefficients a_i ($i = 0,1,2$) are adopted from Forrestal and Tzou, 1997 (see Eq. (4.1.55)).

Models of the types *1&2*, *1&3* and *2&2* are considered below as particular cases of the general *2&3* model.

4.1.2.2 *1&2 models*

4.1.2.2.1 General case of *1&2* model

The general *2&3* model is reduced to a *1&2* model by setting

$$a_1 = 0 \tag{4.1.18}$$

in Eq. (4.1.5), i.e., using two-term model given by Eq. (1.6.9) for the second stage of penetration and assuming that

$$D_* = 0, \tag{4.1.19}$$

whereas Eq. (4.1.3) is ignored. Then Eq. (1.7.3) implies that

$$A_1 = 0 \tag{4.1.20}$$

and Eqs. (4.1.13)–(4.1.16) can be written as follows:

$$c = c(v_{imp}) = \frac{m(v_{imp}^2 - v_0^2)}{h_0^2} = \frac{m(A_2 v_{imp}^2 + A_0)}{h_0(m + A_2 h_0)}, \tag{4.1.21}$$

$$v_0 = v_0(v_{imp}) = \sqrt{\frac{m v_{imp}^2 - A_0 h_0}{m + A_2 h_0}}, \tag{4.1.22}$$

$$v_{imp} \geq \sqrt{\frac{A_0 h_0}{m}}. \tag{4.1.23}$$

The expression for the DOP can be obtained from Eq. (4.1.17) or adding h_0 to the right-hand side of Eq. (1.6.22):

$$H = h_0 + \frac{m}{2A_2} \ln\left(1 + \frac{A_2}{A_0} v_0^2\right), \tag{4.1.24}$$

where A_0 and A_2 are determined by Eq. (1.6.10) and (1.6.15). Eq. (4.1.21) implies that $c > 0$ and, consequently, the model is consistent with the physics of penetration process.

4.1.2.2.2 *Sandia special 1&2 models*

The above model recovers the model suggested by Forrestal *et al.*,1994 for non-truncated ogive-nosed impactors if

$$\mu_{fr} = 0, \tag{4.1.25}$$

$$h_0 = 4R, \tag{4.1.26}$$

$$a_2 = \rho_{sh}, \tag{4.1.27}$$

$$a_0 = sf_c', \tag{4.1.28}$$

$$A_0 = \pi a_0 R^2, \quad A_2 = \pi a_2 R^2 \Psi_2, \tag{4.1.29}$$

where f_c' is the unconfined compressive strength of the material of a shield, Ψ_2 is determined by Eq. (1.6.39) and s is a parameter determined from experimental data. Taking into account Eqs. (4.1.25)–(4.1.29), Eqs. (4.1.22)–(4.1.24) can be written as follows (Forrestal *et al.*,1994):

$$v_0 = \sqrt{\frac{mv_{imp}^2 - 4\pi R^3 sf_c'}{m + 4\pi R^3 \rho_{sh} \Psi_2}}, \quad v_{imp} \geq 2R\sqrt{\frac{\pi R sf_c'}{m}}, \tag{4.1.30}$$

$$H = 4R + \frac{m}{2\pi \rho_{sh} R^2 \Psi_2} \ln\left(1 + \frac{\rho_{sh} \Psi_2}{sf_c'} v_0^2\right). \tag{4.1.31}$$

These equations allow us to determine s as a function of H and v_{imp} (Forrestal *et al.*,1994):

$$s = \frac{\rho_{sh} v_{imp}^2 \Psi_2 / f_c'}{(1 + 4\pi \rho_{sh} R^3 \Psi_2 / m)\exp[2\pi \rho_{sh} R^2 \Psi_2 (H - 4R)/m] - 1}. \tag{4.1.32}$$

The latter equation is suggested for calculating values of s using the experimental data for H. It is recommended to use the average value of s for predicting the DOP.

Several modifications of the model suggested by Forrestal *et al.*,1994 are proposed.

Frew *et al.*, 1998 proposed the following relationship for s:

$$s = 82.6(10^6/f_c')^{0.544}, \tag{4.1.33}$$

rather than estimating s from the experimental data. Li and Chen, 2003 put forward another approximation for s:

$$s = 72.0 \cdot 10^3/\sqrt{f_c'}. \tag{4.1.34}$$

Li and Chen, 2002, 2003 proposed the following formula for h_0 instead of Eq. (4.1.26):

$$h_0 = 1.414R + L. \tag{4.1.35}$$

Forrestal and Tzou, 1997 suggested, among others, the incompressible, elastic-plastic two-term model that implies the following formulas for a_0 and a_2:

$$a_0 = 5.18 f_c', \quad a_2 = 3.88\rho_{sh}, \tag{4.1.36}$$

instead of the expressions in Eqs. (4.1.27)–(4.1.28).

Frew *et al.*, 2000 (for limestone shields), Forrestal *et al.*, 2003 and Frew *et al.*, 2006 have returned to the approach suggested by Forrestal *et al.*,1994 and preferred to consider the resistance parameter $sf' = a_0$ as the measure of the shield resistance which can be determined from experiments using Eq. (4.1.32). Frew *et al.*, 2000 showed that this parameter depends on the projectile shank diameter and proposed the equation that describes the dependence of the resistance parameter on diameter. Frew *et al.*, 2006 studied the effect of the shield diameter on deceleration and penetration depth of 3.0 caliber-radius-head ogive-nosed projectiles, and found negligible changes in the penetration depth and only small decrease of the deceleration magnitude as the shield diameters were reduced.

Teland and Sjøl, 2000 studied the boundary effects in penetration, namely, the effects associated with the finite size of the shield in the direction normal to the penetration direction, and introduced a correcting parameter into the model of Forrestal *et al.*,1994 that was supposed to take into account these effects. Although they did not succeed to derive a formula for the correction coefficient, they found that boundary effects can be neglected in most cases if the ratio of the diameter of the shield to the diameter of the projectile is larger than 15.

Apart from the analysis conducted by the authors, some additional evaluations of the special Sandia *1&2* models based on the experimental data were performed by Forrestal *et al.*, 1996; Frew *et al.*, 1998; Tham, 2006; Vahedi *et al.*, 2008.

Gomez and Shukla, 2001 extended the *1&2* model of Forrestal *et al.*, 1994, 1996; Frew *et al.*, 1998 to multiple impacts introducing in this model an empirical coefficient that is a function of the number of impacts. On the basis of the same *1&2* model, Choudhury *et al.*, 2002; Siddiqui *et al.*, 2002 and Siddiqui *et al.*, 2009 derived expressions for the DOP in a buried shield and applied sensitivity analysis to study the influence of various random variables on projectile reliability and shield safety.

4.1.2.2.3 *1&2 model for small impact velocities*

If

$$v_{imp} < \sqrt{\frac{A_0 h_0}{m}}, \tag{4.1.37}$$

then the radicand in Eq. (4.1.22) is negative, and the model in its original form cannot be used.

Consider motion of the impactor at the first stage of penetration described by Eq. (4.1.8) with $D_* = 0$,

$$m(v_{imp}^2 - v^2) = ch^2, \tag{4.1.38}$$

where c is determined by Eq. (4.1.21). During penetration of a projectile into a shield the velocity of the projectile decreases, and Eqs. (4.1.37)–(4.1.38) yield:

$$h^2 = \frac{m(v_{imp}^2 - v^2)}{c} \le \frac{mv_{imp}^2}{c} = \frac{h_0(m + A_2 h_0)}{(A_2 + A_0/v_{imp}^2)} \le h_0^2 . \qquad (4.1.39)$$

Hence Eq. (4.1.38) can be considered as the equation of motion of the impactor at the first stage of penetration in the case when the second stage does not occur. The formula for the DOP can be obtained by setting $h = H$ and $v = 0$ in Eq. (4.1.38):

$$H = \sqrt{\frac{m}{c}} v_{imp} = \sqrt{\frac{h_0(m + A_2 h_0)}{A_2 v_{imp}^2 + A_0}} v_{imp} . \qquad (4.1.40)$$

Special model can be obtained using Eqs. (4.1.26) and (4.1.27) for h_0 and a_2, correspondingly. Corresponding formula generalizing the model of Forrestal *et al.*,1994 and Frew *et al.*, 1998 is derived by Li and Chen, 2003.

4.1.2.2.4 Other versions of *1&2* model

Li and Chen, 2003 (see also Li and Chen, 2002) found that the generalized model of Forrestal *et al.*, 1994 and Frew *et al.*, 1998 allows determining the dimensionless DOP as a function of two dimensionless parameters.

In a more general case of the *1&2* model, it is convenient to introduce similar dimensionless parameters as follows:

$$N = \frac{\pi m}{4dA_2}, \quad I = \frac{\pi m v_{imp}^2}{4dA_0}, \quad k = \frac{h_0}{d} = \frac{h_0}{2R} . \qquad (4.1.41)$$

Then Eqs (4.1.24) and (4.1.40) recovers the same relationships:

$$\frac{H}{d}=\begin{cases}\sqrt{\dfrac{1+0.25k\pi/N}{1+I/N}\dfrac{4k}{\pi}I} & if \quad \dfrac{H}{d}\le k, \quad or \quad I\le 0.25k\pi \\ k+\dfrac{2N}{\pi}\ln\left(\dfrac{1+I/N}{1+0.25k\pi/N}\right) & if \quad \dfrac{H}{d}>k, \quad or \quad I>0.25k\pi \end{cases} \tag{4.1.42}$$

which have been obtained by Li and Chen, 2003 when the parameters of the model are selected using the model of Forrestal *et al.*,1994 and Frew *et al.*, 1998:

$$N=\frac{m}{8\rho_{sh}d}\Bigg/\int_0^L\frac{\Phi\Phi_x^3}{\Phi_x^2+1}dx, \quad I=\frac{mv_{imp}^2}{82.6f_c'(10^6/f_c')^{0.544}d^3}. \tag{4.1.43}$$

In the limiting case,

$$N>>1, \quad I/N<<1, \tag{4.1.44}$$

Li and Chen, 2003 suggested the following simplified formulas:

$$\frac{H}{d}=\begin{cases}\sqrt{\dfrac{4k}{\pi}I} & if \quad \dfrac{H}{d}\le k \\ \dfrac{k}{2}+\dfrac{2I}{\pi} & if \quad \dfrac{H}{d}>k \end{cases} \tag{4.1.45}$$

Apart from the analysis performed by Li and Chen, 2003, comparison of this model with the results of experiments was conducted also by Shiu *et al.*, 2008.

In the case when the second stage of penetration does occur (the basic model by Forrestal *et al.*,1994 and Frew *et al.*, 1998), similar dimensionless variables were suggested, and the accuracy of the dimensionless formula for the DOP was studied by Sjøl and Teland, 2000; Teland and Sjøl, 2000 and Sjøl *et al.*, 2002 (see also Teland and Sjøl, 2004).

Note that using the following dimensionless variables:

$$\omega_1=\frac{m}{A_2h_0}, \quad \omega_2=\frac{mv_{imp}^2}{A_0h_0}, \tag{4.1.46}$$

allows us to obtain a more simple formula for the DOP:

$$\frac{H}{h_0} = \begin{cases} \sqrt{\dfrac{\omega_2(\omega_1+1)}{\omega_1+\omega_2}} & if \quad 0<\omega_2 \le 1 \\ 1+\omega_1 \ln\left(\dfrac{\omega_1+\omega_2}{\omega_1+1}\right) & if \quad \omega_2 > 1 \end{cases} \tag{4.1.47}$$

4.1.2.3 *1&1 model*

For relatively small impact velocities, Forrestal *et al.*, 2003 used the *1&1* model:

$$D^{(1)}(h) = c\,h, \quad D^{(2)}(v) = A_0 . \tag{4.1.48}$$

Substituting $A_2 = 0$ into Eqs. (4.1.21) and (4.1.22) we can obtain the simplified expressions for c and v_0:

$$c = \frac{A_0}{h_0}, \quad v_0 = \sqrt{\frac{mv_{imp}^2 - A_0 h_0}{m}} . \tag{4.1.49}$$

Formula for the DOP is obtained by passing to the limit $A_2 \to 0$ in Eq. (4.1.24) and taking into account Eq. (4.1.22). Applying the L'Hospital rule we find that

$$H = 0.5h_0 + \frac{mv_{imp}^2}{2A_0} . \tag{4.1.50}$$

4.1.2.4 *1&3 models*

Models of this type can be obtained from the general *2&3* model when Eq. (4.1.19) is satisfied. Similar to the *2&3* model, this model can be applied for describing penetration by projectiles without flat bluntness.

For this model Eqs. (4.1.13) and (4.1.14) read:

$$c = c(v_{imp}) = \frac{m(v_{imp}^2 - v_0^2)}{h_0^2} , \tag{4.1.51}$$

$$v_0 = v_0(v_{imp}) = \frac{\sqrt{A_1^2 + 4(A_2 + \lambda)(\lambda v_{imp}^2 - A_0)} - A_1}{2(A_2 + \lambda)}, \quad (4.1.52)$$

and Eqs. (4.1.17) and (4.1.23) are satisfied.

Clearly, the model is mechanistically sound only if $c > 0$, or $v_0 < v_{imp}$. Taking into account Eq. (4.1.52), the latter inequality reduces to the following constraint:

$$\sqrt{A_1^2 + 4(A_2 + \lambda)(\lambda v_{imp}^2 - A_0)} < 2(A_2 + \lambda)v_{imp} + A_1 . \quad (4.1.53)$$

After some algebra, this constraint can be rewritten as follows:

$$A_2 v_{imp}^2 + A_1 v_{imp} + A_0 > 0 . \quad (4.1.54)$$

This proves the validity of the model since the latter condition is satisfied.

Forrestal and Tzou, 1997 suggested the *1&3* model with

$$a_i = f_c' \left(\frac{\rho_{sh}}{f_c'} \right)^{i/2} \tilde{a}_i, \quad i = 0,1,2, \quad (4.1.55)$$

where a_i are coefficients in Eq. (4.1.5), $\tilde{a}_i$ are the known dimensionless coefficients which determine the model of the shield material (see Table 4.1.1). Forrestal and Tzou, 1997 assumed that Eqs. (4.1.25)–(4.1.26) are valid and considered non-truncated ogive-nosed impactors for which Eqs. (1.6.39) and (1.7.8) must be used for calculating A_2 and A_1 .

Table 4.1.1. Coefficients for the different models of the shield material (Forrestal and Tzou, 1997).

Number of the mode	Characteristic of the model	$\tilde{a}_0$	$\tilde{a}_1$	$\tilde{a}_2$
1	Incompressible, elastic-plastic	5.18	0.0	3.88
2	Incompressible, elastic-cracked-plastic	4.05	1.36	3.51
3	Compressible, elastic-plastic	4.50	0.75	1.29
4	Compressible, elastic-cracked-plastic	3.45	1.60	1.12

4.1.2.5 *2&2 models*

4.1.2.5.1 *Version 1 of 2&2 model*

Consider the *2&2* model as a particular case of the general *2&3* model for which $a_1 = 0$ in Eq. (1.7.1) is valid. Then

$$A_1 = B_1 = 0, \quad D_* = B_2 v_{imp}^2 + B_0, \tag{4.1.56}$$

and taking into account Eq. (4.1.15), the Eqs. (4.1.13) and (4.1.14) can be rewritten as follows:

$$c = \frac{m(v_{imp}^2 - v_0^2) - 2(B_2 v_{imp}^2 + B_0)h_0}{h_0^2} = \frac{(mA_2 - mB_2 - 2h_0 A_2 B_2)v_{imp}^2 + mA_0 - mB_0 - 2h_0 B_0 A_2}{h_0(m + h_0 A_2)}, \tag{4.1.57}$$

$$v_0 = \sqrt{\frac{\eta}{m + h_0 A_2}}, \quad \eta = (m - h_0 B_2)v_{imp}^2 - h_0(A_0 + B_0), \tag{4.1.58}$$

$$\eta \geq 0. \tag{4.1.59}$$

The resistance force at the first stage of penetration as a function of h, $D^{(1)}(h)$, is given by Eq. (4.1.2).

In order to obtain the function $D^{(2)}(h)$, let us write the solution of Eq. (4.1.9) ($A_1 = 0$) with the initial condition $v(h_0) = v_0$ in the following form:

$$\ln\left(\frac{A_2 v_0^2 + A_0}{A_2 v^2 + A_0}\right) = \frac{2A_2}{m}(h - h_0). \tag{4.1.60}$$

Since $A_2 v^2 + A_0$ is a resistance force, we obtain the following formula for $D^{(2)}(h)$:

$$D^{(2)}(h) = (A_2 v_0^2 + A_0)\exp\left(-\frac{2A_2}{m}(h - h_0)\right), \quad h \geq h_0. \tag{4.1.61}$$

The depth of penetration (DOP) is determined by Eq. (4.1.24).

Ben-Dor *et al.*, 2010a took notice to the fact that this model allows positive and negative values of the parameter c. Analysis shows that $c<0$ for relatively large values of the parameter $\bar{r}=r/R$. This situation is illustrated in Fig. 4.1.2 where

$$\xi=\frac{h_0(m+h_0B_2)}{\pi m a_0 R^2}c=2P\theta_1-\bar{r}^2+1-2\theta_1(\theta_1+1)(\bar{r}^2+2P)/\theta_2\,, \qquad (4.1.62)$$

a projectile has a shape of a truncated cone with the equation of the generatrix,

$$\Phi(x)=r+\frac{r-R}{L}x\,, \qquad (4.1.63)$$

and

$$P=\frac{A_2-B_2}{2\pi a_2 R^2}=\frac{1}{R^2}\int_0^L\frac{\Phi\Phi_x^3dx}{\Phi_x^2+1}=\frac{(1-\bar{r})^3(\bar{r}+1)}{2[(L/R)^2+(1-\bar{r})^2]}, \qquad (4.1.64)$$

$$\theta_1=\frac{a_2}{a_0}v_{imp}^2,\quad \theta_2=\frac{m}{\pi h_0 R^2 a_2}. \qquad (4.1.65)$$

In this model the positive value of c is associated with the increase of the resistance force acting on the projectile at the initial stage of penetration. This behavior of the drag force is quite understandable, at least for sharp-nosed impactors, because the resistance force at the moment of impact equals zero. Concerning the case $c<0$, the model predicts the decrease of the resistance at this penetration stage for projectiles having a relatively large radius of the flat bluntness. It is conceivable to suggest that this conclusion is correct although we have not found data which confirm or disprove this prediction.

The model suggested by Teland and Sjøl, 2004 is recovered when the parameters are determined by Eqs. (1.6.35)–(1.6.36) and

$$h_0=L\,. \qquad (4.1.66)$$

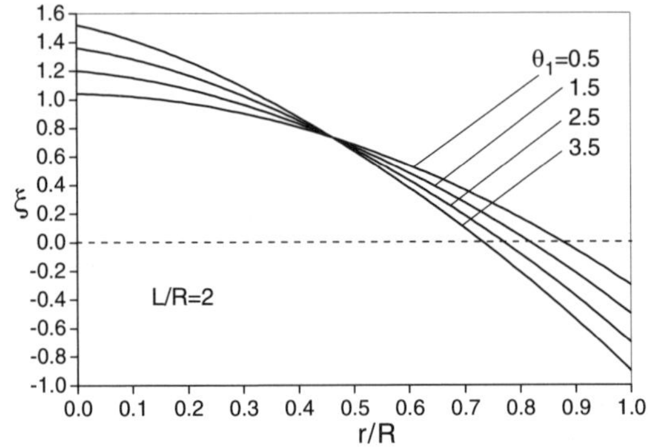

Fig. 4.1.2. Dependence of the parameter ξ vs. the normalized radius of flat thickness.

Teland and Sjøl, 2004 described their model for two classes of the projectile shapes. In the case of truncated ogive nosed impactors, Eqs. (1.6.35)–(1.6.36) can be used for calculation of A_2.

We are not aware of the experiments which allow estimating the accuracy of the linear approximation of the function $D^{(1)}(h)$. However, there exist indirect arguments in favor of the linear model, and these arguments are commonly accepted for substantiating the empirical models. First of all, Teland and Sjøl, 2004 validated the model by comparing the results of calculations based on this model with the experimental data on the DOP. Secondly, a linear model is a simple and natural generalization of the model suggested by Forrestal *et al.*, 1994. This generalization does not require introducing additional empirical parameters and employs the same method for determining the value of only one coefficient. Finally, introducing this model preserves the hierarchy of the models whereby in a particular case of a sharp impactor, the model of Teland and Sjøl, 2004 recovers the model of Forrestal *et al.*, 1994 (with another choice of h_0).

If Eq. (4.1.59) is valid then the impactor penetrates at the depth $h = h_0$, and the above conditions of continuity at $h = h_0$ have a real physical meaning. A different situation arises when $\eta < 0$, and the models suggested by Forrestal *et al.*, 1994 and Teland and Sjøl, 2004 cannot be directly applied in this case. Li and Chen, 2003 solved a similar problem for the $1\&2$ model by using the same expression for c

(see Section 4.1.2.2.3). Ben-Dor *et al.*, 2010a employed this approach for the *2&2* model as shown below.

Eq. (4.1.8) with D_* given by Eq. (4.1.56) yields the following relation that is valid at the first stage of penetration:

$$m(v_{imp}^2 - v^2) = ch^2 + 2(B_2 v_{imp}^2 + B_0)h, \quad 0 \le h \le h_0 . \qquad (4.1.67)$$

If

$$\eta < 0 \qquad (4.1.68)$$

the DOP, $h = H$, can be determined by substituting $v = 0$ into Eq. (4.1.67):

$$cH^2 + 2(B_2 v_{imp}^2 + B_0)H - mv_{imp}^2 = 0 . \qquad (4.1.69)$$

The model is applicable (i.e., Eq. (4.1.69) has a real, positive root) if

$$\gamma_1^2 + \gamma_0 \ge 0, \qquad (4.1.70)$$

where

$$\gamma_0 = mv_{imp}^2 / c, \quad \gamma_1 = (B_2 v_{imp}^2 + B_0) / c . \qquad (4.1.71)$$

This root is given by the following formula:

$$H = \begin{cases} \sqrt{\gamma_1^2 + \gamma_0} - \gamma_1 & if \quad c > 0 \\ -\sqrt{\gamma_1^2 + \gamma_0} - \gamma_1 & if \quad c \le 0 \end{cases} . \qquad (4.1.72)$$

4.1.2.5.2 <u>Version 2 of 2&2 model</u>

Lixin *et al.*, 2000 suggested the penetration model for truncated ogive-nosed impactors. They assumed that the resistance force is determined by Eq. (4.1.1) where

$$D^{(1)}(h) = CC'(h + L_\times), \qquad (4.1.73)$$

$$D^{(2)}(v) = C' D_\times^{(2)}(v), \quad D_\times^{(2)}(v) = A_2 v^2 + A_0 , \qquad (4.1.74)$$

$$C' = 1 + \kappa_1 (r/R)^2, \tag{4.1.75}$$

$$h_0 = \kappa_0 R, \tag{4.1.76}$$

C is a constant that is determined below while κ_0 ($3 \le \kappa_0 \le 5$) and κ_1 are parameters which are determined from experiments. The resistance force of the auxiliary sharp ogive-nosed impactor with the length $L + L_\times$ (see Fig. 1.6.1b), $D_\times^{(2)}$, is calculated on the basis of the model given by Eq. (1.6.14) using Eqs. (4.1.27), (4.1.29) and (4.1.33) for determining parameters A_0 and A_2.

Following the same procedure for determining C as was used above, it can be easily shown that the equation of motion of the impactor at the first stage of penetration implies the following relationship:

$$m(v_{imp}^2 - v_0^2) = CC'(h_0^2 + 2h_0 L_\times), \tag{4.1.77}$$

where v_0 is velocity of the impactor when $h = h_0$. The conditions of continuity of the resistance force and impactor's velocity at $h = h_0$ yield:

$$CC'(h_0 + L_\times) = C'(A_2 v_0^2 + A_0). \tag{4.1.78}$$

Eqs. (4.1.77) and (4.1.78) allow us to determine C and v_0:

$$C = \frac{A_2 v_0^2 + A_0}{h_0 + L_\times} > 0, \tag{4.1.79}$$

$$v_0 = \sqrt{\frac{m v_{imp}^2 (h_0 + L_\times) - A_0 C' h_0 (h_0 + 2L_\times)}{m(h_0 + L_\times) + A_2 C' h_0 (h_0 + 2L_\times)}}, \tag{4.1.80}$$

where

$$v_{imp} \ge \frac{A_0 C' h_0 (h_0 + 2L_\times)}{m(h_0 + L_\times)}. \tag{4.1.81}$$

The expression for the DOP reads:

$$H = h_0 + \frac{m}{2A_2C'}\ln\left(1+\frac{A_2}{A_0}v_0^2\right). \tag{4.1.82}$$

4.1.3 *One-stage models*

Luk and Forrestal, 1987 proposed the spherical cavity expansion model for the material described by locked hydrostats and constant shear strength. This model can be labeled as a *0&2* model that is determined by Eq. (1.4.2) without friction where

$$a_0 = (2/3)Y(1-\ln\eta_*), \tag{4.1.83}$$

$$a_2 = \frac{\dfrac{3Y}{E} + \eta_*\left(1-\dfrac{3Y}{2E}\right)^2 + \dfrac{3\eta_*^{2/3} - \eta_*(4-\eta_*)}{2(1-\eta_*)}}{\left[\left(1+\dfrac{Y}{2E}\right)^3 + \eta_* - 1\right]^{2/3}}\rho_{sh} \tag{4.1.84}$$

Y is the yield stress, E is Young modulus, ρ_{sh} is the initial density of shield material, and η_* is the locked volumetric strain.

DOP can be calculated using Eq. (1.6.22) where A_0 and A_2 are given be Eqs. (1.6.12) and (1.6.15). In the case of ogive-nosed impactors that was considered by Luk and Forrestal, 1987, Eq. (1.6.15) must be replaced by Eqs. (1.6.35) and (1.6.39).

Luk and Forrestal, 1989 modified the expression for the DOP in Luk and Forrestal, 1987 by taking into account friction, and showed that a proper choice of friction coefficient improves the agreement between the experimental results and theoretical predictions. However, taking into account the current level of understanding of the mechanism of interaction of projectiles with concrete shields, this friction coefficient can be considered only as a fudge factor for improving the predictions of the model.

Xu *et al.*, 1997 suggested a *0&2* elastic-cracked spherical cavity expansion model.

Chen and Li, 2002 did not consider the first stage of penetration (cratering) and applied $0\,\&\,2$ models with the parameters which are determined either by the model of Forrestal *et al.*, 1994 or the model of Luk and Forrestal, 1987. In this case, Eq. (1.6.22) allows us to calculate the DOP. Using the dimensionless variables determined by Eq. (4.1.41) or (4.1.43) the DOP can be written as follows:

$$\frac{H}{d}=\frac{2N}{\pi}\ln\left(1+\frac{I}{N}\right)=\frac{2I}{\pi}\varphi(\zeta),\quad \varphi(\zeta)=\frac{\ln(1+\zeta)}{\zeta},\quad \zeta=\frac{I}{N}. \tag{4.1.85}$$

The function $\varphi(\zeta)$ for $0<\zeta<1$ can be expanded into converging alternating-sign power series:

$$\varphi(\zeta)=\sum_{n=1}^{\infty}(-1)^{n+1}\frac{\zeta^{n-1}}{n}. \tag{4.1.86}$$

If $I<<N$ ($\zeta<<1$) the following approximate formula for the DOP can be written keeping only the first term in this series (Chen and Li, 2002):

$$\frac{H}{d}\approx\frac{2}{\pi}I=0.637I. \tag{4.1.87}$$

Rosenberg and Dekel, 2010a proposed the *0&1*model that is based on the assumption that motion of the impactor inside the shield occurs with a constant deceleration, a. Since $D=ma$, under this assumption the solution of the equation of motion, $H=0.5v_{imp}^2/a$, can be written as follows:

$$H=0.5\rho_{imp}L_{eff}v_{imp}^2/R_{sh}, \tag{4.1.88}$$

where R_{sh} is the resisting stress of the shield, ρ_{imp} is the density of the impactor, L_{eff} is the effective length of the impactor and

$$R_{sh}=\rho_{imp}L_{eff}a,\quad L_{eff}=m/(\pi R^2\rho_{imp}). \tag{4.1.89}$$

Eq. (4.1.88) is recommended for calculating the DOP , where L_{eff} is found from Eq. (4.1.89), while the following formula is suggested for determining R_{sh} (in GPa) as a function of f_c' (in MPa):

$$R_{sh} = 0.22\ln(f_c') - 0.285\,. \tag{4.1.90}$$

It is assumed that $v_{imp} < v_c$, where the critical velocity v_c is determined as follows:

$$v_c = \sqrt{R_t / (K_{shape}^{(0)} \rho_{sh})}\,, \tag{4.1.91}$$

and $K_{shape}^{(0)}$ is the nose shape factor (for instance, $K_{shape}^{(0)} = 0.5$ for a hemispherical nose).

Wen, 2002a suggested the model in which the normal stress at the contact surface between the penetrator and a concrete shield is a linear function of the impact velocity.

Teland, 2001 adapted the two-term "averaged" model (see Section 1.6.4.2) to predrilled shields penetrated by ogive-nosed projectile. In this case the DOP is calculated using Eq. (1.6.22), where

$$A_0 = \pi a_0 R^2 (1 - \eta^2), \quad \eta = d_0 / d\,, \tag{4.1.92}$$

$$A_2 = \frac{\pi a_2 R^2 [(8 - 24R^2 - 16R^3) K_{CRH} - 3R^4 + 8R^3 + 6R^2 - 1]}{24 K_{CRH}^2}\,, \tag{4.1.93}$$

d_0 is the diameter of the hole, K_{CRH} is the caliber radius head, coefficients a_0 and a_2 are determined by the model of Forrestal *et al.*, 1994, 1996.

He *et al.*, 2011 proposed a spherical cavity expansion penetration model to predict the penetration and perforation of concrete shields by ogive nosed projectiles. Shear dilatancy as well as compressibility of the material in comminuted region are considered by introducing a dilatant-kinematic relation. On the basis of solution of the cavity expansion problem, coefficients of approximation in Eq. (4.1.5) were determined and applied directly for modeling penetration.

Sun and Yuan, 2012 adapted the model by Yankelevsky and Adin, 1980 for the case of penetration into concrete.

4.2 Shield Having a Finite Thickness

4.2.1 *Brief review*

As a rule, analytical models describing penetration and perforation of finite width shields employ various modifications and generalizations of the model suggested by Forrestal *et al.*, 1994.

Li and Tong, 2003 (see also Chen *et al.*, 2004, Li *et al.*, 2006 and Li *et al.*, 2005), assume that perforation occurs after plug formation, and projectile motion prior plug formation is described, as in the case of semi-infinite shields, by *1&2* model of Forrestal *et al.*,1994, Frew *et al.*, 1998 and Li and Chen, 2003. A modification of this model was suggested by Dai *et al.*, 2005. Chen *et al.*, 2007, 2008a applied a similar approach to a reinforced concrete shield. Ben-Dor *et al.*, 2010a suggested the model that is a combination of the models by Li and Tong, 2003 and Teland and Sjøl, 2004 and is applicable for impactors with a flat bluntness (see Section 4.2.2).

A modified, discussed above in the context of semi-infinite shields, three-phase model by Sjøl and Teland, 2001, 2003, is devised for shields having a finite thickness. He *et al.*, 2011 suggested applying their model not only to semi-infinite shields (see Section 4.1.3) but also to the shield having a finite thickness.

4.2.2 *Models for estimation of perforation thickness*

As a basis we use a *2&2* model for semi-infinite shields which is described in Section 4.1.2.5.1. For convenience, some of the formulas from Section 4.1.2.4 are rewritten here.

If $\eta \geq 0$, where η is determined by Eq. (4.1.58), penetration involves two stages, the DOP, $H = H^{(2)} \geq h_0$, is determined by Eq. (4.1.24) and the relationship for the resistance force reads:

$$D = \begin{cases} D^{(1)}(h) & if \quad 0 \leq h < h_0 \\ D^{(2)}(h) & if \quad h \geq h_0 \end{cases}, \tag{4.2.1}$$

where

$$D^{(1)}(h) = ch + D_*, \quad D_* = B_2 v_{imp}^2 + B_0, \tag{4.2.2}$$

$$D^{(2)}(h) = (A_2 v_0^2 + A_0)\exp\left(-\frac{2A_2}{m}(h - h_0)\right), \tag{4.2.3}$$

c and v_0 are given by Eqs. (4.1.57) and (4.1.58). The parameters A_i and B_i $(i = 0,2)$ are determined by Eqs. (1.6.12), (1.6.15) and (1.6.16). Hereafter formulas (4.1.25)–(4.1.28) for the parameters μ_{fr}, h_0, a_0, a_2 are used in the calculations.

If $\eta < 0$ and the inequality in Eq. (4.1.70) is satisfied, then the DOP, $H = H^{(1)} < h_0$, is determined by Eq. (4.1.72), and formula for the resistance force reads:

$$D(h) = D^{(1)}(h), \quad h < H^{(1)}. \tag{4.2.4}$$

In other cases the model is inapplicable.

Let us consider now a shield having a finite thickness, b, and assume that the above model is valid as long as the following conditions are satisfied:

$$D(h) > G(h,b), \quad h \le \min(b, H), \tag{4.2.5}$$

where

$$G(h,b) = \tau_f S \cos\alpha = \pi\tau_f [(b-h)\tan\alpha + 2R](b-h), \tag{4.2.6}$$

S and α are the lateral surface area and the angle of the conical-shaped plug, correspondingly (see Fig. 4.2.1), $\tau_f = f_c'/\sqrt{3}$ is the shear strength of concrete, D and G are the penetration resistance and the shear resistance force, respectively.

Regarding the selection of α, different recommendations can be found in the literature. Li and Tong, 2003 used the value of $\alpha = 70^\circ$ with the reference to the experimental results by Dancygier, 1998. Yankelevsky and Leibowitz, 1999 estimated the value of α within the range of $58^\circ - 65^\circ$ (for the model when the cratering phase is ignored)

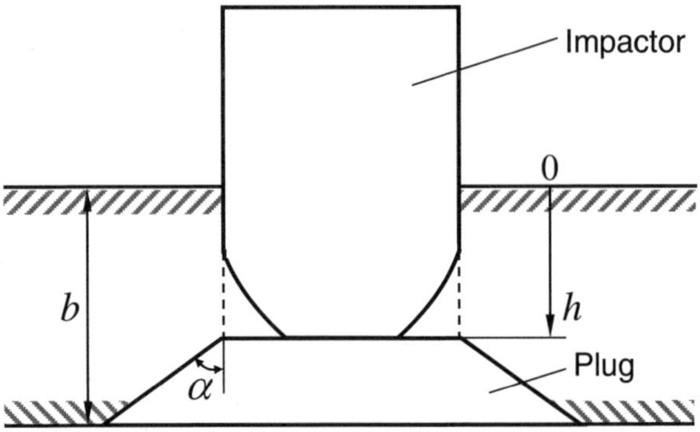

Fig. 4.2.1. Plug formation model.

while Li *et al.*, 2006 used $\alpha = 60^\circ$. Li and Tong, 2003 found that the perforation limit is insensitive to the selection of the value of α, and our calculations confirmed their conclusion.

Following Li and Tong, 2003 we assume that plugging occurs when the penetration resistance force is equal to the shear resistance force. Consequently, the condition for perforation of the shield with a given thickness is as follows. If $D(h) \le G(h,b)$ for some $h \le \min(b,H)$ then a plug is formed, and perforation occurs when the nose of the projectile is located at the depth $h = \min(b,H)$. If $D(h) > G(h,b)$ for $h \le b$ and $b \le H$ then a plug is not formed but perforation still occurs.

Let us determine the perforation thickness of the shield for a given impact velocity, i.e., the minimum thickness, b_{perf}, for which perforation does not occur. For this end we consider a semi-infinite shield and investigate how its thickness can be reduced without perforation.

First, let us consider the case $\eta \ge 0$. In this case $H \ge h_0$, function $D(h)$ is determined by Eqs. (4.2.1)–(4.2.3) and Eq. (4.1.24) is used for determining H.

Fig. 4.2.2 shows the behavior of the functions $D(h)$ and $G(h,b)$. Function $F = D(h)$ is represented by the segment N_1N_2 and by the arc N_2N_3. Parabolas, $F = G(h,b)$, where b is considered as a parameter, intersect the h-axis in the points $h = b$ and can be obtained by translation of one of these parabolas in the horizontal direction. Denote by b_x the value of b that implies intersection of the arcs

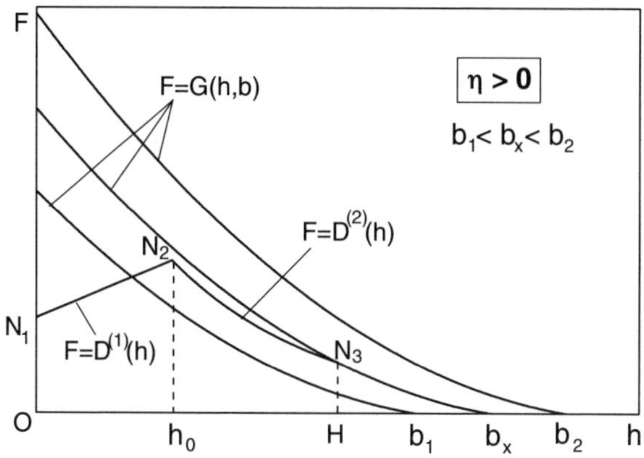

Fig. 4.2.2. Determining perforation thickness.

$F = G(h,b)$ and $F = D^{(2)}(h)$ in the point N_3. Since $D^{(2)}(H) = A_0$, the equation $D^{(2)}(H) = G(H, b_\times)$ can be rewritten as follows:

$$(b_\times - H)^2 + 2\beta_1(b_\times - H) - \beta_0 = 0, \quad \beta_0 = \frac{A_0}{\pi\tau_f \tan\alpha}, \quad \beta_1 = \frac{R}{\tan\alpha}. \qquad (4.2.7)$$

The solution of Eq. (4.2.7) reads:

$$b_\times = H + \sqrt{\beta_1^2 + \beta_0} - \beta_1 > H. \qquad (4.2.8)$$

Clearly, any parabola $F = G(h,b)$ with $b < b_\times$ intersects the arc $ON_1N_2N_3$. Intersection of the parabola with the segment ON_1 means that a plug is formed at the beginning of penetration and the shield is perforated. If the parabola intersects either the segment N_1N_2 or the arc N_2N_3, then a plug is formed at the first stage or at the second stages of penetration, correspondingly, and the shield is perforated as well.

Therefore $b_{perf} = b_\times$ if the following conditions are satisfied:

$$G(h,b_\times) > D^{(1)}(h), \ 0 \le h \le h_0; \quad G(h,b_\times) > D^{(2)}(h), h_0 \le h < H. \qquad (4.2.9)$$

If only one of the latter conditions is violated then the perforation thickness is larger than $b_\times$ and b_{perf} is determined as the maximum b for which either the equation

$$G(h,b) = D^{(1)}(h), 0 \le h \le h_0, \tag{4.2.10}$$

or the equation

$$G(h,b) = D^{(2)}(h), \quad h_0 \le h < H \tag{4.2.11}$$

has a root.

Assume now that $\eta < 0$ and the inequality in Eq. (4.1.70) is valid. Then $H < h_0$, $D(h) = D^{(1)}(h)$ is a linear function, and H is determined by Eq. (4.1.70). The equation $D^{(1)}(H) = G(H, b_\times)$ can be rewritten as follows:

$$(b_\times - H)^2 + 2\beta_1(b_\times - H) - \beta_2 = 0, \quad \beta_2 = \frac{cH + B_2 v_{imp}^2 + B_0}{\pi\tau_f \tan\alpha}. \tag{4.2.12}$$

The solution of Eq. (4.2.12) reads:

$$b_\times = H + \sqrt{\beta_1^2 + \beta_2} - \beta_1 > H. \tag{4.2.13}$$

If

$$G(h, b_\times) > D^{(1)}(h), 0 \le h < H, \tag{4.2.14}$$

then $b_{perf} = b_\times$. The latter relation is always valid when $c > 0$.

If the condition given by Eq. (4.2.14) is violated, then the perforation thickness, b_{perf}, that is larger than $b_\times$ is determined as the maximum b for which the equation

$$G(h,b) = D^{(1)}(h), 0 \le h \le h_0 \tag{4.2.15}$$

has a root.

The predictions of the model and the experimental data are compared in Table 3.29.3 and Table 3.29.4 in Section 3.29.2, where the described above model by Ben-Dor *et al.*, 2010a is called an analytical model; the adopted for the comparison procedure is also described in Section 3.29.2.

In the calculations we used the following values of parameters: $\rho_{sh} = 2200\ kg/m^3$, $\alpha = 70°$, $h_0 = 4R$. The value $k = h_0/R = 4$

is chosen following recommendation by Forrestal *et al.*, 1994. Our calculations have shown that varying k in a wide range practically does not affect the calculated performance indices of the model.

Inspection of the tables reveals that the analytical model does not provide a better accuracy than the best empirical models. However, it must be noted that the experimental data were also partially used in devising the empirical models. Consequently, to a certain extent, the errors of the empirical models are the errors of approximation, while the errors of the analytical models are the errors of predictions. Note also that the model by Li and Tong, 2003 and Li *et al.*, 2006 showed practically the same accuracy as the described above model by Ben-Dor *et al.*, 2010a when the same experimental data were used for their analysis.

Part 3

Penetration into Metallic Shields

A dedicated review by Børvik *et al.*, 1998 deals with penetration into metal shields. Surveys on this topic can be also found in reviews and original papers by Backman and Goldsmith, 1978; Jonas and Zukas, 1978; Neilson, 1985; Brown, 1986; Anderson and Bodner, 1988; Aptukov, 1990; Amde *et al.*, 1996; Corbett *et al.*, 1996; Ben-Dor *et al.*, 2005b; Aly and Li, 2008 and in the books by Zukas, 1982; Recht, 1990; Zukas and Walters, 1990; Bulson, 1997; Bangash and Bangash, 2006; Grigoryan, 2006; Carlucci and Jacobson, 2008; Bangash, 2009; Szuladziński, 2010. Some of the studies compare predictions of the models with experimental results.

This Part 3 comprises Chapters 5 and 6.

In Chapter 5, we have tried to include the maximum number of the suggested empirical models including those that did not appear in the previous reviews.

Chapter 6 is dedicated to the analytical models. Along with well known models much attention is given to the models which have been suggested during recent years and have not yet been included in the published surveys. Chapter 6 also contains a comprehensive description of penetration models suggested in former USSR and Russia which are not well known in the West because of the language barrier or for the lack of easily accessible publications.

Bibliography to Part 2 includes about 200 items.

Chapter 5

Empirical Models

Nomenclature

b	thickness of a shield;
b_{perf}	perforation thickness;
b_{scab}	scabbing thickness;
BLV	ballistic limit velocity;
d	maximum diameter of impactor;
DOP	depth of penetration;
H	depth of penetration ;
$\overline{H}$	$=H/d$;
$K_{shape}^{(2)}$	shape coefficient of the nose of impactor, determined in Eq. (3.9.2) ;
L_{sh}	unsupported shield panel span;
L_{imp}	length of impactor;
m	initial mass of the impactor;
m_{res}	residual mass of impactor;
RHA	rolled homogeneous armor;
v	instantaneous velocity of impactor;
v_{imp}	impact velocity;
$\widetilde{v}_{imp}$	$=v_{imp}/v_{bl}$;
v_0	reference velocity, $1000\,m/s$;
v_{res}	residual velocity of impactor ;
$\widetilde{v}_{res}$	$=v_{res}/v_{bl}$;
v_{bl}	ballistic limit velocity;

$\alpha_1,...,\alpha_{14}$	parameters of models;
θ	angle between a normal vector to a shield surface and the direction of impact;
ρ_{imp}	material density of impactor;
ρ_{sh}	material density of a shield;
σ_u	ultimate tensile strength of shield material.

5.1 Early Relations for DOP

Backman and Goldsmith, 1978 referred to three well known early models for determining the DOP into a semi-infinite shields. These models are described below using the notations in the table at the beginning of this chapter. Additional information on early models may be found in the survey by Herrmann and Jones, 1961.

5.1.1 *Morin equation*

$$\overline{H} = \alpha_1 \frac{mv_{imp}^2}{d^3}. \tag{5.1.1}$$

5.1.2 *Dideon equation*

$$\overline{H} = \alpha_2 \rho_{sh} \ln(1 + \alpha_3 v_{imp}^2). \tag{5.1.2}$$

5.1.3 *Helié equation*

$$\overline{H} = \alpha_4 \frac{mv_{imp}^2}{\rho_{sh} d^3} \ln(1 + \alpha_5 v_{imp}^2). \tag{5.1.3}$$

5.2 De Marre Formula and its Modifications

Applied to a normal impact against steel shields with a finite thickness, the De Marre formula reads (Herrmann and Jones, 1961):

Table 5.2.1. Modifications of the De Marre formula for wrought iron (Herrmann and Jones, 1961).

Model	Author
$v_{bl} = 0.660 \cdot 10^5 \, d^{1/2} b / m^{1/2}$	Fairbairn
$v_{bl} = 0.816 \cdot 10^4 \, d^{1/3} b^{2/3} / m^{1/3}$	Tresidder
$v_{bl} = 0.393 \cdot 10^5 \, d^{5/6} b^{2/3} / m^{1/2}$	Krupps
$v_{bl} = 0.322 \cdot 10^5 \, d^{3/4} b^{0.65} / m^{1/2}$	De Marre

$$v_{bl} = 0.431 \cdot 10^5 \, d^{0.75} b^{0.7} / m^{0.5} \, . \tag{5.2.1}$$

Børvik *et al.*, 1998 give the following ranges of validity of the De Mare formula: $0.1kg < m < 50kg$, $200m/s < v_{bl} < 900m/s$ and note that this formula is recommended in engineering handbooks for blunt-nose projectile penetrating into steel plates.

Modifications of the De Marre formula for wrought iron (Herrmann and Jones, 1961) are presented in Table 5.2.1.

All formulas from Herrmann and Jones, 1961 are converted to SI units.

5.3 Charters and Locke Equation

On the basis of their experiments on high-speed impact of small spheres manufactured from different materials into thick lead and copper plates, Charters and Locke, 1958 suggested the following formula:

$$\overline{H} = 2.28 \left(\frac{\rho_{imp}}{\rho_{sh}} \frac{v_{imp}}{v_{sound}} \right)^{0.69} , \tag{5.3.1}$$

where v_{sound} is speed of sound in shield material.

The formula is recommended for $0.5 \le v_{imp} / v_{sound} \le 2.0$.

5.4 THOR Equations

THOR equations for metallic and nonmetallic shields were suggested in THOR, 1961, 1963. These results were summarized by Greenspon, 1976, and presented in DOE, 1981 (see also Dusenberry, 2010). THOR equations in SI units can be written as follows (Crull and Swisdak, 2005):

$$v_{bl} = 0.3048 \cdot 10^{C_1} (61024bA)^{\alpha_1} (15432.4m)^{\beta_1} (\sec\theta)^{\gamma_1}, \quad (5.4.1)$$

$$v_{res} = v_{imp} - 0.3048 \cdot 10^{C_0} (61024bA)^{\alpha_0} (15432.4m)^{\beta_0} (\sec\theta)^{\gamma_0} (3.28084 v_{imp})^{\lambda_0}, \quad (5.4.2)$$

Table 5.4.1. THOR constants (DOE, 1981)

Parameter	Magnesium	Aluminum (2024T-3	Cast Iron	Titanium	Face Hardened Stee	Mild Homogeneous Steel	Hard Homogeneous Steel	Copper	Lead
C_0	6.904	7.047	4.84	6.292	4.356	6.399	6.475	2.785	1.999
α_0	1.092	1.029	1.042	1.103	0.674	0.889	0.889	0.678	0.499
β_0	-1.17	-1.072	-1.051	-1.095	-0.791	-0.945	-0.945	-0.73	-0.502
γ_0	1.05	1.251	1.028	1.369	0.989	1.262	1.262	0.846	0.655
λ_0	-0.087	-0.139	0.523	0.167	0.434	0.019	0.019	0.802	0.818
C_1	6.349	6.185	10.153	7.552	7.694	6.523	6.601	14.065	10.955
α_1	1.004	0.903	2.186	1.325	1.191	0.906	0.906	3.476	2.735
β_1	-1.076	-0.941	-2.204	-1.314	-1.397	-0.963	-0.963	-3.687	-2.753
γ_1	0.966	1.098	2.156	1.643	1.747	1.286	1.286	4.27	3.59
C_2	-5.945	-6.663	-9.703	-2.318	-1.195	-2.507	-2.264	-5.489	-1.856
α_2	0.285	0.227	0.162	1.086	0.234	0.138	0.346	0.34	0.506
β_2	0.803	0.694	0.673	0.748	0.744	0.835	0.629	0.568	0.35
γ_2	-0.172	-0.361	2.091	1.327	0.469	0.143	0.327	1.422	0.777
λ_2	1.519	1.901	2.71	0.459	0.483	0.761	0.88	1.65	0.934

Table 5.4.2. Range of variables in THOR equations (Dusenberry, 2010).

Material	Target Thickness, b, mm	Impact velocity v_{imp}, m/s	Impact mass, m_{imp}, g
Magnesium	$1.3 \div 76.0$	$152 \div 3200$	$0.97 \div 16$
Aluminum (2024T-3)	$0.5 \div 51.0$	$366 \div 3353$	$0.32 \div 16$
Cast Iron	$4.8 \div 14.0$	$335 \div 1859$	$0.97 \div 16$
Titanium	$1.0 \div 30.0$	$213 \div 3170$	$0.19 \div 16$
Face Hardened Steel	$3.6 \div 13.0$	$762 \div 2987$	$0.97 \div 16$
Homogeneous Steel	$8.0 \div 25.0$	$183 \div 3658$	$0.32 \div 53$
Copper	$1.5 \div 25.0$	$335 \div 3475$	$0.97 \div 16$
Lead	$1.8 \div 25.0$	$152 \div 3170$	$0.97 \div 16$

$$m_{res} = m - 0.648 \cdot 10^{C_2 - 4}(61024bA)^{\alpha_2}(15432.4m)^{\beta_2}(\sec\theta)^{\gamma_2}(3.28084v_{imp})^{\lambda_2}, \quad (5.4.3)$$

where $C_v, \alpha_v, \beta_v, \gamma_v, \lambda_v$ are empirical constants that depend on the shield material (see Table 5.4.1), A is average impact area of the fragment, other notations are given in the table at the beginning of this chapter. Standard fragment with the length d is assumed to have the shape of a cylinder with the hemispherical nose with the diameter $0.5d$ (DOE, 1981).

The ranges for the variables where the THOR equations are valid are shown in Table 5.4.2.

5.5 SRI Equation

Stanford Research Institute (SRI) correlation (1963) for flat-nosed projectiles reads (Corbett *et al.*, 1996):

$$v_{bl} = \alpha_6 b\sqrt{\frac{\sigma_u d}{m}\left(42.7 + \frac{L_{sh}}{b}\right)}, \quad \alpha_6 = 0.44. \quad (5.5.1)$$

Formula (5.5.1) is valid within validity the following ranges of variables: $0.1 < b/d < 0.6$, $0.002 < b/L_{sh} < 0.05$, $10 < L_{imp}/d < 50$, $5 < L_{sh}/d < 8$, $L_{sh}/b < 100$, $21m/s < v_{bl} < 122m/s$.

Amde *et al.*, 1996 recommended to use $\alpha_6 = 0.40$ in Eq. (5.5.1).

5.6 Fuchs Model and its Modification

Fuchs, 1963 proposed the following formula for the resistance force acting on the impactor, D, as a function of the instantaneous velocity of impactor, v:

$$D = \beta_0 + 2\sqrt{\beta_0\beta_2}\,v + \beta_2 v^2 = (\sqrt{\beta_0} + \sqrt{\beta_2}\,v)^2\,, \qquad (5.6.1)$$

where

$$\beta_0 = \frac{\pi d^2}{4}K_{BHN}\,, \quad \beta_2 = \frac{\pi d^2}{4}\frac{\rho_{sh}}{2}C_{shape}\,, \qquad (5.6.2)$$

K_{BHN} can be approximated by the shield Brinell hardness number multiplied by $9.8 \cdot 10^8$ to obtain the result in Pa; C_{shape} is impactor shape factor (similar to drag force coefficient in aerodynamics) for which the following values are recommended: 1.0 for cube or inscribed cylinder, $2/3$ for sphere, and $1/3$ for pointed projectile, like armor-breaking artillery shell.

Eq. (5.6.1) implies the following formula for the DOP:

$$H = \frac{m}{\beta_2}\left[\ln\left(1 + \sqrt{\frac{\beta_2}{\beta_0}}v_{imp}\right) - \frac{\sqrt{\beta_2}\,v_{imp}}{\sqrt{\beta_0} + \sqrt{\beta_2}\,v_{imp}}\right]. \qquad (5.6.3)$$

Zook, 1977 suggested the following generalized, as compared with Eq. (5.6.1), relationship for D:

$$D = 0.25\pi d^2(C_0 K_{BHN} + C_1\sqrt{\rho_{sh}K_{BHN}}\,v + C_2\rho_{sh}v^2)\,, \qquad (5.6.4)$$

where coefficients C_0, C_1 and C_2 are determined using experimental data.

DOP is found on the basis of the formulas similar to Eq. (1.7.9)

5.7 BRL Formula

Ballistic Research Laboratories (BRL) model (1968) reads (Corbett and Reid, 1993):

$$v_{bl} = \alpha_7 \cdot 10^4 \frac{(bd)^{0.75}}{\sqrt{m}}, \quad \alpha_7 = 5.37. \tag{5.7.1}$$

Aly and Li, 2008 indicate that this model can be recommended for flat nose projectiles and present the ranges of projectile and shield parameters where the model is valid: $57\,m/s < v_{bl} < 270\,m/s$, $1.25 \le L_{imp}/d \le 8$, $0.1 \le b/d \le 1.0$, $8 \le L_{sh}/d \le 35$, $315\,MPa \le \sigma_u \le 500\,MPa$.

Amde *et al.*,1996 recommended to use $\alpha_7 = 4.88$ in Eq. (5.7.1).

5.8 Grabarek Formula

Grabarek, 1971 proposed an empirical correlation for penetration of a rod projectile into RHA that can be written in the dimensionless form as follows:

$$\tilde{v}_{res} = \begin{cases} 0 & if \quad 0 \le \tilde{v}_{imp} \le 1 \\ \dfrac{\alpha_{15} z^2 + \alpha_{16} z + \alpha_{17}\sqrt{z}}{z+1} & if \quad 1 < \tilde{v}_{imp} \le 2.5 \end{cases}, \tag{5.8.1}$$

where

$$z = \tilde{v}_{imp} - 1 \tag{5.8.2}$$

and $\alpha_{15} = 1.1, \alpha_{16} = 0.8, \alpha_{17} = 2.0$.

The model is based on the results of experiments with rods where $2\,g \le m \le 900\,g$ and $v_{imp} < 2000\,m/s$.

Anderson *et al.*, 1999 used the model given by Eq. (5.8.1) with the coefficients $\alpha_{15} = 0.9, \alpha_{16} = 1.3, \alpha_{17} = 1.6$.

5.9 Healey and Weissman Model

Healey and Weissman, 1974 formulas in SI units read (Amde *et al.*, 1996, 1997):

$$b_{perf} = \alpha_8 m^{0.33} v_{imp}^{1.22}, \tag{5.9.1}$$

$$v_{res} = v_{imp}\left(1 - \frac{b}{b_{perf}}\right)^{0.82} = \left(v_{imp}^{1.22} - \frac{b}{\alpha_8 m^{0.33}}\right)^{0.82}, b \le b_{perf}, \tag{5.9.2}$$

where $\alpha_8 = 23.2 \cdot 10^{-6}$ for armor piercing missile penetrating into mild steel shield, and $\alpha_8 = 16.1 \cdot 10^{-6}$ for mild steel missile penetrating into mild steel shield.

Clearly, Eq. (5.9.1) implies that

$$v_{bl} = \alpha_9 b^{0.82} / m^{0.27}, \quad \alpha_9 = 1/\alpha_8^{0.82}. \tag{5.9.3}$$

TM 5-855-1, 1986 considers penetration of the standard fragments (for details, see Section 3.26). Eq. (5.9.1) with $\alpha_8 = 16.1 \cdot 10^{-6}$ is recommended for shield manufactured from the common mild structural steel for which the Brinell Hardness Number, μ_{sh}, is $\mu_{sh}^* = 150$. For other types of steels the parameter α_8 in Eq. (5.9.1) is determined as follows:

$$\begin{aligned}&\alpha_8 = 16.1 \cdot 10^{-6} \times \\ &\exp\left[8.77 \cdot 10^{-6}(\mu_{sh}^{*2} - \mu_{sh}^2) - 5.41 \cdot 10^{-3}(\mu_{sh}^* - \mu_{sh})\right]\end{aligned} \tag{5.9.4}$$

The following formula is proposed for the residual velocity:

$$v_{res} = v_{imp}\left(1 - \frac{b}{b_{perf}}\right)^{0.5} \Big/ \left(1 + \frac{17.5b}{m^{0.33}}\right), b \le b_{perf}. \tag{5.9.5}$$

5.10 Lambert and Jonas Approximation

Lambert and Jonas, 1976 suggested the following relationship between the impact, the residual and the ballistic limit velocities for summarizing the results of ballistic tests:

$$v_{res} = \begin{cases} 0 & if \quad 0 \le v_{imp} \le v_{bl} \\ a(v_{imp}^p - v_{bl}^p)^{1/p} & if \quad v_{imp} > v_{bl} \end{cases}, \tag{5.10.1}$$

or

$$\tilde{v}_{res} = \begin{cases} 0 & if \quad 0 \le \tilde{v}_{imp} \le 1 \\ a(\tilde{v}_{imp}^p - 1)^{1/p} & if \quad \tilde{v}_{imp} > 1 \end{cases}, \tag{5.10.2}$$

where $0 \le a \le 1$ and $p > 1$ are constants determined from the experiments.

Lambert, 1978 (see also Zukas, 1982) suggested formulas for calculating a, p and v_{bl} in Eq. (5.10.1) which can be written in SI units as follows:

$$v_{bl} = 31.62\alpha_{10}\left(\frac{L_{imp}}{d}\right)^{0.15}\sqrt{f(z)d^3/m}, \tag{5.10.3}$$

$$a = 1\Big/\left(1 + \frac{\pi\rho_{sh}d^3 z}{12m}\right), \tag{5.10.4}$$

$$p = 2 + z/3, \tag{5.10.5}$$

where

$$z = (b/d)(\sec\theta)^{0.75}, \quad f(z) = z + \exp(z) - 1 \tag{5.10.6}$$

and α_{10} is a constant depending on the shield material (the values $k_{10} = 1750$ for aluminum and $\alpha_{10} = 4000$ for rolled homogeneous armor (RHA) are recommended).

The experiments for validating these formulas were conducted in the following ranges of parameters: $0.5\ g \le m \le 3.63\ kg$, $2\,mm \le d \le 50\,mm$, $4 \le L_{imp}/d \le 30$, $6\,mm \le b \le 150\,mm$, $0 \le \theta \le 60^\circ$, $7800\,kg/m^3 \le \rho_{imp}$ $\le 19000\,kg/m^3$.

The proposed standardization is widely used in penetration mechanics to describe penetration into metallic shields (Børvik *et al.*, 1999, Børvik *et al.*, 2002a, Børvik *et al.*, 2009a, Børvik *et al.*, Børvik *et al.*, 2004, Børvik *et al.*, 2003, Børvik *et al.*, 2001; Dey *et al.*, 2004, Dey *et al.* 2006, Dey *et al.* 2007; Kane *et al.*, 2008) as well as into concrete (Riedel *et al.*, 2010), composites (Garcıa-Castillo *et al.*, 2011; Sabet *et al.*, 2011; Gama and Gillespie, 2008; Zhou and Stronge, 2008; Kasano and Abe, 1997; Yaghoubi and Liaw, 2012), and shields having a complex structure (Børvik *et al.*, 2005; Xue *et al.*, 2010).

Nixdorff, 1983; Mileiko *et al.*, 1994; Ben-Dor *et al.*, 1998b, 2001b found classes of models describing penetration phenomenon which imply Lambert-Jonas correlation.

Ben-Dor *et al.*, 2002d compared the Lambert–Jonas correlation Eq. (5.10.1), where a and p were assumed to be determined from the experimental data, with the similar correlation where $p = 2$ and a is determined from the same data. They showed that the accuracy of both correlations is almost the same in a wide range of variables.

5.11 Kar Formulas

Kar, 1979 proposed the following formula for the perforation thickness:

$$b_{perf} = \frac{3.87 \cdot 10^{-3} K_{shape}^{(2)} \alpha_{11} v_{imp}^{1.65}}{d^{2.03} \sqrt{\sigma_u (e_T / e_T^*)}}, \tag{5.11.1}$$

where shape coefficient of penetrator, $K_{shape}^{(2)}$, is determined by Eq. (3.9.2), e_T, e_T^* are energies determined from Charpy V-notch test for shield material tested at ambient temperature and at $21^\circ C$, respectively. The parameter

$$\alpha_{11} = 0.292 \cdot 10^{-6} (\mu_{imp} + 150) \rho_{imp}, \qquad (5.11.2)$$

where μ_{imp}, is the Brinell Hardness Number of the material of impactor.

Eq. (5.11.1) implies that

$$v_{bl} = \left[\frac{d^{2.03} b \sqrt{\sigma_u (e_T / e_T^*)}}{3.87 \cdot 10^{-3} K_{shape}^{(2)} \alpha_{11}} \right]^{0.606}. \qquad (5.11.3)$$

For the residual velocity, Kar, 1979 proposed the following formula:

$$v_{res} = (v_{imp}^{1.65} - v_{bl}^{1.65})^{1/1.65}, \quad v_{imp} \geq v_{bl}. \qquad (5.11.4)$$

5.12 SwRI Equation

5.12.1 *Model 1*

The staff of Southwest Research Institute (SwRI) prepared the manual (Baker *et al.*, 1980; see also Liang *et al.*, 2003) where the equations for penetration of steel plates by compact steel fragments were described. These equations for the Standard fragment (Fig. 3.26.1) and for the Alternative fragment (Fig. 5.12.1) in SI units can be written as follows:

$$v_{bl} = 0.205 \frac{\beta_1}{\sqrt{m}} S^{\beta_2} (39.37 b \sec\theta)^{\beta_3}, \qquad (5.12.1)$$

$$v_{res} = \frac{v_{bl}}{\sqrt{1 + \rho_{sh} S b / m}} \left(\frac{1.12 z^2 + 0.52 z + 1.29 \sqrt{z}}{z + 1} \right), \qquad (5.12.2)$$

where

$$S = 1.33 \left(\frac{m}{k} \right)^{2/3}, \quad z = \frac{v_{imp}}{v_{bl}} - 1, \qquad (5.12.3)$$

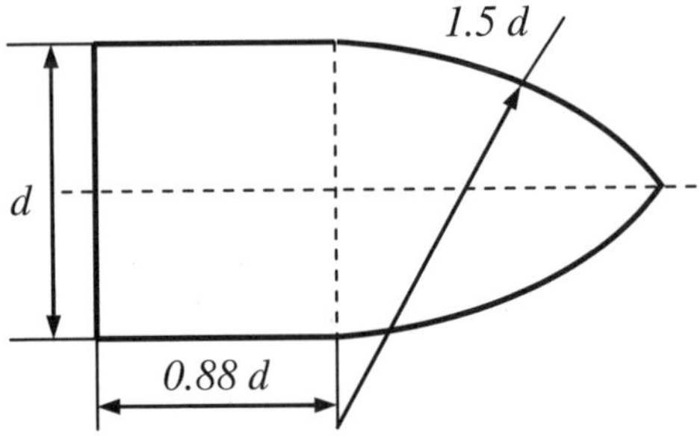

Fig. 5.12.1. Alternative fragment.

Table 5.12.1. Parameters β_1, β_2, β_3 for Eq. (5.12.1).

$\zeta = b/\sqrt{S}$	β_1	β_2	β_3
$0 < \zeta \le 0.46$	1414	0.295	0.910
$0.46 < \zeta \le 1.06$	1936	0.096	1.310
$\zeta > 1.06$	2039	0.064	0.430

$k = 0.186$ for the Standard fragment and $k = 0.34$ for the Alternative fragment, parameters β_1, β_2, β_3 are given in Table 5.12.1.

5.12.2 *Model 2*

This model is based on an examination of data on fragments and hailstones impact upon metal sheets and is described by Baker *et al.*, 1980. For a spherical impactor the equation for the BLV reads:

$$v_{bl} = 7.35 \frac{b\sqrt{Y\rho_{sh}}}{\rho_{imp} d}, \quad \frac{b}{d} \le 1.1, \tag{5.12.4}$$

where Y is the yield stress of the shield material. Here we presented an analytical description of this model instead of the graphical presentation of the model by Baker *et al.*, 1980.

Eq. (5.12.4) can be applied in the case of non-spherical impactor with

$$d = 2\left(\frac{3m}{4\pi\rho_{imp}}\right)^{1/3}. \tag{5.12.5}$$

5.13 Ohte *et al.* Formula

Ohte *et al.*, 1982 suggested the model that can be written in SI units as follows:

$$v_{bl} = 7.67\cdot 10^4 \frac{(bd)^{0.75}}{\sqrt{m}} \tag{5.13.1}$$

for the impactors with a flat and a hemispherical nose, and for cone-nosed impactors if $b\xi > d$, or

$$v_{bl} = 7.67\cdot 10^4 \frac{b^{1.5}\xi^{0.75}}{\sqrt{m}}. \tag{5.13.2}$$

The model can be applied for the cone-nosed impactors if $b\xi \le d$, where

$$\xi = 1 + 2.9(\tan\vartheta)^{2.1} \tag{5.13.3}$$

and ϑ is a semi-vertex angle of a cone.

The ranges of parameters where the model is valid are as follows: $25\,m/s < v_{bl} < 180\,m/s$, $3\,kg \le m \le 50\,kg$, $490\,MPa \le \sigma_u \le 637\,MPa$, $10^\circ \le \vartheta \le 45^\circ$, $7\,mm \le b \le 38\,mm$, $L_{sh}/b \ge 39$, $66\,mm \le d \le 160\,mm$.

5.14 Neilson Formula

Neilson, 1985 suggested the following equation for long flat-nosed penetrators:

$$v_{bl} = \alpha_{12} d\sqrt{\frac{\sigma_u d}{m}\left(\frac{b}{d}\right)^{0.85}\left(\frac{L_{sh}}{d}\right)^{0..3}}, \quad \alpha_{12} = 1.67, \tag{5.14.1}$$

where $L_{imp}/d \geq 13$, $0.14 \leq b/d \leq 0.64$, $4 \leq L_{sh}/d \leq 22$.

5.15 Jowett Formula

Jowett, 1986 proposed the following formula for flat nose short projectiles (Corbett *et al.*, 1996; Aly and Li, 2008):

$$v_{bl} = \begin{cases} 1.62\omega d\sqrt{\dfrac{\sigma_u d}{m}}\left(\dfrac{b}{d}\right)^{0.87} & if \quad 0.1 < \dfrac{b}{d} < 0.25 \\ 0.87\omega d\sqrt{\dfrac{\sigma_u d}{m}}\left(\dfrac{b}{d}\right)^{0.42} & if \quad 0.25 \leq \dfrac{b}{d} < 0.64 \end{cases}, \quad (5.15.1)$$

where

$$\omega = \begin{cases} (L_{sh}/d)^{0.305} & if \quad L_{sh}/d \leq 12 \\ 12.0 & if \quad L_{sh}/d > 12 \end{cases}. \quad (5.15.2)$$

This model is valid in the following ranges of parameters: $2 \leq L_{imp}/d \leq 8$, $315\,MPa \leq \sigma_u \leq 483\,MPa$, $40\,m/s < v_{bl} < 200\,m/s$, $0.1 \leq b/d \leq 0.64$.

Corbett *et al.*, 1996 notes that this formula is called the AEA (for the United Kingdom Atomic Energy Authority) short missile equation.

5.16 Wen and Jones Formulas

Wen and Jones, 1992 proposed the following correlation for low velocity impact by flat-nosed projectiles (Børvik *et al.*, 1998):

$$v_{bl} = 2d\sqrt{\frac{\sigma_y d}{m}\left[0.25\pi\left(\frac{b}{d}\right)^2 + \left(\frac{b}{d}\right)^{1.47}\left(\frac{L_{sh}}{d}\right)^{1.21}\right]} \quad . \quad (5.16.1)$$

The latter formula is valid in the following ranges of parameters: $0.4 < b/d < 1.6$, $25 < L_{sh}/b < 100$, $340\,MPa \leq \sigma_u \leq 440\,MPa$, $L_{sh}/d = 40$, $v_{bl} < 20\,m/s$

In Eq. (5.16.1), σ_y is the yield stress of the shield material; for steel $\sigma_y/\sigma_u \approx 1.35$.

5.17 Modified SRI and Neilson Formulas

Corbett and Reid, 1993 adapted the SRI and Neilson models for hemispherical-nosed impactors. These versions of the models assume, correspondingly, the form of Eq. (5.5.1) with $\alpha_6 = 0.39$ and Eq. (5.14.1) with $\alpha_{12} = 1.34$. The validity ranges of these equation are as follows: $b/d > 0.2$, $31 \le L_{sh}/d \le 83$, $100 m/s < v_{bl} < 250 m/s$.

5.18 Jones and Kim formulas

Jones and Kim, 1997a, 1997b extended the study by Wen and Jones, 1992 to higher velocities and proposed the following equation:

$$v_{bl} = (9.5 \cdot 10^{-3} + \sqrt{9.025 \cdot 10^{-5} + 2m/E_s})(E_s/m), \qquad (5.18.1)$$

where

$$E_s = \sigma_u b^2 d \left[2.075 + 0.0485 L_{sh}/b\right] \qquad (5.18.2)$$

and the impact velocity varies from $7.7 m/s$ to $12.3 m/s$ and from $28 m/s$ to $119 m/s$.

5.19 Aly and Li Formulas

Aly and Li, 2008 advanced the following modification of the Jowett formula (see Section 5.15):

$$v_{bl} = \begin{cases} 1.79d\sqrt{\dfrac{\sigma_u d}{m}}\left(\dfrac{b}{d}\right)^{0.87}\left(\dfrac{L_{sh}}{d}\right)^{0.305} & if \quad 0.1 < \dfrac{b}{d} < 0.25 \\ 1.72d\sqrt{\dfrac{\sigma_u d}{m}}\left(\dfrac{b}{d}\right)^{0.42}\left(\dfrac{L_{sh}}{d}\right)^{0.35} & if \quad 0.25 \le \dfrac{b}{d} < 0.64 \end{cases} \qquad (5.19.1)$$

for $L_{sh}/d \le 12$, and

$$v_{bl}=\begin{cases}3.34d\sqrt{\dfrac{\sigma_u d}{m}}\left(\dfrac{b}{d}\right)^{0.78} & if \quad 0.1<\dfrac{b}{d}<0.25\\ 1.72d\sqrt{\dfrac{\sigma_u d}{m}}\left(\dfrac{b}{d}\right)^{0.41} & if \quad 0.25\le\dfrac{b}{d}<0.64\end{cases} \tag{5.19.2}$$

for $L_{sh}/d>12$.

This formulas are valid in the same ranges of parameters as the Jowett formula.

5.20 Some Other Models and Related Problems

Grigoryan, 2006 presented (without referring to the original paper) the following formula that was proposed in the Central Research Institute of Chemistry and Mechanics for calculating perforation thickness of monolithic armor impacted by elongated penetrator:

$$b_{perf}=\left(\frac{mL_{imp}}{d\rho_{sh}}\right)^{0.5}\left[\alpha_{13}\left(\frac{v_{imp}}{1000}\right)^{2/3}-\alpha_{14}\right]\cos^{2/3}\theta, \tag{5.20.1}$$

where α_{13} and α_{14} are empirical constants depending on the material of the shield. In particular, for the tungsten-iron-nickel alloy VNZh-90 (on Russian classification), $\alpha_{13}=27.7$ and $\alpha_{14}=21.5$.

The best accuracy of this formula is achieved in the range $1200\,m/s\le v_{imp}\le 2000\,m/s$ and $0.45\le b/L_{imp}\le 1.0$.

Amde *et al.*, 1996 proposed modified versions of several well known empirical models.

Gupta *et al.*, 2001 proposed the following approximate formula for the BLV:

$$v_{bl}=\sqrt{\frac{kdY}{m}}b^n, \tag{5.20.2}$$

where k and n are empirical constants, and Eq. (6.3.1) is suggested for determining the residual velocity.

Baker *et al.*, 1991 performed a similitude analysis of penetration into metal shields and showed that well known empirical formulas validated the results of this analysis.

Chapter 6

Analytical Models

Nomenclature

a_0, a_2	parameters defining the model, Eq. (6.4.1);
b	thickness of the shield;
BLV	ballistic limit velocity;
CCE	cylindrical cavity expansion;
d	maximum diameter of impactor;
DOP	depth of penetration;
E	Young modulus;
h	instantaneous distance which impactor has gone after beginning of penetration;
H	depth of penetration;
k_{plast}	constant plastic modulus, Eq. (6.2.3);
K	bulk modulus of the shield material;
K_{CRH}	caliber radius head of ogive nose defined in Eq. (1.6.28) ;
LIM	localized interaction model;
L	length of the nose of the impactor, Fig. 1.1.2;
L_0	length of a cylindrical part of the impactor, Fig. 1.1.2;
L_{imp}	length of the impactor;
m	mass of the impactor;
m_{plug}	mass of the plug;
m_{res}	residual mass of the impactor;
n	coefficient, Eq. (6.3.27);
R	shank radius of impactor;
SCE	spherical cavity expansion;
u	see Eq. (1.3.24);

v	instantaneous velocity of impactor;
v_{imp}	impact velocity;
$\tilde{v}_{imp}$	$= v_{imp} / v_{bl}$;
v_{res}	residual velocity of impactor;
$\tilde{v}_{res}$	$= v_{res} / v_{bl}$;
v_{bl}	ballistic limit velocity ;
V	cavity velocity ;
w	$= v^2$;
W	work done during perforation, minimum perforation energy;
Y	yield stress of the shield material;
α_c	parameters of CCE model, Eqs. (6.5.2), (6.5.22), (6.6.5);
α_s	parameters of SCE model, Eqs. (6.5.1), (6.5.3), (6.5.16), (6.7.1);
β_c	parameters of CCE model, Eqs. (6.5.2), (6.5.22), (6.6.5);
β_s	parameters of SCE model, Eqs. (6.5.1), (6.5.3), (6.5.16), (6.7.1);
γ_c	parameters of CCE model, Eq. (6.5.22);
ε	true strain;
$\dot{\varepsilon}$	true strain rate;
η_c	parameters of CCE model, Eq. (6.6.5);
ϑ	semi-vertex angle of cone-nosed impactor ;
θ, Θ	functions determining the impactor-shield contact surface, Eqs. (1.2.2), (1.2.4);
κ	error control parameter, Eq. (6.2.13);
μ_{fr}	friction coefficient;
ν	Poisson ratio of shield material;
ρ_{imp}	material density of impactor;
ρ_{sh}	material density of a shield;
σ_u	ultimate tensile strength of the shield material;
σ	true stress;
σ_c^{stat}	static component of normal stress in a CCE model;

σ_s^{stat}	static component of normal stress in a SCE model;
σ_r	normal stress, taken positive in compression;
σ_t	tangential stress, taken positive in compression;
τ_s	shear stress;
Φ	function determining the shape of the impactor, Fig.1.1.2;
Φ_x	$= d\Phi/dx$
χ_s	parameters of SCE model, Eq. (6.5.16);
Ψ	function determined by Eq. (6.5.7).

6.1 Introduction

The general criteria for including a particular study in this survey is the following: either the study contains explicit formulas for determining integral characteristics of penetration (BLV, DOP, residual velocity) or allows calculating these characteristics using standard procedures.

Along with the described models some simplified models are used in practical calculations but are not considered here. These models are characterized by one or several of the following features: (i) model does not yield explicit formulas for determining ballistic characteristics which can be determined by solving numerically algebraic or differential equations; (ii) model includes either some unspecified parameters which cannot be easily estimated in advance or empirical coefficients which are defined ambiguously; (iii) model is based on a certain set of experimental data and is in essence a best-fit approximation of the experimental results.

Regarding cavity expansion models, we presented only formulas for the normal stress since integral ballistic characteristics can be determined using methods or explicit formulas described in Chapters 1 and 2.

If the original study contains expression for the minimum penetration energy which can be easily transformed into formula for the BLV, we usually present the expression from the original study.

In order to unify notations all formulas with bulk modulus, K, are converted using the following relation:

$$K = \frac{E}{3(1-2\nu)}, \tag{6.1.1}$$

where E and ν are Young modulus and Poisson ratio, correspondingly.

In the following we present a slightly simplified for engineering applications thickness-of-plate based classification of shields that was suggested by Backman and Goldsmith, 1978. According to this classification we distinguish between the following types of shields: semi-infinite shields when the rear surface does not affect penetration (in the limiting case a shield can be viewed as a half-space); thick (non-thin) shields when the rear surface affects penetration either from the very beginning or after penetration at a certain depth; and thin shields when stresses and deformations in the shield during penetration are independent of the instantaneous depth of penetration.

The main perforation mechanisms (Backman and Goldsmith, 1978) that are approximately described by analytical models are as follows (see Fig. 6.1.1): hole enlargement (semi-infinite and thick shields), plugging (thick and thin shields), petalling (thin shields), and combination of hole enlargement and plugging (thick shields). Plug formation usually occurs during penetration by blunt-nosed projectiles.

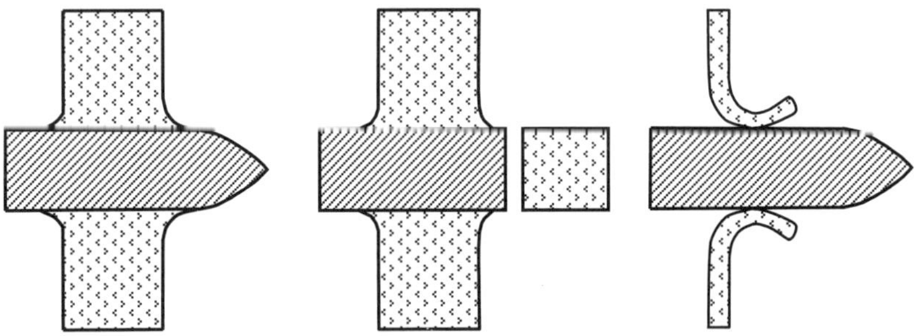

a) Ductile hole enlargement b) Plugging c) Petalling

Fig. 6.1.1. Main perforation mechanisms.

6.2 Early Static Cavity Expansion Models

Static cavity models are widely used in penetration mechanics either directly or as elements of dynamic cavity expansion models.

Classical results from the early works by Bishop *et al.*, 1945 and Hill, 1950a employed von Mises yield condition.

Bishop *et al.*, 1945 proposed the following static SCE and CCE models for incompressible in the plastic region, elastic-plastic, linear strain-hardening materials:

$$\sigma_r = \sigma_s^{stat} = \frac{2Y}{3}\left[1 + \ln\left(\frac{E}{(1+\nu)Y}\right)\right] + \frac{2\pi^2}{27}k_{plast}, \quad (6.2.1)$$

$$\sigma_r = \sigma_c^{stat} = \frac{Y}{\sqrt{3}}\left[1 + \ln\left(\frac{\sqrt{3}E}{2(1+\nu)Y}\right)\right] + \frac{\pi^2}{18}k_{plast}, \quad (6.2.2)$$

where k_{plast} is constant plastic modulus in the linear strain-hardening law:

$$\sigma = \begin{cases} E\varepsilon & if \quad \varepsilon \le Y/E \\ Y + k_{plast}\varepsilon & if \quad \varepsilon > Y/E \end{cases}. \quad (6.2.3)$$

In the case of incompressible in the plastic region, elastic-perfectly plastic materials when

$$\sigma = \begin{cases} E\varepsilon & if \quad \varepsilon \le Y/E \\ Y & if \quad \varepsilon > Y/E \end{cases} \quad (6.2.4)$$

Eqs. (6.2.1) and (6.2.2) can be rewritten for $k_{plast} = 0$ as follows:

$$\sigma_r = \sigma_s^{stat} = \frac{2Y}{3}\left[1 + \ln\left(\frac{E}{(1+\nu)Y}\right)\right], \quad (6.2.5)$$

$$\sigma_r = \sigma_c^{stat} = \frac{Y}{\sqrt{3}}\left[1 + \ln\left(\frac{\sqrt{3}E}{2(1+\nu)Y}\right)\right]. \quad (6.2.6)$$

In the case of incompressible over the whole stress range elastic-perfectly plastic materials ($\nu = 0.5$), Eqs. (6.2.5) and (6.2.6) read:

$$\sigma_r = \sigma_s^{stat} = \frac{2Y}{3}\left[1 + \ln\left(\frac{2E}{3Y}\right)\right], \qquad (6.2.7)$$

$$\sigma_r = \sigma_c^{stat} = \frac{Y}{\sqrt{3}}\left[1 + \ln\left(\frac{E}{\sqrt{3}Y}\right)\right]. \qquad (6.2.8)$$

Finally, in the case of incompressible over the whole stress range linear strain-hardening materials ($\nu = 0.5, k_{plast} > 0$), Eqs. (6.2.1) and (6.2.2) read:

$$\sigma_r = \sigma_s^{stat} = \frac{2Y}{3}\left[1 + \ln\left(\frac{2E}{3Y}\right)\right] + \frac{2\pi^2}{27} k_{plast}, \qquad (6.2.9)$$

$$\sigma_r = \sigma_c^{stat} = \frac{Y}{\sqrt{3}}\left[1 + \ln\left(\frac{E}{\sqrt{3}Y}\right)\right] + \frac{\pi^2}{18} k_{plast}. \qquad (6.2.10)$$

Hill, 1950a proposed static SCE and CCE models for compressible over the whole stress range elastic-perfectly plastic materials:

$$\sigma_r = \sigma_s^{stat} = \frac{2Y}{3}\left[1 + \ln\left(\frac{E}{3(1-\nu)Y}\right)\right], \qquad (6.2.11)$$

$$\sigma_r = \sigma_c^{stat} = \frac{Y}{\sqrt{3}}\left[1 + \ln\left(\frac{\sqrt{3}E}{(5-4\nu)Y}\right)\right]. \qquad (6.2.12)$$

Regarding the expansion of a non-thin layer (shell), Hill, 1950a recommends using models which are valid for incompressible materials, i.e., the models determined by Eqs. (6.2.7)–(6.2.8) and (6.2.9)–(6.2.10).

Masri and Durban, 2007 developed compressible cavity expansion models using a controlled error approach. Introducing an error control parameter, κ , they arrived at the expression for the cavitation pressure which depends only on one parameter. The optimal value of κ is such that allows predicting the cavitation pressure with a high accuracy. A number of CCE models were proposed in the absence of strain-hardening, namely, plane-strain models for von Mises and Tresca yield conditions, respectively (Masri and Durban, 2007):

$$\sigma_r = \sigma_c^{stat} = \frac{Y}{\sqrt{3}}\left\{1+\ln\left[\frac{\sqrt{3}E}{[3+(1-2\kappa)(1-2\nu)]Y}\right]\right\}, \tag{6.2.13}$$

$$\sigma_r = \sigma_c^{stat} = \frac{Y}{2}\left\{1+\ln\left[\frac{E}{2(1-\nu^2)Y}\right]\right\}. \tag{6.2.14}$$

Masri and Durban, 2007 used the value $\kappa = -0.4725$ in Eq. (6.2.13). Note that Eq. (6.2.13) coincides with Eq. (6.2.12) for $\kappa = -0.5$.

Marsi *et al.*, 2010 suggested the following plane-stress models for von Mises and Tresca yield conditions, correspondingly:

$$\sigma_r = \sigma_c^{stat} = 2.04Y\left[1-(2e-1)\frac{Y}{E}\right], \tag{6.2.15}$$

$$\sigma_r = \sigma_c^{stat} = 2Y\left[1-(2e-1)\frac{Y}{E}\right]. \tag{6.2.16}$$

6.3 Momentum and Energy Balance Approach

For a relatively thin plate ($b/L_{imp} < 0.5,\ b/d < 0.5$) penetrated by a blunt fragment, Recht and Ipson, 1963 proposed the following formula for the residual velocity that is based on momentum and energy balance:

$$v_{res} = a(v_{imp}^2 - v_{bl}^2)^{1/2}, \quad v_{imp} \geq v_{bl}, \tag{6.3.1}$$

or

$$\tilde{v}_{res} = a(\tilde{v}_{imp}^2 - 1)^{1/2}, \quad \tilde{v}_{imp} \geq 1, \tag{6.3.2}$$

where

$$a = 1\Big/\left(1+\frac{m_{plug}}{m}\right) = 1\Big/\left(1+\frac{\rho_{sh} b d_{plug}^2}{\rho_{imp} L_{imp} d^2}\right), \tag{6.3.3}$$

$$\tilde{v}_{imp} = v_{imp} / v_{bl} , \quad \tilde{v}_{res} = v_{res} / v_{bl} , \tag{6.3.4}$$

m_{plug} and d_{plug} are the mass and the diameter of the plug, correspondingly.

For the BLV, Recht and Ipson, 1963 suggested the following formula:

$$v_{bl} = \frac{4\zeta_1 \rho_{sh} b^2 \psi \eta}{L_{imp} d} \left[1 + \sqrt{\left(1 + \frac{L_{imp} \rho_{imp}}{b \rho_{sh}}\right)\left(1 + \frac{\zeta_2 d}{4\rho_{sh} b \psi^2 \eta}\right)} \right], \tag{6.3.5}$$

where

$$\psi = \frac{1}{\rho_{imp} C_{imp}} + \frac{1}{\rho_{sh} C_{sh}}, \tag{6.3.6}$$

C_{sh}, C_{imp} are the longitudinal acoustic-wave velocities in the projectile and shield materials, respectively, ζ_1, ζ_2 are constants depending on chosen units for parameters, η is a coefficient related to the dynamic shear strength if the shield material properties are assumed to be constant. Recht and Ipson, 1963 demonstrated that their model successfully predicts penetration of mild steel plate with $\eta = 1.76 \cdot 10^5\, psi = 1.21\, MPa$ by a cylindrical projectile.

In the case of a thick plate penetrated by a cylinder, Recht and Ipson, 1963 suggested the following formula for the parameter a in Eqs. (6.3.1) and (6.3.2):

$$a = 1 \Big/ \sqrt{\left(1 + \frac{\hat{m}_{plug}}{m}\right)\left(1 + \frac{\rho_{sh} b d_{plug}^2}{\rho_{imp} L_{imp} d^2}\right)}, \tag{6.3.7}$$

where $\hat{m}_{plug}$ is the experimental value of the ejected plug mass.

The model by Recht and Ipson, 1963, for the case of penetration by a sharp impactor without plug formation, is based on the energy conservation law that can be written as follows:

$$0.5 m v_{imp}^2 - 0.5 m v_{res}^2 = W , \tag{6.3.8}$$

where W is the work performed during perforation.

Assumption that W does not depend on the impact velocity implies Eqs. (6.3.1) and (6.3.2) with $a = 1$:

$$v_{res} = (v_{imp}^2 - v_{bl}^2)^{1/2}, \quad v_{imp} \geq v_{bl}, \tag{6.3.9}$$

$$\tilde{v}_{res} = (\tilde{v}_{imp}^2 - 1)^{1/2}, \quad \tilde{v}_{imp} \geq 1 \tag{6.3.10}$$

and the following formula for the BLV:

$$v_{bl} = \sqrt{2W/m}\,. \tag{6.3.11}$$

The formula for the work, W_{cav}, that is required for expanding the hole in a ductile material up to the radius R reads (for details, see Zaid *et al.*, 1973 and Corbett *et al.*, 1996):

$$W_{cav} = \pi\, k_W R^2 bY\,, \tag{6.3.12}$$

where k_W equals to 2.0 (Bethe, 1941), 1.33 or 0.5, depending of the deformation mode (Taylor, 1948), 1.92 (Hill, 1950b). Associating W_{cav} with the energy loss of the rigid, point-nosed impactor, W, and assuming that $W = W_{cav}$ one can obtain formulas for the residual velocity and the BLV by substituting W_{cav} from Eq. (6.3.12) instead of W in Eq. (6.3.11).

Thomson, 1955 proposed the model that takes into account a dynamic component of the work required for perforation of a thin plate. This model can be written as follows (Sodha and Jain, 1958, Nixdorff, 1987b):

$$W(h) = 0.5\pi bY\Phi^2(h) + \pi b\rho_{sh} v^2 I(h)\,, \tag{6.3.13}$$

where h is the instantaneous penetration depth of the impactor into the plate after beginning of penetration with a constant velocity v, W is the power that is needed to overcome the resistance of the plate, function $\Phi(x)$ determines the shape of a sharp nose of the impactor (see Fig. 1.1.2), and

$$I(h) = \int_0^h [\Phi\Phi_{xx} + 2\Phi_x^2]\Phi\Phi_x dx\,. \tag{6.3.14}$$

Following a quasi-dynamic approach assume that Eq. (6.3.13) is valid for $0 \le h \le L$, where L is the length of the nose of impactor. Then energy conservation equation yields the following formula for the instantaneous impactor velocity v (Sodha and Jain, 1958):

$$W(h) = 0.5m(v_{imp}^2 - v^2)\,. \tag{6.3.15}$$

Substituting $v = v_{res}, h = L$ and $\Phi(L) = R$ in Eqs. (6.3.13) and (6.3.15) yields:

$$v_{res} = \sqrt{\frac{mv_{imp}^2 - \pi bYR^2}{m + 2\pi b\rho_{sh}I(L)}}, \quad v_{bl} = R\sqrt{\pi bY/m}, \quad v_{imp} \ge v_{bl}\,. \tag{6.3.16}$$

Gupta *et al.*, 2001 considered a thin plate and took into account the work required in the formation of crater by ogive nosed projectile and energy of the radial stretching in the following form:

$$W = \pi R^2 b\left[0.5Y + 0.62\rho_{sh}\left(\frac{v_{imp}R}{L}\right)^2\right] + \frac{\pi bYw_c^2(1+2kl)\exp(-2kl)}{4\sqrt{1-\nu+\nu^2}} \tag{6.3.17}$$

where l is length of the crack. Gupta *et al.*, 2001 used the following empirical correlation between the radial plate deflection, w, the central deflection of the plate, w_c, and the distance from the point of impact, $\tilde{r}$, (Calder and Goldsmith, 1971; Levy and Goldsmith, 1984):

$$w = w_c \exp(-k\tilde{r})\,, \tag{6.3.18}$$

where a constant k can be evaluated from the profile of the plate. The residual velocity and the BLV can be determined from Eqs. (6.3.8) and (6.3.11), respectively.

Zhang and Mu, 2000 considered perforation of a thin shield by a rigid sphere or hemispherical-nosed impactor and developed an analytical model describing perforation in a dicing mode. Their approach is based on the estimation of the energy dissipations during three stages of penetration: bulging, fracture, and hole enlargement.

Pol *et al.*, 2009 suggested the model whereby the work required for perforation of a thin plate by ogive-nosed projectile comprises the work required for plastic deformation, the work for transferring the matter to new position, and the work for bending of the petals.

Zaid and Paul, 1957 and Paul and Zaid, 1958 investigated normal penetration of pointed and truncated cone-nosed and ogive-nosed impactors into thin ductile plate using momentum balance and obtained a number of relationships which allow estimating characteristics of penetration. In particular, their analysis implies the following formula for a pointed cone-nosed impactor with a semi-vertex angle ϑ:

$$v_{res} = v_{imp}(1 - \pi b \rho_{sh} R^2 \sin\vartheta / m) . \qquad (6.3.19)$$

Brown, 1964 applied a quasi-dynamic theory to describe penetration of truncated conical-nosed impactor into plastically deformed thin plate. In order to estimate the thickness of the shield that allows containing the impactor, he derived formulas for the energy of the impactor.

Woodward, 1978a proposed a model for conical-nosed impactors taking into account ductile and dishing failure modes that yields the following formula for the minimum perforation energy:

$$W = \frac{\pi bYR^2}{2}\left[1 + 0.5\pi\left(\frac{b}{R}\right) + 0.867\left(\frac{b}{R}\right)^2\right]. \qquad (6.3.20)$$

Woodward, 1978b also suggested the following formulas for W which can be applied for thick and thin shields, correspondingly:

$$W = 2\pi R^3 Y\left(\frac{b}{R} + \frac{1}{3\tan\vartheta} - \sqrt{3}\right). \qquad (6.3.21)$$

$$W = \frac{2\pi b^3 Y}{\sqrt{3}\tan\vartheta}, \tag{6.3.22}$$

where the difference between thick and thin plates is determined by the value of $b = \sqrt{3}R$.

Forrestal *et al.*, 1987 and Rosenberg and Forrestal, 1988 developed an engineering model for "shields that exhibit ductile hole-growth" which was applied for conical-nosed rods penetrating into aluminum plate having a finite thickness:

$$v_{bl} = R\sqrt{2\pi b\sigma_c^{stat}/m}, \quad v_{res} = v_{bl}\sqrt{(v_{imp}/v_{bl})^2 - 1}, \quad v_{imp} \geq v_{bl}, \tag{6.3.23}$$

$$\sigma_r = \sigma_c^{stat} = \frac{Y}{2}\left[1 + \ln\left(\frac{2E}{(5-4\nu)Y}\right)\right], \tag{6.3.24}$$

where σ_c^{stat} corresponds to the static CCE model for elastic-perfectly plastic material with compressible plastic zone and the Tresca yield condition (Forrestal, 1986; Satapathy, 1997). The second equation in Eqs. (6.3.23) for v_{res} recovers formula suggested by Recht and Ipson, 1963.

Dikshit and Sundararajan, 1992 (see also Liaghat and Malekzadeh, 1995; Diskshit and Sundararajan, 1995) proposed an analytical model describing penetration of thick steel plates by ogive shaped projectiles. Liaghat and Malekzadeh, 1999 modified this model by introducing an analytical equation instead of an empirical relationship for determining the size of the plastic zone around the projectile inside the shield.

The approach used by Landkof and Goldsmith, 1993 and Wierzbicki, 1999 for thin plates is based on an energy balance including the energies required for crack propagation, petal bending, and plate dishing. Wierzbicki, 1999 emphasizes that the main advantage of his model is determining these three energies simultaneously while Landkof and Goldsmith, 1993 assume that these fractional energies are independent of each other. Wierzbicki, 1999 proposed the following formula for the BLV of plate penetrated by a sharp conical-nosed impactor:

$$v_{bl} = 4.22\left(\frac{\sigma_u Y}{1+n}\right)^{0.5}\left(\frac{\delta_b}{b}\right)^{0.1}\left(\frac{R\rho_{sh}}{m}\right)^{0.7} b^{0.8}, \tag{6.3.25}$$

where n is the exponent in the power-type stress-strain law, δ_b is the crack tip opening displacement parameter (CTOD). For simplicity, it can be assumed that $\delta_b/b=1$.

Masri, 2010 suggested a simple model for penetration resistance of pointed nose projectiles, D, that is based on power balance between the power supplied by the projectile during steady-state penetration and elasto-plastic power required for expanding the penetration hole:

$$D = \frac{\pi d^2}{4}\tilde{D}, \quad \tilde{D} = \left(\frac{2S}{Ld}\right)\left[\hat{\alpha}\sigma_s^{stat} + \left(\frac{2L}{d}\right)\frac{k_{fr}Y}{\sqrt{3}}\right], \tag{6.3.26}$$

where S and L are the meridional cross section area and the length of the projectile nose, correspondingly, d is a projectile shank diameter, k_{fr} is a friction factor (Durban and Fleck, 1992), $0 \le k_{fr} < 1$, $\hat{\alpha}$ is an empirical parameter, σ_s^{stat} is static spherical cavitation pressure. For the hardening power law with the parameter n (Chakrabarty, 2010),

$$\sigma = \begin{cases} E\varepsilon & if \quad \varepsilon \le Y/E \\ Y(E\varepsilon/Y)^n & if \quad \varepsilon > Y/E \end{cases}. \tag{6.3.27}$$

Masri, 2010 proposed the following approximate formula for σ_s^{stat}:

$$\sigma_s^{stat} = \frac{2Y}{3}\left\{1 - \ln 2 - \ln(1-\nu) + \frac{1}{n}\left[\left(\frac{2E}{3Y}\right)^n - 1\right] + 0.73n\left(\frac{2E}{3Y}\right)^n\right\}. \tag{6.3.28}$$

In particular, formula for penetration resistance force for ogive-nosed projectile reads:

$$\tilde{D} = f(K_{CRH})\left[\hat{\alpha}\sigma_s^{stat} + \sqrt{4K_{CRH} - 1}\frac{k_{fr}Y}{\sqrt{3}}\right], \tag{6.3.29}$$

where

$$f(K_{CRH}) = 1 + 2K_{CRH}\left[\frac{\sin^{-1}(\omega)}{\omega} - 1\right], \omega = \frac{\sqrt{4K_{CRH} - 1}}{2K_{CRH}}. \quad (6.3.30)$$

Solving the equation of projectile motion, $mv\,dv/dh = -\tilde{D}$, where v is velocity of the projectile, we arrive at the following formula for the depth of penetration (DOP):

$$H = \frac{\tilde{m}}{2\tilde{D}} v_{imp}^2, \quad \tilde{m} = m/(0.25\pi d^2). \quad (6.3.31)$$

Using results of experiments from the literature, Masri, 2010 concluded that the empirical parameter $\hat{\alpha}$ is practically independent of the shield material characteristics but depends on the shape of the projectile nose, whereas $\hat{\alpha}$ varies in a narrow range and is slightly smaller than for ogival nose projectiles.

It was found that experimental data on penetration into aluminum shields by spherical-nose projectiles can be fairly well described (up to some critical impact velocity) by the following formula:

$$H = \frac{\tilde{m}}{\pi\sigma_s^{stat}} v_{imp}^2, \quad (6.3.32)$$

which is obtained from Eqs. (6.3.28)–(6.3.31) for $K_{CRH} = 0.5$, $\hat{\alpha} = 1$ and $k_{fr} = 0$. Penetration of ogival-nosed projectile with $K_{CRH} = 0.5$ into aluminum shields can be described by these equations with $\hat{\alpha} = 1$ and $0.2 \le k_{fr} \le 0.6$; $k_{fr} = 0.4$ is suggested as a representative value for all the experiments.

Marsi *et al.*, 2010 and Cohen *et al.*, 2010 examined the validity of static CCE models for predicting the BLVs and the residual velocities on the basis of Eq. (6.3.23) with various relations for σ_c^{stat}. They used experimental data for the shields manufactured from aluminum alloys and Weldox steels as well as numerical simulations for steel. Different versions of the models corresponding to Tresca or von Mises conditions, plane-stress or plane-strain patterns, were compared and the parameters of the models were determined.

Using the energy balance approach, Srivathsa and Ramakrishnan, 1997, 1998 derived a ballistic performance index to estimate and compare the ballistic quality of metal materials. This index is a function of the commonly determined mechanical properties of the material and the impact velocity of the projectile. In their study Srivathsa and Ramakrishnan, 1999 presented these indexes as maps.

6.4 Non Cavity Expansion Models

To the best of our knowledge, the first model in the penetration mechanics which belongs to the class of the localized interaction models (LIMs) was suggested by Nishiwaki, 1951. This is a two-term model described by Eq. (1.4.2):

$$\Omega_n(u,v) = a_0 + a_2 u^2 v^2 , \qquad (6.4.1)$$

where a_0 is the "static contact pressure" and $a_2 = \rho_{sh}$. This model takes into account the impactor-shield friction. Using his experimental results for conical impactors penetrating into aluminum shields, Nishiwaki, 1951 arrived at the conclusion that a_0 is proportional to the thickness of the perforated plate. He developed a relationship between the impact velocity and the residual velocity of a cone-shaped impactor, taking into consideration the change in the impactor-plate contact surface during penetration.

Vitman and Stepanov, 1959 investigated experimentally penetration of cone-nosed projectiles with impact velocities up to $1000\,m/s$ into various metal shields. They found that the ratio of the resistance force to the shank area exhibits a linear dependence on the squared impactor velocity, provided that the penetration depth exceeds the length of the conical nose. Analysis of these linear correlations for penetrators by projectiles having different cone angles yielded two interesting observations. The first is that the slope of these lines is approximately the same as the slope of the drag force dependence for a conical projectile moving in a gaseous medium and calculated according to the Newton model (Hayes and Probstein, 1959). The second is that all straight lines obtained in the experiments on penetration into the shields manufactured from the same material intersect at the same point. These two observations imply that the formula for a drag force for a conical impactor may be written as follows:

$$D_{cone} = \pi R^2 (a_0 + a_2 \sin^2 \vartheta \, v^2) , \qquad (6.4.2)$$

where $a_2 = \rho_{sh}$, $a_0 = \tilde{H}_{sh}$ is the "dynamic hardness of the metal for impact velocities of $\tilde{v} \sim 10\,\mathrm{m/s}$", ϑ is the half angle of the apex of the cone. Some values of $\tilde{H}_{sh}$ adopted from the study by Vitman and Ioffe, 1948 are as follows: $350\,MPa$, $1850\,MPa$, $910\,MPa$ and $1330\,MPa$ for aluminum, soft steel, copper and duraluminum, respectfully.

Initially, Vitman and Stepanov, 1959 used the following correlation:

$$a_0 = \tilde{H}_{sh} (v/\tilde{v})^{\alpha} , \qquad (6.4.3)$$

but they subsequently decided that the dependence of a_0 on the impactor velocity can be neglected because $0 < \alpha << 1$.

Let us assume that the local interaction force between the projectile and the host medium is equal to the force acting at the surface of the tangent cone at this location, when the projectile velocity and the host medium are the same in both cases. Then Eq. (6.4.2) for the cone-nosed impactors implies the model given by Eq. (6.4.1) with $a_2 = \rho_{sh}$ and $a_0 = \tilde{H}_{sh}$ for sharp impactors of different shapes. Hence the result obtained using the Vitman and Stepanov, 1959 formula is often referred as a LIM, although describing the local interaction between the medium and the impactor contact surface was not the direct goal of their study.

Different versions of the above assumption are known in aerodynamics as "methods of tangent cones" (Chernyi, 1969; Hayes and Probstein, 1959). Similar approach in penetration mechanics was used by Recht, 1990.

Golubev and Medvedkin, 2000, 2001 studied penetration of a rigid rods with conical and hemispherical heads into a thick plate manufactured from mild low-carbon steel at velocities of up to $600\,\mathrm{m/s}$. On the basis of the obtained experimental data they proposed a three-stage penetration model. At the initial stage of penetration ($h \le R$), the linear dependence between the resistance to penetration and the impactor velocity is used. At the third stage ($h > 2R$), the resistance force is assumed to be constant. At the intermediate stage ($R < h \leq 2R$), the resistance force is determined by linear interpolation between the values of the resistance force at the initial and final stages.

Landgrov and Sarkisyan, 1984 proposed the following model:

$$\Omega_n(u,v) = E(\hat{\sigma}_0/E)^{1-uv/c}, \quad c = \sqrt{E/\rho_{sh}}, \tag{6.4.4}$$

where $\hat{\sigma}_0$ is experimentally determined parameter that is close to the value of the Mayer hardness.

Yarin *et al.*, 1995 studied penetration of a rigid projectile (ovoid of Rankine) into an elastic-plastic shield. This particular shape of the impactor was selected because it implied a reasonably simple velocity field that exactly satisfied the continuity equation and the condition of impenetrability of the projectile. Although equation of motion of projectile can be generally solved numerically, analytical formulas for the DOP, the residual velocity and the BLV were derived by including some additional assumptions. Particularly, for deep penetration into a semi-infinite shield, the DOP is given by the following formula:

$$H = \frac{3m + \pi\rho_{sh}R^3}{6\pi YR^2 \ln\left(4^{5/9}\dfrac{3G}{4Y}\right)} v_{imp}^2, \tag{6.4.5}$$

where G is the shear modulus of the material of the shield. Yarin *et al.*, 1995 also described the procedure that allows applying the suggested approach to a projectile having a tip of arbitrary shape.

Levy and Goldsmith, 1984 considered normal impact of thin plates by hemispherical tip cylindrical hard-steel projectiles and derived an expressions for the resistance force, D, as a function of time, t. The authors note that "the theory is based on a lumped-parameter system that includes the radial variation in axial target displacement". For the non-perforation case (the impact velocity is less than the BLV) the theory implies the following relationship:

$$D(t) = \frac{1.33m^{2.5}\sqrt{Yb}}{(m+m_{eq})^2} v_{imp} \Big/ \exp\left(\frac{1.33\sqrt{mYb}}{m+m_{eq}} t\right), \quad m_{eq} = \frac{\pi\rho_{sh}b}{2k^2}, \tag{6.4.6}$$

where t is time, k is the same as in Eq. (6.3.18). For the perforation case (the impact velocity is larger than the BLV) the sub-model "requires either measured information from the plug separation process or corresponding reasonable geometric and kinematic assumptions for a solution in the perforation domain". The detailed analysis of the model by Levy and Goldsmith, 1984 may be found in Corbett *et al.*, 1996.

Wen, 2002a, 2002b suggested the model in which the normal stress acting on the projectile-metal shield contact surface is a linear function of the impact velocity.

Shaw, 2012 suggested the model of the type given by Eq. (6.4.1) with $a_0 = \sigma_r$ and $a_2 = \rho_{sh}G(v_{imp})$, where σ_r is determined in Eq. (6.2.5) while G is a function that strongly depends on the impact velocity in addition to the shield material properties. The function $G(v_{imp})$ is determined on the basis of experiments; it is a decreasing function that tends to a constant value of 0.5 when the impact velocity is very large. Shaw, 2012 gave two examples of approximations for $G(v_{imp})$, for 0.4 GPa aluminum and 0.8 GPa aluminum, correspondingly:

$$G = 0.0019v_{imp}^4 - 0.0373v_{imp}^3 + 0.2860v_{imp}^2 - 1.0059v_{imp} + 1.8603, \tag{6.4.7}$$

$$G = 0.0092v_{imp}^4 - 0.1369v_{imp}^3 + 0.7877v_{imp}^2 - 2.1346v_{imp} + 2.8697. \tag{6.4.8}$$

On the basis of numerical simulations Rosenberg and Dekel, 2009a, 2010b proposed the following formula for the BLV of sharp-nosed projectiles penetrating into elastic-plastic shields:

$$v_{bl} = R\sqrt{2\pi b\sigma_*/m}, \tag{6.4.9}$$

where

$$\frac{\sigma_*}{Y} = \begin{cases} 2/3 + 4\bar{b} & if \quad 0 < \bar{b} \le 1/3 \\ 2.0 & if \quad 1/3 < \bar{b} \le 1.0 \\ 2.0 + 0.8\ln\bar{b} & if \quad \bar{b} > 1.0 \end{cases} \tag{6.4.10}$$

$$\bar{b} = b/d . \tag{6.4.11}$$

6.5 Quasi-Dynamic Cavity Expansion Models

Spherical and cylindrical quasi-dynamic models of penetration dynamics are generally based on cavity expansion laws described by Eqs. (2.2.1) and (2.3.5) and can be reduced to LIMs using Eqs. (2.2.2) and (2.3.8). Most of the cavity expansion laws are written in one of the following forms:

$$\sigma_r = \alpha_s + \beta_s V^2, \quad \alpha_s = \sigma_s^{stat}, \tag{6.5.1}$$

$$\sigma_r = \alpha_c + \beta_c V^2, \quad \alpha_c = \sigma_c^{stat}, \tag{6.5.2}$$

or

$$\sigma_r = \alpha_s + \gamma_s V + \beta_s V^2 . \tag{6.5.3}$$

Eqs. (6.5.1), (6.5.2) and (6.5.3) imply LIMs given by Eqs. (2.4.3), (2.4.4) and (2.4.6), correspondingly. Formulas that allow calculating the resistance force, the DOP, the BLV for different LIMs are presented in Chapter 1.

Consider now some particular quasi-dynamic models.

Forrestal *et al.*, 1988 suggested quasi-dynamic SCE and CCE models for compressible elastic-perfectly plastic material similar to those given by Eqs. (6.5.1) and (6.5.2), respectively, and used Eqs. (6.2.11) and (6.1.12) for the static part of the normal stress while coefficients β_S and β_c were determined by curve-fitting of the corresponding results by Forrestal and Luk, 1988 and Forrestal, 1986. For the 7075-T651 shield material ($K = 69GPa$, $\nu = 0.33$, $Y = 400MPa$, $\rho_{sh} = 2710kg/m^3$), the values of these coefficients are $\beta_s/\rho_{sh} = 1.041$ and $\beta_c/\rho_{sh} = 1.837$.

Forrestal *et al.*, 1990 developed a CCE model for rate independent, power law strain hardening material that is described by Eq. (6.5.2) with

$$\sigma_c^{stat} = \frac{Y}{\sqrt{3}}\left[1 + \left(\frac{E}{\sqrt{3}Y}\right)^n \Psi(1-\eta, n)\right], \tag{6.5.4}$$

$$\beta_c = \frac{\rho_{sh}}{2}\left(\frac{1+\ln 2}{1-\nu} + \eta - \ln\eta - 1\right), \tag{6.5.5}$$

where

$$\eta = \frac{2(\nu+1)Y}{\sqrt{3}E}, \tag{6.5.6}$$

$$\Psi(z,n) = \int_0^z \frac{(-\ln\varsigma)^n}{1-\varsigma} d\varsigma, \quad 0 < z < 1. \tag{6.5.7}$$

It was assumed that the plastic zone is incompressible while the elastic region is compressible. Parameter n (which is generally less than 0.5) is the coefficient in empirical stress-strain relation given by Eq. (6.3.27). For calculating the integral with singularity at $\varsigma = 0$ in Eq. (6.5.7), Forrestal and Romero, 2007 proposed the following formula:

$$\Psi(z,n) = 2.994 - 3.742n + 3.511n^2 + \frac{0.05^n - (1-z)^n}{n} + \frac{n[0.05^{n+1} - (1-z)^{n+1}]}{2(n+1)}. \tag{6.5.8}$$

Luk and Amos, 1991 developed numerical CCE model for compressible, elastic-plastic, rate independent materials with power-law strain-hardening. For the 5083-H131 aluminum shield material at a strain rate of $10^0\, s^{-1}$ ($E = 70.3\,GPa$, $\nu = 0.33$, $Y = 276\,MPa$, $\rho_{sh} = 2710\,kg/m^3$, $n = 0.084$), this model is described by Eq. (6.5.2) with $\sigma_c^{stat}/Y = 4.090$ and $\beta_c/\rho_{sh} = 2.610$. Luk and Amos, 1991 compared numerical results obtained using their model with the results of analytical model by Forrestal *et al.*, 1990 for incompressible materials and concluded that application of the compressible model improved predictions for the thicker plates and $v_{imp} >> v_{bl}$.

Luk *et al.*, 1991 proposed dynamic SCE model for elastic-plastic, rate independent material with power-law hardening. For the incompressible material, the closed-form solution was obtained in the form of Eq. (6.5.1) where

$$\sigma_s^{stat} = \frac{2Y}{3}\left[1 + \frac{\Psi(1-\eta,n)}{\eta^n}\right], \tag{6.5.9}$$

$$\beta_s = 1.5\rho_{sh}, \qquad (6.5.10)$$

$$\eta = 1.5Y/E. \qquad (6.5.11)$$

Application of the compressible model requires numerical solution of differential equations.

In order to verify these models Forrestal *et al.*, 1991 conducted experiments with spherical-nose rods penetrating into semi-infinite steel and aluminum shields. Penetration model for the incompressible material given by Eqs. (6.5.9)–(6.5.11) was used along with the results of numerical solutions for compressible material obtained by Luk *et al.*, 1991. For aluminum shields ($E = 68.9GPa$, $\nu = 0.33$, $Y = 276MPa$, $n = 0.051$, $\rho_{sh} = 2710kg/m^3$), the compressible model is described by Eq. (6.5.1) with $\sigma_s^{stat}/Y = 4.407$ and $\beta_s/\rho_{sh} = 1.133$ while $\sigma_s^{stat}/Y = 4.595$ and $\beta_s/\rho_{sh} = 1.5$ for the incompressible model.

Forrestal *et al.*, 1992 performed similar analysis for ogive-nosed rods. For the 7075-T651 shield material ($E = 73.1GPa$, $\nu = 0.33$, $Y = 448MPa$, $n = 0.089$, $\rho_{sh} = 2710kg/m^3$), the compressible model is described by Eq. (6.5.2) with $\sigma_s^{stat}/Y = 4.418$ and $\beta_s/\rho_{sh} = 1.068$ while $\sigma_s^{stat}/Y = 4.609$ and $\beta_s/\rho_{sh} = 1.5$ for the incompressible model.

Warren and Forrestal, 1998 proposed the set of SCE models for 6061-T651 aluminum shields which are determined by Eq. (6.5.3) where

$$\alpha_s = \tilde{\alpha}_s Y, \quad \gamma_s = \tilde{\gamma}_s\sqrt{\rho_{sh}Y}, \quad \beta_s = \tilde{\beta}_s\rho_{sh}, \qquad (6.5.12)$$

and the coefficients $\tilde{\alpha}_s$, $\tilde{\gamma}_s$, $\tilde{\beta}_s$ are given in Table 6.5.1.

Table 6.5.1. Coefficients of models by Warren and Forrestal, 1998.

Model	$\tilde{\alpha}_s$	$\tilde{\gamma}_s$	$\tilde{\beta}_s$
Incompressible without strain rate effects	4.8376	0.0	1.5
Incompressible with strain rate effects	5.5468	0.6255	1.4377
Compressible without strain rate effects	4.4534	0.4680	0.9926
Compressible with strain rate effects	5.0394	0.9830	0.9402

Warren, 1999 developed a CCE model for elastic-plastic rate-dependent material with power-law strain hardening assuming that the plastic zone is incompressible while the elastic region is compressible:

$$\sigma_r = \sigma_c^{stat} + \frac{\rho_{sh} V^2}{2}\left[\frac{1}{(1-\nu)\sqrt{1-\lambda^2}}\ln\left(\frac{1+\sqrt{1-\lambda^2}}{\lambda}\right) - \ln\left(\frac{2(1+\nu)Y}{\sqrt{3}E}\right) - 1\right] + \frac{\tilde{\alpha}}{\sqrt{3}\tilde{m}}\left(\frac{2V}{\sqrt{3}a_{cav}\dot{\varepsilon}_0}\right)^{\tilde{m}}\left[1-\left(\frac{2(1+\nu)Y}{\sqrt{3}E}\right)^{\tilde{m}}\right] \tag{6.5.13}$$

where σ_c^{stat} and η are determined by Eqs. (6.5.4) and (6.5.6), correspondingly, a_{cav} is the radius of the cavity surface, $\tilde{\alpha}$, $\tilde{m}$, $\dot{\varepsilon}_0$ are parameters in the stress-strain relation (Chakrabarty, 2010; Yadav *et al.*, 1995):

$$\sigma = \begin{cases} E\varepsilon & if \quad \sigma \le Y_d \\ Y(E\varepsilon/Y)^n + \tilde{\alpha}(\dot{\varepsilon}/\dot{\varepsilon}_0)^{\tilde{m}} & if \quad \sigma > Y_d \end{cases}, \tag{6.5.14}$$

σ, ε, n are the same as in Eq. (6.3.27), $\dot{\varepsilon}$ is the strain rate, $\dot{\varepsilon}_0$ is a reference strain rate, $\tilde{\alpha}$ is the curve-fitting parameter having the units of stress, Y_d is the dynamic yield stress given by the following formula:

$$Y_d = Y + \tilde{\alpha}(\dot{\varepsilon}/\dot{\varepsilon}_0)^{\tilde{m}}. \tag{6.5.15}$$

Masri and Durban, 2005 derived a closed SCE analytical model for compressible, elastic/perfectly plastic material including strain-hardening that does not require application of best fit procedure to the numerical data:

$$\sigma_r = \sigma_s^{stat} + \beta_s V^2 + \chi_s V^3, \tag{6.5.16}$$

where σ_s^{stat} is determined by Eq. (6.2.11) and

$$\beta_s = \rho_{sh}\left[\frac{3}{2} - 0.160\frac{(1-2\nu)(13-5\nu)}{(1-\nu)^{5/3}}\left(\frac{Y}{E}\right)^{1/3}\right], \tag{6.5.17}$$

$$\chi_s = -\frac{2}{9}\sqrt{(1-2\nu)\left(\frac{1+\nu}{1-\nu}\right)^5 \frac{\rho_{sh}^3}{E}}. \tag{6.5.18}$$

This model also includes the case where the expansion velocity is much smaller than the speed of sound in the solid, i.e., $V << \sqrt{E/\rho_{sh}}$. The authors note that the term $\chi_s V^3$ can often be omitted when the model is applied to penetration mechanics.

Masri and Durban, 2006 proposed the CCE model for incompressible, elastic-plastic, linear strain-hardening materials which is determined by Eq. (6.5.2) with

$$\sigma_c^{stat} = \frac{Y}{\sqrt{3}}\left[1+\ln\left(\frac{E}{\sqrt{3}Y}\right)\right] + \frac{\pi^2}{18}\eta_{hard}, \tag{6.5.19}$$

$$\beta_c = \frac{\rho_{sh}}{2}\ln\left(\frac{E}{4\sqrt{3}Y}\right). \tag{6.5.20}$$

The following stress-strain equation is used:

$$\sigma = \begin{cases} E\varepsilon & if \quad \varepsilon \le Y/E \\ \overline{\eta}_{hard}E\varepsilon + (1-\overline{\eta}_{hard})Y & if \quad \varepsilon > Y/E \end{cases}, \tag{6.5.21}$$

where $\overline{\eta}_{hard} = \eta_{hard}/E$, η_{hard} is a constant hardening modulus, $0.001 < \overline{\eta}_{hard} < 0.01$.

Masri and Durban, 2009 suggested the CCE model for plastic orthotropic strain hardening materials that is described by Eq. (6.5.16). In order to simplify application of the model for penetration mechanics, the model was approximated by Eq (6.5.1). Similar procedure was applied for a SCE model.

On the basis of the numerical simulations of expansion of spherical and cylindrical cavities in elastic-perfectly plastic materials (aluminum, steel, lead) Rosenberg and Dekel, 2008 proposed approximate SCE and CCE models. The SCE model is described by Eq. (6.5.1) with σ_s^{stat} given by Eq. (6.2.11) and $\beta_s = 1.1875\rho_{sh}$. The formula for CCE model reads:

$$\sigma_r = \alpha_c + \gamma_c V + \beta_c V^2, \tag{6.5.22}$$

where

$$\alpha_c = \frac{Y}{\sqrt{3}}\left[1 + \ln\left(\frac{\sqrt{3}E}{6(1-\nu)Y}\right)\right], \quad \gamma_c = 0.29\rho_{sh}, \quad \beta_c = 0.29\rho_{sh}. \tag{6.5.23}$$

Employing similar approach, Rosenberg and Dekel, 2009c suggested a SCE model that is determined by Eq. (6.5.16) with σ_s^{stat} given by Eq. (6.2.11) and

$$\begin{aligned} \beta_s &= K\left[1.5 - (1.3 - 0.53\nu - 1.19\nu^2)\left(\frac{Y}{K}\right)^{1/3}\right], \\ \chi_s &= -K\left[[2/3 - \zeta(\nu)]\left(\frac{Y}{K}\right)^{1/3}\right], \end{aligned} \tag{6.5.24}$$

where K is the bulk modulus, while ζ is equal to 2.36, 2.237, 2.066, 1.88 for ν equal to $0.1, 0.2, 0.3, 0.4$, respectively.

Rosenberg and Dekel, 2009b analyzed the results of numerical simulations of penetration of rigid long rods having ogive, spherical, conical and flat nose shapes into semi-infinite aluminum and steel shields having different strengths. They found that for the impact velocities, v_{imp}, less than some critical velocity, v_c, the deceleration (resistance force) of the rods is practically constant and is independent on the impact velocity, while for the velocities larger than v_c a velocity dependent term must be added to the static resistance term.

Mathematically, the model by Rosenberg and Dekel, 2009b can be described as follows. The cavity expansion approach implies the following expression for the impactor's resistance force, D:

$$\frac{D}{\pi R^2} = \sigma^{stat} + \chi\rho_{sh}v^2, \tag{6.5.25}$$

where σ^{stat} is static normal stress determined from a SCE or CCE model, χ depends on the nose shape of the rod.

Instead of the model given by Eq. (6.5.25), the following model is used:

$$\frac{D}{\pi R^2} = \begin{cases} \sigma^{stat} & if \quad v \le v_c \\ 0.5(\sigma^{stat} + \chi\rho_{sh}v^2) & if \quad v > v_c \end{cases}, \tag{6.5.26}$$

where

$$v_c = \sqrt{\sigma^{stat}/(\chi\rho_{sh})} \,. \tag{6.5.27}$$

It is convenient to denote as "basic model" and "suggested model" the models determined by Eqs. (6.5.25) and (6.5.26), respectively, written in a dimensionless form:

$$\overline{D} = 1 + \overline{v}^2 \,, \tag{6.5.28}$$

$$\overline{D} = \begin{cases} 1 & if \quad \overline{v} \le 1 \\ 0.5(1 + \overline{v}^2) & if \quad \overline{v} > 1 \end{cases}, \tag{6.5.29}$$

where

$$\overline{D} = \frac{D}{\pi R^2 \sigma^{stat}}, \quad \overline{v} = \frac{v}{v_c} \,. \tag{6.5.30}$$

Solving the equation of motion of impactor allows determining the DOP, H. For basic and suggested models, the expressions for the dimensionless DOP, $\overline{H}$, read:

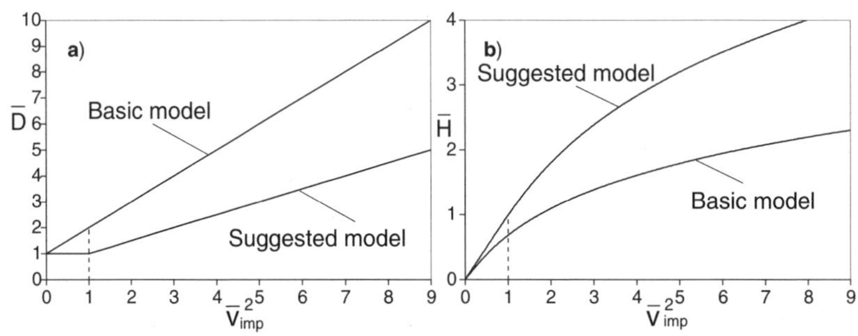

Fig. 6.5.1. Basic and the suggested by Rosenberg and Dekel, 2009b models.

$$\overline{H} = \ln(1+\overline{v}_{imp}^2), \tag{6.5.31}$$

$$\overline{H} = \begin{cases} \overline{v}_{imp}^2 & if \quad \overline{v}_{imp} \le 1 \\ 2\ln(1+\overline{v}_{imp}^2) - c & if \quad \overline{v}_{imp} > 1 \end{cases}, \tag{6.5.32}$$

where

$$\overline{H} = \frac{2\pi R^2 \sigma^{stat}}{m v_c^2} H, \quad \overline{v}_{imp} = \frac{v_{imp}}{v_c}, \quad c = 2\ln 2 - 1. \tag{6.5.33}$$

Comparison of the basic and the suggested by Rosenberg and Dekel, 2009b models is shown in Fig. 6.5.1.

6.6 Dynamic Cavity Expansion Models

Hopkins, 1960 noted that "important aspects of this type of theory were first considered by R. Hill" referring to Hill unpublished report of 1948, and suggested the following dynamic SCE model for incompressible elastic-plastic material:

$$\sigma_r = \sigma_r(y, \dot{y}, \ddot{y}) = \sigma_s^{stat} + \sigma_s^{dyn}, \tag{6.6.1}$$

where

$$\sigma_s^{dyn} = \sigma_s^{dyn}(y, \dot{y}, \ddot{y}) = \rho_{sh}(y\ddot{y} + 1.5\dot{y}^2), \tag{6.6.2}$$

y is the radius of the hole, σ_s^{stat} is determined by Eq. (6.2.7). On the basis of the Eq. (6.6.2), Goodier, 1965 proposed the following formula for the normal stress at the surface of the hemispherical-nose impactor:

$$p = \sigma_s^{stat} + \sigma_s^{dyn}(R, v, \dot{v})\cos\upsilon = \sigma_s^{stat} + \rho_{sh}(R\dot{v} + 1.5v^2)\cos\upsilon, \tag{6.6.3}$$

where υ is the angle between the direction of motion of impactor and the local vector normal to the surface, R is the radius of the hemisphere. The equation of motion of the impactor reads (Bernard and Hanagud, 1975):

$$(m + m_{add})\dot{v} = -\pi R^2(\sigma_s^{stat} + \rho_{sh}v^2), \quad m_{add} = \pi\rho_{sh}R^3. \tag{6.6.4}$$

Eq. (6.6.4) implies that m_{add} can be neglected if $m_{add} << m$ and, consequently, the term $\rho_{sh} R\dot{v}\cos\upsilon$ in Eq. (6.6.3) can be omitted. Then the quasi-dynamic SCE model can be considered whereby a solution of the equation of motion of impactor is used for determining penetration characteristics (Bernard and Hanagud, 1975; Forrestal *et al.*, 1995). Bernard and Hanagud, 1975 also extended the approach suggested by Goodier, 1965 to impactors with non-hemispherical nose shapes.

Sagomonyan, 1975 considered CCE in incompressible elastic-plastic material and proposed the following model:

$$\sigma_r = \sigma_r(y, \dot{y}, \ddot{y}) = \alpha_c + \beta_c \dot{y}^2 + \eta_c\, y\ddot{y}, \tag{6.6.5}$$

where

$$\begin{aligned} &\alpha_c = \frac{Y}{2}[1 + \ln(1+\omega)], \quad \omega = \frac{2E}{3Y}, \\ &\beta_c = \frac{\rho_{sh}}{2}\left[\ln(1+\omega) - \frac{\omega}{\omega+1}\right], \quad \eta_c = \frac{\rho_{sh}}{2}\ln(1+\omega) \end{aligned} \tag{6.6.6}$$

Then Eq. (2.3.4) implies the following formula for the normal stress at the surface of the impactor:

$$p = \alpha_c + (\beta_c \Phi'^2 + \eta_c \Phi\Phi'')v^2 + \eta_c \Phi\Phi'\dot{v}, \tag{6.6.7}$$

where the function $\Phi(x)$ describes the shape of the impactor nose (see Fig. 2.3.1). Sagomonyan, 1975 considered conical-nosed impactors in more detail (see also Ben-Dor *et al.*, 2000d). Generally, Eq. (6.6.7) implies the equation of motion of the impactor that is similar to the equation obtained for two-term model described in Chapter 1.

Sagomonyan, 1988 developed a model describing the expansion of a cylindrical cavity inside an incompressible elastic-plastic medium, starting from the non-zero radius and outlined the applications of this model for investigating penetration of an impactor having a plane bluntness that is accompanied by plug formation. Penetration was considered as two simultaneous processes, namely, expansion of the cavity in the shield and motion of the plug. Ben-Dor *et al.*, 2006a extended this approach and determined the closed-form solutions for the

BLV of arbitrary body of revolution without any additional simplifications.

Aptukov *et al.*, 1992 derived an approximate solution for the expansion of cylindrical hole in a compressible elastic-plastic medium in the form of Eq. (6.6.5) with

$$\beta_c = \hat{\beta}_c / v_*^2, \quad \eta_c = \hat{\eta}_c / v_*^2, \tag{6.6.8}$$

where values of the coefficients α_c, $\hat{\beta}_c$, $\hat{\eta}_c$ for several metals are given in Table 6.6.1, while v_* is equal to $5040 m/s$, $5160 m/s$ and $4650 m/s$ for aluminum, titanium and steel, correspondingly.

Aptukov, 1991c determined an approximate solution for the expansion of spherical hole in a compressible elastic-plastic medium in the form of Eq. (6.6.5) and proposed formulas for calculating the coefficients of the model.

Warren and Forrestal, 1998 suggested a SCE model that includes power-law strain hardening and strain-rate sensitivity of an incompressible target material:

$$\sigma_r = \sigma_s^{stat} + \rho_{sh}(1.5V^2 + y\dot{V}) + \frac{2\tilde{\alpha}}{3\tilde{m}}\left(\frac{2V}{y\dot{\varepsilon}_0}\right)^{\tilde{m}}, \tag{6.6.9}$$

where σ_s^{stat} is determined by Eq. (6.5.9).

Li *et al.*, 2007b and Chen *et al.*, 2008b, Chen *et al.*, 2008c considered the class of SCE models of the following type:

$$\sigma_r = \alpha_s + \xi_s V + \beta_s V^2 + \varsigma_s \dot{V}, \tag{6.6.10}$$

Table 6.6.1. Parameters $\alpha_c, \hat{\beta}_c, \hat{\eta}_c$.

Parameter	Aluminum, Y (GPa)			Titanium, Y (GPa)			Steel, Y (GPa)		
	0.1	0.3	0.5	0.4	0.6	0.8	0.8	1.0	1.2
α_c (GPa)	0.41	1.10	1.70	1.49	2.12	2.70	2.94	3.57	4.18
$\hat{\beta}_c$ (GPa)	76	71	64	122	118	112	170	162	157
$\hat{\eta}_c$ (GPa)	55	50	48	88	85	83	121	118	117

$$\sigma_r = \alpha_s + \xi_s V + \beta_s V^2 + \varsigma_s \dot{V}, \tag{6.6.10}$$

where $\alpha_s, \xi_s, \beta_s, \varsigma_s$ are shield's material coefficients. They derived also a dimensionless formula for the DOP as a function of three parameters for general convex shapes of a projectile.

6.7 Oversimplified Models

6.7.1 *Shields having a finite thickness*

Chen and Li, 2003a; 2011 started from a quasi-dynamic SCE model of the type given by Eq. (2.4.1):

$$\sigma_r = \alpha_s + \beta_s V^2, \quad \sigma_t = \mu_{fr}\sigma_r, \quad \alpha_s = \tilde{\alpha}_s Y, \quad \beta_s = \tilde{\beta}_s \rho_{sh}, \tag{6.7.1}$$

where $\tilde{\alpha}_s$ and $\tilde{\beta}_s$ are dimensionless constants depending on the chosen model of shield material (such models were discussed above). They showed that this quasi-dynamic SCE model implies the following formulas for the residual velocity and the BLV:

$$v_{res}^2 = (v_{imp}^2 - v_{bl}^2)\exp\left(-\frac{\pi b}{2dN}\right), \quad v_{bl}^2 = \frac{\tilde{\alpha}_s Y N_1}{\tilde{\beta}_s \rho_{sh} N_2}\left[\exp\left(\frac{\pi b}{2dN}\right) - 1\right], \tag{6.7.2}$$

where

$$N_1 = 1 + 2\mu_{fr} K_{CRH}^2 (\pi - 2\phi - \sin 2\phi), \tag{6.7.3}$$

$$N_2 = N^* + \mu_{fr} K_{CRH}^2 \left[\frac{\pi}{2} - \phi - \frac{1}{3}\left(2\sin 2\phi + \frac{\sin 4\phi}{4}\right)\right], \tag{6.7.4}$$

$$N^* = \frac{8K_{CRH} - 1}{24K_{CRH}^2}, \quad \phi = \sin^{-1}\left(1 - \frac{1}{2K_{CRH}}\right), \quad K_{CRH} \geq \frac{1}{2}. \tag{6.7.5}$$

for ogive-nosed projectiles, and

$$N_1 = 1 + \mu_{fr} \cot\vartheta, \quad N_2 = \frac{1 + \mu_{fr} \cot\vartheta}{1 + \cot^2\vartheta} \tag{6.7.6}$$

for conical-nosed projectiles.

For sufficiently sharp and slender projectiles when the parameter

$$N = \frac{m}{\tilde{\beta}_s \rho_{sh} d^3 N_2} \tag{6.7.7}$$

is sufficiently large, Chen and Li, 2003a proposed the following simplified version of formulas in Eq. (6.7.2):

$$v_{res}^2 = v_{imp}^2 - v_{bl}^2, \quad v_{bl}^2 = \frac{\pi b \tilde{\alpha}_s Y N_1 d^2}{2m}. \tag{6.7.8}$$

Forrestal and Warren, 2009 started from a quasi-dynamic CCE model of the type described by Eq. (2.4.2),

$$\sigma_r = \alpha_c + \beta_c V^2, \quad \sigma_\tau = 0, \quad \alpha_c = \sigma_c^{stat}, \quad \beta_c = \tilde{\beta}_c \rho_{sh}, \tag{6.7.9}$$

where $\tilde{\beta}_c$ is a dimensionless constant depending on the chosen model of shield material, σ_c^{stat} is determined by Eq. (6.5.4). This model implies the following formulas for the residual velocity and the BLV:

$$v_{res} = \sqrt{v_{imp}^2 - v_{bl}^2}\,\exp(-C), \quad v_{bl} = \left(\frac{\sigma_c^{stat}}{\tilde{\beta}_c \rho_{sh} N_*}\right)^{1/2} \sqrt{\exp(2C) - 1}, \tag{6.7.10}$$

where

$$C = \frac{\tilde{\beta}_c b \rho_{sh} N_*}{(L_0 + k_{shape} L) \rho_{imp}}, \tag{6.7.11}$$

$$N_* = 8K_{CRH}^2 \ln\left(\frac{2K_{CRH}}{2K_{CRH} - 1}\right) - 4K_{CRH} - 1, \tag{6.7.12}$$

for ogive-nose projectiles, and

$$N_* = \tan^2 \vartheta \tag{6.7.13}$$

for conical-nosed projectiles.

Forrestal and Warren, 2009 assumed the following formula for the mass of the projectile:

$$m = \frac{\pi d^2 \rho_{imp}}{4}(L_0 + k_{shape} L), \tag{6.7.14}$$

where

$$k_{shape} = 4K_{CRH}^2 - \frac{4}{3}K_{CRH} + \frac{1}{3} - \frac{4K_{CRH}^2(2K_{CRH}-1)}{\sqrt{4K_{CRH}-1}}\sin^{-1}\left(\frac{\sqrt{4K_{CRH}-1}}{2K_{CRH}}\right) \tag{6.7.15}$$

for ogival-nosed strikers, and

$$k_{shape} = 1/3 \tag{6.7.16}$$

for conical-nosed strikers.

If the nose of the projectile is sufficiently sharp then keeping several first terms in Taylor series expansion of the exponent in Eq. (6.7.10) yields the following simplified formulas for the residual velocity and the BLV:

$$v_{res} = \sqrt{v_{imp}^2 - v_{bl}^2}\left(1 - C + 0.5C^2\right)^{1/2}, \tag{6.7.17}$$

$$v_{bl} = \sqrt{\frac{2\sigma_c^{stat} b}{(L_0 + k_{shape} L)\rho_{imp}}}\left(1 + C + \frac{2}{3}C^2\right)^{1/2}. \tag{6.7.18}$$

The latter equations with $C = 0$ can be also considered as a version of a simplified penetration model.

Forrestal and Warren, 2009; Forrestal *et al.*, 2010; Børvik *et al.*, 2010 analyzed the accuracy of the model given by Eqs. (6.7.17) and (6.7.18) using experimental results.

Chen and Li, 2011 compared the models suggested by Chen and Li, 2003a and by Forrestal and Warren, 2009 and concluded that there are the differences in their predictions are quite small. Note that Chen and Li, 2011 used the simplified (as compared with Eq. (6.7.17)) formula for the residual velocity (Forrestal and Warren, 2009):

$$v_{res} = \sqrt{v_{imp}^2 - v_{bl}^2}\left(1 - 2C + 2C^2\right)^{1/2}. \tag{6.7.19}$$

This formula is obtained by raising Eq. (6.7.10) to the second power and keeping only the first three terms in Taylor series expansion of $\exp(-2C)$.

6.7.2 *Semi-infinite shields*

Forrestal and Warren, 2009 started from a quasi-dynamic SCE model given by Eqs. (6.5.1), (6.5.9)–(6.5.11). This model implies the following formula for the DOP of ogive-nosed projectiles:

$$\frac{H}{L_0 + k_{shape}L} = \frac{\rho_{imp}}{3\rho_{sh}N^*}\ln\left(1 + \frac{3\rho_{sh}N^*}{2\sigma_s^{stat}}v_{imp}^2\right), \tag{6.7.20}$$

where k_{shape} and N^* are determined by Eq. (6.7.15) and (6.7.5), respectively.

Assuming that $1.5\rho_{sh}N^* v_{imp}^2 / \sigma_s^{stat}$ is much smaller than 1 Eq. (6.7.20) can be written in two simplified forms by replacing the logarithm by one or two first terms of the Taylor series expansion:

$$\frac{H}{L_0 + k_{shape}L} = \frac{\rho_{imp}v_{imp}^2}{2\sigma_s^{stat}}, \tag{6.7.21}$$

$$\frac{H}{L_0 + k_{shape}L} = \frac{\rho_{imp}v_{imp}^2}{2\sigma_s^{stat}}\left(1 - \frac{3\rho_{sh}N^*}{4\sigma_s^{stat}}v_{imp}^2\right). \tag{6.7.22}$$

Comparison with the results of experiments showed that Eq. (6.7.22) is very accurate for striking velocities smaller than $1300\,m/s$.

The model similar to Eq. (6.7.21) was also proposed by Chen and Li, 2002.

6.8 Plugging and Multi-Stage Models

Most of models taking into account plug formation, for instance, suggested by Awerbuch, 1970; Awerbuch and Bodner, 1974; Chen and Li, 2003a, 2003b; Goldsmith and Finnegan, 1971; Hsueh, 1988, 1989; Jenq *et al.*, 1988; Krishna Teja Palleti *et al.*, 2012; Liss *et al.*, 1983; Ning *et al.*, 2005; Shoukry *et al.*, 1991; Ravid and Bodner, 1983, 1994; Teng and Wierzbicki, 2005; Wu and Batra, 1996 (most of which is surveyed by Corbett *et al.*, 1996) generally do not allow deriving explicit formulas for determining the residual velocity or the BLV. Two models considered below provide a possibility to determine the residual velocity and the BLV.

6.8.1 *Basic simplified model*

The simplest model of plugging for a cylindrical impactor (see e.g., Sagomonyan, 1988; Holt *et al.*, 1993) is based on the assumption that the cylindrical plug with the radius $R = r$, height b and mass

$$m_{plug} = \pi\rho_{sh}r^2 b \tag{6.8.1}$$

is formed at the beginning of penetration, and that this plug moves together with the impactor under the action of the resistance force associated with the plug:

$$D_{plug}(h) = 2\pi\tau_s r(b-h), \quad h \le b\,, \tag{6.8.2}$$

where h is the instantaneous depth of penetration and τ_s is a constant shear stress. If the impactor penetrates at the depth $h = b$ then perforation occurs. More complicated models of plug formation when the plate is penetrated by cylindrical impactor were suggested by Sagomonyan, 1977; Sagomonyan *et al.*, 1982.

Ben-Dor *et al.*, 2000a, 2006a, 2007c considered impactors with arbitrary generatrix having a flat bluntness, and took into account the resistance force at the lateral surface of the impactor, $D_{lat} = D_{lat}(h,\dot{h},\ddot{h})$. In this case, the impactor is subjected to the resistance forces D_{plug} and

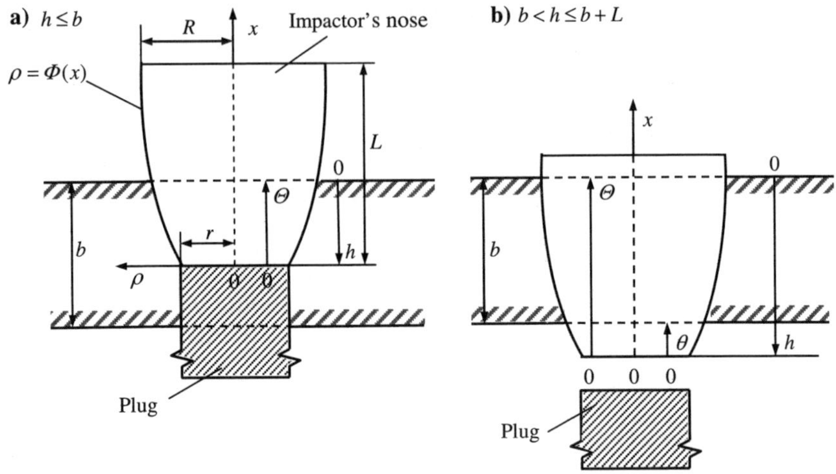

Fig. 6.8.1. Model of penetration with plug formation.

D_{lat} until $h \le b$ (see Fig. 6.8.1a) and the resistance force D_{lat} when the plug breaks apart at the depth $h = b$ (see Fig. 6.8.1b). Therefore, equations describing the motion of the impactor read:

$$(m + m_{plug})\ddot{h} + D_{lat}(h, \dot{h}, \ddot{h}) + D_{plug}(h) = 0, \quad 0 \le h < b, \qquad (6.8.3)$$

$$m\ddot{h} + D_{lat}(h, \dot{h}, \ddot{h}) = 0, \quad b \le h \le b + L. \qquad (6.8.4)$$

Considering $w = v^2 = \dot{h}^2$ as a function of the independent variable h, Eqs. (6.8.3) and (6.8.4) can be replaced by the following equation:

$$\frac{1}{2}[m + m_{plug}\delta(h)]\frac{dw}{dh} + D_{lat}\left(h, \sqrt{w}, \frac{1}{2}\frac{dw}{dh}\right) + D_{plug}(h)\delta(h) = 0, \qquad (6.8.5)$$

where the function $\delta(h)$ is determined by Eq. (1.3.9).

In most of analytical approximate models, D_{lat} can be written in the following form:

$$D_{lat}(h, \dot{h}, \ddot{h}) = \chi_0(h) + \chi_1(h)\dot{h}^2 + \chi_2(h)\ddot{h}, \qquad (6.8.6)$$

where functions χ_0, χ_1, χ_2 depend upon h, upon the shape of the impactor and upon characteristics of the material of the shield. Substituting D_{lat} from Eq. (6.8.6) into Eq. (6.8.5) yields the following linear ordinary differential equation:

$$\frac{1}{2}[m + m_{plug}\delta(h) + \chi_2(h)]\frac{dw}{dh} + \chi_1(h)w + D_{plug}(h)\delta(h) + \chi_0(h) = 0, \quad (6.8.7)$$

which can be solved analytically to obtain formulas for the BLV and the residual velocity.

6.8.2 *Slezkin model*

N. A. Slezkin in 1944 (see Grigoryan, 2006; Sagomonyan, 1974) suggested a simple model based on the assumption that only shear associated with plug formation causes the resistance force. We describe this model in the generalized form, for an arbitrary impactor (body of revolution) with flat bluntness.

The two-stage model is illustrated in Fig. 6.8.2. The impactor and the plug move together as one body. The expression for the resistance force acting on the body comprising the impactor and the plug reads:

Motion of this composite body having a variable mass at the first stage of penetration ($0 \le h \le L$) is described by the equation:

$$\frac{d}{dt}[(1 + k_1\Phi^2)v] = -k_2\Phi, \quad (6.8.10)$$

where t is time and

$$k_1 = \pi b\rho_{sh}/m, \quad k_2 = 2\pi b\tau_s/m. \quad (6.8.11)$$

After multiplying by $(1 + k_1\Phi^2)vdt$ both sides of Eq. (6.8.10) and taking into account that $vdt = dh$, we obtain that

$$0.5d[(1 + k_1\Phi^2)v]^2 = -k_2(1 + k_1\Phi^2)\Phi dh. \quad (6.8.12)$$

Motion of this composite body having a variable mass at the first stage of penetration ($0 \le h \le L$) is described by the equation:

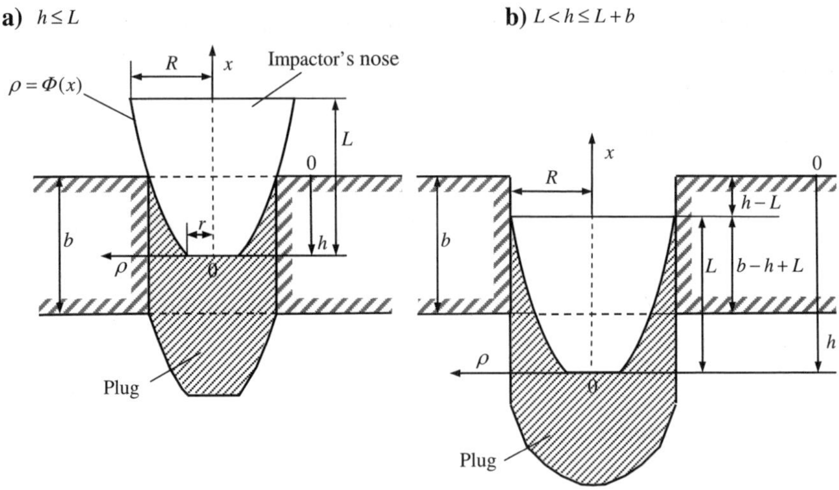

Fig. 6.8.2. Slezkin model of penetration with plug formation.

$$\frac{d}{dt}[(1+k_1\Phi^2)v] = -k_2\Phi\,, \tag{6.8.10}$$

where t is time and

$$k_1 = \pi b\rho_{sh}/m, \quad k_2 = 2\pi b\tau_s/m\,. \tag{6.8.11}$$

After multiplying by $(1+k_1\Phi^2)vdt$ both sides of Eq. (6.8.10) and taking into account that $vdt = dh$, we obtain that

$$0.5d[(1+k_1\Phi^2)v]^2 = -k_2(1+k_1\Phi^2)\Phi dh\,. \tag{6.8.12}$$

The latter equation in total differentials implies that

$$0.5d[(1+k_1\Phi^2)v]^2\Bigg|_{h=0,\, v=v_{imp}}^{h=L,\, v=v_*} = -k_2\int_0^L(1+k_1\Phi^2)\Phi dh\,, \tag{6.8.13}$$

where v_* is velocity of the impactor when $h = L$. After some algebra, Eq. (6.8.13) can be written as follows:

$$(1+k_1r^2)^2v_{imp}^2 - (1+k_1R^2)^2v_*^2 = 2k_2(J_1+k_1J_3)\,, \tag{6.8.14}$$

where

$$r = \Phi(0),\ R == \Phi(L),\ J_1 = \int_0^L \Phi(h)dh,\ J_3 = \int_0^L \Phi^3(h)dh\,. \quad (6.8.15)$$

Motion of the composite body having a constant mass that comprises the impactor and the plug, at the second stage of penetration ($L \le h \le L+b$) is described by the following equation:

$$(1+k_1R^2)\frac{dv}{dt} = -k_2\frac{R}{b}(b+L-h)\,. \qquad (6.8.16)$$

Since $dv/dt = v\,dv/dh$, Eq. (6.8.16) yields:

$$(1+k_1R^2)\int_{v_*}^{v_{res}} vdv = -k_2\frac{R}{b}\int_L^{L+b}(b+L-h)dh\,. \qquad (6.8.17)$$

After calculation of integrals in Eq. (6.8.17) we arrive at the following formula:

$$(1+k_1R^2)(v_*^2 - v_{res}^2) = k_2Rb\,. \qquad (6.8.18)$$

Excluding v_* from Eqs. (6.8.14) and (6.8.18), we obtain the relationship for the residual velocity when the impact velocity is sufficiently high for perforating the shield:

$$v_{res} = \frac{\sqrt{(1+k_1r^2)^2 v_{imp}^2 - 2k_2(J_1+k_1J_3) - k_2Rb(1+k_1R^2)}}{(1+k_1R^2)}\,. \qquad (6.8.19)$$

Substituting $v_{res} = 0$ and $v_{imp} = v_{bl}$ into Eq. (6.8.19) we obtain the following formula for the BLV:

$$v_{bl} = \frac{\sqrt{2k_2(J_1+k_1J_3) + k_2Rb(1+k_1R^2)}}{(1+k_1r^2)}\,. \qquad (6.8.20)$$

6.9 Some Other Models and Related Problems

Using energy and momentum balance equations and assuming that the shield is thin and that the absorbed energy is small in comparison with the impact energy, Giere, 1964 derived the following formula:

$$v_{res}/v_{imp} = m_{plug}/(m_{plug} + m_{imp}). \qquad (6.9.1)$$

Sagomonyan, 1974 applied the "model of normal sections" (see Section 8.1.2) for modeling penetration of cones with non small vertex angle. This approach was used also by Bagdoev *et al.*, 1997.

In the studies of Partom, 1996 and Littlefield *et al.*, 1997, cavity expansion models were modified to account for the finite size of the shield in the direction normal to the direction of penetration.

A number of solutions based on elastic-plastic isotropic models of shield material were obtained by Vantsyan, 1981; Bagdoev and Vantsyan, 1981a, 1981b; Rahmatulin and Bagdoev, 1982. Works by Bagdoev and Vantsyan, 1983, 1987; Bagdoev *et al.*, 1988, Bagdoev *et al.*, 1989, Bagdoev *et al.*, 1997; Asatryan *et al.*, 1996 (see also Sagomonyan, 1988) were devoted to modeling of translational penetration into anisotropic media. Bagdoev and Vantsyan, 1989 offered a solution for the problem of penetration into an elastic transversely isotropic medium when the impactor rotates or twists with constant angular velocity.

Huang *et al.*, 2011a, Huang *et al.*, 2011b applied cavity expansion theory for calculating the DOP in the case when the impact velocity is smaller than the BLV and the projectile only partially penetrates the plate. They found that for relatively low velocities nose shape of the projectile has a stronger influence on the penetration depth while the reverse dependence is observed for relatively high velocities.

Dienes and Miles, 1977 proposed a membrane model that requires numerical calculation of an integral for determining the BLV of thin plate.

Tirosh *et al.*, 2009 suggested a closed-form assessment of residual cavity depth left after a rigid hemispherical nosed cylinder impacts a rigid-plastic shield.

Teland and Moxnes, 2003 conducted numerical simulations using a hydrocode in order to compare analytical results from cavity expansion models with results of "exact" numerical calculations.

Wijk *et al.*, 2005; Wijk, 2005 suggested a model that can be described as follows. The drag force acting on the projectile is assumes to be a sum of the drag force acting on the nose of a projectile and the friction drag force acting at the body of a projectile. Initially the nose drag increases with the penetration depth until it attains a constant value which corresponds to lateral displacement of the shield material along the trajectory of a projectile. The projectile body friction force also increases at the initial stage of penetration and becomes constant when the trailing edge of the projectile passes through the front face of the shield. It is assumed that when the leading edge of the projectile is close to the rear surface of the shield, the remaining bulk of shield material ahead of the projectile crushes and forms fragments. At this penetration depth the force required for crushing is equal to the force required for continued lateral displacement of the shield material, and the fragments are ejected with the same velocity as the projectile.

Quite a few engineering models on penetration of metal plates were published in Chinese (Chen *et al.*, 2009; Jiang *et al.*, 2004, 2005a, 2005b, 2007a, 2007b; Song *et al.*, 2005; Wang *et al.*, 2005b, Wang *et al.*, 2010; Wang and Zhu, 2010; Yan *et al.*, 2008; Zhou and Wen, 2006).

Part 4

Penetration into Geological Shields

Dedicated surveys by Thompson and Mitchell, 1971; Adeli *et al.*, 1986; Heuzé, 1990 are devoted to engineering models describing penetration in geological media. Papers and surveys by Backman and Goldsmith, 1978; Brown, 1986; Corbett *et al.*, 1996; Wen and Xu, 2004; Ben-Dor *et al.*, 2005b and monographs by Bulson, 1997; Carlucci and Jacobson, 2008; Bangash, 2009; Szuladziński, 2010; Seah *et al.*, 2011 summarize some of the results obtained in the field. Performance of a number of publications include comparisons between the results obtained with different models and experiments.

This Part 4 includes Chapters 7 and 8 that dedicate to empirical models and analytical models, correspondingly.

Along with well known models much attention is given to the less familiar outside the former USSR models which have been suggested by Russian (Soviet) scientists.

A reference list to Part 4 includes more than 60 items.

Chapter 7

Empirical Models

Nomenclature

d	shank diameter of impactor;
D	resistance force;
DOP	depth of penetration;
f_c'	unconfined compressive strength;
H	depth of penetration;
$\overline{H}$	$= H/d$;
k_0, k_1, k_2	empirical coefficients, Eq. (7.1.1);
$\hat{k}_0, \hat{k}_1$	coefficients in Eq. (7.10.3);
K_{CRH}	caliber radius head defined in Eq. (1.6.28);
m	mass of impactor;
P	pressure at the surface of impactor nose;
v	instantaneous velocity of impactor;
v_{imp}	impact velocity;
$\overline{v}_{imp}$	$= v_{imp}/1000$;
$\gamma_1, \gamma_2 \ldots, \gamma_{17}$	parameters of models;
θ	angle between the normal vector to the front surface of a shield and the direction of impact;
ϑ	semi-vertex angle of cone-nosed impactor;
ρ_{imp}	material density of an impactor;
ρ_{sh}	material density of a shield.

7.1 Early Relations for DOP

Thompson and Mitchell, 1971; Backman and Goldsmith, 1978 and Corbett *et al.*, 1996 cited three best known early models for the DOP which are based on the assumption that the relation between the velocity of impactor, v, and the resistance force, D, is given by a polynomial:

$$D = k_0 + k_1 v + k_2 v^2, \tag{7.1.1}$$

where k_0, k_1, k_2 are empirical constants depending on the shape of the impactor and mechanical properties of the shield; the values of the constants for some soils can be found in Backman and Goldsmith, 1978. Formulas for the DOP are obtained by integrating the Newton second law equation. Hereafter we use the following notations: m is mass of the impactor, v_{imp} is the impact velocity, H is the depth of penetration.

7.1.1 *Robins (1742) and Euler (1745) equation*

$$H = \frac{m v_{imp}^2}{2k_0}, \quad D = k_0. \tag{7.1.2}$$

7.1.2 *Ponselet equation (1829)*

$$H = \frac{m}{2k_2} \ln\left(1 + \frac{k_2 v_{imp}^2}{k_0}\right), \quad D = k_0 + k_2 v^2. \tag{7.1.3}$$

Discussion on applicability of this model for sand with the constants k_0 and k_2 obtained from the experimental data, may be found in Sierakowski *et al.*, 1977.

7.1.3 *Resel equation (1895)*

$$H = \frac{m}{k_2} \ln\left(1 + \frac{k_2 v_{imp}}{k_1}\right), \quad D = k_1 v + k_2 v^2. \tag{7.1.4}$$

7.2 Petri Formula (1910)

According to Bulson, 1997 the Petri formula reads:

$$H = 0.552 \cdot 10^{-3} \frac{m}{d^2} K_{Pet} \ln(1 + 0.5 \cdot 10^{-4} v_{imp}^2), \tag{7.2.1}$$

where K_{Pet} depends on soil properties and impact velocity. For typical soils, the following approximation was proposed:

$$K_{Pet} = K_{Pet}^{(0)} - K_{Pet}^{(1)} v_{imp}, \tag{7.2.2}$$

where coefficients are given in Table 7.2.1.

Note that all formulas in Bulson, 1997 are converted into SI units.

Table 7.2.1. Coefficients in Petri formula.

Soil	$K_{Pet}^{(0)}$	$K_{Pet}^{(1)}$
Sand	154	0.230
Sandy foam	190	0.295
Loam	227	0.360
Clay	341	0.623

7.3 Some Formulas Suggested in the 1960s

Adeli *et al.*, 1986 mentioned a few simple models for the DOP which were suggested in the 1960s.

7.3.1 *Hermann et al. equation (1963)*

$$H = \gamma_1 d \left(\frac{\rho_{imp}}{\rho_{sh}} \right)^{0.67} v_{imp}^{0.66}, \tag{7.3.1}$$

where d is shank diameter of impactor, γ_1 is a coefficient that depends on shield material and is obtained from the experimental data.

7.3.2 *Rohani equation (1965)*

$$H = \gamma_2 d v_{imp}^{\gamma_5}, \tag{7.3.2}$$

where γ_2 and γ_3 are empirical coefficients.

7.4 Young Equations

The overview of the model by Young, 1997 is given in Section 3.23. Modifications of this model for the case of penetration into soil, rock and ice, are described in Sections 7.4.1–7.4.3.

7.4.1 *Penetration equations for rock*

In this case Eqs. (3.23.1)–(3.23.8) and explanations to these equations are still valid while Eq. (3.23.9) must be replaced by the following relationship:

$$K_S = 2.7/(f_c' K_Q)^{0.3}, \tag{7.4.1}$$

Table 7.4.1. Descriptive terms for rock quality, K_Q (Young, 1997).

Descriptive Terms	K_Q
Massive	0.9
Interbedded	0.6
Joint Spacing < 0.5m	0.3
Joint Spacing > 0.5m	0.7
Fractured, blocky, or fissured	0.4
Highly fractured or jointed	0.2
Slightly weathered	0.7
Moderately weathered	0.4
Highly weathered	0.2
Frost shattered	0.2
Rock Quality, very good/excellent	0.9
Rock Quality, good	0.7
Rock Quality, fair	0.5
Rock Quality, poor	0.3
Rock Quality, very poor	0.1

where f_c' is an unconfined compressive strength and the values of the parameter K_Q are presented in Table 7.4.1.

The model is recommended for $m > 5kg, \overline{H} = H/d \geq 3, v_{imp} < 1220 m/s$.

7.4.2 *Penetration equations for soil*

In this case Eqs. (3.23.1)–(3.23.5), (3.23.7)–(3.23.8) and explanations to these equations are still stand while Eq. (3.23.6) must be replaced by the following relationship:

$$K_m = \begin{cases} 0.27m^{1.1} & if \quad m < 27 \\ m^{0.7} & if \quad m \geq 27 \end{cases}. \qquad (7.4.2)$$

Parameter K_S for typical soils is given in Table 7.4.2.

The model is recommended for $m > 2kg$, $\overline{H} \geq 3$, $v_{imp} < 1220 m/s$.

Table 7.4.2. Parameter K_S for typical soils (Young, 1997).

Shield Description	K_S
Dense, dry, cemented sand. Dry caliche. Massive gypsite and selenite deposits.	2 – 4
Gravel deposits. Sand, without cementation. Very stiff and dry clay. Moderately dense to loose sand, not cementation, water content not important.	4 – 6
Soil fill material, with K_S range depending on compaction.	8 – 10
Silt and clay, low to medium moisture content, stiff. Water content dominates penetrability.	5 – 10
Silt and clay, moist to wet. Topsoil, loose to very loose.	10 – 20
Very soft, saturated clay. Very low shear strength.	20 – 30
Clay marine sediments, either currently (Gulf of Mexico) or recent geologically (mud deposits near Wendover, Utah).	30 – 60

7.4.3 *Penetration equations for ice and frozen soil*

After some algebra, the Young equation for ice and frozen soil (Young, 1997) can be written as follows:

$$\overline{H} = \frac{K_S K_{shape} K_m}{d^{2.2}} P(\overline{v}_{imp}), \quad \overline{H} = \frac{H}{d}, \quad \overline{v}_{imp} = \frac{v_{imp}}{1000}, \tag{7.4.3}$$

where

$$P(\overline{v}_{imp}) = \begin{cases} 0.28 \cdot 10^{-3} \ln(1 + 205\overline{v}_{imp}^2) & if \quad \overline{v}_{imp} < 0.061 \\ 0.52 \cdot 10^{-2} (\overline{v}_{imp} - 0.0305) & if \quad \overline{v}_{imp} \geq 0.061 \end{cases} \tag{7.4.4}$$

$$K_m = m^{0.6} \ln(50 + 0.29m^2), \tag{7.4.5}$$

and coefficient K_{shape} is determined by Eq. (3.23.7).

Regarding the coefficient K_S, Young, 1997 recommends the following values: $K_S = 4.5 \pm 0.25$ for fresh water ice and sea ice, and $K_S = 2.75 \pm 0.5$ for completely frozen saturated soil. The magnitude of K_S for the partially frozen soil may be as high as 7.0.

The model is recommended for $m > 5kg$, $\overline{H} \geq 3$, $v_{imp} < 1220 m/s$.

7.5 Berezan' Formula

The name of this formula originates from the island Berezan' where ballistic experiments on penetration into soils and concrete shields were conducted in 1912.

The Berezan' formula reads (Efimov, 1939; Sagomonyan, 1974):

$$\overline{H} = \frac{\gamma_4 m \overline{v}_{imp} \cos\theta}{100 d^3}, \tag{7.5.1}$$

where θ is the angle between the normal to the shield surface and the direction of the impact while parameter γ_4 for typical soils is given in Table 7.5.1.

Table 7.5.1. Parameter γ_4 for typical soils.

Shield material	γ_4
Loose earth	1.3
Clay soil	1.0
Loose sand	0.9
Dense clay	0.7
Dense earth	0.65
Loam soil	0.6
Sandy loam	0.5
Sand	0.45
Limestone	0.2
Granite	0.16

7.6 Modified Berezan' Formula

Berezan' formula was improved in Artillery Research Institute (former USSR). The modified version of this formula reads (Balagansky and Merzhievsky, 2004):

$$\overline{H} = \frac{\gamma_5 \gamma_6 m \overline{v}_{imp}}{1000 d^3} \frac{\cos(\xi\theta)}{\sqrt{\cos\theta}}, \tag{7.6.1}$$

Table 7.6.1. Parameter γ_6 for typical soils.

Shield material	γ_6
Loose earth	13–17
Regular soil	11–13
Clay having medium density	7–10
Frozen soil	3.5
Loam soil	6–8
Sand with regular water content	4.5–7
Saturated sand	6–9
Sandy loam	5–7
Limestone	1.8–2
Granite	1.4–1.7

where

$$\gamma_5 = (2.8\sqrt[3]{d} - 1.3\sqrt{d})[0.5 + 0.4(L/d)^{2/3}], \qquad (7.6.2)$$

$\xi = 1.82$ for the long-range projectiles and $\xi = 2.62$ for non long-range projectiles, the angle $\theta \le 0.5\pi/(2\xi - 1)$, $\bar{v}_{imp} \le 0.4$ and γ_6 for typical soils is given in Table 7.6.1.

7.7 Zabudsky Formula

The Zabudsly formula (Balagansky and Merzhievsky, 2004; Efimov, 1939) for the case of normal penetration of blunt projectiles with impact velocity less than $400\,m/s$ is given by the following relation between the dimensionless instantaneous velocity, $\bar{v}$, and the instantaneous depth of penetration of the impactor, $\bar{h}$:

$$\bar{h} = \gamma_7 \ln\left(\frac{1 + \gamma_9 \bar{v}_{imp}^2}{1 + \gamma_9 \bar{v}^2}\right), \qquad (7.7.1)$$

where

$$\gamma_7 = \frac{2m\cos\theta}{\pi\gamma_8\gamma_9\gamma_{10}d^2L}, \quad \bar{h} = \frac{h}{d}, \quad \bar{v} = \frac{v}{1000}, \qquad (7.7.2)$$

L is length of the nose of the projectile, coefficient γ_{10} varies in the range from 1.15 to 1.20, values of the coefficients γ_8 and γ_9 for typical soils are given in Table 7.7.1.

Table 7.7.1. Parameter γ_8 and γ_9 for typical soils.

Shield material	γ_8	γ_9
Loose earth	4.5	60
Clay having a medium density	10.2	35
Sand with regular water content	4.3	20

Eq. (7.7.1) implies the following formulas for the dimensionless DOP, $\overline{H}$, BLV, $\overline{v}_{bl}$, and the residual velocity of the impactor, $\overline{v}_{bl}$:

$$\overline{H} = \gamma_7 \ln(1 + \gamma_9 \overline{v}_{imp}^2), \tag{7.7.3}$$

$$\overline{v}_{bl} = \sqrt{\frac{1}{\gamma_9}[\exp(\overline{b}/\gamma_7) - 1]}, \tag{7.7.4}$$

$$\overline{v}_{res} = \frac{v_{res}}{1000} = \sqrt{\frac{1}{\gamma_9}[(1 + \gamma_9 \overline{v}_{imp}^2)\exp(-\overline{b}/\gamma_7) - 1]}, \quad \overline{v}_{imp} \le \overline{v}_{bl}, \tag{7.7.5}$$

where $\overline{b} = b/d$ is the dimensionless thickness of the shield.

7.8 Kar Formula

Using the dimensionless variables $\overline{H}$ and $\overline{v}_{imp}$ and SI units for the dimensional parameters the formula by Kar, 1977, that was mentioned in Adeli *et al.*, 1986, can be written as follows:

$$G(\overline{H}) = \gamma_{11} \overline{v}_{imp}^{1.25}, \tag{7.8.2}$$

where

$$\gamma_{11} = 20.5 \frac{K_{shape}^{(5)} m}{d^{2.31}\sqrt{f_c'}} \left(\frac{E}{E_{steel}} \right), \tag{7.8.3}$$

E and E_{steel} are Young modules of the material of the projectile and of the D6A-C steel, respectively, $E_{steel} = 207\,GPa$. Nose shape parameter for an ogive-nosed impactor, $K_{shape}^{(5)}$, is calculated as follows:

$$K_{shape}^{(5)} = \min(0.72 + 10^{-3} K_{CRH}^{2.72}, 1.45), \tag{7.8.4}$$

$$G(\overline{H}) = \begin{cases} 0.25\overline{H}^2 & if \quad \overline{H} \le 2 \\ \overline{H} - 1 & if \quad \overline{H} > 2 \end{cases} \tag{7.8.5}$$

Using Eq. (7.8.5) , the Kar formula can be rewritten as follows:

$$\overline{H} = \begin{cases} 2\sqrt{\gamma_{11}}\overline{v}_{imp}^{0.625} & if \quad \overline{v}_{imp} \le 1/\gamma_{11}^{0.8} \\ \gamma_{11}\overline{v}_{imp}^{1.25} + 1 & if \quad \overline{v}_{imp} > 1/\gamma_{11}^{0.8} \end{cases} . \tag{7.8.6}$$

For the case of perforation of shield having a finite thickness, b, Kar, 1979 recommended the following formula for calculating the dimensionless residual velocity, $\overline{v}_{res} = v_{res}/1000$:

$$\overline{v}_{res} = (\overline{v}_{imp}^{1.25} - \overline{v}_{bl}^{1.25})^{1/1.25}, \tag{7.8.7}$$

where

$$\overline{v}_{bl} = [G(\overline{b})/\gamma_{11}]^{1/1.25}, \quad \overline{b} = b/d . \tag{7.8.8}$$

7.9 Adeli–Amin–Sierakowski Model

Adeli *et al.*, 1986 proposed the model that can be written after conversion to SI units as follows:

$$H = \begin{cases} 0.09348 + 0.1925 \cdot 10^{-3}\omega & if \quad 69\,kPa \le f_c' < 552\,kPa \\ 0.1185 + 0.9622 \cdot 10^{-3}\omega & if \quad 5529\,kPa \le f_c' < 965\,kPa \end{cases}, \tag{7.9.1}$$

where

$$\omega = 2.68 \frac{K_{shape}^{(5)} m v_{imp}^2}{f_c' d^3} . \tag{7.9.2}$$

Adeli *et al.*, 1986 recommended the following ranges of applicability of their model: $40\,mm \le d \le 216\,mm, 3.3\,kg \le m \le 206\,kg, 69\,kPa \le f_c' \le 965\,kPa$, $18\,m/s \le v_{imp} \le 107\,m/s$.

7.10 WES models for penetration into rock

Three models, which are presented below, were suggested at the U. S. Army Engineer Waterways Experiment Station (WES). Analysis of the performance of the models conducted by their authors revealed that the first two models have almost the same accuracy while the accuracy of the third model is somewhat better.

7.10.1 *Bernard model (1977)*

Bernard, 1977 proposed the following model for penetration into rock:

$$H = 0.255\frac{mv_{imp}}{d^2\sqrt{f_c'\rho_{imp}}}\left(\frac{100}{K_{RQD}}\right)^{0.8}, \tag{7.10.1}$$

where ρ_{imp} is material density of impactor, K_{RQD} is the rock quality in % (Jaeger, 1979). Table 7.10.1 shows the relationship between K_{RQD} and rock quality (Deere, 1968).

Bernard, 1977 estimated the accuracy of prediction of the model as $\pm 20\%$ at best and formulated some restrictions on the validity of the model. Eq. (7.10.1) is valid only when $\overline{H} > 3$. For nearly intact rock ($K_{RQD} > 90$), the model appears to be applicable for the projectiles having the shank diameter in the interval $3cm \le d \le 30cm$. For rock having a rock quality parameter $K_{RQD} < 90$, the model has not been verified for the projectiles having shank diameter outside the interval, $3cm \le d \le 30cm$. If $K_{RQD} < 20$, the model should not be used at all. The effect of nose shape seems to be weak, but the model is not recommended for blunt or near-blunt ($K_{CRH} < 1.5$) projectiles.

Murphy, 1984 generalized this model to shields with a predrilled hole:

Table 7.10.1. Relationship between K_{RQD} and rock quality.

Rock quality	K_{RQD} (%)
Very poor	0–25
Poor	25–50
Fair	50–75
Good	75–90
Excellent	90–100

$$H = 0.255 \frac{mv_{imp}}{(d^{2.} - d_0^2)\sqrt{f_c' \rho_{imp}}} \left(\frac{100}{K_{RQD}} \right)^{0.8}, \tag{7.10.2}$$

where d_0 is hole diameter.

7.10.2 *Bernard model (1978)*

Bernard, 1978 proposed a model which is based upon the relation between the velocity of the impactor, v, and the resistance force, D:

$$D = \frac{\pi d^2}{4} (\hat{k}_0 + \hat{k}_1 v), \tag{7.10.3}$$

where

$$\hat{k}_0 = 1.6 f_c' \left(\frac{K_{RQD}}{100} \right)^{1.6}, \quad \hat{k}_1 = 3.6 \sqrt{f_c' \rho_{imp}} \left(\frac{K_{RQD}}{100} \right)^{0.8}. \tag{7.10.4}$$

Integrating the Newton second law equation yields the following formula for the DOP:

$$H = \frac{4m}{\pi d^2} \left[\frac{v_{imp}}{\hat{k}_1} - \frac{\hat{k}_0}{\hat{k}_1^2} \ln \left(1 + \frac{\hat{k}_1}{\hat{k}_0} v_{imp} \right) \right]. \tag{7.10.5}$$

7.10.3 *Bernard and Creighton model (1979)*

Bernard and Creighton, 1979 proposed a modification of the model by Bernard, 1978 in which the resistance force for the ogive-nosed and conical-nosed (with half angle ϑ) impactors is determined by Eq. (7.10.3) where

$$\hat{k}_0 = \frac{4 f_c'}{K_{shape}^{(6)}} \left(\frac{K_{RQD}}{100} \right)^{0.2}, \quad \hat{k}_1 = \frac{3\sqrt{f_c' \rho_{imp}}}{K_{shape}^{(6)}} \left(\frac{K_{RQD}}{100} \right)^{0.1}, \tag{7.10.6}$$

$$K_{shape}^{(6)} = \begin{cases} 0.863\left(\dfrac{4K_{CRH}^2}{4K_{CRH}-1}\right)^{0.25} & \textit{for ogive nose} \\ 0.805/\sqrt{\sin\vartheta} & \textit{for conical nose} \end{cases} \quad (7.10.7)$$

and the DOP is determined by Eq. (7.10.5).

7.11 WES Model for Penetration into Soil

Crull *et al.*, 1999 proposed the following equation for the DOP of a fragment penetrating into soils (after conversion to SI units):

$$H = 0.165\gamma_{12}m^{1/3}\log(1+0.50\cdot 10^{-4}v_{imp}^2), \quad (7.11.1)$$

where the coefficient γ_{12} is given in Table 7.11.1.

Table 7.11.1. Relationship between γ_{12} and soil type.

Soil type	γ_{12}
Limestone	0.775
Sandy soil	5.29
Soil containing vegetation	6.95
Gray soil	10.6

7.12 Allen–Mayfield–Morrison Model for Sand

Allen *et al.*, 1957a, 1957b proposed the model that is based on the piecewise linear dependence between the squared velocity and the resistance force acting at the impactor:

$$\frac{D}{m} = \begin{cases} \gamma_{15} + \gamma_{14}v^2 & \textit{if} \quad 0 < v < v_* \\ \gamma_{13}v^2 & \textit{if} \quad v_* < v < v_{imp} \end{cases} \quad (7.12.1)$$

where $\gamma_{13}, \gamma_{14}, \gamma_{15}, v_*$ are empirical constants.

7.13 DAFL Model for Soil

The Differential Area Force Law (DAFL) model (Hadala, 1975; Heuzé, 1990; Bernard and Creighton, 1979) was proposed by the AVCO Corporation in the early 1970s and was used mainly for computer simulation of motion of impactor in soils after high speed oblique impact.

It was assumed that the normal stress (pressure) distribution at a given location at the surface of impactor, P, after conversion to SI units is described by the following formula (Bernard and Creighton, 1979):

$$P = 0.178 \cdot 10^7 \frac{\sqrt{u}}{\tilde{K}_s \Phi(x)} + 3.69 \cdot 10^6 \frac{v\sqrt{u\Phi(x)}}{\tilde{K}_s} + 1.45 \cdot 10^8 \frac{h_{loc} u}{\tilde{K}_s^2}, \quad (7.13.1)$$

if $u > 0$ and $P = 0$ if $u \leq 0$, where the values of the parameter $\tilde{K}_s$ are given in Table 7.13.1 (Young, 1972) which is an early version of the Table 7.4.2 (Young, 1997).

Table 7.13.1. Parameter $\tilde{K}_s$ for typical soils (Young, 1972).

Materials	$\tilde{K}_s$
Frozen silt or clay, saturated, very hard. Rock, weathered, low strength, fractured. Sea or freshwater ice more than 10 feet (3.0 m) thick.	1–2
Massive gypsite deposits (White Sands Missile Range, New Mexico.). Well-cemented coarse sand and gravel. Caliche, dry. Frozen, moist silt or clay.	2–3
Sea or freshwater ice from 1 to 3 feet (from 0.3 to 0.9 m) thick. Medium dense, medium to coarse sand, no cementation, wet or dry. Hard, dry, dense silt or clay (Tonopah Test Range, Nevada, dry lake playas). Desert alluvium.	4–6
Very loose, fine sand, excluding topsoil. Moist, stiff clay or silt, medium dense, less that about 50 percent sand.	8–12
Moist topsoil, loose, with some clay or silt. Moist, medium stiff clay, medium dense, with some sand.	10–15
Loose, moist topsoil with humus material, mostly sand and silt. Moist to wet clay, soft, low shear strength.	20–30
Very loose, dry, sandy topsoil (Eglin AFB). Saturated, very soft clay and silts, with very low shear strengths and high plasticity (Great Salt Lake Desert and bay mud at Skaggs Island). Wet lateritic clays.	40–50

Generally, Eq. (7.13.1) is valid for an arbitrary orientation of an impactor relative to the front surface of a shield. The particular location is determined by the abscissa x and the ordinate ρ in the coordinate system associated with the impactor (see Fig. 1.1.2), where $\rho = \Phi(x)$ is the equation of the generatrix of the impactor. Parameter u is the cosine of the angle between the outer normal vectors and the velocity vector at a given location at the projectile surface, $\vec{v}$; h_{loc} is the distance between the considered location and the front surface of the shield.

7.14 Three-Term Models

Khromov, 2004 used a square-law dependence between the local normal stress at the impactor-shield contact surface, σ_n, and the local normal component of the velocity of impactor, v_n,

$$\sigma_n = \tilde{k}_0 + \tilde{k}_1 v_n + \tilde{k}_2 v_n^2, \tag{7.14.1}$$

and developed the procedure for determining the coefficients $\tilde{k}_0, \tilde{k}_1, \tilde{k}_2$ on the basis of experimental data. The values of these coefficients are given in Table 7.14.1.

Veldanov, 1989; Ostapenko *et al.*, 1994 also applied three-term models although they did not consider methods for determining the coefficients in the models.

Table 7.14.1. Coefficients in Eq. (7.14.1) (Khromov, 2004).

Material	$\tilde{k}_0\,(MPa)$	$\tilde{k}_1\,(MPa \cdot s/m)$	$\tilde{k}_2\,(MPa \cdot s^2/m^2)$
Sand, humidity 5%	4.680	0.9237	0.00626
Sand, humidity 10%	4.972	1.2345	0.00647
Clay	2.664	0.8379	0.00763

7.15 ConWer-FOI model

Hansson, 2005 proposed the following generalization of the model by Hyde, 1991:

$$\overline{H} = \gamma_{16}(\gamma_{17}\overline{v}_{imp}^{1.8} + 1), \quad \overline{v}_{imp} > \left(\frac{2-\gamma_{16}}{\gamma_{17}\gamma_{16}}\right)^{0.56}, \tag{7.15.1}$$

where

$$\gamma_{17} = \frac{2.95\hat{K}_{nose}m}{d^{2.8}\sqrt{f'_{c*}}}, \tag{7.15.2}$$

$$\gamma_{16} = \begin{cases} 1.25 & for \quad 3 \le \overline{L}_{imp} \le 7, \quad granite/gneiss\ aggregate \\ 1.07 & for \quad 7 < \overline{L}_{imp} \le 10, \quad granite/gneiss\ aggregate \\ 1.25 & for \quad 3 \le \overline{L}_{imp} \le 10, \quad \lim estone\ aggregate \end{cases} \tag{7.15.3}$$

$$f'_{c*} = \begin{cases} f'_c & if \quad 30\,MPa \le f'_c \le 65\,MPa \\ 65 \cdot 10^6 & if \quad f'_c > 65\,MPa \end{cases} \tag{7.15.4}$$

L_{imp} is length of the impactor, $\overline{L}_{imp} = L_{imp}/d$.

The model by Hyde, 1991 is determined by Eqs. (7.15.1) and (7.15.2) with $\gamma_{16} = 1$ and $f'_{c*} = f'_c$ (Hansson, 2005).

7.16 Some Other Models and Related Problems

Gagin *et al.*, 1979 (see also Malanin and Pensky, 2007; Pensky, 2007) suggested the formula for the DOP for penetration into sand that can be written as follows:

$$\overline{H} = 0.324\frac{m\overline{v}_{imp}}{d^{2.2}\vartheta^{0.1}}, \tag{7.16.1}$$

where ϑ is a semi-vertex angle of cone-nosed impactor in degrees.

Zhang *et al.*, 2006b conducted a series of experiments on normal penetration into granite shields and used the results of these experiments for analysis of the known empirical models and validation of the proposed new equation for determining the DOP.

Adeli *et al.*, 1986 compared Kar model, Adeli–Amin–Sierakowski model and Young model (early version of the model from Young, 1967) and concluded that Young model is the best. Comparison of different models can be found also in Malanin and Pensky, 2007; Pensky, 2007.

Chapter 8

Analytical Models

Nomenclature

CCE	cylindrical cavity expansion;
d	maximum diameter of impactor, $d = 2R$;
D	resistance force acting on impactor;
DOP	depth of penetration;
E	Young's modulus of shield material;
f_c'	unconfined compressive strength of shield material;
h	instantaneous depth of penetration;
H	depth of penetration;
K	bulk modulus of shield material ;
m	mass of impactor;
P	pressure at the surface of impactor nose;
R	shank radius of impactor;
s	parameter, Eq. (8.3.12);
$\tilde{s}$	parameter, Eq. (8.1.5);
SCE	spherical cavity expansion;
v	instantaneous velocity of impactor;
v_{imp}	impact velocity;
V	velocity at the surface of a cavity;
y	radius of a cavity;
Y	yield stress of the shield material;
α_c	parameter in CCE model, Eqs. (8.1.1);
α_s	parameter in SCE model, Eqs. (8.3.16);
β_c	parameter in CCE model, Eqs. (8.1.1);
β_s	parameter in SCE model, Eqs. (8.3.16);
η_c	parameter in CCE model, Eq. (8.1.1);
ϑ	semi-vertex angle of cone-nosed impactor;

κ	apparent cohesion;
μ	parameter determining mechanical properties of soil, Eq. (8.3.2);
$\tilde{\mu}$	parameter determining mechanical properties of soil, Eq. (8.1.9);
ν	Poisson ratio of shield material;
ρ_{sh}	density of shield material;
ρ_{sh}^{*}	density of soil after compression by a shock wave;
σ_n	normal failure stress (compressive stress is assumed to be positive);
σ_t	circumferential stress (positive in compression);
σ_r	radial stress (positive in compression);
τ	failure shear stress of soil;
τ_0	parameter determining mechanical properties of soil, Eq. (8.3.2);
$\tilde{\tau}_0$	parameter determining mechanical properties of soil, Eq. (8.1.9);
φ	angle of shearing resistance (angle of internal friction);
ω	parameter, Eq. (8.1.6).

8.1 Moscow State University Models

8.1.1 *Dynamic CCE model for sharp slender impactors*

The dynamic CCE model, which is described below, was developed in Moscow State University (MSU) at the end of 1950s and documented in the publications by Rahmatulin *et al.*, 1964; Sagomonyan, 1960, 1974, 1988.

Rahmatulin *et al.*, 1964 (see also Sagomonyan, 1974,1988) proposed the dynamic CCE model that can be written in a simplified form as follows:

$$\sigma_r = \sigma_r(y, \dot{y}, \ddot{y}) = \alpha_c + \beta_c \dot{y}^2 + \eta_c y\ddot{y}\,, \qquad (8.1.1)$$

where

$$\alpha_c = \frac{\tilde{\tau}_0}{2\tilde{\mu}}(\omega^{-\tilde{s}/2} - 1), \quad \eta_c = \frac{\rho_{sh}}{\tilde{s}(1-\omega)}(\omega^{-\tilde{s}/2} - 1), \tag{8.1.2}$$

$$\beta_c = \frac{\rho_{sh}}{(1-\omega)(\tilde{s}-2)\omega^{\tilde{s}/2}}\left[1 + \frac{2}{\tilde{s}}(\omega^{\nu/2} - 1) - \omega + (\tilde{s}-2)(1-\omega)\right] \tag{8.1.3}$$

for $\tilde{\mu} > 0$,

$$\alpha_c = -\frac{\tilde{\tau}_0}{2}\ln\omega, \ \eta_c = -\frac{\rho_{sh}}{2(1-\omega)}\ln\omega, \ \beta_c = \frac{\rho_{sh}}{2(1-\omega)}(1-\omega-\ln\omega), \tag{8.1.4}$$

for $\tilde{\mu} = 0$,

$$\tilde{s} = 2\tilde{\mu}/(\tilde{\mu}+1), \tag{8.1.5}$$

ρ_{sh} is material density of the shield, y is radius of the cavity.

It was assumed that the material density of the shield does not change during unloading and the density of soil compressed by a shock wave, ρ^*_{sh}, is a constant. Consequently,

$$\omega = 1 - \frac{\rho_{sh}}{\rho^*_{sh}} \tag{8.1.6}$$

in Eqs.(8.1.2), (8.1.3) and (8.1.4) is a constant. In order to take into account compressibility of the shield material, compressed by a shock wave, the authors proposed to use iterative procedures.

The model given by Eqs. (8.1.1)–(8.1.6) is based on the Coulomb soil model (Scott, 1980):

$$\tau = \sigma_n \tan\varphi + \kappa, \tag{8.1.7}$$

that can be expressed in terms of Mohr-Coulomb failure condition (the authors of the model use the term Prantl condition):

$$\sigma_r - \sigma_t = \tilde{\tau}_0 + \tilde{\mu}(\sigma_r + \sigma_t), \tag{8.1.8}$$

where

$$\tilde{\tau}_0 = 2\kappa\cos\varphi, \ \tilde{\mu} = \sin\varphi, \tag{8.1.9}$$

τ is the shear stress in the soil at failure, σ_n is the normal stress at failure (compressive stress is considered positive), κ is the apparent cohesion, φ is the angle of shearing resistance (the angle of internal friction).

Eq. (8.1.8) yields:

$$\sigma_r = \frac{\tilde{\tau}_0}{1-\tilde{\mu}} + \frac{\tilde{\mu}+1}{1-\tilde{\mu}}\sigma_t. \qquad (8.1.10)$$

On the basis of the above model Rahmatulin *et al.*, 1964 obtained formulas for the DOP of an impactor (an arbitrary sharp thin body of revolution) taking into account variation of the contact interaction surface between the impactor nose and the shield at the initial stage of penetration.

Earlier, Sagomonyan, 1960 comprehensively investigated penetration of a cone-nosed impactors using the same model.

8.1.2 *Models for non slender impactors*

Sagomonyan, 1974 proposed several simple penetration models for impactors having the shapes different from those described in the previous section. Some of these models are described below.

The expression for the resistance force, D acting on a cylinder having radius R and moving along the axis h, can be written as follows:

$$D = \frac{\pi R^2 \rho_{sh}}{\omega}(\dot{h}^2 + h\ddot{h}), \quad \dot{h} = v, \ddot{h} = \dot{v}, \qquad (8.1.11)$$

where v is the velocity of the impactor.

In the case of cone-nosed impactor with the cone half angle ϑ close to a right angle, the resistance force is given by the following expression:

$$D = \begin{cases} kh^2(\dot{h}^2 + h\ddot{h}/3) & \text{if} \quad h \le R\cot\vartheta \\ kR^2\cot^2\vartheta(\dot{h}^2 + R\cot\vartheta\ddot{h}/3) & \text{if} \quad h > R\cot\vartheta \end{cases}, \qquad (8.1.12)$$

where

$$k = \pi \rho_{sh} \tan^2 \vartheta / \omega . \tag{8.1.13}$$

Models determined by Eqs. (8.1.11) and (8.1.12) yield a linear ordinary differential equation with respect to the squared instantaneous velocity, v^2, which is considered as a function of the instantaneous penetration depth, h. This ODE for impactor motion can be readily solved.

In the case of impactor having a hemispherical nose having radius R, two stages of penetration can be identified.

At the first stage of penetration separation of the shield material from the surface of a sphere does not occur. Separation develops at $\psi = \psi_*$, where ψ is the angle between the radius of the hemisphere passing through a considered point and the direction of the penetration and

$$\psi_* = \cos^{-1}(\sqrt{0.43}) = 49^\circ . \tag{8.1.14}$$

Consequently, penetration is considered until the depth $h = h_* = R(1-\cos\psi_*)$, whereas the pressure, P, at the surface of impactor nose is given by the following relation:

$$P = \frac{1.75 \rho_{sh} \dot{h}^2}{1-\omega} \left[(1-\omega)\omega^{1-\tilde{s}} - \frac{1-\omega^{2-\tilde{s}}}{2-\tilde{s}} + \frac{1-\omega^{1-\tilde{s}}}{1-\tilde{s}} \right] \tag{8.1.15}$$
$$\times (\cos^2 \psi - 0.43), \quad 0 \le \psi \le \psi_h,$$

where

$$\psi_h = \cos^{-1}(1 - h/R) . \tag{8.1.16}$$

The second stage of penetration begins when the penetration depth $h > h_*$, and the contact surface area between the projectile and the shield remains constant. Pressure at this stage is determined by Eq. (8.1.15) with $\psi_h = \psi_*$.

The model suggested by Sagomonyan, 1997 for penetration into perfect plastic compressible media can be formulated similarly. In this case

$$\psi_* = \sin^{-1}\left(\frac{2\sqrt{1-\omega^2}}{3-\omega} \right) \tag{8.1.17}$$

and pressure at the first stage of penetration is calculated as follows:

$$\frac{2P}{\rho_{sh}} = \left[1+\omega-\frac{(3-\omega)^2}{1-\omega}\sin^2\psi\right]\dot{h}^2 + R\cos\psi\ddot{h}, \quad 0 \le \psi \le \psi_h. \tag{8.1.18}$$

At the second stage of penetration, Eq. (8.1.18) is used with additional term $2gh$, where g is the acceleration of gravity and $\psi_h = \psi_*$.

Grigoryan, 1993 analyzed the qualitative features of penetration of a blunt impactor into soil and determined the approximate solution of a problem. This solution comprises explicit formulas which allow calculating all parameters of the motion of impactor. Formulas for the DOP read:

$$\frac{\overline{H}}{k_1} = \frac{H}{k_1 d} = \begin{cases} \ln[1+k_2^2(v_{sound}/v_\times)^2] & if \quad k_2 \le 1 \\ \ln k_2/\omega + \ln[1+(v_{sound}/v_\times)^2] & if \quad k_2 > 1 \end{cases}, \tag{8.1.19}$$

where

$$k_1 = \frac{m}{\pi d^3 \rho_{sh} C_x (0.25+\mu L/d)}, \quad k_2 = \frac{v_{imp}}{v_{sound}}\left(1-\frac{\pi k_0 d^3 \rho_{sh}}{4m}\right), \tag{8.1.20}$$

$$v_\times = K\sqrt{2(gH_{av} + \tau_0/\rho_{sh})}, \tag{8.1.21}$$

v_{sound} is the velocity of sound in the undisturbed soil, C_x is a shape coefficient of impactor nose ($C_x \sim 0.4 \div 06$), L is the length of the nose of impactor, k_0, K are coefficients ($k_0 \sim 1 \div 3$, $K \sim 5 \div 10$), τ_0 and μ are the characteristics of soil in the relationship between the shear failure stress in soil, τ, and the normal failure stress, σ_n, i.e., $\tau = \tau_0 + \mu p$, where p is the hydrostatic pressure.

The iteration procedure is proposed for determining H, but the author notes that these iterations are redundant if the exact (not simplified) equation of motion of impactor is used. This equation permits an explicit analytical solution although a bulky one.

The "model of normal sections" (Sagomonyan, 1974) for non slender impactors (e.g., cones with non small vertex angle) is based on the assumption that after contact with impactor particles of the shield move in the direction of the local normal vector to the impactor surface. Even for conical impactors, this model yields more complicated formulas for the resistance force as compared with the "model of plane sections" (cylindrical cavity expansion model).

8.2 Ross and Hanagud Model for Ice

Ross and Hanagud, 1969 developed a model based on SCE theory that describes penetration into a homogeneous semi-infinite isotropic shield by a spherical-nosed projectile. The material of the shield is modeled as a locked-plastic, locked-elastic medium. According to this model the DOP is determined by the following formula (Richmond, 1980):

$$\overline{H} = \frac{H}{d} = \frac{3m}{\pi\rho_* \beta_2 d^3} + \frac{\beta_1}{4\beta_2 d} \ln\left(1 + \frac{2\beta_2 \rho_* v_{imp}^2}{3\beta_3 E}\right), \qquad (8.2.1)$$

where

$$\rho_* = \rho_{sh} \exp(\Sigma_p), \qquad (8.2.2)$$

$$\beta_1 = 1 - \delta^{1/3}, \quad \beta_2 = 1.5 - \delta^{1/3}[2 + 0.5\delta - \exp(-\Sigma_p)], \qquad (8.2.3)$$

$$\beta_3 = \frac{4}{9}[1 - \exp(-3\beta)] - \frac{2Y \ln\delta}{3E} + \frac{2E_p}{9E}\left(\frac{\pi^2}{3} - 2\eta\right), \qquad (8.2.4)$$

$$\delta = [1 - \exp(-\Sigma_p)]\exp(-3\beta), \quad \beta = \frac{Y}{2E} - \frac{\Sigma_e}{3}, \quad \eta = \sum_{i=1}^{\infty} \frac{\delta^i}{i^2}, \qquad (8.2.5)$$

Σ_p is the plastic volumetric strain (%), Y is yield strength of the shield material, E is the Young's modulus of the shield material, E_e is the elastic volumetric strain of the shield material (%), E_p is the plastic modulus of deformation of the shield material.

8.3 Sandia Research Laboratories Models

Quite a few CEA models were developed in the Sandia Research Laboratories (SRL), sometimes in cooperation with other research establishments.

8.3.1 *Forrestal–Norwood–Longcope dynamic model*

Forrestal *et al.*, 1981b proposed the CCE model that is based on the expression for the hydrostatic pressure, p:

$$p = (\sigma_r + 2\sigma_t)/3 \tag{8.3.1}$$

and on the shear failure-pressure relation:

$$\tau = \sigma_r - \sigma_t = \tau_0 + \mu p\ , \tag{8.3.2}$$

where τ_0 and μ are the characteristics of soil.

Eqs. (8.3.1) and (8.3.2) yield:

$$\sigma_r = \frac{3}{3-\mu}\tau_0 + \frac{2\mu+3}{3-\mu}\sigma_t\ . \tag{8.3.3}$$

On the basis of the above relations, Forrestal *et al.*, 1981b proposed the dynamic CCE model in the form of Eq. (8.1.1) where the coefficients are determined for $\mu > 0$ by the following formulas:

$$\alpha_c = \frac{\tau_0}{\mu}(\omega^{-\nu/2} - 1),\quad \eta_c = \frac{\rho_{sh}}{q(1-\omega)}(\omega^{-\nu/2} - 1)\ , \tag{8.3.4}$$

$$\beta_c = \frac{\rho_{sh}}{(1-\omega)\omega^{\nu/2}}\left[1 + \frac{1-\omega^{\nu/2}}{q} + \frac{\omega - \omega^{\nu/2}}{2-q} - \omega\right], \tag{8.3.5}$$

$$q = 3\mu/(2\mu+3)\ . \tag{8.3.6}$$

The constant ω is determined by Eq. (8.1.6).

Comparison of Eqs. (8.1.10) and (8.3.3) reveals that in the case when $\mu = 0$, the model by Forrestal *et al.*, 1981b coincides with the model suggested by Rahmatulin *et al.*, 1964, and is given by Eq. (8.1.4) after replacing $\tilde{\tau}_0$ by τ_0.

Forrestal *et al.*, 1981b also analyzed comprehensively a quasi-dynamic version of the model for conical- and ogive-nosed impactors.

8.3.2 *Forrestal static models*

Forrestal, 1986 considered conical-nosed impactors and proposed a static elastic-plastic, elastic-cracked-plastic and elastic-cracked CCE models assuming that the plastic region of response is described by the following relationships:

$$p = K(1-\rho_{sh}/\rho) = K\eta\,, \tag{8.3.7}$$

$$\tau = \sigma_r - \sigma_t = \tau_0 + \mu p, \quad \tau_0 = (1-\mu/3)f_c'\,, \tag{8.3.8}$$

$$\sigma_t \geq -Y\,, \tag{8.3.9}$$

where p is a hydrostatic pressure; ρ_{sh} and ρ are the densities in the non deformed and deformed materials, correspondingly; η is a volumetric strain; σ_r and σ_t are radial and circumferential stress components (positive in compression); μ and τ_0 are the failure criteria; f_c' is an unconfined compressive strength; Y is the tensile strength. The elastic region is characterized by material constants K and Poisson ratio ν. If a cracked region exists, it is considered as an elastic region with $\sigma_t = 0$.

The static solution is described below. For the elastic-plastic model, formula for the radial stress at the cavity wall, σ_r, reads:

$$\frac{\sigma_r}{\tau_0} = \begin{cases} z_1^{-s/2}/s - 1/\mu & if \quad \mu \neq 0 \\ 0.5(1-\ln z_1) & if \quad \mu = 0 \end{cases}, \tag{8.3.10}$$

where

$$z_1 = \frac{\tau_0}{K}\left[\frac{s}{\mu(2-s)} + \frac{1+\nu}{3(1-2\nu)}\right], \tag{8.3.11}$$

$$s = 2\mu/(\mu+1)\,. \tag{8.3.12}$$

For the elastic-cracked-plastic model, σ_r is determined by the following formula:

$$\frac{\sigma_r}{\tau_0} = \begin{cases} \dfrac{K(2+\mu)}{Y(2-\mu)} z_2^{-s} - 1/\mu & if \quad \mu \neq 0 \\ 1 - \ln z_2 & if \quad \mu = 0 \end{cases}, \tag{8.3.13}$$

where

$$z_2 = \frac{2\tau_0}{(2-\mu)Y}\sqrt{\frac{5-4\nu}{3(1-2\nu)}}\,. \tag{8.3.14}$$

For the elastic-cracked model, formula for σ_r reads:

$$\frac{\sigma_r}{Y} = \sqrt{\frac{3(1-2\nu)K}{(5-4\nu)Y}}\,. \tag{8.3.15}$$

8.3.3 *Forrestal and Luk quasi-dynamic models*

Forrestal and Luk, 1992 proposed two quasi-dynamic SCE models of the following type:

$$\sigma_r = \sigma_r(V) = \alpha_s + \beta_s V^2, \quad V = \dot{y}\,. \tag{8.3.16}$$

8.3.3.1 *Model for a Mohr-Coulomb material*

The first model is based on Eqs. (8.3.1) and (8.3.2). The coefficients in Eq. (8.3.16) are determined by the following formulas:

$$\frac{\alpha_s}{\tau_0} = \frac{1}{q}\left(\frac{1+0.5\kappa}{\varsigma}\right)^{2\nu} - \frac{1}{\mu}, \tag{8.3.17}$$

$$\frac{\beta_s}{\rho_{sh}} = \frac{3}{(1-\omega)(1-2q)(2-q)} + \frac{1}{\varsigma^2}\left(\frac{1+0.5\kappa}{\varsigma}\right)^{2\nu}$$
$$\times\left\{3\kappa + \omega(1-1.5\kappa)^2 - \frac{\varsigma^3[2(1-\omega)(2-q)+3\varsigma^3]}{(1-\omega)(1-2q)(2-q)(1+0.5\kappa)^4}\right\} \qquad (8.3.18)$$

if $\mu \neq 0.75, \quad \mu > 0$;

$$\alpha_s/\tau_0 = 2(1+0.5\kappa)/\varsigma - 4/3, \qquad (8.3.19)$$

$$\frac{\beta_s}{\rho_{sh}} = -\frac{2\ln\varsigma}{1-\omega} + \frac{(1+0.5\kappa)[3\kappa+\omega(1-1.5\kappa)^2}{\varsigma^3}$$
$$-\frac{2}{3(1+0.5\kappa)^3} + \frac{2\ln(1+0.5\kappa)}{1-\omega} \qquad (8.3.20)$$

if $\mu = 0.75$;

$$\frac{3\alpha_s}{2\tau_0} = 1 - \ln\left[1 - \frac{1-\omega}{(1+0.5\kappa)^3}\right], \qquad (8.3.21)$$

$$\frac{\beta_s}{\rho_{sh}} = \frac{3}{2(1-\omega)} + \frac{3\kappa+\omega(1-1.5\kappa)^2}{[(1+0.5\kappa)^3+\omega-1]^{2/3}}$$
$$-\frac{[(1+0.5\kappa)^3+\omega-1]^{1/3}}{2(1+0.5\kappa)^4}\left[1+\frac{3(1+0.5\kappa)^3}{1-\omega}\right] \qquad (8.3.22)$$

if $\mu = 0$, where

$$\varsigma = [(1+0.5\kappa)^3+\omega-1]^{1/3}, \quad \kappa = \tau_0/E. \qquad (8.3.23)$$

For the case $\kappa << 1$, Forrestal and Luk, 1992 proposed simplified formulas which were obtained when $\kappa \to 0$:

$$\frac{\alpha_s}{\tau_0} = \frac{1}{q\omega^{2\nu/3}} - \frac{1}{\mu}, \qquad (8.3.24)$$

$$\frac{\beta_s}{\rho_{sh}} = \frac{3-\omega^{(1-2\nu)/3}[(2-q)(1+2q)+\omega(1-2q)(1-q)]}{(1-\omega)(1-2q)(2-q)} \quad (8.3.25)$$

if $\mu \neq 0.75$, $\mu > 0$;

$$\frac{\alpha_s}{\tau_0} = \frac{2}{\omega^{1/3}} - \frac{4}{3}, \quad \frac{3\beta_s}{\rho_{sh}} = 1 - \frac{2\ln\omega}{1-\omega}, \quad (8.3.26)$$

if $\mu = 0.75$;

$$\frac{3\alpha_s}{2\tau_0} = 1 - \ln(\omega), \quad \frac{\beta_c}{\rho_{sh}} = \omega^{1/3} + \frac{3-\omega^{1/3}(4-\omega)}{2(1-\omega)}, \quad (8.3.27)$$

if $\mu = 0$.

8.3.3.2 *Mohr-Coulomb-Tresca-limit yield model*

The second approach is based on the so called Mohr-Coulomb Tresca-limit yield constitutive model in which the shear strength τ attains a constant value, τ_*, for sufficiently high pressure p_*:

$$\tau = \sigma_r - \sigma_t = \begin{cases} \tau_0 + \mu p & if \quad 0 \le p \le p_* \\ \tau_* & if \quad p > p_* \end{cases} \quad (8.3.28)$$

where

$$p_* = (\tau_* - \tau_0)/\mu. \quad (8.3.29)$$

This model including Eq. (8.3.28) required three solutions (Forrestal and Luk, 1992). For $V < V_{\min}$, the stress state in the plastic region lies on the Mohr-Coulomb line given by Eq. (8.3.2), and the solutions are described above. For $V_{\min} < V < V_{\max}$, there are stress states in the plastic region that lie on the Mohr-Coulomb and Tresca-limit yield lines. If $V > V_{\max}$, the entire stress state in the plastic region lies on the Tresca-limit line. The expressions for $V_{\min}$ and $V_{\max}$ read:

$$V_{\min} = \sqrt{\frac{1}{q\beta_c}\left[\tau_* - \tau_0\left(\frac{1+0.5\kappa}{\varsigma}\right)^{2\nu}\right]}, \quad (8.3.30)$$

$$V_{\max} = \varsigma\sqrt{\frac{3\mu\tau_* - (3+2\mu)\nu\tau_0}{3q\mu\rho_{sh}[3\kappa + \omega(1-1.5\kappa)^2]}}. \qquad (8.3.31)$$

For $V_{\min} < V < V_{\max}$,

$$\alpha_s = \frac{3 + 2\nu\ln[1+(1-\omega)(\chi/\varsigma)^3]}{3\nu}\tau_* - \frac{\tau_0}{\mu}, \qquad (8.3.32)$$

$$\beta_s = \frac{\rho_0}{2(1-\omega)}\left\{3 - \frac{3+4(1-\omega)(\chi/\varsigma)^3}{[1+(1-\omega)(\chi/\varsigma)^3]^{4/3}}\right\}, \qquad (8.3.33)$$

χ is found numerically as the solution of the following equation:

$$\frac{(1+0.5\kappa)^{2\nu}\tau_0}{[(1-\omega)\chi^3 + \varsigma^3]^{2\nu/3}} + \frac{\psi\nu\rho_{sh}V^2}{\varsigma^2} - \tau_* = 0, \qquad (8.3.34)$$

where

$$\psi = \frac{\varsigma^3[2(1-\omega)(2-q)\chi^3 + 3\varsigma^3]}{(1-\omega)(1-2q)(2-q)[(1-\omega)\chi^3 + \varsigma^3]^{4/3}} + \frac{(1+0.5\kappa)^{2\nu}}{[(1-\omega)\chi^3 + \varsigma^3]^{2\nu/3}}$$
$$\times\left\{3\kappa + \omega(1-1.5\kappa)^2 - \frac{\varsigma^3[2(1-\omega)(2-s) + 3\varsigma^3]}{(1-\omega)(1-2q)(2-q)(1+0.5\kappa)^4}\right\} \qquad (8.3.35)$$

In this case, coefficients in Eq. (8.3.16) depend on the cavity expansion velocity, V.

For $V > V_{\max}$,

$$\alpha_s = \frac{2}{3}\tau_0 + 2\tau_*\ln\left(\frac{1+0.5\kappa}{\varsigma}\right), \qquad (8.3.36)$$

$$\frac{\beta_s}{\rho_{sh}} = \frac{3}{2(1-\omega)} + \frac{3\kappa + \omega(1-1.5\kappa)^2}{\varsigma^2} - \frac{\varsigma[1-\omega+3(1+0.5\kappa)^3]}{2(1+0.5\kappa)^4(1-\omega)}. \qquad (8.3.37)$$

8.3.4 *Frew–Forrestal–Hanchak Model for a Limestone Shield*

Frew *et al.*, 2000 adapted the model by Forrestal *et al.*, 1994 (Section 4.1.2.2.2) for limestone shields as follows: sf_c' in Eqs. (4.1.31) and (4.1.32) was replaced by the unknown "strength constant", a_0. Values of parameter a_0 are determined from experiments conducted for different impact velocities v_{imp} (other characteristics of the shield and the impactor are the same in all the experiments) on the basis of modified Eq. (4.1.32). The obtained values of parameter a_0 are averaged, and the average value is used in the predictive model when v_{imp} varies.

Frew *et al.*, 2000 conducted three sets of experiments with the limestone shields and 3.0 caliber-radius-head (CRH) ogive-nose steel projectiles which have the length-to-diameter ratios of ten while the impactor shank diameters are different in every set. The authors found that the average a_0 can be expressed as a linear function of the impactor shank diameter.

8.4 Some Other Models and Related Problems

Ross *et al.*, 1971 (see also references therein) applied dynamic SCE approach for mathematical modeling of penetration into two-layer snow-ice media.

Song *et al.*, 1985 studied penetration of a projectile into a semi-infinite plastic shield on the basis of dynamic and quasi-dynamic CCE models.

Yankelevsky and Adin, 1980; Yankelevsky and Gluck, 1980 proposed the disk model that belongs to the class of cylindrical cavity expansion models. Yankelevsky, 1983a adapted this model for the case of shields with a predrilled hole.

Forrestal *et al.*, 1981a developed penetration model for conical-nosed impactors when the material of the shield is described by a linear hydrostatic and by a linear shear failure-pressure relations. The equations determining the model were solved numerically, and the results of this solution were used for calculating a parameter of the empirical model of the type $\sigma_r \sim V$. Siddiqui and Abbas, 2002 extended the model by

Forrestal *et al.*, 1981a for ogive-nosed impactors and took into account friction.

Kravchenko *et al.*, 1994 suggested a set of engineering models for calculating penetration into different soils. For non-rock soils (sand, clay, loam) the cavity expansion approximation approach for plastic medium and assumption about step-like dependence of compressibility were used, while for rock and concrete shields the empirical relationship between the local normal stress at the surface of the impactor and the instantaneous velocity was employed.

Warren *et al.*, 2004; Longcope *et al.*, 1999 took into account the influence of the free surface. To this end, they used spherical cavity expansion models and assumed that radial stress vanishes at some sphere that was viewed as the "free surface". A spherical cavity expansion model was also used by Macek and Duffey, 2000 in order to take into account the near-surface effects and layering. Free surface effect was taken into account by Aptukov, 1991b in his spherical cavity expansion model.

Kotov *et al.*, 2002; Bazhenov *et al.*, 2003; Kotov, 2008 analyzed the applicability of some simplified models using the results of numerical simulation and experiments.

Hankins *et al.*, 2006 suggested a penetration model of the type given by Eq. (8.3.16) for coarse sand that employs different values of the friction coefficient at the impactor nose surface and at the lateral surface of the cylindrical part of impactor. Similar approach, when different sub-models are used for the nose and cylindrical part of the penetrator, was used by Aptukov and Fonarev, 2010 for modeling piles driving into the earth.

Wen, 2002a suggested the model where the normal stress at the impactor-soil contact surface is a linear function of the impact velocity.

Based on the experimental observations, Wen and Xu, 2004 suggested the function describing dependence of the shear strength on projectile diameter and unconfined compressive strength of a shield and verified this correlation using penetration data for the limestone shields impacted by ogival-nosed projectiles.

Part 5

Some Special Inverse Problems

Inverse problems are often encountered in various fields of science and are a subject of numerous studies (see e.g. Denisov, 1994; Ramm, 2005 and references therein). Solution of the inverse problems for ordinary differential equations (ODEs) implies determining either the right-hand-side of ODEs or their coefficients (including the case when they are functions of the independent variable), using the known solution. However, less common inverse problems of different type are also encountered in applied sciences, and, in particular, in penetration mechanics It is convenient to describe the essence of these problems without dwelling on a particular application.

Let $y = \zeta(x, y_0)$ be a solution of the first order ODE

$$dy/dx = \Gamma(x, y) \tag{5.1}$$

with the initial condition $y = y_0$ for $x = 0$. Varying y_0, yields a set of different solutions. Suppose that we are interested in the solutions with a prescribed value $\tilde{y}$ of the function y. Let us denote by x_1 the value of argument x when this condition is satisfied. Clearly, x_1 and y_0 are related by the formula $\Omega(x_1, y_0) = 0$, where $\Omega(x_1, y_0) = \zeta(x_1, y_0) - \tilde{y}$.

Let us consider the case when the function $\Gamma(x, y)$ in Eq. (5.1) is unknown but the empirical dependence $\Omega(x_1, y_0) = 0$ for given $\tilde{y}$ is known. Let us consider the following inverse problem: to recover the ODE (to determine the function $\Gamma(x, y)$) which implies the dependence $\Omega(x_1, y_0) = 0$. Generally, this inverse problem does not have a unique solution.

In the case when ODE describes penetration of a striker into a semi-infinite shield, x and y are the instantaneous depth of penetration and the instantaneous velocity of the striker, respectively, and formula $\Omega(x_1, y_0) = 0$ relates the initial velocity of the striker, y_0, and the depth of penetration (DOP), x_1, i.e., the depth at which the striker stops ($\tilde{y} = 0$). The goal is determining the dependence of the resistance force acting at the striker as a function of the instantaneous velocity and the instantaneous depth of penetration that is associated with the function $\Gamma(x, y)$.

For the simplest particular case when it is assumed that the resistance force depends only of the instantaneous velocity ($\Gamma = \Gamma(y)$), the inverse problem was solved by Bivin *et al.*, 1982; Roy *et al.*, 1987; Eisler *et al.*, 1998; Pilyugin, 2004; Malanin and Pensky, 2007, Pensky, 2007. A number of studies associated with solving inverse problems of penetration mechanics are discussed in the following. In a more general case, the above problem was studied by Ben-Dor *et al.*, 2008c, 2009f, 2010c.

Part 5 includes two chapters. In Chapter 9 we present the theoretical basis of the method while Chapter 10 contains examples of the applications of the theory to penetration mechanics.

Chapter 9

Theoretical Basis of the Method

Nomenclature

D	dimensionless drag force acting on projectile, Eq. (9.1.4);
D_0, D_1	functions, Eqs. (9.3.1), Eq. (9.4.1);
D_h	function, Eq. (9.2.1);
D_w	function determined by Eq. (9.2.1);
DOP	depth of penetration;
F	function that determines empirical dependence, $H_{\max} = F(V_{imp})$;
F_w	function that determines empirical dependence $h_{\max} = F_w(w_{imp})$;
F_h	function that determines empirical dependence $w_{imp} = F_h(h_{\max})$;
G	function determined by Eqs. (9.2.29) and (9.2.30);
h	$= H/H_0$, dimensionless instantaneous depth of penetration;
$h_{\max}$	$= H_{\max}/H_0$, dimensionless DOP;
H	instantaneous DOP;
$H_{\max}$	depth of penetration (DOP);
H_0	characteristic length;
m	mass of projectile;
p	parameter, Eq. (9.2.18);
q	parameter, Eq. (9.2.23);
R	drag force (resistance force) acting on projectile;
v	$= V/V_0$, dimensionless instantaneous velocity of projectile;
V	instantaneous velocity of projectile;

V_0	characteristic velocity of projectile;
V_{imp}	impact velocity;
w	$= (V/V_0)^2 = v^2$;
w_{imp}	$= (V_{imp}/V_0)^2 = v_{imp}^2$;
Θ	function, Eqs. (9.2.29), (9.2.30);
λ	parameter, Eq. (9.4.1);
φ	function, Eqs. (9.3.2), (9.4.3);
Ψ	function, Eqs. (9.2.12).

9.1 Formulation of the Problem

Consider normal penetration of a rigid projectile into a semi-infinite shield in the direction of the coordinate axis, H. The coordinate H is the instantaneous depth of penetration which is defined as the distance between some reference point at the impactor and the front surface of the shield. The effects associated with the stage of the incomplete immersion of the nose of the impactor into the shield are neglected.

Assuming that the total force acting on the impactor, R, depends on the velocity of the impactor, V, and the instantaneous coordinate of the impactor, H, Newton second law describing the motion of the impactor with mass m can be written as follows:

$$mV\,dV/dH = -R(H,V). \tag{9.1.1}$$

Let $V = \xi(V_{imp}, H)$ be the solution of Eq. (9.1.1) with the initial condition $V(0) = V_{imp}$, where V_{imp} is the impact velocity. Then the DOP, a depth whereby the velocity of the penetrator vanishes, is determined by the following formula: $\xi(V_{imp}, H_{\max}) = 0$.

Assume that the empirical dependence, $H_{\max} = F(V_{imp})$, is known and

$$F(0) = 0, \quad F(V_{imp}) > 0, \quad F'(V_{imp}) > 0, \quad V_{imp} > 0. \tag{9.1.2}$$

Let us consider now the problem of determining the function R that implies this empirical dependence.

Using the dimensionless variables, Eq. (9.1.1) can be rewritten as follows:

$$dw/dh = -D(h,w)\,, \tag{9.1.3}$$

where

$$h = \frac{H}{H_0},\ v = \frac{V}{V_0},\ w = v^2 = \frac{V^2}{V_0^2},\quad D(h,w) = \frac{2H_0}{mV_0^2} R(H_0 h, V_0\sqrt{w})\,, \tag{9.1.4}$$

V_0 is some reference velocity and H_0 is the characteristic length. The empirical dependence between the squared dimensionless impact velocity, w_{imp}, and the dimensionless DOP, $h_{\max}$, can be expressed in the following two forms:

$$h_{\max} = F_w(w_{imp}),\quad F_w(0) = 0 \quad or \quad w_{imp} = F_h(h_{\max}),\quad F_h(0) = 0\,, \tag{9.1.5}$$

where F_h and F_w are the known, mutually inverse functions,

$$w_{imp} = (V_{imp}/V_0)^2,\quad h_{\max} = H_{\max}/H_0\,. \tag{9.1.6}$$

We consider formulas for $D(h,w)$ that allow solving the equation of motion by quadratures.

9.2 Equation with Separable Variables

9.2.1 *General solution*

Consider the case when the function D can be written as follows:

$$D(h,w) = D_h(h)\,D_w(w)\,, \tag{9.2.1}$$

where D_h and D_w are some functions. Separating the variables in Eq. (9.1.3) and using the initial conditions, $w = w_{imp}$ for $h = 0$, we obtain the equation of motion of the impactor:

$$\int\limits_{w_{imp}}^{w} \frac{d\tilde{w}}{D_w(\tilde{w})} = -\int\limits_0^h D_h(\tilde{h})d\tilde{h}\,. \tag{9.2.2}$$

Equation (9.2.2) for the terminal point of the trajectory can be written by substituting $w = 0$ and $h = h_{\max}$:

$$\int_0^{w_{imp}} \frac{d\tilde{w}}{D_w(\tilde{w})} = \int_0^{h_{\max}} D_h(\tilde{h})d\tilde{h} \,. \tag{9.2.3}$$

Considering Eq. (9.2.3) as a definition of the functional dependence $h_{\max}$ vs. w_{imp} and differentiating both sides of Eq. (9.2.3) with respect to w_{imp} yields:

$$D_w(w_{imp})D_h(h_{\max})(dh_{\max}/dw_{imp}) = 1 \,. \tag{9.2.4}$$

Substituting $h_{\max} = F_w(w_{imp})$ into Eq. (9.2.4) reduces Eq. (9.2.4) to the identity, $D_w(w_{imp})D_h(F_w(w_{imp}))F_w'(w_{imp}) = 1$, which must be satisfied for arbitrary w_{imp}. It is convenient to rewrite this identity using the variable w instead of w_{imp}:

$$D_w(w)D_h(F_w(w))F_w'(w) = 1 \,. \tag{9.2.5}$$

If function $D_h(h)$ in Eq. (9.2.1) is known, then Eq. (9.2.5) implies the following expression for the unknown function, $D_w(w)$:

$$D_w(w) = 1/[D_h(F_w(w))F_w'(w)] \,. \tag{9.2.6}$$

Using the dependence between w_{imp} and $h_{\max}$ in the second equation in Eqs. (9.1.5), $w_{imp} = F_h(h_{\max})$, Eq. (9.2.6) can be rewritten after the change of variables, $F_w(x) = F_h^{-1}(x)$, as $D_w(F_h(h))D_h(h) - F_h'(h) = 0$. If $D_w(w)$ is known then

$$D_h(h) = F_h'(h)/D_w(F_h(h)) \,. \tag{9.2.7}$$

Eqs. (9.2.6)–(9.2.7) imply that the function $D(h,w)$ cannot be determined uniquely on the basis of the known correlation between the impact velocity and the DOP. Consequently, one can select either the function $D_w(w)$ or the function $D_h(h)$ in order to determine $D(h,w)$.

Substituting $D_w(w)$ from Eq. (9.2.6) or $D_h(h)$ from Eq. (9.2.7) into Eq. (9.2.1) we obtain two expressions for D:

$$D(h,w)=\frac{D_h(h)}{D_h(F_w(w))\,F_w'(w)},\quad D(h,w)=\frac{D_w(w)F_h'(h)}{D_w(F_h(h))}. \tag{9.2.8}$$

In order to write the equation of motion of the impactor using only one unknown function, let us substitute $D_w(w)$ from Eq. (9.2.6) into Eq. (9.2.2):

$$\int_w^{w_{imp}} D_h(F_w(\tilde{w}))\,F_w'(\tilde{w})d\tilde{w}=\int_0^h D_h(\tilde{h})d\tilde{h}\,. \tag{9.2.9}$$

After change of variable in the left-hand-side, $F_w(\tilde{w})=\hat{h}$, Eq. (9.2.9) can be rewritten as follows:

$$\int_{F_w(w)}^{F_w(w_{imp})} D_h(\hat{h})\,d\hat{h}=\int_0^h D_h(\tilde{h})d\tilde{h}\,. \tag{9.2.10}$$

Equation (9.2.10) implies that

$$\Psi(F_w(w_{imp}))-\Psi(F_w(w))=\Psi(h)\,, \tag{9.2.11}$$

where

$$\Psi(z)=\int_0^z D_h(\tilde{h})d\tilde{h}\,. \tag{9.2.12}$$

Functions $D_h(z)$ and $D_w(z)$ can be replaced by the function $\Psi(z)$. Clearly (see Eq. (9.2.12)),

$$D_h(z)=\Psi'(z)\,, \tag{9.2.13}$$

and Eqs. (9.2.6) and (9.2.13) imply that

$$D_w(z)=1/[\Psi'(F_w(z))\,F_w'(z)]\,. \tag{9.2.14}$$

Therefore, formulas for the instantaneous depth of penetration as a function of the instantaneous velocity of the impactor as well as the instantaneous drag force can be written using the function Ψ as follows:

$$h(w) = \Psi^{-1}\left(\Psi(F_w(w_{imp})) - \Psi(F_w(w))\right), \tag{9.2.15}$$

$$D(h, w) = \frac{\Psi'(h)}{\Psi'(F_w(w))\, F_w'(w)}, \tag{9.2.16}$$

where Ψ is an arbitrary function that satisfies the following conditions:

$$\Psi(0) = 0, \quad \Psi(z) > 0, \quad \Psi'(z) > 0, \quad z > 0. \tag{9.2.17}$$

Using Eq. (9.2.15), the drag force D could be written as a function of a single argument, h or w, but corresponding expressions are quite involved and are not presented here.

9.2.2 *Some sub-classes of model*

9.2.2.1 *The model* $\Psi(z) = z^p$

Assume that

$$\Psi(z) = z^p. \tag{9.2.18}$$

where $p \geq 1$ is a constant.

Eq. (9.2.13) implies that the latter assumption is equivalent to the following formula:

$$D_h(z) = pz^{p-1}. \tag{9.2.19}$$

Since $\Psi^{-1}(z) = z^{1/p}$, Eqs. (9.2.15) and (9.2.16) can be written as follows:

$$h(w) = [F_w(w_{imp})^p - F_w(w)^p]^{1/p}, \tag{9.2.20}$$

$$D(h, w) = \frac{h^{p-1}}{F_w(w)^{p-1} F_w'(w)}. \tag{9.2.21}$$

Substituting h from Eq. (9.2.20) into Eq. (9.2.21) we obtain expression for D as a function of w:

$$D(w) = \frac{[F_w(w_{imp})^p - F_w(w)^p]^{(p-1)/p}}{F_w(w)^{p-1} F_w'(w)}. \tag{9.2.22}$$

9.2.2.2 *The model* $\Psi(z) = F_h(z)^q$

If

$$\Psi(z) = F_h(z)^q = F_w^{-1}(z)^q, \tag{9.2.23}$$

where $q \geq 1$ is a constant, then

$$\Psi'(F_w(w))F_w'(w)\Big|_{\Psi(z)=F_w^{-1}(z)} = \frac{d\Psi(F_w(w))}{dw}\Big|_{\Psi(z)=F_w^{-1}(z)} \tag{9.2.24}$$

$$= \frac{dw^q}{dw} = qw^{q-1}, \quad \Psi^{-1}(z) = F_w(z^{1/q})$$

and Eqs. (9.2.15) and (9.2.16) can be written as follows:

$$h(w) = F_w\left((w_{imp}^q - w^q)^{1/q}\right), \tag{9.2.25}$$

$$D(h,w) = F_h(h)^{q-1}F_h'(h)w^{1-q}. \tag{9.2.26}$$

Eq. (9.2.24) implies that

$$w(h) = [w_{imp}^q - F_h(h)^q]^{1/q} \tag{9.2.27}$$

and the expression for the drag force D as a function of h reads:

$$D(h) = -dw/dh = [w_{imp}^q - F_h(h)^q]^{(1-q)/q}F_h(h)^{q-1}F_h'(h)w^{1-q}. \tag{9.2.28}$$

9.2.2.3 *A particular case*

Assume that the dependence between w_{imp} and $h_{\max}$ is given in the implicit form:

$$\Theta(w_{imp}) = G(h_{\max}), \tag{9.2.29}$$

where Θ and G are positive, increasing functions for positive arguments and

$$\Theta(0) = G(0) = 0. \tag{9.2.30}$$

In a similar manner Eq. (9.2.3) can be considered as a definition of the functional dependence $h_{\max}$ vs. w_{imp}. Hence Eqs. (9.2.3) and (9.2.29) imply that

$$\int_0^{w_{imp}} \frac{d\tilde{w}}{D_w(\tilde{w})} = \Theta(w_{imp}), \quad \int_0^{h_{\max}} D_h(\tilde{h})d\tilde{h} = G(h_{\max}), \tag{9.2.31}$$

and, consequently,

$$D_w(w) = 1/\Theta'(w), \quad D_h(h) = G'(h), \tag{9.2.32}$$

$$D(h,w) = G'(h)/\Theta'(w). \tag{9.2.33}$$

Using Eqs. (9.2.12), (9.2.29) and (9.2.32) we find that

$$\Psi(z) = G(z), \quad F_w(w) = G^{-1}(\Theta(w)). \tag{9.2.34}$$

Combining Eqs. (9.2.15) and (9.2.34) we arrive at the following relation:

$$h(w) = G^{-1}(\Theta(w_{imp}) - \Theta(w)). \tag{9.2.35}$$

Note that Eq. (9.2.31) yields only one of the possible models of penetration. The complete set of models can be obtained using the general method outlined in Section 9.2.1 and applying it to the dependence, $F_w(w) = G^{-1}(\Theta(w))$.

9.2.2.4 *The model* $D = D_w(w)$

Assume that the resistance force does not depend on the instantaneous depth of penetration of the impactor, i.e.,

$$D_h(h) = 1. \tag{9.2.36}$$

Then Eq. (9.2.12) implies that $\Psi(z) = z$, and Eqs. (9.2.15)–(9.2.16) yield:

$$h(w) = F_w(w_{imp}) - F_w(w), \tag{9.2.37}$$

$$D = D_w(w) = \frac{1}{F_w'(w)}. \tag{9.2.38}$$

9.2.2.5 *The model* $D = D_h(h)$

In this case

$$D_w(w) = 1, \tag{9.2.39}$$

and Eq. (9.2.7) implies that

$$D = D_h(h) = F_h'(h). \tag{9.2.40}$$

Substituting $D_h(h)$ from Eq. (9.2.40) to Eq. (9.2.12) we obtain that

$$\Psi(z) = F_h(z). \tag{9.2.41}$$

Taking into account that $F_h\left(F_w(z)\right) = z$, Eq. (9.2.15) yields:

$$h(w) = F_h^{-1}(w_{imp} - w), \tag{9.2.42}$$

9.3 Linear Equation

Assume that the expression for the resistance force D can be written as follows:

$$D(h,w) = D_1(h)\,w + D_0(h), \tag{9.3.1}$$

where $D_0(h)$ and $D_1(h)$ are some functions. The solution of Eq. (9.1.3) with the initial condition $w(0) = w_{imp}$ reads:

$$w(h) = \left(w_{imp} - \int_0^h [D_0(\tilde{h}) / \varphi(\tilde{h})] d\tilde{h} \right) \varphi(h), \quad \varphi(h) = \exp\left(-\int_0^h D_1(\tilde{h})\, d\tilde{h} \right). \tag{9.3.2}$$

Setting $w = 0$ and $h = h_{\max}$ in Eq. (9.3.2) we conclude that this model implies the following relationship between the DOP, $h_{\max}$, and the impact velocity:

$$w_{imp} = \int_0^{h_{max}} [D_0(\tilde{h})/\varphi(\tilde{h})]d\tilde{h} \,. \tag{9.3.3}$$

Considering w_{imp} as a known function of h_{max}, $w_{imp} = F_h(h_{max})$, differentiating Eq. (9.3.3) with respect to h_{max} and replacing h_{max} by h yields:

$$D_0(h) = \varphi(h)F_h'(h) \,. \tag{9.3.4}$$

The second equation in Eqs. (9.3.2) implies that

$$D_1(h) = -\varphi'(h)/\varphi(h) \,. \tag{9.3.5}$$

Therefore, the functions $D_0(h)$ and $D_1(h)$ can be determined by selecting the function $\varphi(h)$, where

$$\varphi(h) > 0, \quad \varphi(0) = 1 \,. \tag{9.3.6}$$

Substituting $D_0(h)$ and $D_1(h)$ from Eqs. (9.3.4) and (9.3.5) into Eqs. (9.3.1) and (9.3.2) we obtain:

$$D(h,w) = -[\varphi'(h)/\varphi(h)]\,w + \varphi(h)F_h'(h) \,, \tag{9.3.7}$$

$$w(h) = \left(w_{imp} - F_h(h)\right)\varphi(h) \,. \tag{9.3.8}$$

Substituting w from Eq. (9.3.8) into Eq. (9.3.7) yields:

$$D(h) = -\varphi'(h)\left(w_{imp} - F_h(h)\right) + \varphi(h)F_h'(h) = -dw/dh \tag{9.3.9}$$

9.4 Bernoulli Equation

Assume that the expression for the resistance force D can be written as follows:

$$D(h,w) = D_1(h)\,w + D_0(h)w^{1-\lambda}, \quad 0 < \lambda < 1 \,, \tag{9.4.1}$$

where $D_0(h)$ and $D_1(h)$ are some functions. This dependence implies that the equation of motion of the striker is the equation of Bernoulli type. The solution of this equation with the initial condition $w(0) = w_{imp}$ reads (Korn and Korn, 1968):

$$[w(h)]^{\lambda} = \left(w_{imp}^{\lambda} - \lambda \int_{0}^{h} [D_0(\tilde{h}) / \varphi(\tilde{h})] d\tilde{h} \right) \varphi(h) , \qquad (9.4.2)$$

where

$$\varphi(h) = \exp\left(-\lambda \int_{0}^{h} D_1(\tilde{h})\, d\tilde{h} \right) . \qquad (9.4.3)$$

Eq. (9.3.2) implies the following relationship between the DOP, $h_{\max}$, and the impact velocity:

$$w_{imp}^{\lambda} = \lambda \int_{0}^{h_{\max}} [D_0(\tilde{h}) / \varphi(\tilde{h})] d\tilde{h} . \qquad (9.4.4)$$

Considering w_{imp} as a known function of $h_{\max}$, $w_{imp} = F_h(h_{\max})$, and Eq. (9.4.4) can be written as

$$[F_h(h_{\max})]^{\lambda} = \lambda \int_{0}^{h_{\max}} [D_0(\tilde{h}) / \varphi(\tilde{h})] d\tilde{h} . \qquad (9.4.5)$$

Differentiating Eq. (9.4.5) with respect to $h_{\max}$ and replacing $h_{\max}$ by h yields:

$$D_0(h) = \varphi(h) [F_h(h)]^{\lambda - 1} F_h'(h) . \qquad (9.4.6)$$

Eq. (9.4.3) implies that

$$D_1(h) = -\frac{\varphi'(h)}{\lambda \varphi(h)} . \qquad (9.4.7)$$

Therefore the functions $D_0(h)$ and $D_1(h)$ can be determined by selecting the function $\varphi(h)$, where

$$\varphi(h) > 0, \quad \varphi(0) = 1 . \qquad (9.4.8)$$

Substituting $D_0(h)$ and $D_1(h)$ from Eqs. (9.4.6) and (9.4.7) into Eqs. (9.4.1) and (9.4.2) we obtain:

$$D(h,w) = -\frac{\varphi'(h)}{\lambda\varphi(h)}w + \varphi(h)[F_h(h)]^{\lambda-1}F_h'(h)w^{1-\lambda}, \qquad (9.4.9)$$

$$w(h) = \left\{\left(w_{imp}^{\lambda} - [F_h(h)]^{\lambda}\right)\varphi(h)\right\}^{1/\lambda}. \qquad (9.4.10)$$

Relation for D as a function of h can be obtained by substituting w from Eq. (9.4.10) into Eq. (9.4.9).

Clearly, the considered here model recovers the model analyzed in Section 9.4 for $\lambda = 1$.

Chapter 10

Application to Penetration Mechanics

Nomenclature

d	maximum diameter of impactor;
D	dimensionless drag force acting on projectile, Eq. (9.1.4);
D_0, D_1	functions determined by Eq. (10.3.5);
DOP	depth of penetration;
F_w	function determining empirical dependence, $h_{\max} = F_w(w_{imp})$;
F_h	function determining empirical dependence, $w_{imp} = F_h(h_{\max})$;
G	function, Eqs. (9.2.29)–(9.2.30);
h	$= H/H_0$, dimensionless instantaneous depth of penetration;
$h_{\max}$	$= H_{\max}/H_0$, dimensionless DOP;
H	instantaneous DOP;
$H_{\max}$	depth of penetration (DOP);
H_0	characteristic length;
k	parameter in Young model, Eqs. (10.2.11)–(10.2.17);
m	mass of projectile;
n	parameter of the model, Eq. (10.2.2);
p	parameter in Eq. (9.2.18);
q	parameter in Eq. (9.2.23);
R	drag force (resistance force) acting on projectile;

v	$= V/V_0$, dimensionless instantaneous velocity of projectile;
$\bar{v}_0$, $\bar{v}_*$	parameter in Young model, Eqs. (10.2.11)–(10.2.17);
V	instantaneous velocity of projectile;
V_0	characteristic velocity of impactor;
V_{imp}	impact velocity;
w	$= (V/V_0)^2 = v^2$;
w_{imp}	$= (V_{imp}/V_0)^2 = v_{imp}^2$;
α_1, α_2	parameters in Young model, Eq. (10.2.11);
β_0, β_2	parameters in Young model, Eq. (10.2.17);
Θ	function, Eqs. (9.2.29)–(9.2.30);
φ	function, Eq. (9.3.2);
Ψ	function, Eq. (9.2.12).

10.1 Introduction

In this chapter, we illustrate application of the proposed method by several examples in the field of penetration mechanics. However, the scope of applications of the method is considerably wider since it can be used in modeling of different phenomena described by ODEs. The suggested approach allows extending applications of the method to other types of ODEs and other kinds of relationships that are assumed to be known.

Non-uniqueness in the choice of the model which implies the known empirical relationship, affords a wide range of possibilities for devising an appropriate model of the considered phenomenon. We developed a general procedure for devising a variety of models for choosing a particular model. The choice of the particular model is another problem which is outside the scope of this study. It must be emphasized that the suggested method on its own does not guarantee the adequacy of the obtained equations.

10.2 High Speed Penetration

10.2.1 *Inverse problem for NDRC formula for concrete*

10.2.1.1 *Initial relationships*

The widely known modified National Defense Research Committee (NDRC) formula for penetration of a rigid projectile into a semi-infinite concrete shield is presented in Section 3.8. This model can be formulated in the form given by Eq. (9.2.29),

$$\Theta(w_{imp}) = G(h_{\max}), \tag{10.2.1}$$

with

$$\Theta(w_{imp}) = w_{imp}^{n}, \quad n = 0.9, \tag{10.2.2}$$

$$G(h_{\max}) = \begin{cases} (h_{\max}/2)^2 & if \quad 0 \le h_{\max} \le 2 \\ h_{\max} - 1 & if \quad h_{\max} > 2 \end{cases} \tag{10.2.3}$$

$$H_0 = d, \quad V_0 = 1000/\mu^{5/9}, \tag{10.2.4}$$

where d is the diameter of a projectile, μ is determined by Eq. (3.8.2). Hereafter in this chapter, variables and functions that are introduced in Chapter 9 are not described in the text but are presented in the table in the beginning of the chapter; regarding the notations H_0 and V_0 see also Eq. (9.1.4).

Eq. (10.2.1) allows us to write the following expression for the functions F_w and F_h:

$$F_w(z) = G^{-1}(\Theta(z)) = \begin{cases} 2z^{n/2} & if \quad 0 \le z \le 1 \\ z^n + 1 & if \quad z > 1 \end{cases} \tag{10.2.5}$$

$$F_h(z) = F_w^{-1}(z) = \begin{cases} (z/2)^{2/n} & if \quad 0 \le z \le 2 \\ (z-1)^{1/n} & if \quad z > 2 \end{cases} \tag{10.2.6}$$

10.2.1.2 *Sub-model* $\Psi(z) = z^p$

In the case of the model with $\Psi(z) = z^p$, Eqs. (9.2.20)–(9.2.21), after substituting F_w from Eq. (10.2.5), yield:

$$h(w) = \begin{cases} 2(w_{imp}^{pn/2} - w^{pn/2})^{1/p} & \text{if} \quad 0 \le w \le w_{imp},\ w_{imp} < 1 \\ [(w_{imp}^n + 1)^p - (2w^{n/2})^p]^{1/p} & \text{if} \quad 0 \le w \le 1,\ w_{imp} \ge 1 \\ [(w_{imp}^n + 1)^p - (w^n + 1)^p]^{1/p} & \text{if} \quad 1 \le w \le w_{imp},\ w_{imp} \ge 1 \end{cases} \tag{10.2.7}$$

$$D(h,w) = \begin{cases} (h/2)^{p-1} w^{(2-pn)/2} / n & \text{if} \quad 0 \le w \le 1 \\ h^{p-1} \dfrac{w^{1-n}}{n(w^n + 1)^{p-1}} & \text{if} \quad w > 1 \end{cases} \tag{10.2.8}$$

where $1 \le p < 2/n$.

10.2.1.3 *Sub-model* $\Psi(z) = F_h(z)^q$

When $\Psi(z) = F_h(z)^q$, Eqs. (9.2.25)–(9.2.26) after substituting F_w from Eq.(10.2.6) yield the following relations:

$$h(w) = \begin{cases} 2(w_{imp}^q - w^q)^{n/(2q)} & \text{if} \quad 0 \le w \le w_{imp},\ w_{imp} < 1 \\ (w_{imp}^q - w^q)^{n/q} + 1 & \text{if} \quad 0 \le w \le (w_{imp}^q - 1)^{1/q},\ w_{imp} \ge 1 \\ 2(w_{imp}^q - w^q)^{n/(2q)} & \text{if} \quad (w_{imp}^q - 1)^{1/q} \le w \le w_{imp}, w_{imp} \ge 1 \end{cases} \tag{10.2.9}$$

$$D(h,w) = \begin{cases} (h/2)^{(2q-n)/n} w^{1-q} / n & \text{if} \quad 0 \le h \le 2 \\ (h-1)^{(q-n)/n} w^{1-q} / n & \text{if} \quad h > 2 \end{cases} \tag{10.2.10}$$

where $n/2 < q \le 1$.

In the particular case $q = n = 0.9$ the model recovers the equation suggested by Kennedy, 1976. The method suggested by Kennedy, 1976 for determining equation of motion can be interpreted as realization of the approach described in Section 9.2.2.3. It must be emphasized that the method described in Section 9.2.2.3 allows determining a single equation of motion while the suggested general approach allows determining a set of these equations which yield different laws of motion of a projectile. The latter assertion is illustrated in the next section.

10.2.1.4 *Results of calculations*

Figs. 10.2.1 and 10.2.2 show the dependences $w(h)$ and $D(h)$ for the models described by Eqs. (10.2.7)–(10.2.8) (solid curves) and by Eqs. (10.2.9)–(10.2.10) (dash curves); the dependencies $D(h)$ are obtained by excluding w from Eqs. (10.2.7)–(10.2.8) and (10.2.9)–(10.2.10). Inspection of these figures shows that a large variety of different dependencies can be obtained by selecting the function Ψ .

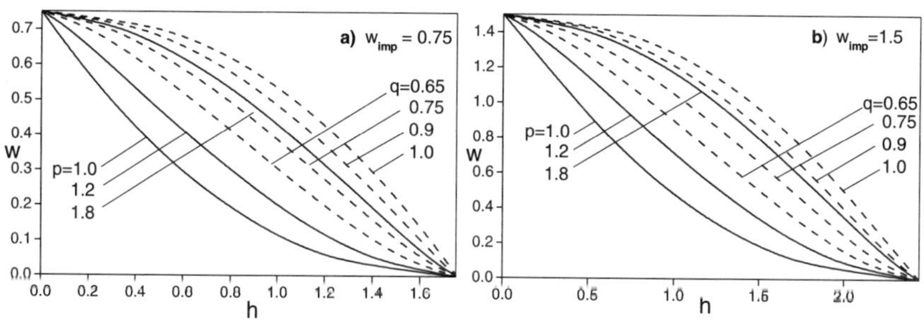

Fig. 10.2.1. Instantaneous squared velocity of impactor vs. instantaneous depth of penetration (dimensionless variables, penetration into concrete) for models described by a) Eq. (10.2.7) - solid curves; b) Eq. (10.2.9) - dash curves.

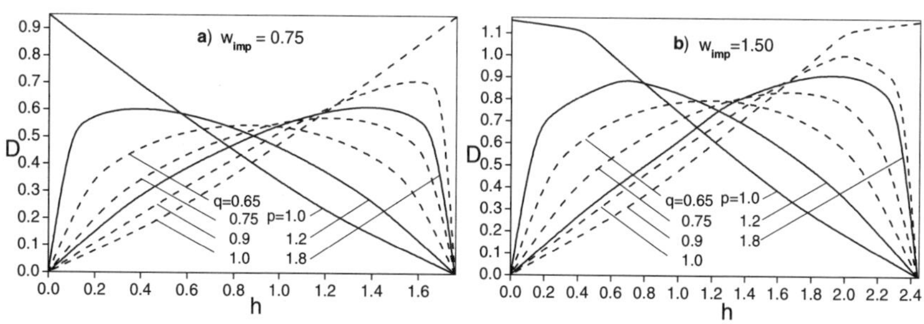

Fig. 10.2.2. Instantaneous resistance force vs. instantaneous depth of penetration of impactor (dimensionless variables, penetration into concrete) for models described by a) Eqs. (10.2.7)–(10.2.8) - solid curves; b) Eqs. (10.2.9)–(10.2.10) - dash curves.

10.2.2 *Inverse problem for Young equations*

The Young equation for concrete (see Section 3.23.1) can be written in the form $h_{\max} = F_w(w_{imp})$ with the function F_w is determined by the following formulas:

$$F_w(w_{imp}) = \begin{cases} \alpha_1 \ln(1+\alpha_2 w_{imp}) & if \quad w_{imp} < \bar{v}_*^2 \\ k(\sqrt{w_{imp}} - \bar{v}_0) & if \quad w_{imp} \geq \bar{v}_*^2 \end{cases} \tag{10.2.11}$$

where

$$H_0 = \frac{K_S K_{shape} K_m}{d^{1.4}}, \quad V_0 = 1000\, m/s, \tag{10.2.12}$$

parameters K_S, K_{shape}, K_m are described in Section 3.23.1. In the original Young model, parameters α_1, α_2, k, $\bar{v}_0$, $\bar{v}_*$ are given by Eqs. (3.23.3)–(3.23.5).

Assume that the resistance force depends on the instantaneous velocity of penetration and does not depends on the instantaneous depth. This case is considered in Section 9.2.2.4, and Eqs. (9.2.37) and (9.2.38) allow determining formulas for $h(w)$ and $D(w)$.

Application of the method to the original Young model is not straightforward because function $F_w(w_{imp})$ and its derivative are discontinuous at the point $w_{imp} = \bar{v}_*^2$ when α_1, α_2, k, $\bar{v}_0$, $\bar{v}_*$ are selected from Eqs. (3.23.3)–(3.23.5). Therefore we use modifications of the Young model proposed in Sections 3.23.2.1 and 3.23.2.2 which provide the necessary smoothness of the function $F_w(w_{imp})$.

In the first modification (Section 3.23.2.1), new values of α_1, α_2, $\bar{v}_*$ given by Eqs. (3.23.19) and (3.23.20) are used. Hence Eqs. (9.2.37) and (9.2.38) yield:

$$h(w) = F_w(w_{imp}) - F_w(w) = \alpha_1 \ln\left(\frac{1+\alpha_2 w_{imp}}{1+\alpha_2 w}\right), \tag{10.2.13}$$

$$D_w(w) = \frac{1}{F_w'(w)} = \frac{2\sqrt{w}}{k}, \tag{10.2.14}$$

for $w_{imp} \leq \bar{v}_*^2$, and

$$h(w)=\begin{cases}\alpha_1\ln\left(\dfrac{1+\alpha_2 w_{imp}}{1+\alpha_2 w}\right) & if \quad w<\bar{v}_*^2\\ \alpha_1\ln(1+\alpha_2 w_{imp})-k(\sqrt{w}-\bar{v}_0) & if \quad w\ge\bar{v}_*^2\end{cases} \tag{10.2.15}$$

$$D_w(w)=\begin{cases}\alpha_1\alpha_2/(1+\alpha_2 w) & if \quad w<\bar{v}_*^2\\ 2\sqrt{w}/k & if \quad w\ge\bar{v}_*^2\end{cases} \tag{10.2.16}$$

for $w_{imp}>\bar{v}_*^2$.

In the case of the second modification of the Young model (Section 3.23.2.2),

$$F_w(w)=\begin{cases}(\beta_2 w+\beta_0)w & if \quad w<\bar{v}_*^2\\ k(\sqrt{w}-\bar{v}_0) & if \quad w\ge\bar{v}_*^2\end{cases} \tag{10.2.17}$$

where $\beta_0,\beta_2,\bar{v}_*$ are given by Eq. (3.23.27) while k and $\bar{v}_0$ assume the same values as in the original Young model. Hence Eqs. (9.2.37) and (9.2.38) yield:

$$h(w)=(\beta_2 w_{imp}+\beta_0)w_{imp}-(\beta_2 w+\beta_0)w, \tag{10.2.18}$$

$$D_w(w)=1/(2\beta_2 w+\beta_0), \tag{10.2.19}$$

for $w_{imp}\le\bar{v}_*^2$, and

$$h(w)=\begin{cases}(\beta_2 w_{imp}+\beta_0)w_{imp}-(\beta_2 w+\beta_0)w & if \quad w<\bar{v}_*^2\\ (\beta_2 w_{imp}+\beta_0)w_{imp}-k(\sqrt{w}-\bar{v}_0) & if \quad w\ge\bar{v}_*^2\end{cases} \tag{10.2.20}$$

$$D_w(w)=\begin{cases}1/(2\beta_2 w+\beta_0) & if \quad w<\bar{v}_*^2\\ 2\sqrt{w}/k & if \quad w\ge\bar{v}_*^2\end{cases} \tag{10.2.21}$$

10.3 Low Speed Penetration into Granular Media

This example is of particular significance because the associated inverse problem was analyzed in a number of studies. In these studies a single solution was obtained by trial and error method while the approach suggested in Chapter 9 allows generating many solutions using a simple procedure. The history of the search for the analytical dependence for the penetration drag for granular media can be found in Katsuragi and Durian, 2007. We are interested in this problem only from the point of view of formulating and solving the inverse problems.

Uehara *et al.*, 2003a, 2003b found the following scaling for the DOP of a ball penetrating into a granular bed:

$$H_{\max} = 0.14 \frac{1}{\mu_{fr}} \left(\frac{\rho_{imp}}{\rho_g} \right)^{1/2} d^{2/3} \left(\frac{V_{imp}^2}{2g} + H_{\max} \right)^{1/3}, \tag{10.3.1}$$

where ρ_g and ρ_{imp} are densities of grains and of the impacting ball, respectively, $\mu = \tan\theta$ is the grain-grain friction coefficient, θ is the repose angle of the grains, d is the diameter of the ball, and g is the acceleration of gravity, other notations are specified at the beginning of this chapter. Ambroso *et al.*, 2005 assumed that the resistance force is described by a linear ODE and by trial and error method arrived at the analytical formula for the resistance force that implies Eq. (10.3.1). We will present this result below as a particular case of our general analysis.

In our analysis, we consider a more general empirical model:

$$\frac{H_{\max}}{d} = s \left(\frac{V_{imp}^2}{2gd} + \frac{H_{\max}}{d} \right)^{\alpha}, \tag{10.3.2}$$

where parameters s and α are determined by the choice of a particular model.

In the model suggested by Uehara *et al.*, 2003a, 2003b, $s = (0.14/\mu_{fr})\sqrt{\rho_{imp}/\rho_g}$, $\alpha = 1/3$ and Eq. (10.3.2) recovers Eq. (10.3.1).

In the model considered by Seguin *et al.*, 2008 which takes into account the influence of the confinement on penetration, $s = A_\infty A_* (\rho_{imp}/\rho_g)^\beta$, $\alpha = \alpha_\infty \alpha_*$, where parameters $A_\infty > 0$ and $0 < \alpha_\infty < 1$ correspond to the unbounded case, $0 < A_* < 1$ and $0 < \alpha_* < 1$ are the corrections for the confinement, and $\beta > 1$ is a parameter.

We consider the case when the expression for the total resistance force, R, reads:

$$R(H,V) = R_1(H)\,V^2 + R_0(H)\,, \tag{10.3.3}$$

where R_0 and R_1 are some functions. Using the dimensionless variables, Newton second law, Eq. (9.1.1), can be rewritten in the form given by Eqs. (9.1.3) and (9.3.1):

$$dw/dh = -[D_1(h)w + D_0(h)]\,, \tag{10.3.4}$$

where

$$D_0(h) = \frac{R_0(H_0 h)}{mg}, \quad D_1(h) = \frac{2H_0}{m} R_1(H_0 h), \tag{10.3.5}$$

$H_0 = d_{imp} s^{1/(1-\alpha)}$ is the initial collapse depth of the ball prior to the beginning of penetration from the state of rest ($V_{imp} = 0$), variables $h, v, w, w_{imp}, h_{\max}$ are determined by Eqs. (9.1.4) and (9.1.6) and $V_0 = \sqrt{2gH_0}$. Using these dimensionless variables, Eq. (10.3.2) can be rewritten as follows:

$$w_{imp} = F_h(h_{\max}), \quad F_h(z) = z^{\gamma} - z, \quad \gamma = 1/\alpha \tag{10.3.6}$$

and Eqs. (9.3.8)–(9.3.9) imply the following dependencies for the resistance force and squared instantaneous velocity vs. the instantaneous depth of penetration:

$$D(h) = -(w_{imp} - h^{\gamma} + h)\varphi'(h) + \varphi(h)(\gamma h^{\gamma-1} - 1)\,, \tag{10.3.7}$$

$$w(h) = \left(w_{imp} - h^{\gamma} + h\right)\varphi(h)\,, \tag{10.3.8}$$

where the function $\varphi(h)$ must satisfy the constraints given by Eq. (9.3.6).

Since the considered problem has a non unique solution, it is feasible, as a rule, to generate different models having the same accuracy by selecting different functions $\varphi(h)$. In order to illustrate this point, several

Table 10.3.1. Three models that imply Eq. (10.3.6) for $\gamma = 3$.

Model	$\varphi(h)$	**Value of the parameter**
1	$\exp(-c_1 h)$	$c_1 = 0.86$
2	$1/(1+c_2 h^{3/2})$	$c_2 = 1.38$
3	$(1-c_3 h)^2$	$c_3 = 0.30$

versions of $\varphi(h)$ are selected (see Table 10.3.1, columns 1-2) including the model 1 by Ambroso *et al.*, 2005 that can be readily obtained in the framework of the suggested approach.

In comparison with the experiment, we use the same approach that was suggested in Ambroso *et al.*, 2005 for their model. This approach is based on the following relationship that can be obtained from Eqs. (10.3.6) and (10.3.8):

$$\left(\frac{V}{V_{imp}}\right)^2 = \left(1 - \frac{\tilde{H}^\gamma - \eta^{\gamma-1}\tilde{H}}{1-\eta^{\gamma-1}}\right)\varphi(\tilde{H}/\eta), \tag{10.3.9}$$

where $\tilde{H} = h/h_{\max} = H/H_{\max} \le 1, \eta = 1/h_{\max} = H_0/H_{\max} \le 1$. Following Ambroso *et al.*, 2005, we determine parameters of the models 2 and 3 using the relationship given by Eq. (10.3.9) on the basis of the same experimental data (495 points) for the same value $\gamma = 3$ and an average value $\lambda = 0.36$. The best one-parameter fits to Eq. (10.3.9) for the models 2 and 3 are shown in Fig. 10.3.1, as well as, for comparison, the curve corresponding to the model 1 suggested by Ambroso *et al.*, 2005; the values of the parameters are presented in Table 10.3.1 (column 3). Inspection of this figure demonstrates that the experimental data are faithfully approximated by each of the models and the models have the same accuracy.

Note the in the case of Ambroso *et al.*, 2005 model when $\alpha = 1/3$, cubic algebraic equation $h_{\max}^3 - h_{\max} = w_{imp}$ allows to determine $h_{\max}$ as a function of w_{imp}, $h_{\max} = F_w(w_{imp})$, in the following explicit form:

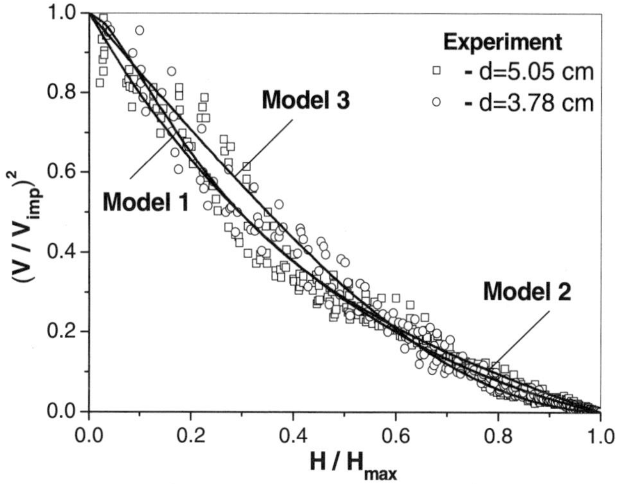

Fig. 10.3.1. Comparison of models with experimental data for penetration into granular media. Squares and rhombs – experiments by Ambroso *et al.*, 2005.

$$F_w'(w_{imp}) = \begin{cases} \dfrac{2}{\sqrt{3}}\cos\left(\dfrac{1}{3}\arccos \overline{w}_+\right) & if \quad 0 \le \overline{w}_+ \le 1 \\ \dfrac{1}{\sqrt{3}}\left(\sqrt[3]{\overline{w}_\times} + \dfrac{1}{\sqrt[3]{\overline{w}_\times}}\right) & if \quad \overline{w}_+ \ge 1 \end{cases}, \quad (10.3.10)$$

where $\overline{w}_+ = 1.5\sqrt{3}\, w_{imp}$, $\overline{w}_\times = \overline{w}_+ + \sqrt{\overline{w}_+^2 - 1}$.

Part 6

Method of Basic Impactors for Prediction of Penetration and Perforation

In next two chapters we describe an engineering method (Ben-Dor *et al.*, 2009b, 2009d) that allows using the known (e.g., from the experiments) data on the depth of penetration of several high-speed impactors for predicting the depth of penetration of the projectiles having different shapes. The suggested method is based on the conjecture that a complete description of the impactor-shield interaction model for determining the depth of penetration is redundant.

This method is an outgrowth of the approach that was suggested earlier (Bunimovich and Dubinsky, 1995; Dubinsky and Elperin, 1997) for calculating aerodynamic characteristics (ACs) of flying bodies when ACs are expressed as integrals depending on the shape of a projectile. This method allows determining the ACs of projectiles having different shapes using the known (e.g., measured) ACs for several basic projectiles. The essence of the approach is that it is based on the assumption about the local character of the projectile-host medium interaction (see Eqs. (1.1.1)-(1.1.6)) but does not require complete specification of the model and its parameters (functions Ω_n and Ω_τ).

Chapter 11

Simplified Version of the Method

Nomenclature

a_0	parameters determining the model, Eqs. (11.1.3), (11.1.4);
$b^{(i)}$	thickness of shield;
BLV	ballistic limit velocity;
DOP	depth of penetration;
$H^{(i)}$	depth of penetration (DOP);
$I^{(i)}$	integral in Eq. (11.1.5);
$L^{(i)}$	length of the nose of impactor;
$m^{(i)}$	mass of impactor;
n	number of basic impactors;
$R^{(i)}$	shank radius of impactor;
$u^{(i)}$	$=1/\Phi^{(i)\prime}(x^{(i)})$;
u_*, u_{**}	cotangents of the angle between the tangent to the generatrix and the axis of a projectile at the leading and trailing edges of the projectile, respectively (the same for all projectiles);
$v_{imp}^{(i)}$	impact (initial) velocity of impactor;
$v_{bl}^{(i)}$	ballistic limit velocity (BLV) of impactor;
$x^{(i)}(u^{(i)})$, $y^{(i)}(u^{(i)})$	equation of generatrix of impactor in a parametric form;
$\alpha^{(i)}$	parameter in Eq. (11.3.1);
$\beta^{(i)}$	coefficients, Eq. (11.2.7);
μ_{fr}	friction coefficient;
$\Xi^{(i)}$	impactor with an ordinal number i;
$\sigma^{(i)}$	shank cross-sectional area of impactor;

$\Phi^{(i)}(x^{(i)})$	function determining the shape of impactor;
ω	function determining the particular model for impactor-shield interaction;

Superscripts

(0)	denotes parameters associated with the impactor with the unknown characteristics (sometime omitted);
(i)	denotes the ordinal number of the basic impactor, $i = 1,2,...,n$.

11.1 Formulation of the Problem

Consider n impactors, $\Xi^{(1)}, \Xi^{(2)}, ..., \Xi^{(n)}$ (hereafter in this chapter, the superscript in parentheses denotes the ordinal number of the impactor), having the shape of bodies of revolution without a flat bluntness. The generatrix of the nose of the i-th impactor is determined, in the system of coordinates $x^{(i)}O^{(i)}y^{(i)}$ associated with the impactor, by the equation $y^{(i)} = \Phi^{(i)}(x^{(i)})$, where the origin of the coordinates coincides with the leading edge of the projectile, axis $O^{(i)}x^{(i)}$ is directed along the axis of the projectile, and axis $O^{(i)}y^{(i)}$ is perpendicular to the axis $O^{(i)}x^{(i)}$. We assume that only the nose of the impactor with the length $L^{(i)}$ interacts with a shield when the impactor penetrates into the shield, and the functions $\Phi^{(i)}$ satisfy the following constraints:

$$\Phi^{(i)}(0) = 0, \quad \Phi^{(i)\prime}(x^{(i)}) > 0, \quad \Phi^{(i)\prime\prime}(x^{(i)}) < 0, \quad 0 \le x^{(i)} \le L^{(i)}, \tag{11.1.1}$$

$$\Phi^{(i)\prime}(0) = 1/u_*, \quad \Phi^{(i)\prime}(L^{(i)}) = 1/u_{**} \quad i = 1,2,...,n, \tag{11.1.2}$$

where u_* and u_{**} are the cotangents of the angle between the tangent to the generatrix and the axis of a projectile at the leading and trailing edges of the projectile, respectively. Each impactor is considered as a rigid body penetrating normally into a shield having the same mechanical properties along the axis h, where the coordinate h, the instantaneous location of the penetrator, is defined as the distance between the leading edge of the impactor and the frontal surface of the shield. Effects associated with the stage of the incomplete immersion of the nose of the impactor into the shield are neglected.

The subsequent analysis is based on the model presented in Section 1.6. In particular, impactor-shield interaction at a given location at the impactor surface which is in contact with the shield is determined by Eq. (1.6.9) with $\mu_{fr} = 0$. Eqs. (1.6.22) and (1.6.7) written for i th impactor give the relationships for the depth of penetration (DOP) into a semi-infinite shield, $H^{(i)}$, and for the ballistic limit velocity (BLV) for a shield having a finite thickness, $v_{bl}^{(i)}$, correspondingly:

$$H^{(i)} = \varphi(I^{(i)}); \quad \varphi(I^{(i)}) = \frac{m^{(i)}}{4\pi I^{(i)}} \ln\left(1 + \frac{2\pi (v_{imp}^{(i)})^2}{a_0 \sigma^{(i)}} I^{(i)}\right), \tag{11.1.3}$$

$$v_{bl}^{(i)} = \psi(I^{(i)}); \quad \psi(I^{(i)}) = \sqrt{\frac{a_0 \sigma^{(i)}}{2\pi I^{(i)}} \left[\exp\left(\frac{4\pi b^{(i)}}{m^{(i)}} I^{(i)}\right) - 1\right]}, \tag{11.1.4}$$

where

$$I^{(i)} = \int_0^{L^{(i)}} \omega(u^{(i)}) \Phi^{(i)} \Phi^{(i)\prime} dx^{(i)}, \quad u^{(i)} = 1/\Phi^{(i)\prime}(x^{(i)}),$$
$$\sigma^{(i)} = 2\pi \int_0^{L^{(i)}} \Phi^{(i)} \Phi^{(i)\prime} dx^{(i)} = \pi \Phi^{(i)2}(L^{(i)}), \tag{11.1.5}$$

$m^{(i)}, v_{imp}^{(i)}, \sigma^{(i)}$ are the mass of the impactor, the impact velocity and the cross-sectional area of the shank, respectively, $b^{(i)}$ is the thickness of the shield. The function $\omega(u^{(i)}) > 0$ determines the particular model for impactor-shield interaction; it depend also upon the parameters which describe the properties of the material of the shield and are not written as the arguments. Parameter a_0 has a particular and well defined meaning in engineering models, many of which are described in Parts 2-4 of the monograph, and the magnitude of a_0 is assumed to be known.

We consider the problem associated with semi-infinite shields while the problem for the shields having a finite thickness can be analyzed in a similar manner. Suppose that the DOPs of the impactors $\Xi^{(1)},\Xi^{(2)},...,\Xi^{(n)}$ (the values $H^{(1)},H^{(2)},...,H^{(n)}$) are known. Consider an impactor $\Xi^{(0)}$ for which the above penetration model is valid, i.e., in the corresponding formulas the superscript i can be replaced by the superscript 0. Our goal is to determine a set of the impactors $\Xi^{(0)}$ such that their DOPs can be determined on the basis of the known DOPs, $H^{(1)},H^{(2)},...,H^{(n)}$, independently of the particular impactor-shield interaction model (function ω). In the analysis of shields having a finite thickness the DOPs are replaced by BLVs.

11.2 Solution of the Problem

The main idea of the suggested method is illustrated in Figs. 11.2.1a-b, where the tilde denotes some value of the corresponding variable. The equation of the generatrix of the impactor is written in a parametric form whereby the variable u_i which varies in the same range for all impactors, is selected as a parameter. This allows us, after transformation of the variables, $x^{(i)} \to u^{(i)}$, to write all integrals $I^{(i)}$ as the integrals with the same lower and upper limits for the same integration variable u. Mathematically this approach can be realized as follows.

Choosing $u^{(i)} = 1/\Phi^{(i)\prime}(x^{(i)})$ as a parameter, the equation of the generatrix of the impactor can be written in a parametric form as follows:

$$x^{(i)} = x^{(i)}(u^{(i)}),\ y^{(i)} = y^{(i)}(u^{(i)}), \tag{11.2.1}$$

where

$$u^{(i)} = \dot{x}^{(i)}(u^{(i)})/\dot{y}^{(i)}(u^{(i)}),\ \dot{x}^{(i)} = dx^{(i)}/du^{(i)},\ \dot{y}^{(i)} = dy^{(i)}/du^{(i)}. \tag{11.2.2}$$

Since $u^{(i)}$ varies in the range from u_* to u_{**} for all impactors, the variables in the integrals $I^{(i)}$ in Eq. (11.1.5) can be replaced by the same variable, u:

$$I^{(i)} = \int_{u_*}^{u_{**}} \omega(u)y^{(i)}(u)\dot{y}^{(i)}(u)du = \frac{1}{2}\int_{u_*}^{u_{**}} \omega(u)\frac{dy^{(i)2}(u)}{du}du,\ i=1,2,...,n. \tag{11.2.3}$$

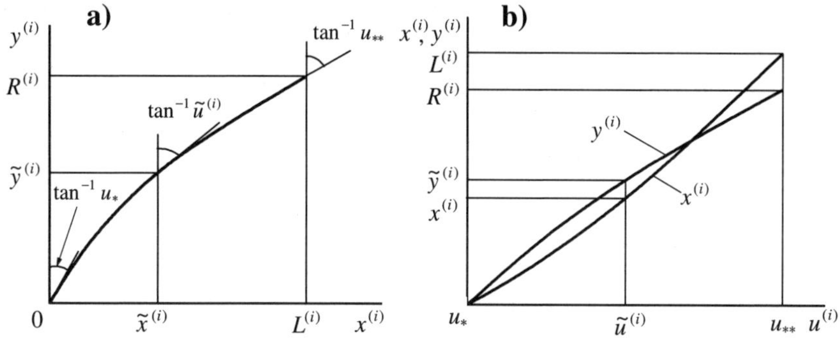

Fig. 11.2.1. Generatrix of the impactors: a) original representation as the dependence, $y^{(i)} = \Phi^{(i)}(x^{(i)})$, and b) representation in a parametric form, $x^{(i)} = x^{(i)}(u^{(i)})$, $y^{(i)} = y^{(i)}(u^{(i)})$.

Consider the impactor $\Xi^{(0)}$ with the generatrix determined by the equations $x^{(0)} = x^{(0)}(u^{(0)})$, $y^{(0)} = y^{(0)}(u^{(0)})$, $u_* \le u^{(0)} \le u_{**}$, where

$$y^{(0)}(u) = \xi(u), \quad x^{(0)}(u) = \int_{u_*}^{u} U \frac{d\xi(U)}{dU} dU = u\xi(u) - \int_{u_*}^{u} \xi(U) dU \,, \qquad (11.2.4)$$

$$\xi(u) = \sqrt{\sum_{i=1}^{n} \beta^{(i)} y^{(i)2}(u)},\ y^{(i)}(u_*) = 0,\ x^{(0)}(u_*) = 0,\ u = \frac{dx^{(0)}/du}{dy^{(0)}/du}, \qquad (11.2.5)$$

$\beta^{(1)}, \beta^{(2)}, ..., \beta^{(n)}$ are some coefficients. Using Eqs. (11.2.4) and (11.2.5) the integral $I^{(0)}$ which is associated with the impactor $\Xi^{(0)}$ and has a similar form as the integrals in Eq. (11.2.3), can be written as follows:

$$I^{(0)} = \frac{1}{2} \int_{u_*}^{u_{**}} \omega(u) \frac{dy^{(0)2}(u)}{du} du = \frac{1}{2} \sum_{i=1}^{n} \beta^{(i)} \int_{u_*}^{u_{**}} \omega(u) \frac{dy^{(i)2}(u)}{du} du \,. \qquad (11.2.6)$$

Equations (11.2.3) and (11.2.6) imply that the following formula:

$$I^{(0)} = \sum_{i=1}^{n} \beta^{(i)} I^{(i)} \,, \qquad (11.2.7)$$

regardless of the choice of the function ω. Using Eq. (11.2.4) allows determining the length of the impactor $\Xi^{(0)}$, $L^{(0)}$, its shank radius, $R^{(0)}$, and shank cross-sectional area, $\sigma^{(0)}$:

$$L^{(0)} = x^{(0)}(u_{**}), \quad R^{(0)} = y^{(0)}(u_{**}), \quad \sigma^{(0)} = \sum_{i=1}^{n} \beta^{(i)} \sigma^{(i)}. \quad (11.2.8)$$

Suppose that the shapes of the impactors $\Xi^{(1)}, \Xi^{(2)}, ..., \Xi^{(n)}$ and the values $H^{(i)}, v_{imp}^{(i)}, m^{(i)}$ ($i = 1,2,...,n$) and a_0 are known. Hence the following procedure allows determining the shape of the impactor $\Xi^{(0)}$ and its DOP, $H^{(0)}$, for given coefficients $\beta^{(1)}, \beta^{(2)}, ..., \beta^{(n)}$.

Step 1. Determining $I^{(i)}$ by solving the equations $\varphi(I^{(i)}) = H^{(i)}$ ($i = 1,2,...,n$). Function $\varphi(I^{(i)})$ decreases in the range from $\varphi = \mu = m^{(i)} (v_{imp}^{(i)})^2 / (2a\sigma^{(i)})$ to 0 when $I^{(i)}$ increases from 0 to $+\infty$. Therefore the equation $\varphi(I^{(i)}) = H^{(i)}$ has a single root if $H^{(i)} \le \mu$. If $H^{(i)} > \mu$ then these data are excluded from the analysis. Note that several experiments with different $m^{(i)}$ and $v_{imp}^{(i)}$ for the same impactor can be used to obtain the averaged values of $I^{(i)}$ on the basis of several values of $H^{(i)}$.

Step 2. Formulating equations of the generatrices of the impactors $\Xi^{(1)}, \Xi^{(2)}, ..., \Xi^{(n)}$ in the parametric form and formulating the equation of the generatrix for the impactor $\Xi^{(0)}$ using Eqs. (11.2.4) and (11.2.5).

Step 3. Calculating integral $I^{(0)}$ using Eq. (11.2.7).

Step 4. Determining the DOP, $H^{(0)}$, using Eq. (11.1.3) for $i = 0$.

Varying coefficients $\beta^{(i)}$, allows calculating the DOPs for a large number of projectiles having different shapes.

11.3 Example

Consider as impactors $\Xi^{(1)}$ and $\Xi^{(2)}$ ($n = 2$) power–shaped bodies with the same shank radius, $R = R^{(1)} = R^{(2)}$, and let us illustrate the most involved *Step 2* of the outlined above procedure.

Equations of the generatrices of the impactors can be written as follows:

$$y^{(i)} = \Phi^{(i)}(x^{(i)}) = R(x^{(i)}/L^{(i)})^{\alpha^{(i)}}, 0 < \alpha^{(i)} < 1, i = 1,2, \quad (11.3.1)$$

where $\alpha^{(i)}$ are given. Since $\Phi^{(i)\prime}(x^{(i)}) = \alpha^{(i)} R x_i^{\alpha_i - 1}/L_i^{\alpha_i}$, Eq. (11.1.2) implies that $u_* = 1/\Phi^{(1)\prime}(0) = 1/\Phi^{(2)\prime}(0) = 0$ and

$$u_{**} = 1/\Phi^{(i)\prime}(L^{(i)}) = L^{(i)}/(\alpha^{(i)} R), \quad i = 1,2,...,n, \quad (11.3.2)$$

where Eq. (11.3.2) describes the constraints concerning the shapes of the impactors, $L^{(1)}/(\alpha^{(1)}R) = L^{(2)}/(\alpha^{(2)}R)$. Choosing

$$u = 1/\Phi^{(i)\prime}(x^{(i)}) = L^{(i)\alpha^{(i)}} x^{(i)1-\alpha^{(i)}} / (\alpha^{(i)} R) \quad (11.3.3)$$

as a parameter and substituting $L^{(i)} = \alpha^{(i)} R u_{**}$ we obtain the equations of the generatrices of the impactors in a parametric form:

$$x^{(i)}(u) = \alpha^{(i)} R u_{**}^{\alpha^{(i)}/(\alpha^{(i)}-1)} u^{1/(1-\alpha^{(i)})},$$
$$y^{(i)}(u) = R(u/u_{**})^{\alpha^{(i)}/(1-\alpha^{(i)})}, 0 < u < u_{**} \quad (11.3.4)$$

In order to elucidate the obtained formulas, let us consider a special case with the parameters $\alpha^{(1)} = 1/2, \alpha^{(2)} = 2/3$. Hence Eqs. (11.2.4) and (11.2.5) imply the following equations of the generatrix for the impactor $\Xi^{(0)}$:

$$\frac{3}{Ru_{**}} x^{(0)}(u) = \eta(u)\left[2\left(\frac{u}{u_{**}}\right)^2 - \frac{\beta^{(1)}}{\beta^{(2)}}\right] + \frac{\beta^{(1)3/2}}{\beta^{(2)}}$$
$$\frac{y^{(0)}(u)}{R} = \frac{u}{u_{**}}\eta(u), \quad (11.3.5)$$

where

$$\eta(u) = \sqrt{\beta^{(1)} + \beta^{(2)}(u/u_{**})^2}, \quad 0 \le u \le u_{**}. \quad (11.3.6)$$

The length and the shank radius of the impactor $\Xi^{(0)}$ are as follows:

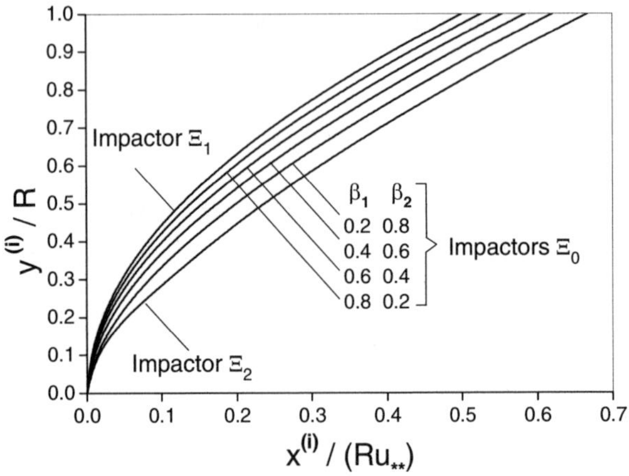

Fig. 11.3.1. Shape of the impactors.

$$L^{(0)} = x^{(0)}(u_{**}) = \frac{Ru_{**}}{3}\left[\sqrt{\beta^{(1)}+\beta^{(2)}}\left(2-\frac{\beta^{(1)}}{\beta^{(2)}}\right)+\frac{\beta^{(1)\,3/2}}{\beta^{(2)}}\right], \tag{11.3.7}$$

$$R^{(0)} = y^{(0)}(u_{**}) = R\sqrt{\beta^{(1)}+\beta^{(2)}}$$

Excluding u in Eqs. (11.3.5), the equation of the generatrix of the impactor $\Xi^{(0)}$ can be written as the dependence between $x^{(0)}$ and $y^{(0)}$:

$$3\bar{x}^{(0)}\beta^{(2)}/u_{**} = (\xi - 2\beta^{(1)})\sqrt{(\beta^{(1)}+\xi)/2} + \beta^{(1)\,3/2}, \tag{11.3.8}$$

where

$$\xi = \sqrt{\beta^{(1)2} + 4\beta^{(2)}\bar{y}_0^2},$$
$$\bar{x}^{(0)} = x^{(0)}/R, \quad \bar{y}^{(0)} = y^{(0)}/R, \quad 0 \le \bar{y}^{(0)} \le \sqrt{\beta^{(1)}+\beta^{(2)}} \tag{11.3.9}$$

If $\beta^{(1)}+\beta^{(2)}=1$, then all impactors, $\Xi^{(0)}, \Xi^{(1)}, \Xi^{(2)}$, have the same shank radius. Some versions of the shape of impactor $\Xi^{(0)}$ for different $\beta^{(1)}$ and $\beta^{(2)}$, as well as the shape of the "basic" impactors $\Xi^{(1)}$ and $\Xi^{(2)}$, are shown in Fig. 11.3.1.

Chapter 12

Complete Version of the Method

Nomenclature

DOP	depth of penetration;
E_k	function, Eq. (12.2.24);
$f^{(i)}$	function, Eq. (12.1.5);
F_k	function, Eq. (12.2.19);
G	function, Eq. (12.2.10);
$I_j^{(i)}$	integral with ordinal number j, Eq. (12.1.6);
$L^{(i)}$	upper limit for $x^{(i)}$, the length of the nose of impactor;
$m^{(i)}$	mass of impactor;
n	number of basic objects (impactors);
N	number of integral functionals which are arguments of function $f^{(i)}$ in Eq. (12.1.5);
p	parameter, Eq. (12.2.16);
$p^{(i)}$	characteristic parameter of object (impactor);
$r^{(i)}$	$=\Phi^{(i)}(0)$;
$R^{(i)}$	$=\Phi^{(i)}(L^{(i)})$, shank radius of i-th impactor;
s	number of subintervals in partition of the interval $[u_*^{\min}, u_{**}^{\max}]$;
$T^{(i)}$	known function for i-th object (impactor), Eq. (12.1.6);
$u_*^{(i)}$	$=1/\Phi^{(i)\prime}(0)$, cotangents of the angle between the tangent to the generatrix and the axis of i-th projectile at the leading edges of the projectile;

$u_{**}^{(i)}$ — $=1/\Phi^{(i)\prime}(L^{(i)})$, cotangents of the angle between the tangent to the generatrix and the axis of i-th projectile at the trailing edges of the projectile;

$u_*^{\min}$, $u_{**}^{\max}$ — parameters determined by Eq. (12.1.4);

U_k — end points of the subintervals in partition of the interval $[u_*^{\min}, u_{**}^{\max}]$, $k = 0,1,...,s$;

$v_{imp}^{(i)}$ — impact velocity of i-th impactor;

$v_{bl}^{(i)}$ — ballistic limit velocity (BLV) of impactor;

$x^{(i)}(u^{(i)})$, $y^{(i)}(u^{(i)})$ — functions determining object (generatrix of the impactor) in a parametric form;

$\alpha^{(i)}$ — parameter, Eq. (12.3.8);

$\beta^{(i)}$ — coefficient in the relationship between characteristics;

Δ — function, Eq. (12.2.8);

θ — function that determines a correspondence between the superscript i in $u_*^{(i)}$ and the subscript k in U_k, $u_*^{(i)} = U_{\theta(i)}$;

Θ — function that determines a correspondence between the superscript i in $u_{**}^{(i)}$ and the subscript k in U_k, $u_{**}^{(i)} = U_{\Theta(i)}$;

μ_{fr} — friction coefficient;

$\Xi^{(i)}$ — object (impactor) with ordinal number i;

$\Phi^{(i)}(x^{(i)})$ — function determining object (shape of impactor),

χ_j — unknown functions determining the model for the object, Eq. (12.1.6);

$\Psi_j^{(i)}$ — functions determined by Eq. (12.2.4)

ω, ω_0 — functions determining the particular model for impactor-shield interaction, Eq. (12.1.9);

Superscripts

(0) — denotes parameters associated with the impactor with unknown characteristics (sometime omitted);

(i) — denotes the ordinal number of basic object (impactor), $i = 1,2,...,n$.

12.1 Formulation of the Problem

In this chapter we present the generalized version of the method. Since the range of the applicability of the method is far beyond the field of penetration mechanics, hereafter the method is described for general mathematical objects. The method is illustrated by examples from penetration mechanics.

The generalization of the method in comparison with the preceding chapter is performed in two principal directions: (i) Formulas for calculating the desired characteristics can include dependencies on several integral functionals. In penetration mechanics this generalization implies that the model of projectile-shield interaction can include several functions; (ii) Constraint that all basic objects – bodies of revolution – have the same angle between the tangent to the generatrix and the axis of the projectile at the leading and trailing edges, is removed.

Consider n objects $\Xi^{(1)}, \Xi^{(2)}, ..., \Xi^{(n)}$. As in the previous chapter, the superscript in parentheses indicates the ordinal number of the object. Assume that function $y^{(i)} = \Phi^{(i)}(x^{(i)})$, which is defined for $0 \le x^{(i)} \le L^{(i)}$, determines the object $\Xi^{(i)}$ and has the following properties:

$$\Phi^{(i)}(0) \ge 0, \quad \Phi^{(i)\prime}(x^{(i)}) > 0, \Phi^{(i)\prime\prime}(x^{(i)}) < 0, x^{(i)} > 0, \quad i = 1,2,...,n \,. \tag{12.1.1}$$

Introduce the following notations:

$$u_*^{(i)} = 1/\Phi^{(i)\prime}(0), \quad u_{**}^{(i)} = 1/\Phi^{(i)\prime}(L^{(i)}) \quad i = 1,2,...,n \,. \tag{12.1.2}$$

Hereafter it is assumed that

$$\bigcup_{i=1}^{n} [u_*^{(i)}, u_{**}^{(i)}] = [u_*^{\min}, u_{**}^{\max}], \tag{12.1.3}$$

where

$$u_*^{\min} = \min_{0 \le i \le n} u_*^{(i)}, \quad u_{**}^{\max} = \max_{0 \le i \le n} u_{**}^{(i)}, \tag{12.1.4}$$

i.e., there are no gaps between the intervals, $[u_*^{(i)}, u_{**}^{(i)}]$, $i = 1,2,...,n$.

For each object, $\Xi^{(i)}$, we consider a characteristic parameter, $P^{(i)}$, that can be written in the following form:

$$P^{(i)} = f^{(i)}(I_1^{(i)}, I_2^{(i)}, ..., I_N^{(i)}), \tag{12.1.5}$$

where

$$I_j^{(i)} = \int_0^{L^{(i)}} T^{(i)}(x^{(i)}, \Phi^{(i)}, \Phi^{(i)'})\chi_j(\Phi^{(i)'})dx, \ \Phi^{(i)'} = d\Phi^{(i)}/dx^{(i)}, \tag{12.1.6}$$

$T^{(i)}$ and $f^{(i)}$ are given functions while χ_j are the unknown functions, N is a number of integral functionals which are the arguments of the functions $f^{(i)}$. Functions $f^{(i)}$ may depend also on some known parameters that are not shown in the list of arguments.

Let us consider the following problem. Suppose that series of experiments are performed for each object $\Xi^{(i)}$ in which the parameters vary while the function $\Phi^{(i)}(x^{(i)})$ remains the same. We assume that the results of these experiments allow determining parameters $P^{(i)}$ ($i = 1,2,...,n$). Consider an object $\Xi^{(0)}$ for which the above model is valid, i.e., in the corresponding relationships the subscript i can be replaced by the subscript 0. Our goal is to determine a set of the objects $\Xi^{(0)}$ such that their characteristic parameter $P^{(0)}$ can be determined on the basis of the parameters, $P^{(i)}, P^{(i)}, ..., P^{(n)}$. It must be emphasized that functions χ_j are unknown and, consequently, the characteristic parameters of the objects cannot be calculated directly.

This general formulation of the problem can be readily illustrated in a particular case when penetrating impactor and the depth of penetration (DOP) into a shield are considered as an object and as a characteristic parameter, correspondingly.

Let objects $\Xi^{(1)}, \Xi^{(2)}, ..., \Xi^{(n)}$ be projectiles having the shape of bodies of revolution and, as in the previous chapter, the generatrix of the nose of the i th projectile is determined by the equation $y^{(i)} = \Phi^{(i)}(x^{(i)})$. Parameters $u_*^{(i)}$ and $u_{**}^{(i)}$ are the cotangents of the angle between the tangent to the generatrix and the axis of a projectile at the leading and trailing edges of the projectile, respectively. These parameters have the same meaning as u_* and u_{**} in the previous chapter, although in the general case they are different for different basic projectiles.

Consider the set of all different values of $u_*^{(i)}$ and $u_{**}^{(i)}$ ($i = 1,2,...,n$) and denote them by U_k, whereby the ordinal numbers are chosen such that $U_0 < U_1 < ... < U_s$, where $U_0 = u_*^{\min}$, $U_s = u_{**}^{\max}$ and $s \leq 2n$ (see Fig. 12.2.1). Introduce the function that determines a correspondence between the superscript i in $u_*^{(i)}$ and the subscript k in U_k, so that $u_*^{(i)} = U_{\theta(i)}$. Similarly, the function Θ determines a correspondence between the superscript i in $u_{**}^{(i)}$ and the subscript k in U_k, so that $u_{**}^{(i)} = U_{\Theta(i)}$.

Along with the direct problem studied above, we can consider an inverse problem of determining the DOP of the impactor Ξ_0 with a given shape. This problem has, generally, only an approximate solution and is reduced to the optimal selection of the coefficients, $\beta_1, \beta_2, ..., \beta_n$.

The DOP is considered as the characteristic parameter, and the general average two-term model (see Section 1.6), where functions Ω_n and Ω_τ are determined by Eq. (1.6.1), is adopted as a model of impactor-shield interaction. Hence Eqs. (1.6.22), (1.6.3), (1.6.4), written for i-th impactor, imply the following relationship for $P^{(i)}$:

$$P^{(i)} = \frac{m^{(i)}}{2I_2^{(i)}} \ln\left(1 + \frac{I_2^{(i)}}{I_1^{(i)}} v_{imp}^{(i)2}\right), \tag{12.1.7}$$

where the integrals $I_1^{(i)}$ and $I_2^{(i)}$ are determined by Eq. (12.1.6) with

$$T^{(i)}(x^{(i)}, \Phi^{(i)}, \Phi^{(i)\prime}) = \Phi^{(i)}, \tag{12.1.8}$$

$$\begin{aligned} \chi_1(\Phi^{(i)\prime}) &= 2\pi\,\omega_0\left(\frac{\Phi^{(i)\prime}}{\sqrt{(\Phi^{(i)\prime})^2+1}}\right)(\Phi^{(i)\prime} + \mu_{fr}) \\ \chi_2(\Phi^{(i)\prime}) &= 2\pi\,\omega\left(\frac{\Phi^{(i)\prime}}{\sqrt{(\Phi^{(i)\prime})^2+1}}\right)(\Phi^{(i)\prime} + \mu_{fr}) \end{aligned} \tag{12.1.9}$$

$m^{(i)}$ and $v_{imp}^{(i)}$ are the mass of the impactor and the impact (initial) velocity, respectively, $\Phi^{(i)}(0) = 0$.

In the considered case, it is assumed that functions ω_0 and ω as well as the friction coefficient, μ_{fr}, that describe the model of projectile-shield interaction are unknown.

12.2 Solution of the Problem

12.2.1 *General case*

Choosing $u = 1/\Phi^{(i)\prime}(x^{(i)})$ as a parameter, the function $y^{(i)} = \Phi^{(i)}(x^{(i)})$ can be written in a parametric form as follows:

$$x^{(i)} = x^{(i)}(u),\ y^{(i)} = y^{(i)}(u), \tag{12.2.1}$$

where

$$u = \dot{x}^{(i)}(u)/\dot{y}^{(i)}(u), \quad \dot{x}^{(i)} = dx^{(i)}/du, \quad \dot{y}^{(i)} = dy^{(i)}/du, \tag{12.2.2}$$

and u varies in the range from $u_*^{(i)}$ to $u_{**}^{(i)}$ for the object $\Xi^{(i)}$. Hence the integrals $I_j^{(i)}$ in Eq. (12.1.6) can be rewritten as follows:

$$I_j^{(i)} = \int_{u_*^{(i)}}^{u_{**}^{(i)}} \Psi_j^{(i)}(u)du, \quad i = 1,2,...,n, \tag{12.2.3}$$

where

$$\Psi_j^{(i)}(u) = T^{(i)}\left(x^{(i)}(u), y^{(i)}(u), u^{-1}\right)\chi_j(u^{-1})\dot{x}^{(i)}(u) \tag{12.2.4}$$

and $i = 1,2,...,n$; $j = 1,2,...,N$.

Assuming that $\Phi^{(0)\prime}(x^{(0)})$ varies from u_*^{min} to u_{**}^{max} when $x^{(0)}$ varies from 0 to $L^{(0)}$, the expression for $I_j^{(0)}$, associated with the object $\Xi^{(0)}$, can be written as follows:

$$I_j^{(0)} = \int_{u_*^{\min}}^{u_{**}^{\max}} \Psi_j^{(0)}(u)du. \tag{12.2.5}$$

Using these notations, Eqs. (12.2.3) and (12.2.5) can be rewritten as follows:

$$I_j^{(i)} = \int\limits_{U_{\theta(i)}}^{U_{\Theta(i)}} \Psi_j^{(i)}(u)du = \sum_{k=\theta(i)+1}^{\Theta(i)} \int\limits_{U_{k-1}}^{U_k} \Psi_j^{(i)}(u)du, \quad i = 1,2,...,n, \tag{12.2.6}$$

$$\begin{aligned} I_j^{(0)} &= \int\limits_{U_o}^{U_s} \Psi_j^{(0)}(u)du = \sum_{k=1}^{s} \int\limits_{U_{k-1}}^{U_k} \Psi_j^{(0)}(u)du \\ &= \sum_{k=1}^{s} \int\limits_{U_{k-1}}^{U_k} T^{(0)}\left(x_k(u), y_k(u), u^{-1}\right)\chi_j(u^{-1})\dot{x}_k(u)du, \end{aligned} \tag{12.2.7}$$

where $x_k(u)$ and $y_k(u)$ denote functions $x^{(0)}(u)$ and $y^{(0)}(u)$, respectively, in the interval $[U_{k-1}, U_k]$.

Consider the following expression:

$$\Delta = \sum_{i=1}^{n} \beta^{(i)} I_j^{(i)} - I_j^{(0)}, \tag{12.2.8}$$

where $\beta^{(i)}$ are some coefficients. Substituting $I_j^{(i)}$ and $I_j^{(0)}$ from Eq. (12.2.6) and (12.2.7) into Eq. (12.2.8), changing the order of summation and integration and taking into account Eqs. (12.2.4) yields:

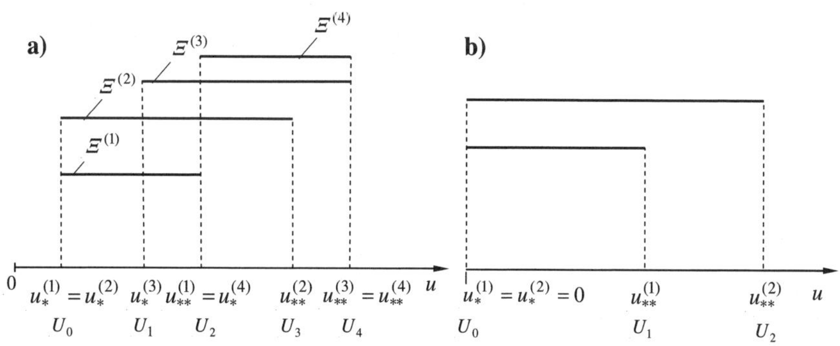

Fig. 12.2.1. Definition of the parameters U_k.

$$\Delta = \sum_{k=1}^{s} \int_{U_{k-1}}^{U_k} \left[G(u) - T^{(0)}\left(x_k(u), y_k(u), u^{-1}\right)\dot{x}_k(u) \right] \chi_j(u^{-1})\,du\,, \tag{12.2.9}$$

where

$$G(u) = \sum_i \beta^{(i)} T^{(i)}\left(x^{(i)}(u), y^{(i)}(u), u^{-1}\right)\dot{x}^{(i)}(u) \tag{12.2.10}$$

and summation is conducted over those i for which $\theta(i)+1 \le k \le \Theta(i)$. Function $G(u)$ has discontinuities at the joint points of the adjacent intervals.

Let us require that $x_k(u)$ and $y_k(u)$ satisfy the system of the following ordinary differential equations (ODEs):

$$T^{(0)}\left(x_k(u), y_k(u), u^{-1}\right)\dot{x}_k(u) = G(u)\,, \tag{12.2.11}$$

$$\dot{x}_k(u) = u\,\dot{y}_k(u)\,,\ k = 1,2,\ldots,m\,. \tag{12.2.12}$$

Then Eq. (12.2.9) implies that $\Delta = 0$, i.e.,

$$I_j^{(0)} = \sum_{i=1}^{n} \beta^{(i)} I_j^{(i)}\,, \tag{12.2.13}$$

for every j and independently of the choice of functions χ_j. The latter assertion allows calculating the characteristic parameter, $P^{(0)}$.

We require that the solution of Eqs. (12.2.11) and (12.2.12) satisfies the following conditions:

$$x_1(u_*^{\min}) = 0, \quad y_1(u_*^{\min}) = r^{(0)}\,, \tag{12.2.14}$$

$$x_{k-1}(U_{k-1}) = x_k(U_{k-1}), \quad y_{k-1}(U_{k-1}) = y_k(U_{k-1}), \quad k = 2,\ldots,m\,, \tag{12.2.15}$$

where Eq. (12.2.14) represents the initial conditions with some selected $r^{(0)}$, while Eq. (12.2.15) describes continuity conditions at the joint points of the adjacent intervals. Note that $\Phi^{(0)\prime}(u) = u^{-1}$ is a continuous function.

For given coefficients, $\beta^{(1)}, \beta^{(2)}, \ldots, \beta^{(n)}$, the application of the method reduces to the following procedure.

Step 1. Determining $I_j^{(i)}$ for every object $\Xi^{(i)}$, using the known values of $P^{(i)}$. For instance, for penetrating impactor, several experiments with different $m^{(i)}$ and $v_{imp}^{(i)}$ for the same impactor can be used for determining the averaged values of $I_j^{(i)}$ on the basis of several values of the DOP, $P^{(i)}$ (see Eq. (12.1.7)).

Step 2. Formulating equations $y^{(i)} = \Phi^{(i)}(x^{(i)})$ for the objects $\Xi^{(1)}, \Xi^{(2)}, ..., \Xi^{(n)}$ in a parametric form (see Eq. (12.2.1). To this end function $x^{(i)}(u)$ can be determined by finding the dependence $x^{(i)}$ vs. u using the expression $u^{-1} = d\Phi^{(i)}/dx^{(i)}$ and the function $y^{(i)}(u) = \Phi^{(i)}(x^{(i)}(u))$.

Step 3. Formulating equation for the function $y^{(0)} = \Phi^{(0)}(x^{(0)})$ for the object $\Xi^{(0)}$ in a parametric form using Eqs. (12.2.11)-(12.2.12) and (12.2.14)-(12.2.15).

Step 4. Calculating integrals $I_j^{(0)}$ using Eq. (12.2.7).

Step 5. Determining parameter $P^{(0)}$ using Eq. (12.1.5) for $i = 0$.

Varying coefficients $\beta^{(i)}$, the characteristic parameters $P^{(0)}$ for a large number of various objects can be calculated.

12.2.2 *Class of solutions*

Solution of Eqs. (12.2.11), (12.2.12), (12.2.14) and (12.2.15) can be found for a number of important particular cases in the explicit form. For instance, if

$$T^{(0)}(x^{(0)}, \Phi^{(0)}, \Phi^{(0)\prime}) = \Phi^{(0)p} \qquad (12.2.16)$$

the Eqs. (12.2.11) and (12.2.12) imply the following equations for $y_k(u)$:

$$y_k^p \, \dot{y}_k = G(u)/u, \quad k = 1,2,...,m \, . \qquad (12.2.17)$$

The general solution of Eq. (12.2.17) reads:

$$y_k^{p+1} = F_k(U_{k-1}; u) + a_k \, , \qquad (12.2.18)$$

where

$$F_k(c;u) = (p+1)\int_c^u G(t)dt/t, \; U_{k-1} \le c \le U_k, \; U_{k-1} \le u \le U_k \, . \qquad (12.2.19)$$

Substituting conditions given by Eqs. (12.2.14) and (12.2.15) for y_k into Eq. (12.2.18) we obtain:

$$\begin{aligned}
&r^{(0)p+1} = F_1(U_0;U_0) + a_1 \\
&F_1(U_0;U_1) + a_1 = F_2(U_1;U_1) + a_2 \\
&\ldots \\
&F_{k-1}(U_{k-2};U_{k-1}) + a_{k-1} = F_k(U_{k-1};U_{k-1}) + a_k \\
&\ldots \\
&F_{s-1}(U_{s-2};U_{s-1}) + a_{s-1} = F_s(U_{s-1};U_{s-1}) + a_s
\end{aligned} \tag{12.2.20}$$

where a_k are constants. Taking into account that all $F_k(c;c) = 0$, Eqs. (12.2.20) imply that

$$a_k = \sum_{\lambda=1}^{k} F_{\lambda-1}(U_{\lambda-2};U_{\lambda-1}), \quad F_0(U_{-1};U_0) = r^{(0)p+1}, \quad k = 1,2,\ldots s\,. \tag{12.2.21}$$

Therefore

$$y_k(u) = \left[F_k(U_{k-1};u) + \sum_{\lambda=1}^{k} F_{\lambda-1}(U_{\lambda-2};U_{\lambda-1}) \right]^{1/(p+1)}. \tag{12.2.22}$$

When the functions $y_k(u)$ are found, Eq. (12.2.12) allows determining general solutions for $x_k(u)$:

$$x_k = E_k(U_{k-1};u) + b_k\,, \tag{12.2.23}$$

where

$$E_k(c;u) = \int_c^u t\dot{y}_k(t)dt, \quad U_{k-1} \le c \le U_k, \quad U_{k-1} \le u \le U_k \tag{12.2.24}$$

and b_k are constants. Formulas for x_k, similar to those given Eq. (12.2.20),

$$\begin{aligned}
&0 = E_1(U_0;U_0) + b_1, \\
&E_{k-1}(U_{k-2};U_{k-1}) + b_{k-1} = E_k(U_{k-1};U_{k-1}) + b_k,\ k = 2,\ldots,s,
\end{aligned} \tag{12.2.25}$$

imply that

$$b_1 = 0, \quad b_k = \sum_{\lambda=2}^{k} E_{\lambda-1}(U_{\lambda-2};U_{\lambda-1}), \quad k = 2,...s\,. \qquad (12.2.26)$$

Consequently,

$$\begin{aligned} x_1(u) &= \int_{u_*^{\min}}^{u} t\dot{y}_1(t)dt, \\ x_k(u) &= \int_{U_{k-1}}^{u} t\dot{y}_k(t)dt + \sum_{\lambda=2}^{k} \int_{U_{\lambda-2}}^{U_{\lambda-1}} t\dot{y}_{\lambda-1}(t)dt, \quad k = 2,...,s. \end{aligned} \qquad (12.2.27)$$

The integrals in Eq. (12.2.27) can be transformed using integration by parts.

Taking into account Eq. (12.2.17) we find that

$$\frac{d^2 x_k}{dy_k^2} = \frac{d}{dy_k}\left(\frac{dx_k}{dy_k}\right) = \left(\frac{dy_k}{du}\right)^{-1} = \frac{uy_k^p}{G(u)}. \qquad (12.2.28)$$

Therefore $y^{(0)} = \Phi^{(0)}(x^{(0)})$ is a convex function if $G(u) > 0$.

In the particular case when

$$T^{(i)}(x^{(i)}, \Phi^{(i)}, \Phi^{(i)'}) = \Phi^{(i)p}, \quad i = 1,2,...,n\,, \qquad (12.2.29)$$

Eq. (12.2.10) implies that

$$G(u) = \sum_i \beta^{(i)} y^{(i)p} \dot{x}^{(i)} = u \sum_i \beta^{(i)} y^{(i)p} \dot{y}^{(i)}\,, \qquad (12.2.30)$$

the integrals in Eq. (12.2.19) can be calculated:

$$\begin{aligned} &F_k(c;u) = (p+1)\sum_i \beta^{(i)} \int_{y^{(i)}(c)}^{y^{(i)}(u)} y^{(i)p} dy^{(i)} = \\ &\sum_i \beta^{(i)} \left[(y^{(i)}(u))^{p+1} - (y^{(i)}(c))^{p+1}\right] \end{aligned} \qquad (12.2.31)$$

and the relationship between $y_k(u)$ and $y^{(i)}(u)$, given by Eq. (12.2.22), can be written in the explicit form.

12.3 Example

Let us illustrate *Step 2* and the most complicated *Step 3* that are characteristic for the above outlined procedure. As an example, we consider the case when

$$n = 2, \quad T^{(i)} = \Phi^{(i)}, \quad \Phi^{(i)}(0) = 0, \quad \Phi^{(i)\prime}(0) = \infty, \quad i = 0,1,2, \tag{12.3.1}$$

that corresponds to the case of two blunt penetrating impactors.

Assume that

$$\Phi^{(1)\prime}(L^{(1)}) > \Phi^{(2)\prime}(L^{(2)}). \tag{12.3.2}$$

Then

$$u_*^{(1)} = u_*^{(2)} = 0, \quad u_{**}^{(1)} < u_{**}^{(2)}, \quad r^{(0)} = 0 \tag{12.3.3}$$

and, consequently,

$$\begin{aligned} &s = 2, \quad U_0 = 0, \quad U_1 = u_{**}^{(1)}, \quad U_2 = u_{**}^{(2)}, \\ &\theta(1) = \theta(2) = 0, \quad \Theta(1) = 1, \quad \Theta(2) = 2. \end{aligned} \tag{12.3.4}$$

Eq. (12.2.31) implies in this case ($p = 1$) that

$$\begin{aligned} &F_1(c;u) = \beta^{(1)}\left[(y^{(1)}(u))^2 - (y^{(1)}(c))^2\right] + \beta^{(2)}\left[(y^{(2)}(u))^2 - (y^{(2)}(c))^2\right], \\ &F_2(c;u) = \beta^{(2)}\left[(y^{(2)}(u))^2 - (y^{(2)}(c))^2\right]. \end{aligned} \tag{12.3.5}$$

Using Eq. (12.2.22) and taking into account Eqs. (12.3.3)-(12.3.5), we obtain:

$$\begin{aligned} &y_1(u) = \sqrt{\beta^{(1)}(y^{(1)}(u))^2 + \beta^{(2)}(y^{(2)}(u))^2}, 0 \le u \le U_1, \\ &y_2(u) = \sqrt{\beta^{(1)}(y^{(1)}(U_1))^2 + \beta^{(2)}(y^{(2)}(u))^2}, U_1 \le u \le U_2. \end{aligned} \tag{12.3.6}$$

After calculating integrals in Eq. (12.2.27) by parts, the expressions for $x_1(u)$ and $x_2(u)$ can be written as follows:

$$x_1(u) = uy_1(u) - \int_0^u y_1(t)dt, 0 \le u \le U_1,$$

$$x_2(u) = uy_2(u) - \int_{U_1}^u y_2(t)dt + U_1[y_1(U_1) - y_2(U_1)] - \int_0^{U_1} y_1(t)dt, U_1 \le u \le U_2. \tag{12.3.7}$$

Consider in more details the case of power-law functions

$$\Phi^{(i)}(x^{(i)}) = R^{(i)}(x^{(i)}/L^{(i)})^{\alpha^{(i)}}, \quad 0 < \alpha^{(i)} < 1, \quad i = 1,2, \tag{12.3.8}$$

which can be presented in a parametric form as follows (*Step 2*):

$$x^{(i)}(u) = L^{(i)}\left(\frac{u}{u_{**}^{(i)}}\right)^{\frac{1}{1-\alpha^{(i)}}}, y^{(i)}(u) = R^{(i)}\left(\frac{u}{u_{**}^{(i)}}\right)^{\frac{\alpha^{(i)}}{1-\alpha^{(i)}}}, \tag{12.3.9}$$

$$u_{**}^{(i)} = \frac{L^{(i)}}{\alpha^{(i)}R^{(i)}}, i = 1,2.$$

Hence Eq. (12.3.6) can be rewritten as follows:

$$y_1(u) = \sqrt{\beta^{(1)}R^{(1)2}\left(\frac{u}{u_{**}^{(1)}}\right)^{\frac{2\alpha^{(1)}}{1-\alpha^{(1)}}} + \beta^{(2)}R^{(2)2}\left(\frac{u}{u_{**}^{(2)}}\right)^{\frac{2\alpha^{(2)}}{1-\alpha^{(2)}}}}, \tag{12.3.10}$$

$$0 \le u \le U_1 = u_{**}^{(1)},$$

$$y_2(u) = \sqrt{\beta^{(1)}R^{(1)2} + \beta^{(2)}R^{(2)2}\left(\frac{u}{u_{**}^{(2)}}\right)^{\frac{2\alpha^{(2)}}{1-\alpha^{(2)}}}}, \tag{12.3.11}$$

$$u_{**}^{(1)} = U_1 \le u \le U_2 = u_{**}^{(2)}.$$

Substituting $y_1(u)$ and $y_2(u)$ given by Eqs. (12.3.10)-(12.3.11) into Eq. (12.3.7) we arrive at formulas for $x_1(u)$ and $x_2(u)$.

If $\alpha^{(1)} = 2/3$ and $\alpha^{(1)} = 1/2$ then the solution of the problem can be obtained in the explicit form:

$$\frac{x_1(u)}{b} = \Lambda(u), \quad \frac{y_1(u)}{b} = u\sqrt{au^2+1}, \quad 0 \le u \le u_{**}^{(1)}$$

$$\frac{x_2(u)}{b} = \frac{1}{2}u\sqrt{u^2+e} - \frac{e}{2}\ln\left(\frac{u+\sqrt{u^2+e}}{u_{**}^{(1)}+\sqrt{u_{**}^{(1)2}+e}}\right)$$

$$-\frac{1}{2}u_{**}^{(1)}\sqrt{u_{**}^{(1)2}+e} + \Lambda(u_{**}^{(1)}), \tag{12.3.12}$$

$$\frac{y_2(u)}{b} = \sqrt{u^2+e}, \quad u_{**}^{(1)} \le u \le u_{**}^{(2)},$$

where

$$u_{**}^{(1)} = \frac{3}{2\tau^{(1)}}, \quad u_{**}^{(2)} = \frac{2}{\tau^{(2)}}, \quad \tau^{(i)} = \frac{R^{(i)}}{L^{(i)}},$$

$$\Lambda(u) = \frac{(2au^2-1)\sqrt{au^2+1}+1}{3a}, \tag{12.3.13}$$

$$a = \frac{\beta^{(1)}R^{(1)2}u_{**}^{(2)2}}{\beta^{(2)}R^{(2)2}u_{**}^{(1)4}}, \quad b = \frac{\sqrt{\beta^{(2)}}R^{(2)}}{u_{**}^{(2)}}, \quad e = \frac{\beta^{(1)}R^{(1)2}u_{**}^{(2)2}}{\beta^{(2)}R^{(2)2}}.$$

The parameters $L^{(0)}$ and $R^{(0)}$ for the object $\Xi^{(0)}$ are given by the following formulas:

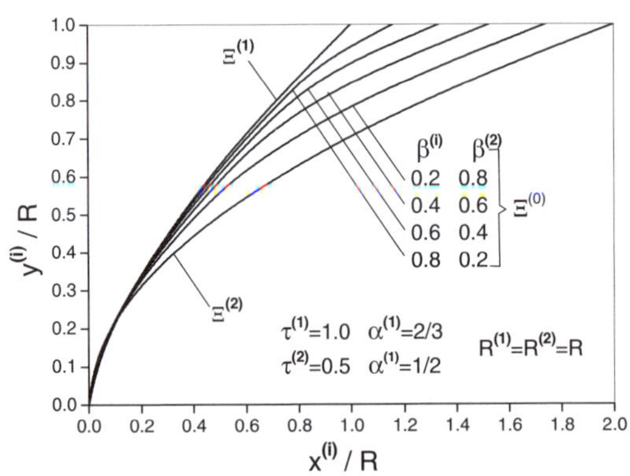

Fig. 12.3.1. Functions $y^{(i)} = \Phi^{(i)}(x^{(i)})$ associated with the object $\Xi^{(0)}$, for "basic" objects $\Xi^{(1)}$ and $\Xi^{(2)}$.

$$L^{(0)} = x_2(u_{**}^{(2)}),\ R^{(0)} = y_2(u_{**}^{(2)}) = \sqrt{\beta^{(1)} R^{(1)2} + \beta^{(2)} R^{(2)2}}. \quad (12.3.14)$$

If $R^{(1)} = R^{(2)}$ and $\beta^{(1)} + \beta^{(2)} = 1$ then $R^{(0)} = R^{(1)} = R^{(2)}$.

Versions of the function $y^{(0)} = \Phi^{(0)}(x^{(0)})$ for the object $\Xi^{(0)}$ (this function describes the shape of the generatrix of the projectile if a penetrating projectile is considered as an object) for different $\beta^{(1)}$ and $\beta^{(2)}$ as well as the functions $y^{(i)} = \Phi^{(i)}(x^{(i)})\ (i = 1,2)$ for the "basic" objects $\Xi^{(1)}$ and $\Xi^{(2)}$ (the shape of the "basic" impactors) are shown in Fig. 12.3.1. Inspection of this figure shows that varying coefficients $\beta_1, \beta_2, ..., \beta_n$ allows obtaining estimations for a wide class of penetrators having different shapes.

The suggested approach which is formulated for applications in penetration mechanics, can be useful when there are no alternative methods for estimating the DOP of an impactor using the available experimental data on the DOPs of the impactors having other shapes.

Part 7

Shape Optimization of Impactors

Part 7 includes Chapters 13–17.

Chapter 13 is devoted to a survey of investigations on shape optimization of penetrating impactors based on approximate models which includes more than 50 studies published since 1980. We analyze publications which employed analytical and numerical method for shape optimization of high-speed penetrators against concrete, metal, fiber-reinforced plastic (FRP) laminate and soil shields. The reviewed studies are summarized in the table where we display the following information: the model; indicate whether the model accounts for or neglects friction at the surface of penetrator; criterion for optimization (depth of penetration (DOP), ballistic limit velocity (BLV), several criteria); class of considered shapes of penetrators (bodies of revolution, different classes of 3-D bodies, etc.); method of solution (analytical or numerical); in comments we present additional information on formulation of the optimization problem. In assessing performance of different models we use a proposed classification of the mathematical models for describing interaction between a penetrator and a shield.

Chapter 14 includes Sections 14.1 and 14.2.

Section 14.1 deals with approximate penetration models that take into account sliding velocity (SV) and pressure dependence of the friction coefficient (FC); this chapter includes the survey of the relevant investigations. It is shown that introducing variable FC in a localized interaction model (LIM) (see Chapter 1) yields a model that belongs to the same class of LIMs. We develop a general method for determining the DOP using the piecewise linear approximation of the impactor generatrix. For some classes of SV dependent friction models, analytical formulas for calculating the DOP are obtained. Using the experimental data available in the literature, the dependencies of FC versus pressure and SV are determined. We also conducted numerical modeling of penetration of a metal striker into metal and concrete shields employing models with variable and constant FC. It is shown that taking into account variable FC strongly affects the DOP when FC changes appreciably for large velocities that are characteristic for the high speed penetration.

Section 14.2 deals with shape optimization of penetrating impactors using the criterion of the maximum DOP when impactor-shield interaction is described by a LIM. The universal numerical optimization procedure that can be employed for arbitrary LIM is suggested. This procedure is applied for shape optimization of impactor against a semi-infinite metallic shield using a version of the LIM with friction coefficient depending on sliding velocity.

Chapter 15 considers numerical solution of the problem of shape optimization of projectiles penetrating into concrete semi-infinite shield. Along with optimal shapes in the class of bodies of revolution we determined optimal truncated cone-shaped and truncated ogive-shaped impactors. We analyzed the effects of selected model and class of the considered bodies on the magnitude of the DOP for the optimal striker.

The problem of shape optimization of a striker against a finite width shield taking into account plug formation and incomplete immersion of the striker into the barrier at the initial and final stages of penetration is solved numerically in Chapter 16.

Chapter 17 deals with the problem of determining the shape of the impactor that provides the minimum BLV against FRP laminates. The suggested approach accounts for the non-unique choice of the phenomenological parameter of the model and for the comparatively weak variation of the optimization criterion in the vicinity of the optimum. Using this approach the shape optimization problem for the strikers having a truncated cone nose is comprehensively studied. Results of the investigation are presented in the form of intervals for the radius of bluntness of the nose of the striker whereby the BLV is in the vicinity of the optimum value.

In writing Part 7 we used the results obtained by Ben-Dor *et al.*, 2007a, 2007b, 2007c, 2009c, 2009h.

Chapter 13

Survey

13.1 Introduction

In this chapter the notion of shape optimization of impactors refers to the choice of the best impactor among the infinite number of possible solutions. Although the term "shape optimization of impactor" is commonly used for brevity, it is actually implied that the problem of determining the optimal shape of the nose of the impactor is considered. The required solution comprises either a function that determines the shape of the projectile (in this case a variational problem must be solved) or geometrical parameters which determine the shape of the projectile among the considered types of bodies, e.g. ogive-shaped projectiles (in this case optimization problem for a function of one or several variables must be solved). We do not include publications where on the basis of numerical or experimental investigations the projectile having the best performance characteristics is selected among several considered projectiles.

All results on shape optimization of penetrators are based on approximate models that describe penetrator-barrier interaction. In most cases these are localized interaction models (LIMs) (see Chapter 1).

The natural criteria for shape optimization of impactors are either the maximum depth of penetration (DOP) into a semi-infinite shield (the depth inside a barrier at which penetrator is brought to rest) or the minimum ballistic limit velocity (BLV), i.e., the minimum impact velocity required for shield perforation, in the case of a shield having a finite thickness. In order to simplify the mathematical formulation of the problem, along with these criteria (particularly at the early stages of investigating the problem) the indirect criteria, which did not have a clear physical meaning, were used. These studies are also included in the survey.

Solutions of the problems related to shape optimization of penetrating projectiles can be interpreted in two different ways. The penetrator having an optimum shape maximizes the efficiency of the attack or, in contrast, determining ballistic parameters of the optimal penetrator allows optimizing a shield for the most unfavorable case.

13.2 Optimization Using of Indirect Criteria

Kucher, 1967 optimized the shape of penetrator using as the criterion the "dynamic work" from Thomson theory (Thomson, 1955) for thin plates. Nixdorff, 1987b compared the efficiency of conical, different power-law and ogival heads and found that there exist impactors that are superior to "Kucher's optimum head", which was determined by solving the corresponding variational problem. This paradox was explained by Ben-Dor *et al.*, 2001a with reference to the correct solution (Miele, 1962a, 1965) of the mathematically similar variational problem in hypersonic aerodynamics, namely, determining a thin head with a minimum drag using the Newton-Busemann model for a projectile-medium interaction. Challenging questions that arise when solving this gas dynamics problem were discussed in the monographs by Hayes and Probstein, 1959 and Miele, 1965.

The first studies of shape optimization of non-thin shields also employed indirect optimization criteria. Using the previously developed disks model (Yankelevsky and Adin, 1980), Yankelevsky, 1983b optimized the shape of a projectile (a body of revolution) penetrating into soil by minimizing the instantaneous resistance force. The optimal shape was found to be determined by a single parameter that depends on the velocity and deceleration of the impactor and the properties of the medium. Gendugov *et al.*, 1984; Bunimovich and Yakunina, 1987a, 1987b, 1989; Ostapenko and Yakunina, 1992; Ostapenko, 1997 determined analytically the shape of the minimum drag body. Similar approach was applied by Jones *et al.*, 1998, for determining the optimum shape numerically. The latter studies employed the *0&2* model. Hereafter we employ the notations *p&q* model, which are described in Section 4.1.1, for non composite shields while *0&q* models denote one-stage models. Generally *1&q* and *2&q* models are used for describing penetration into concrete shields while *0&2* and *0&3* models are used for analysis of penetration into metal and soils.

Baranov and Lopa, 1995, Baranov *et al.*, 1995, Baranov *et al.*, 1996 minimized impactor drag force assuming that the normal stress at the surface of the penetrator is proportional to the normal component of the local velocity and neglecting friction. Khromov, 2003, 2004 used similar model for determining the minimum drag force impactor among the truncated conical nosed impactors, and a model with a constant friction for numerical investigation of the variational problem for bodies of revolution.

13.3 Optimization Using Direct Criteria

Table 13.1 includes the studies which used the technically meaningful criterion for optimization, namely the maximum DOP or the minimum BLV.

In column "Model" the unified notations for the models are used. In the case of semi-infinite shield, the optimization criterion is the maximum DOP, the asterisk at the right indicates the exact model (Section 1.2.1), while the absence of the asterisk implies neglecting incomplete immersion of the projectile in the shield at the initial stage of penetration ("averaged model", Section 1.5–1.7). "Averaged models" are not used for shape optimization of projectiles penetrating into finite-thickness shields (optimization criterion is the BLV); in this case Θ and θ are determined by Eqs. (1.2.2) and (1.2.4). Yankelevsky and Gluck, 1980 employ a special model suggested by Yankelevsky and Adin, 1980.

In the case of fiber-reinforce plastic (FRP) laminates, Ben-Dor and coauthors used models that are based on a linear dependence between the normal stress and the normal component of the impact velocity (Ben-Dor *et al.*, 2002a, 2002b, 2002c) and the normal component of the instantaneous velocity of impactor (Ben-Dor *et al.*, 2002c). Section 17.1 contains a more comprehensive survey of the studies on shape optimization of impactors penetrating into FRP laminates.

Symbol "-" in column «Friction» indicates that friction is not taken into account while "*constant* μ_{fr} " or "*constant* τ_{fr} " indicate that either friction coefficient or friction force per unit surface area are considered constant and non zero; "*variable* μ_{fr} " indicates using the model with friction coefficient depending on sliding velocity, and "*different models*" indicates that analysis is conducted using different friction models (either a model with constant μ_{fr} or a model suggested by Grigoryan, 1979).

In column "Shapes of impactors" we indicate which shapes of the impactors are considered for optimization. Depending on the considered shape either a variational problem or optimization problem for a function of several variables is solved. The typical situation is solving a variational problem in the class of bodies of revolution (indicated as "Bodies of revolution"). Notation "A class (classes) of 3-D bodies" indicates that bodies with complicated shapes are considered.

In column "Method of solution" we indicate the general characteristic of the method: "A" when analytical method is used, and "N" if a numerical method is employed.

In column "Comments" we present additional requirements for the considered bodies. Some clarification is needed here.

Typical situation is when a projectile having a nose (its shape is to be determined) and a cylindrical part is considered. The length of a cylindrical part is varied in order to keep the mass of the projectile constant. This constant is assumed to be specified, and it appears in the equation of motion of the projectile and in the formula for optimization criterion. In this case mass of a projectile is not considered as a constraint for optimization of the nose shape and is not mentioned in the column "Comments".

The situation is quite different when the shape of the whole impactor has to be determined, and there is no provision for the presence of the cylindrical part. The latter approach is employed in the studies by Banichuk and coauthors. Here either the mass of the monolithic projectile or the mass of the thin shell of a hollow striker are the constraints in the optimization problem and they are indicated in "Comments" column.

In the considered variational problems the size of the projectile (the total length or the length of the nose, shank diameter for bodies of revolution) is assumed to be specified. For 3-D impactors, it is a shank area that is commonly assumed to be preseted. In the case of conical 3-D impactors with a given length the latter constraint is equivalent to prescribing the mass of the projectile provided the mass is distributed uniformly over the volume of a striker.

Quite unusual approach is used by Aptukov and Pozdeev, 1982. They considered the minimax problem for determining the shape of impactor that penetrates to the maximum depth under the most unfavorable distribution of the mechanical properties over the thickness of the shield with a prescribed areal density. The two-term model without friction was used with a linear relationship between the parameters of the model.

Table 13.1 Optimization using criteria of maximum DOP and minimum BLV.

Reference	**Model**	**Friction**	**Criterion**	**Shapes of impactors**	**Method of solution**	**Comments**
Yankelevsky and Gluck, 1980	The disk model	-	Max DOP	Ogive nosed impactors	N	For details see main text
Aptukov and Pozdeev, 1982	*0&2*	-	Max DOP	Bodies of revolution	A	For details see main text
Bondarchuk *et al.* 1982	*0&2*	-	Max DOP	A class of 3-D bodies	N	
Ostapenko *et al.*, 1994	*0&3*	Constant μ_{fr}	Max DOP	A class of 3-D bodies	N	Different sets of conditions
Bunimovich and Dubinsky, 1995	*0&2*	-	Max DOP	Sharp 3-D conical bodies	A	
Vedernikov and Shchepanovsky, 1995; Vedernikov *et al.*, 1995	*0&2*	Constant μ_{fr}	Max DOP	Classes of 3-D bodies	N	
Ben-Dor *et al.*, 1997c	*0&2**	-	Max DOP, Min BLV	Sharp 3-D bodies: cones, slender bodies, pyramids	A	Mass of monolithic impactor is preset. Layered shield

Reference	Model	Friction	Criterion	Shapes of impactors	Method of solution	Comments
Ostapenko, 1997	*0&2*	Constant μ_{fr}	Max DOP	Slender 3-D bodies with self-similar cross sections	A	
Ostapenko and Yakunina, 1999	*0&2*	Different models	Max DOP	Slender 3-D bodies with self-similar cross sections	A	
Ben-Dor *et al.*, 2000a	*0&2**	-	Min BLV	Truncated cones	N	
Jones and Rule, 2000	*0&2*	Constant μ_{fr}	Max DOP	Bodies of revolution	A/N	
Rule and Jones, 1999	*0&2*	Constant τ_{fr}	Max DOP	Bodies of revolution	A/N	
Ben-Dor *et al.*, 2001a	*0&2*	-	Max DOP, Min BLV	3-D bodies	A	For details see main text
Ben-Dor *et al.*, 2002a	See main text*	-	Min BLV	Bodies of revolution	A	FRP laminates
Ben-Dor *et al.*, 2002b	See main text*	-	Max DOP	Bodies of revolution	N	FRP laminates
Ben-Dor *et al.*, 2002c	See main text*	-	Max DOP. Min BLV	A class of 3-D bodies	A	FRP laminates
Yakunina, 2001, 2005a, 2005b, 2005c	*0&2*	Different models	Max DOP	3-D bodies	A	
Ben-Dor *et al.*, 2003b	*1&2*	-	Max DOP	Bodies of revolution	A	
Ben-Dor *et al.*, 2003a	*0&2**	Constant μ_{fr}	Max DOP	Bodies of revolution	N	
Ben-Dor *et al.*, 2006c	*1&3*	-	Max DOP	Bodies of revolution	A	
Ben-Dor *et al.*, 2006a	*0&3*	-	Max DOP	Bodies of revolution	A	

Table 13-1 (cont'd)

Reference	Model	Friction	Criterion	Shapes of impactors	Method of solution	Comments
Ben-Dor *et al.*, 2006a	*1&2*	-	Max DOP	Different classes of bodies of revolution	N	
Ben-Dor *et al.*, 2007b	*0&2*	Variable μ_{fr}	Max DOP	Bodies of revolution	N	
Ben-Dor *et al.*, 2007c	*0&2**	-	Min BLV	Bodies of revolution	N	Taking into account plug formation
Banichuk and Ivanova, 2007; Banichuk *et al.*, 2007	*1&3*	-	Max DOP	Bodies of revolution	N	Mass of monolithic impactor or mass of shell is preset
Banichuk and Ivanova, 2008; Banichuk *et al.*, 2008a	*1&2*	-	Max DOP	Sharp or truncated pyramids	A	Mass of monolithic impactor is given
Banichuk *et al.*, 2008b	*1&2*	-	Max DOP	Sharp pyramids	A	Mass of monolithic impactor or mass of shell is given
Banichuk *et al.*, 2 009	*1&2*	-	Max DOP	Bodies of revolution	A	Mass of monolithic impactor or mass of shell is preset
Banichuk *et al.*, 2 009	*1&3*	-	Max DOP	Bodies of revolution	N	Mass of monolithic impactor is given
Ben-Dor *et al.*, 2009h	*2&3, 2&2*	-	Max DOP	Bodies of revolution	N	
Ben-Dor *et al.*, 2009c	See main text*	-	Min BLV	Truncated cones	N	FRP laminates
Ragnedda and Serra, 2010	*1&3*	-	Max DOP	Special class of bodies	N	
Banichuk *et al.*, 2011, 2012	*1&2*	-	Max DOP, min mass of shell, min of volume	Bodies of revolution	A	Multi-objective optimization

Table 13-1 (cont'd)

13.4 Some Methodological Remarks

13.4.1 *Analogy between different optimization problems*

In the case of *0&2* "averaged" model without friction the problems of determining the maximum DOP, minimum BLV and minimum drag for constant penetration velocity for 3-D bodies with preset shank diameter are reduced to minimization of the same functional. The same variational problem is encountered in hypersonic aerodynamics for drag coefficient which is calculated using the Newton model (Ben-Dor *et al.*, 2001a, 2006a). Clearly, the latter assessment is also valid in a particular case of bodies of revolution with a preset shank diameter. Jones and Rule, 2000 were the first to notice the similarity between drag minimization for constant velocity and maximization of the DOP for sharp bodies of revolution. Maximization of the DOP using *1&2* "averaged" model without friction for penetration into concrete shield can be reduced to the maximization of the same functional as in the case of frictionless model for penetration into concrete shields (Ben-Dor *et al.*, 2003b, 2006a). Therefore, solution of the questionable from the physical point of view problem of minimizing drag during projectile motion in dense medium with constant velocity and neglecting friction yields the shape of the striker that penetrates at the maximum depth. However, this similarity is no longer valid when friction is taken into account. Moreover, the shape of the minimum drag projectile depends upon the velocity and parameters of the model.

Due to the described above similarity, instead of solving penetration problems with *0&2* and *1&2* "averaged" model without friction it is sufficient to use the solutions obtained earlier in hypersonic aerodynamics (many of these solutions can be found in Miele, 1965).

It is important to note that correct formulation of the problem of shape optimization of penetrating impactors must include provision for the striker with a flat bluntness. In most cases the impactors having a flat bluntness are the optimal ones. In particular, such solution was obtained by Newton, 1687 in drag minimization for a body of revolution using a *0&2* model. This solution is valid for the maximization of the DOP using *0&2* or *1&2* "averaged" models without friction.

13.4.2 *About optimization of 3-D impactors*

Analysis based on the approximate models of projectile-shield interaction shows that the drag force acting on star-shaped strikers with plane fins tends to zero. Consequently, the depth of penetration tends to infinity when the number of fins grows. Since this shape and similar star-shaped solutions are of no practical importance, the shape optimization problem in conventional setting is ill-posed. The typical method that allows avoiding this problem is to consider optimization for a certain specified classes of projectile shape. For comparing the ballistic properties of 3-D impactors and projectiles having a conventional shape of bodies of revolution, it is important to choose correctly the requirements and constraints for the compared strikers as has been emphasized, e.g. by Bivin, 1999 (see also Bivin and Simonov, 2010).

13.4.3 *Phenomenon of cavitating penetration*

Under certain conditions, which depend on projectile velocity and shape, penetration can be accompanied by cavitation phenomenon, i.e. loss of contact between the the surface of the projectile nose and the surrounding medium. Some additional information regarding this phenomenon and its modeling can be found, e.g., in the studies by Hill, 1980, Yankelevsky, 1985; Simonov, 1993; Mayersak, 2004; Osipenko and Simonov, 2002; Bivin and Simonov, 2010. In a limiting case, the projectile moves inside the cavity without a contact of its nose surface with a barrier. This results in sharp reduction of drag and independence of drag on the projectile shape as long as the projectile remains inside the cavity.

In this context it seems worthwhile to consider shape optimization of the projectiles taking into account the cavitation phenomenon. In particular, it is conceivable to suggest that using the projectile with a jet thruster (see Part 9) for maintaining stable cavitation regime of penetration can result in considerable increase of the DOP.

13.4.4 *Area rules*

In the context of shape optimization problems it is pertinent to mention area rules. Area rules determine the conditions under which the difference between the values of some characteristics of a 3-D projectile and those of a reference projectile having a simpler shape (usually, a body of revolution) is of the higher order of smallness than the difference in their shapes if the difference in their shapes is small. The requirement that the distributions of cross-sectional areas of two equal length projectiles along the longitudinal axis are the same, can serve as an area rule. Consequently, area rule determines a large class of 3-D projectiles having near optimal parameters when the optimal body of revolution is chosen as a reference projectile. The latter assertion is used extensively in aircraft design for different high velocity flight conditions, provided a body of revolution and a drag force are chosen as a reference projectile and parameter, correspondingly.

Area rules have been determined for special ranges of flight conditions for various integral characteristics (not only for the drag force), including the characteristics that can be expressed through functionals of a quite general form. More information on this subject can be found, e.g., in Bunimovich and Dubinsky, 1995 and Apshtein and Titow, 1996.

Ben-Dor *et al.*, 1997b, 1998c determined area rules for penetrating impactors using the assumption of the localized interaction theory. Two versions of these rules are proved for the BLV, namely, when a body of revolution or a projectile with a polygonal cross-section is considered as the reference impactor.

13.4.5 *Concluding remarks*

It must be noted that there are no published studies that compare the obtained optimal solutions with experimental data and with the "exact" numerical simulations. The reason is that dedicated experiments and "exact" numerical calculations are needed for this purpose, because it is necessary to meet the same requirements for the compared shapes as in the theoretical investigations (the same mass, diameter and length of the projectile nose, etc.).

It would be unwarranted and inappropriately optimistic to expect that the obtained solutions of the shape optimization problems can be used directly for selecting the shape of penetrators. Similarly to other situations, the solutions obtained on the basis of approximate models may be used only for elucidating tendencies which must be taken into account in planning experiments and "exact" numerical calculations for design of penetrating projectiles.

Chapter 14

Penetration with Non-Constant Friction

Nomenclature

a_0, a_2	parameters determining the model, Eq. (14.1.16);
D	resistance force acting on projectile;
$\tilde{D}$	$=D/\pi$;
DOP	depth of penetration;
F	functions, Eq. (14.1.12);
FC	friction coefficient;
H	depth of penetration (DOP);
$\tilde{H}$	$=(\pi/m)H$,
H_*	depth of penetration for constant coefficient of friction;
H_0	depth of penetration for variable coefficient of friction;
J_i	integrals, Eq. (14.2.5);
L	length of the nose of impactor;
LIM	localized interaction model;
m	mass of projectile;
M	number of intervals in piecewise linear representation of the generatrix of projectile, Eq. (14.1.7);
N	number of intervals for discretization of velocity range of projectile;
p	$=\Omega_n(u,v)$, local pressure (normal stress);
r	radius of flat bluntness of projectile nose;
R	shank radius of impactor;
SV	sliding velocity;
t_0, t_1	parameter in Eq. (14.2.11);

t_2, t_3	parameter in Eq. (14.2.14);
u	$= \cos \upsilon$;
v_{imp}	impact (initial) velocity of projectile;
v	instantaneous velocity of projectile;
v_i	$= i\Delta v$
w	sliding (tangential) velocity of impactor
w_0	parameters in the model, Eq. (14.2.12)
x_j	nodes in piecewise linear representation of the generatrix of projectile, Eq. (14.1.7);
x	coordinate associated with the impactor, Fig. 1.1.2;
α_i, β_i	parameters, Eq. (14.1.8);
γ_i	parameter, Eq. (14.1.10);
Δv	$= v_{imp}/N$;
ϑ	half-angle of the cone apex
ϑ_j	slope of the j-th segment of polygonal approximation of function $\rho = \Phi(x)$.
θ_j	$= 1/\sqrt{\alpha_j^2 + 1} = \cos \vartheta_j$, Eq. (14.1.15);
κ	$= H_*/H_0$;
λ_i	coefficients in the Newton-Cotes integration formula, Eq. (14.2.2);
μ	function determining FC in terms of FC and SV, or of SV, or of product SV$\times$FC
μ_{fr}	friction coefficient;
μ_{eq}	value of FC whereby the depth of penetration for preset v_{imp} is the same as in the case with a non-constant FC;
μ_0, μ_∞	parameters in the model determined by Eq. (14.2.12);
μ_*, μ_{**}	functions, Eq. (14.1.24);
τ	$= R/L$;
υ	angle between the local inner normal vector at the projectile surface and the vector in the opposite direction to the vector of velocity of projectile;
φ	function in Eq. (1.4.21);
Φ	function determining the generatrix of the body of revolution (projectile);
Φ_i	$= \Phi(x_i)$;

ω	function in Eq. (14.1.5);
Ω_n, Ω_τ	functions determining the LIM.

14.1 Modeling of Ballistic Characteristics

14.1.1 *Introduction*

Modeling of high speed penetration into metals, concrete and soils whereby the penetrator velocity is of the order of hundreds meters per second and pressure (normal stress) approaches hundreds of MPa, requires investigating the behavior of friction coefficient (FC). However, in the tribological studies the emphasis is given to the analysis of friction for relatively low velocities and pressures (Bowden and Tabor, 1964; Kraghelsky, 1965; Kraghelsky *et al.*, 1982; Kraghelsky and Vinogradova, 1962; Rabinowicz, 1995; Martins *et al.*, 1990; Berger, 2002; Oden and Martins, 1985).

There are only a few publications where the influence of high and very high (with respect to the velocities and pressures employed in tribological research) velocities and pressures on friction has been investigated. Ogawa, 1997; Rajagopalan *et al.*, 1999; Rajagopalan and Prakash, 1999; Espinosa *et al.*, 2000a, 2000b investigated a pair of metals at velocities up to 10 m/s and pressures up to 200 MPa. Bowden and Freitag, 1958 and Bowden and Persson, 1961 conducted experiments with the velocities up to 800 m/s and loads less than 0.015 MPa. Pressures up to 3 GPa and speeds up to 30 m/s were used in the experiments conducted by Prakash and Clifton, 1993 and Prakash, 1995,1998. Irfan and Prakash, 2000 and Philippon *et al.*, 2004 studied friction for pressure of the order of 1.5 GPa and pressures varying in the range from 9 MPa to 33 MPa, respectively, and speeds up to 60 m/s. Lim and Ashby., 1987; Lim *et al.*, 1989 analyzed a large number of experimental data on the variation of FC in sliding of steel on steel in a wide range of sliding conditions and presented the results of the analysis in the form of a friction regime map. They concluded that at low speeds (less than 1 m/s) the FC strongly depends on the state of the surface; at higher velocities the FC becomes, increasingly, dependent on the pressure at the surface and the sliding velocity (SV); at very high loads and speeds, the magnitude of FC is very small.

A number of models describing the dependence of FC on sliding conditions are proposed for relatively low velocities (Bo and Pavelescu, 1982; Banerjee, 1968; Kraghelsky, 1965; Kraghelsky and Vinogradova 1962; Molinari *et al.*, 1999). Kraghelsky, 1965; Kraghelsky and Vinogradova 1962 identified the distinctive features of the dependence of FC vs. SV. At large pressures FC is always smaller than at lower pressures. Generally, pressure dependence of FC has a maximum that shifts towards the origin when pressure is increased. Experiments reported by Vukolov and Zharov, 2005 show that this dependence may be more involved.

The behavior of FC at larger velocities and pressures was investigated by Krafft, 1955; Montgomery, 1976; Balakin, 1980; Klepaczko, 2001, 2002; Pelak *et al.*, 2000, 2001; Yuan *et al.*, 2009. Montgomery, 1976 presented the experimental data on FC for different pairs of metals for pressures up to 180 MPa and velocities up to 550 m/s. Data of Balakin, 1980 for steel/steel FCs correspond to pressures up to 18 MPa and velocities up to 500 m/s. Klepaczko, 2001, 2002 suggested a theoretical model for calculating the FC of a steel projectile moving inside a geological or concrete shield with velocities up to 1000 m/s. Yuan *et al.*, 2009, detected experimentally decreasing exponential character of the dependence of FC vs. SV in the range of SV from 30 m/s up to 250 m/s for 7075-T6 AL Alloy/CH Tool steel tribological pair.

According to the data reported in the literature there is a clear evidence that FC decreases (and assumes very low values for pairs metal/metal) with the increase of SV and pressure starting from low (as compared to the velocities encountered in high speed penetration) velocities. The latter conclusion with only a few exceptions (Bowden and Tabor, 1964) is valid for most of the materials.

Some codes (Taylor and Flanagan, 1987; Attaway *et al.*, 1998) allow taking into account the prescribed dependence of the FC on the SV when modeling penetration into a geological shield. The first of these codes was used by Chen, 1989 to compare the model with a constant friction and a hypothetical model with a velocity dependent friction. In terms of the depth of penetration and the peak deceleration, these models yield almost identical results. However, Chen, 1989 concluded that the model with velocity dependent friction should be a preferred model of sliding friction in penetration analysis because it predicts better the behavior of the impactor deceleration.

Jones *et al.*, 2003 and Davis, 2003 developed engineering models for penetration modeling that take into account SV dependence of the FC. They also noticed that this generalization of the model renders the derivation of an explicit formula for the depth of penetration (DOP) in the case of arbitrary dependence of the FC, μ_{fr}, upon SV, w, $\mu_{fr} = \mu(w)$, unfeasible. Jones *et al.*, 2003 derived a formula for the DOP of an ogive-shaped impactor (body of revolution) using the two-term impactor-shield interaction model whereby the dependence of the FC vs. SV is approximated by two linear segments where one segment is the part of the abscissa axis. Generalizing this approach, Davis, 2003; Davis *et al.*, 2003 proposed using a step function approximation to FC as a function of SV.

Baranov *et al.*, 2004; Khromov, 2004 devised a mathematical model describing penetration into soil by rotating impactor when the local normal stress is a polynomial function of the local velocity of the impactor and the dependence of the FC on the SV is described by a linear fractional function. Numerical calculations and experiments showed that the rotation of impactor increases the DOP, and this effect is more pronounced for the relatively small velocities (under 300 m/s).

14.1.2 *LIMs with velocity and pressure dependent friction coefficient*

Consider a general LIM that is determined by Eqs. (1.1.1)–(1.1.3) and assume that the FC, μ_{fr}, depends on the local sliding (tangential) velocity of the impactor, w, and local pressure, p, i. e.,

$$\mu_{fr} = \mu(w, p), \tag{14.1.1}$$

where

$$w = v\sqrt{1-u^2}, \quad p = \Omega_n(u, v). \tag{14.1.2}$$

Hence

$$\Omega_\tau(u,v) = \mu(v\sqrt{1-u^2}, p) \cdot p. \tag{14.1.3}$$

Therefore, the LIM with velocity dependent FC remains in the class of LIM, and general properties of the LIM are valid also in this case. In particular, the latter conclusion concerns the ballistic properties of the spaced shields (Ben-Dor *et al.*, 1998a).

In the case of penetration of a body of revolution, the averaged model for the general LIM that can be obtained similarly as has been done in Sections 1.5-1.7 for some types of models, yields the following formula for the resistance force:

$$\frac{D(v)}{\pi} = \tilde{D}(v) = r^2 \Omega_n(1,v) + 2\int_0^L \omega(\Phi', v)\Phi dx, \tag{14.1.4}$$

where $\Phi = \Phi(x)$ determines the generatrix of the body of revolution (Fig. 1.1.2b) and

$$\omega(\Phi', v) = \Phi' \Omega_n(u,v) + \Omega_\tau(u,v), \quad u = \Phi' \big/ \sqrt{\Phi'^2 + 1}. \tag{14.1.5}$$

The equation of motion of impactor having the mass m, $mv(dv/dh) = -\pi\tilde{D}(v)$, yields the following expression for the DOP, H:

$$\tilde{H} = \frac{\pi}{m} H = \int_0^{v_{imp}} \frac{vdv}{\tilde{D}(v)}. \tag{14.1.6}$$

14.1.3 *Piecewise linear approximation of generatrix*

Expression in Eq. (14.1.6) for the penetration depth becomes quite involved when taking into account Eq. (14.1.4), and is not convenient for calculations. This problem can be eliminated by using a piecewise-linear approximation of the generatrix of a striker.

The equation of a piecewise linear generatrix can be written as follows (see Fig. 14.1.1):

$$\Phi(x)=\begin{cases}\alpha_1 x+\beta_1 & if \quad r=x_0\le x\le x_1\\ & \dots\\ \alpha_j x+\beta_j & if \quad x_{j-1}\le x\le x_j\\ & \dots\\ \alpha_M x+\beta_M & if \quad x_{M-1}\le x\le x_M=L\end{cases} \tag{14.1.7}$$

where $\Phi_j=\Phi(x_j)$, $j=0,1,...,M$ and

$$\alpha_j=\frac{\Phi_j-\Phi_{j-1}}{x_j-x_{j-1}}, \beta_j=\frac{x_j\Phi_{j-1}-x_{j-1}\Phi_j}{x_j-x_{j-1}}, \quad j=1,2,...,M\,. \tag{14.1.8}$$

Using these formulas the integral in Eq. (14.1.4) can be calculated to yield:

$$\widetilde{D}(v)\approx r^2\Omega_n(1,v)+\sum_{j=1}^{M}\gamma_j\omega(\alpha_j,v), \tag{14.1.9}$$

where

$$\gamma_j=(x_j-x_{j-1})[\alpha_j(x_j+x_{j-1})+2\beta_j]. \tag{14.1.10}$$

The appropriate choice of M allows attaining the required accuracy of approximation.

The integral in Eq. (14.1.6) can be calculated by numerical techniques. Using the trapezoidal rule (Korn and Korn, 1968) that is based on the piecewise linear approximation of the integrand we obtain:

$$\widetilde{H}\approx \Delta v\left[\frac{\hat{F}_0+\hat{F}_N}{2}+\sum_{i=1}^{N-1}\hat{F}_i\right], \tag{14.1.11}$$

where N is the number of segments in the polygonal line,

$$F=v/D(v),\quad \hat{F}_i=F(\hat{v}_i),\quad \hat{v}_i=i\Delta v,\quad \Delta v=v_{imp}/N,$$
$$i=0,1,...,N\,. \tag{14.1.12}$$

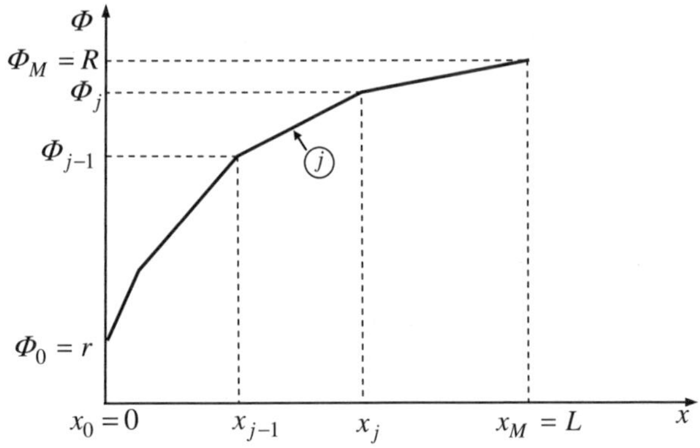

Fig. 14.1.1. Piecewise linear approximation of the generatrix of impactor

The suggested method allows determining the DOP for an arbitrary LIM (in particular, for arbitrary friction law) and for an impactor having an arbitrary shape.

In the case of a conical nosed impactor, $M = 1$, $\alpha_1 = \tan\vartheta$, $\beta_1 = r = 0$, $\gamma_1 = L^2 \tan\vartheta$, where ϑ is a half-angle of the cone apex. Hence $\tilde{D}(v) = \omega(\tan\vartheta, v)L^2 \tan\vartheta$ and Eq. (14.1.6) yields:

$$\frac{\pi L^2 \tan\vartheta}{m} H = \int_0^{v_{imp}} \frac{v dv}{\omega(\tan\vartheta, v)}. \qquad (14.1.13)$$

14.1.4 *Semi-analytical solutions*

14.1.4.1 *General formulas*

Consider now some particular cases when analytical solution can be obtained without approximating the integral in Eq. (14.1.6).

In a general case when FC depends on SV, substitution of Eqs. (14.1.3) into Eq. (14.1.5) yields the following expression for $\omega(\Phi', v)$:

$$\omega(\Phi', v) = \Omega_n(u, v)[\mu(v\sqrt{1-u^2}) + \Phi'] \qquad (14.1.14)$$

and, consequently,

$$\omega(\alpha_j, v) = \Omega_n(\alpha_j \theta_j, v)[\mu(\theta_j v) + \alpha_j],$$
$$\theta_j = 1\big/\sqrt{\alpha_j^2 + 1} = \cos\vartheta_j,\ \alpha_j = \tan\vartheta_j \tag{14.1.15}$$

where ϑ_j — is the slope of the j-th segment of the polygonal line that approximates the function $\rho = \Phi(x)$.

For the most widely used models,

$$\Omega_n(u, v) = a_2 u^2 v^2 + a_0, \quad \Omega_\tau(u, v) = \mu_{fr}\Omega_n(u, v), \tag{14.1.16}$$

and Eq. (14.1.6) can be rewritten as follows:

$$\tilde{H} = \int_0^{v_{imp}} F(v)dv, \tag{14.1.17}$$

where

$$F(v) = \frac{v}{b_0 + b_2 v^2 + \sum_{j=1}^{M} \varphi_j(v)\mu(\theta_j v)},$$
$$\varphi_j(v) = \gamma_j(a_2 \alpha_j^2 \theta_j^2 v^2 + a_0) \tag{14.1.18}$$

$$b_0 = a_0(r^2 + \sum_{j=1}^{M} \gamma_j \alpha_j), \quad b_2 = a_2(r^2 + \sum_{j=1}^{M} \gamma_j \alpha_j^3 \theta_j^2). \tag{14.1.19}$$

Since theory does not specify the type of the function, $\mu(w)$, we can select this function such that it decreases in the range of feasible values of w, and the integral in Eq. (14.1.17) can be calculated analytically.

14.1.4.2 *Approximation 1*

As a first approximation let us select the following dependence:

$$\mu(w) = \mu_0 - kw^2, \quad k < \mu_0 \big/ v_{imp}^2, \tag{14.1.20}$$

where μ_0 and k are constants. Substituting $\mu(w)$ given by Eq. (14.1.20) into Eq. (14.1.18), we obtain:

$$F(v) = \frac{v}{c_0 + c_1 v^2 - c_2 v^4}, \tag{14.1.21}$$

where

$$\begin{aligned} c_0 &= a_0[r^2 + \sum_{j=1}^{M} \gamma_j(\alpha_j + \mu_0)], \\ c_1 &= a_2 r^2 + \sum_{j=1}^{M} \gamma_j \theta_j^2 [a_2(\alpha_j + \mu_0)\alpha_j^2 - a_0 k] \\ c_2 &= a_2 k \sum_{j=1}^{M} \gamma_j \theta_j^4 \alpha_j^2 \end{aligned} \tag{14.1.22}$$

Changing the variables in the integral in Eq. (14.1.17), $v = \sqrt{z}$, we obtain the following expression for $\tilde{H}$:

$$\tilde{H} = \frac{1}{2} \int_0^{v_{imp}^2} \frac{dz}{c_0 + c_1 z - c_2 z^2}. \tag{14.1.23}$$

Integral in Eq. (14.1.23) can be easily calculated (Gradstein and Ryzhik, 1980).

14.1.4.3 *Approximation 2*

We can expand the possibilities of describing the dependence of FC on SV using the following expression for the function $\mu(w)$:

$$\mu(w) = \begin{cases} \mu_*(w) & \text{if} \quad 0 \le w \le w_0 \\ \mu_{**}(w) & \text{if} \quad w \ge w_0 \end{cases} \tag{14.1.24}$$

where $\mu_*(w)$ and $\mu_{**}(w)$ are some functions, w_0 is a positive constant, $\mu_*(w_0) = \mu_{**}(w_0)$.

Since we consider the impactors with the convex noses, $\rho = \Phi(x)$ is a convex upward function:

$$\theta_1 < \theta_2 < \ldots < \theta_j < \ldots < \theta_M , \tag{14.1.25}$$

and

$$\begin{aligned} &0 = v_{M+1} \le v_M \le v_{M-1} \le \ldots \le v_j \le \ldots \le v_1, \\ &v_j = w_0/\theta_j ,\ j = 1,2,\ldots,M \end{aligned} \tag{14.1.26}$$

Equations (14.1.25)–(14.1.26) yield the following inequalities for $v_{s+1} \le v \le v_s$ ($s = 0,1,\ldots,M$ and the parameter $v_0 > v_1$ is determined below):

$$\theta_1 v < \theta_2 v < \ldots < \theta_s v \le w_0, \quad w_0 \le \theta_{s+1} v < \theta_{s+2} v < \ldots < \theta_M v , \tag{14.1.27}$$

which allow determining the location at the surface of impactor where the functional dependence in Eq. (14.1.24) changes. Taking into account Eq. (14.1.27) the integral in Eq. (14.1.17) with the integrand given by Eq. (14.1.18) can be represented as a sum of the integrals:

$$\tilde{H} = \sum_{s=M}^{j_+} I_s , \tag{14.1.28}$$

where

$$I_s = \int_{v_{s+1}}^{v_s} \frac{v dv}{b_0 + b_2 v^2 + \sum_{1 \le j \le s} \varphi_j(v) \mu_*(\theta_j v) + \sum_{s < j \le M} \varphi_j(v) \mu_{**}(\theta_j v)} . \tag{14.1.29}$$

The value of the parameter j_+ is determined as follows. If $v_{imp} \ge w_0/\theta_1$, then $j_+ = 0$ and $v_0 = v_{imp}$. Otherwise, j_+ is determined from the conditions $w_0/\theta_{j_+ +1} \le v_{imp} \le w_0/\theta_{j_+}$ and v_{j_+} is set to $v_{j_+} = v_{imp}$.

Let us consider the FC function $\mu(w)$ in the form given by Eq. (14.1.24) with

$$\mu_*(w) = \mu_0, \quad \mu_{**}(w) = k + (\mu_0 - k) w_0^2 / w^2, \, \mu_0 > 0, k > 0. \qquad (14.1.30)$$

Substituting these expressions in the integrand in Eq. (14.1.28) we obtain:

$$I_s = \int_{v_{s+1}}^{v_s} \frac{v^3 dv}{c_0 + c_1 v^2 + c_2 v^4} = \frac{1}{2} \int_{v_{s+1}^2}^{v_s^2} \frac{z dz}{c_0 + c_1 z + c_2 z^2}, \qquad (14.1.31)$$

where

$$\begin{aligned} c_0 &= a_0 (\mu_0 - k) w_0^2 \sum_{s<j\le M} \gamma_j / \theta_j^2, \\ c_1 &= b_0 + a_0 \mu_0 \sum_{1\le j\le s} \gamma_j + \sum_{s<j\le M} \gamma_j [a_0 k + (\mu_0 - k) a_2 w_0^2 \alpha_j^2] \\ c_2 &= b_2 + a_2 \mu_0 \sum_{1\le j\le s} \gamma_j \alpha_j^2 \theta_j^2 + a_2 k \sum_{s<j\le M} \gamma_j \alpha_j^2 \theta_j^2 \end{aligned} \qquad (14.1.32)$$

The integral in Eq. (14.1.28) can be easily calculated (Gradstein and Ryzhik, 1980).

14.1.5 *Numerical simulations and discussion*

14.1.5.1 *Introductory remarks*

There are two approaches for approximate modeling of interaction of a striker with a shield taking into account a non-constant coefficient of friction.

In the first approach one postulates a model that takes into account variation of a coefficient of friction as a function of SV or other parameters. This model comprises a number of coefficients that are determined as the best fit coefficients to a particular experimental data set. Such approach was adopted in the studies by Jones *et al.*, 2003 and Davis, 2003. The approach is convenient and allows improving the accuracy of the model when penetration conditions and shape of the impactor are close to those used in determining the parameters of the model. However, this approach even in the case of a model with a constant coefficient of friction has a shortcoming that there is no guarantee that the improvement of the model is associated with taking into account friction (for details see Luk and Forrestal, 1989). Since the model is approximate introducing additional coefficients can formally improve the performance of the model for the data set used for determining these coefficients in the model while the predictive performance of the model can deteriorate.

The second approach is based on using data obtained in the experiments for measuring coefficient of friction. This approach seems more attractive although its application is restricted because of the scarcity of the quantitative information about the effect of penetration conditions (SV, etc.) on friction between a striker and a shield. Nevertheless, using the available data, one can employ this approach for the analysis of the problem that was first posed by Chen, 1989: whether the model with a variable FC offers advantages when compared with the model based on a constant coefficient of friction.

14.1.5.2 *Steel shield*

In this series of simulations we used the model determined by Eq. (14.1.16) for mild steel with the coefficients $a_0 = 1850 Mpa$ and $a_2 = 7830 kg/m^3$ (Vitman and Stepanov, 1959; Vitman and Ioffe, 1948), where a_0 and a_2 are "dynamical hardness" and material density of the shield, respectively (for details see Section 6.4). It is assumed that this model describes penetration without friction, and friction is accounted for in devising the computational model. Strictly speaking, since the original model has been developed using experimental data, it takes into account friction (if friction is significant). However, for simulation purposes, it is not important especially because the results of computations show that the obtained conclusions are valid in a wide range of the parameters of the model.

In order to account for the dependence of FC, μ_{fr}, on SV, w, and pressure, p, we used two sets of experimental data obtained by Balakin, 1980 and Montgomery, 1976. These data sets are the most representative among the data sets that are available in the literature although they do not cover the whole range of velocities and pressures

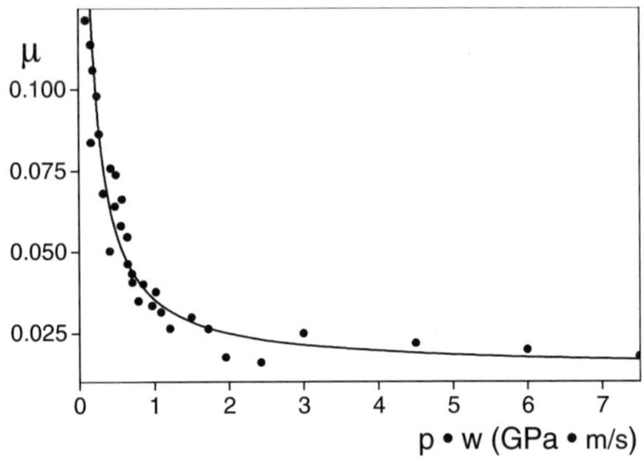

Fig. 14.1.2. FC as a function of SV × pressure. Filled circles denote experimental data of Balakin, 1980.

encountered in high-speed penetration. In describing the dependence of μ_{fr} on w and p we employed the approximation, $\mu_{fr} = \mu(pw)$, where following the suggestion by Montgomery, 1976, the product pw is used as an independent variable. Simulations are performed for conical nose-shaped impactors having half-angle of the apex ϑ equal to $30^\circ, 45^\circ$ and 60°.

In Fig. 14.1.2 we showed the experimental data by Balakin, 1980 and the approximating curve

$$\mu(pw) = \xi_2 + \xi_3/(\xi_1 pw + 1), \tag{14.1.33}$$

where $\xi_1 = 25 \cdot 10^{-9} m \cdot s/N, \xi_2 = 0.014, \xi_3 = 0.556$.

Inspection of Fig.14.1.2 shows that the quality of approximation is good. The limiting values of FC for $w = 0$ and $pw \to \infty$ are equal to $\xi_2 + \xi_3 = 0.57$ and $\xi_2 = 0.014$, respectively.

For illustration, in Fig. 14.1.3 we showed the dependencies of μ_{fr} vs. dimensionless velocity of impactor $\bar{v} = v/v_*$ ($v_* = 1\, m/s$) for strikers with various ϑ; the relationship $pw = (a_2 \sin^2 \alpha v^2 + a_0) v \cos\alpha$ was used ($\alpha = \tan^{-1} \vartheta$). Inspection of Fig. 14.1.3 shows that transition from the maximum value of FC to the minimum asymptotic value occurs for velocities less than $10\, m/s$. Therefore, it can be expected that variation of FC for small velocities of impactor does not affect the accuracy of the model. Indeed, Fig. 14.1.4 confirms this expectation. In this figure we showed the dependence of the performance parameter $\kappa = H_* / H_0$ vs. μ_{fr}, where the penetration depth H_* is determined assuming that coefficient of friction is constant and equals μ_{fr}, and H_0 is the penetration depth determined taking into account that FC varies. Since the magnitude of κ is practically independent of the impact velocity for $v_{imp} > 100 m/s$, the impact velocity is not shown in the figure. With high accuracy it can be assumed that all curves intersect at the point $\mu_{fr} = 0.014$, $\kappa = 1.0$, i.e., introducing the dependence of FC on SV and pressure in the model does not offer advantage over the model with a constant FC equal to its asymptotic value for high velocities.

Experimental data obtained by Montgomery, 1976 have large scatter (see Fig. 14.1.5). These data can be approximated by the following formula:

$$\mu(pw) = \xi_2 + \xi_3 / (\sqrt{\xi_4 pw} + 1) \quad , \tag{14.1.34}$$

where the coefficients ξ_2 and ξ_2 are the same as in Eq. (14.1.33), and $\xi_4 = 67.6 \cdot 10^{-12} m \cdot s / N$. The approximating curve passes through the same limiting points as in the previous case. This simplifies the comparison and does not impair the accuracy of the approximation. Fig.14.1.6 shows that this case differs considerably from the previous one since FC strongly varies in the considered range of impact velocities, $0.1 \div 1000 m/s$. Inspection of Figs. 14.1.7a-c shows that in this case the location of the curves $\kappa = \kappa(\mu)$ depends on the impact velocity, v_{imp}. In Fig. 14.1.8 we showed the dependencies μ_{eq} vs. v_{imp} for different values of α. Here μ_{eq} is the magnitude of FC whereby the depth of penetration for given v_{imp} is the same as in the case with a non-constant FC. Clearly, replacing the non-constant FC by a constant value can result in considerable errors when the shape of the impactor and the impact velocity vary.

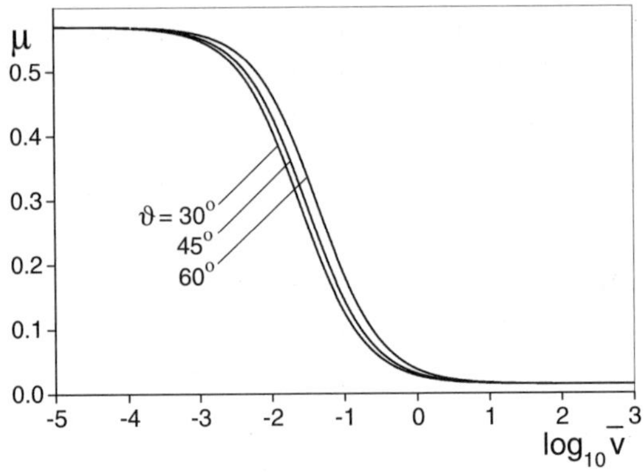

Fig. 14.1.3. FC as a function of velocity of the cone-shaped impactor with half-apex angle ϑ.

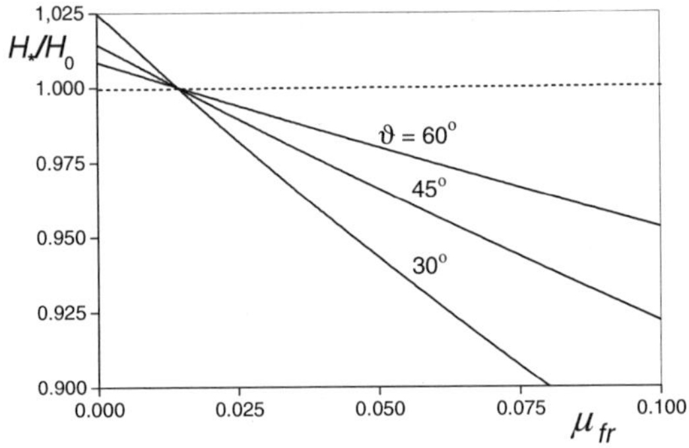

Fig. 14.1.4. Comparison of the DOP, H_0, calculated using a variable FC, and the DOP, H_*, calculated using a constant FC, μ_{fr}.

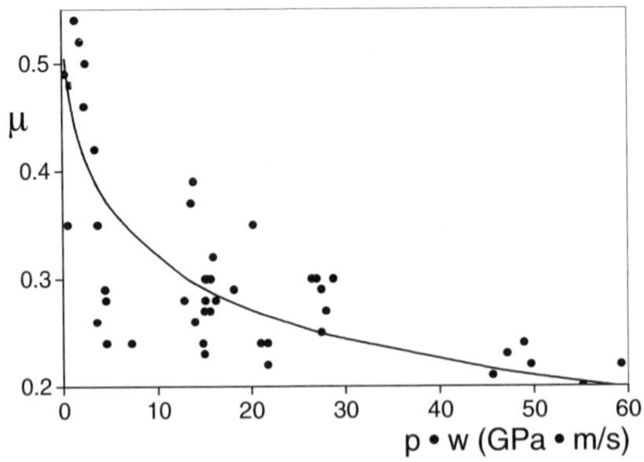

Fig. 14.1.5. FC as a function of SV × pressure. Filled circles denote experimental data of Montgomery, 1976.

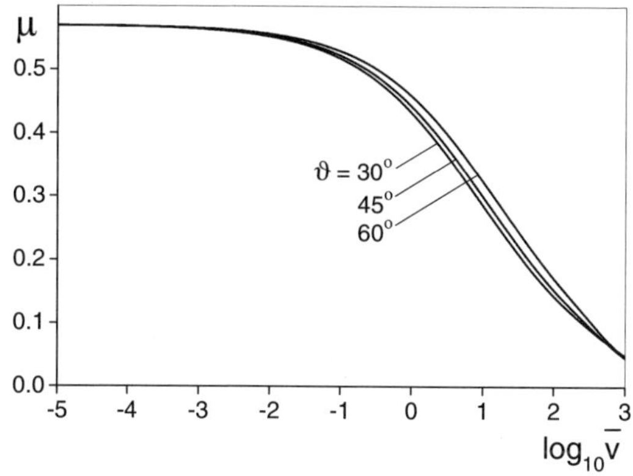

Fig. 14.1.6. FC as a function of velocity of the cone-shaped impactor with half-apex angle ϑ .

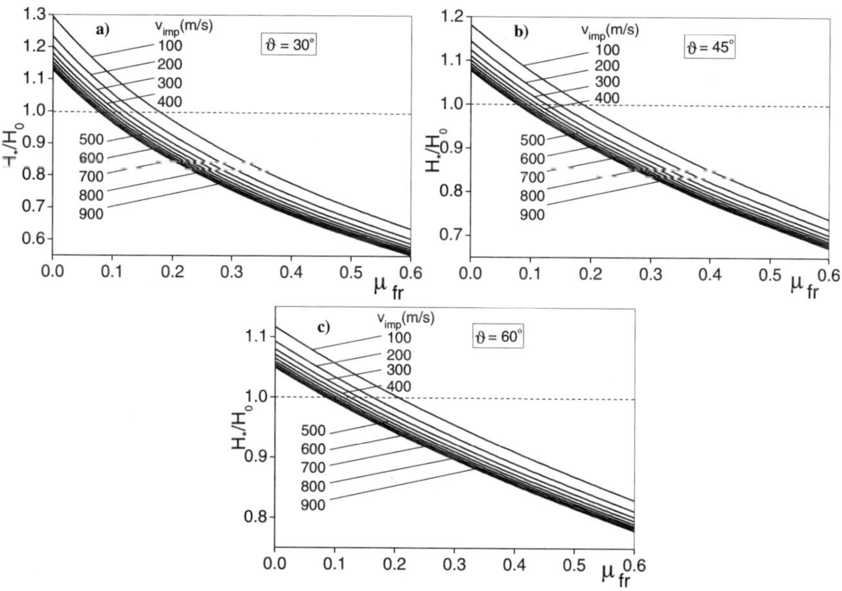

Fig. 14.1.7. Comparison of the DOP, H_0 , calculated using a variable FC, and DOP, H_* , calculated using a constant FC, μ_{fr} .

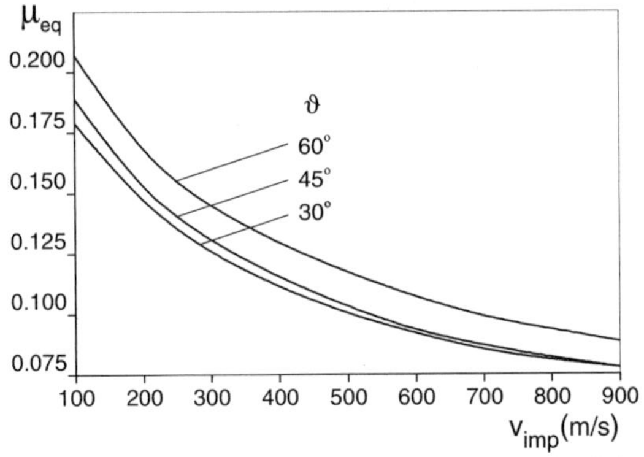

Fig. 14.1.8. Constant FC that provides the same DOP as the variable FC for a given impact velocity.

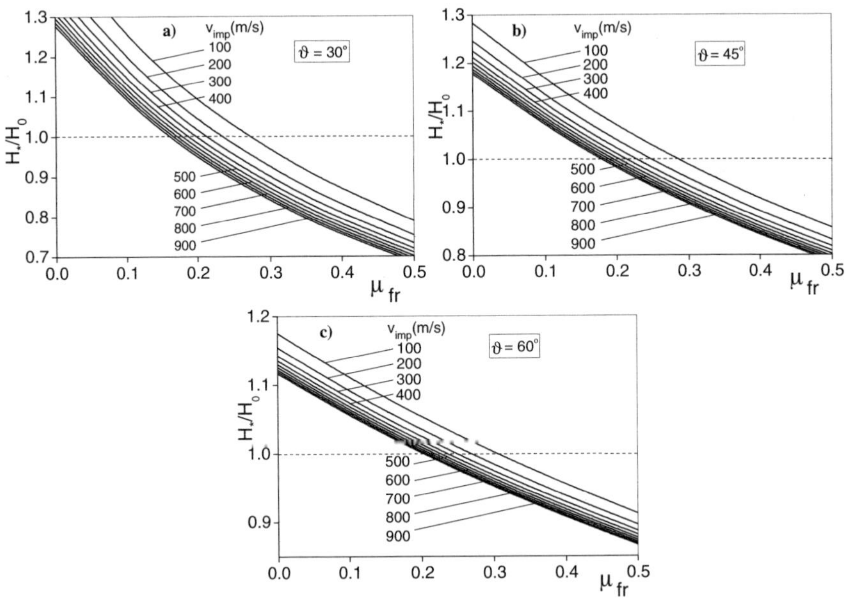

Fig. 14.1.9. Penetration into a concrete shield. Comparison of the DOP, H_0, calculated using the variable FC, and DOP, H_*, calculated using a constant FC, μ_{fr},

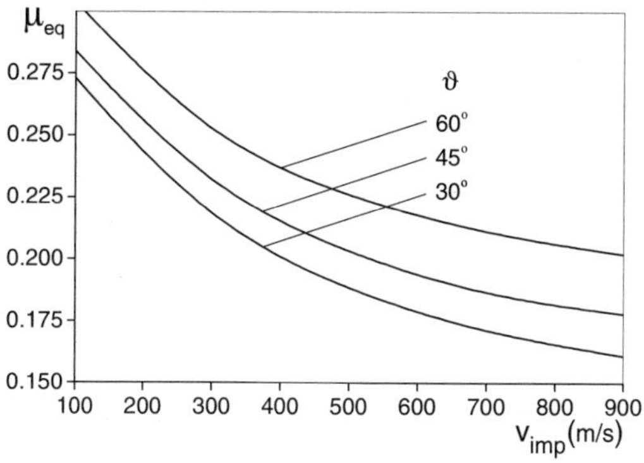

Fig. 14.1.10. Constant FC that provides the same DOP as the variable FC for a given impact velocity

14.1.5.3 *Concrete shield*

In the calculations, we use the model given by Eq. (14.1.16) (cratering is not taken into account) with $a_0 = 100\,MPa$ and $a_2 = 2000\,kg/m^3$ and the following dependence of FC vs. SV (Klepaczko, 2002):

$$\mu(w) = \xi_5[1 - \xi_6 \log_{10}(1+\overline{w})], \quad \overline{w} = w/v_*, \tag{14.1.35}$$

where $\xi_5 = 0.5$ and $\xi_6 = 0.263$. The results of calculations presented in Figs. 14.1.9 and 14.1.10 demonstrate the advantage of taking into account the dependence of FC on velocity.

14.1.5.4 *Concluding remark*

Results of numerical simulations show that taking into account variation of FC has a strong effect on the DOP when strong variation of FC occurs at high velocities that are characteristic for high speed penetration. Therefore, the effect of variable FC as a function of SV and pressure for large penetration velocities and various combinations of shield-striker materials warrants further investigation.

14.2 Shape Optimization of Penetrating Impactors

14.2.1 *Formulation of the problem and method of solution*

Let us consider the problem of maximizing the DOP of an impactor (body of revolution) for a given impact velocity using a general LIM. The model describing impactor/shield interaction is determined by the functions Ω_n and Ω_τ, as well as by parameters depending on the mechanical properties of the shield. Mass of the impactor m, the length L and the shank radius of the nose R are assumed to be given. The problem reduces to the optimization of the functional $\tilde{H}$ in Eq. (14.1.6) where $\tilde{D}(v)$ is given by Eq. (14.1.4). The solution must satisfy the constraints:

$$\Phi(L) = R, \quad \Phi' \geq 0, \quad 0 \leq x \leq L. \qquad (14.2.1)$$

While in the general case this problem cannot be solved analytically, it is possible to suggest a reasonably universal method for the numerical solution of the problem. To this end let us discretize the problem.

The Newton-Cotes formulas for approximate calculation of the integral in Eq. (14.1.6) can be written as follows (Korn and Korn, 1968):

$$\tilde{H} \approx \Delta v \sum_{i=0}^{N} \lambda_i \frac{v_i}{\tilde{D}(v_i)}, \qquad (14.2.2)$$

where $N+1$ is the number of nodes where the values of the integrand are calculated,

$$\Delta v = v_{imp} / N, \quad v_i = i\Delta v, \quad , \quad i = 0,1,\ldots,N. \qquad (14.2.3)$$

The coefficients λ_i depend on the version of the employed Newton-Cotes integration formula. In particular, $\lambda_0 = \lambda_1 = \ldots = \lambda_{N-1} = 1, \lambda_N = 0$ for the rectangular integration formula while $\lambda_0 = \lambda_N = 0.5$, $\lambda_1 = \lambda_2 = \ldots = \lambda_{N-1} = 1$ when the trapezoidal integration formula is used. The appropriate choice of N allows attaining the required accuracy of approximation.

Eqs. (14.1.4), (14.1.5) and (14.2.2) allow representing $\tilde{H}$ as a function of the integral functionals:

$$\tilde{H} = f(J_0, J_1, ..., J_N) = \Delta v \sum_{i=0}^{N} \lambda_i \frac{v_i}{J_i}, \tag{14.2.4}$$

where

$$J_i = \tilde{D}(v_i) = r^2 \Omega_n(1, v_i) + 2\int_0^L \omega(\Phi', v_i)\Phi dx. \tag{14.2.5}$$

An efficient numerical method for solving such problems is the method of local variations (Banichuk *et al.*, 1969; Chernous'ko and Banichuk, 1973). A brief description of this method can be found in Ben-Dor *et al.*, 2006a. For the sake of completeness in the following we summarize the basic concepts of this method. The unknown function $\Phi(x)$ is represented approximately as a piecewise linear function determined by its values in the nodes of interpolation. Hence every integral J_i can be represented as a sum whereby every term depends only on the values of the function $\Phi(x)$ in no more then in two adjacent nodes of interpolation. The essence of the method of local variations is that it determines the rule of sequential variation of the values of the function $\Phi(x)$ in the nodes that maximizes the optimization criterion. The important advantage of the method is that it requires recalculating no more then two terms in the sum when the function varies in any node. Decreasing the function variation increment and increasing the number of nodes of interpolation allow attaining the desired accuracy. Similar to most of the existing numerical methods, the method of local variations allows to find only the local extremum. In order to overcome this shortcoming calculations must be performed with different initial functions, and the obtained results must be compared.

Eqs. (14.2.4) and (14.2.5) imply that the variation of the function in an interior interpolation node requires recalculation of $2(N+1)$ terms in the sums for the integrals $J_0, J_1, ..., J_N$ and calculating the new value of the optimization criterion.

Consider the particular case when the model of impactor-shield interaction is such that the function $\omega(\Phi',v)$ can be written as

$$\omega(\Phi',v)=\sum_{s=1}^{S}\varphi_s(v)\,\psi_s(\Phi'),\qquad(14.2.6)$$

where φ_s, ψ_s ($s=1,2,...,S$) are some functions. In this case Eq. (14.2.5) can be rewritten as follows:

$$J_i=\tilde{D}(v_i)=r^2\Omega_n(1,v_i)+2\sum_{s=1}^{S}\varphi_s(v_i)\,G_s,\quad G_s=\int_0^L\psi_s(\Phi')\Phi dx\,.\qquad(14.2.7)$$

Consequently, variation of the function in an interior interpolation node requires recalculating $2S$ terms in the sums for the integrals $G_1,G_2,...,G_S$ and recalculating the optimization criterion. In this particular case, since usually $S \ll N$, the procedure requires less computational time.

14.2.2 *Mmodel with friction coefficient dependent on sliding velocity*

In the case when the model given by Eq. (14.1.16) and FC depends on the SV, Eq. (14.1.14) yields:

$$\omega(\Phi',v)=(a_2u^2v^2+a_0)[\mu(w)+\Phi']\,.\qquad(14.2.8)$$

Eqs. (14.2.3)–(14.2.5) yield the following expression for the DOP written using the dimensionless variables:

$$H=\frac{m}{\pi}\tilde{H}=t_1\overline{H},\quad \overline{H}=\Lambda(\overline{J}_0,\overline{J}_1,...,\overline{J}_N)=\Delta\overline{v}\sum_{i=1}^{N}\lambda_i\frac{\overline{v}_i}{\overline{J}_i},\qquad(14.2.9)$$

where

$$\bar{J}_i = \bar{r}^2(\bar{v}_i^2 + 1/t_0) + 2\int_0^1 \left(\bar{v}_i^2 \frac{\overline{\Phi'}^2}{\overline{\Phi'}^2+1} + \frac{1}{t_0} \right) \left[\overline{\Phi'} + \mu\left(\frac{v_{imp}\bar{v}_i}{\sqrt{\overline{\Phi'}^2+1}} \right) \right] \overline{\Phi} d\bar{x}\,, \tag{14.2.10}$$

$$t_0 = \frac{a_2}{a_0} v_{imp}^2,\ t_1 = \frac{m}{\pi L^2 a_2},\ \bar{v}_i = \frac{v_i}{v_{imp}} = \frac{i}{N},$$

$$\Delta\bar{v} = \frac{\Delta v}{v_{imp}} = \frac{1}{N},\ \bar{x} = \frac{x}{L},\ \overline{\Phi} = \frac{\Phi}{L},\ \overline{\Phi'} = \frac{d\overline{\Phi}}{d\bar{x}} \tag{14.2.11}$$

Since the theory does not specify the type of the function $\mu(w)$, we use a simple empirical relationship that was proposed by Jones *et al.*, 2003 and Davis, 2003 for modeling high-speed penetration (see Fig. 14.2.1):

$$\mu(w) = \begin{cases} \mu_0 - (\mu_0 - \mu_\infty)\left(\dfrac{w}{w_0} \right) & if \quad 0 \le w \le w_0 \\ \mu_\infty & if \quad w \ge w_0 \end{cases} \tag{14.2.12}$$

where μ_0 and μ_∞ are the values of FC for very small and large SV, respectively, and w_0 is the value of SV when the transition from SV dependent FC to a constant FC occurs.

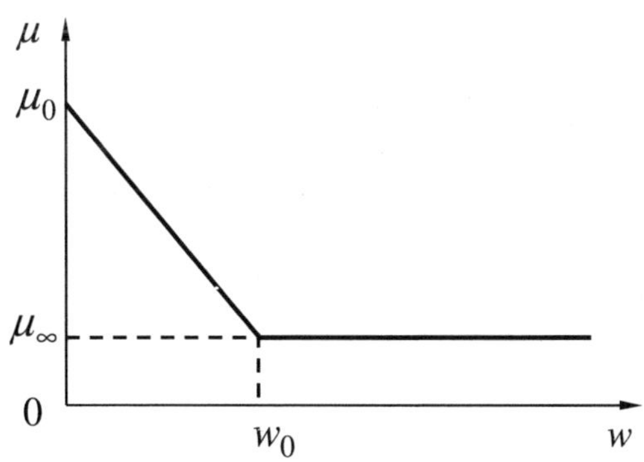

Fig. 14.2.1. Friction coefficient as a function of sliding velocity (Davis, 2003).

Hence the optimization problem in a dimensionless form can be formulated as maximization of the functional $\overline{H}$ subjected to the following constraints:

$$\overline{\Phi}(1) = \tau, \overline{\Phi}'(\overline{x}) \geq 0, \tau\overline{x} \leq \overline{\Phi}(\overline{x}) \leq \tau, 0 \leq \overline{x} \leq 1 , \qquad (14.2.13)$$

where $t_0, t_2 = v_{imp}/w_0$, μ_0, μ_∞ and $\tau = R/L$ are known parameters. The third constraint stipulates that the optimum shape is sought in the range between a sharp cone and a cylinder.

14.2.3 *Numerical results and conclusions*

Davis, 2003 performed calculations for several steels using models determined by Eqs. (14.1.16) and (14.2.12) and compared the obtained results with experimental data for ogive-nosed impactor having the following parameters: $L = 21.06\,mm$, $R = 6.35\,mm$ ($\tau = 0.302$) and mass $m = 65\,g$. The latter parameter was estimated using the plots in Davis, 2003. We conducted the first series of optimization calculations (Figs. 14.2.2 and 14.2.3) for the same maximum sizes (L and R) and mass of the impactor and used the models of Davis, 2003 for Experimental Steel 2 with the following parameters: $a_0 = 473\,MPa$, $a_2 = 1125\,kg/m^3$, $\mu_0 = 0.5770$, $\mu_\infty = 0.0079$, $w_0 = 201\,m/s$ and for AF1410 Steel with the following parameters: $a_0 = 479\,MPa$, $a_2 = 1125\,kg/m^3$, $\mu_0 = 0.5770$, $\mu_\infty = 0.0551$, $w_0 = 479\,m/s$.

Fig. 14.2.2 shows the generatrix of the optimum impactor which coincides with that of a sharp cone for $v_{imp} \leq v^*_{imp}$, where v^*_{imp} is some upper bound value. When $v_{imp} > v^*_{imp}$, increasing v_{imp} results in the increase of the convexity of the optimum generatrix. For large v_{imp}, when friction is not taken into account, the solution for the optimum generatrix recovers the optimum solution of Newton (for details see, e.g., Ben-Dor *et al.*, 2006a). The sharp cone is also the optimum impactor for all considered impact velocities if FC is assumed to be equal to the maximum value, $\mu = \mu_0$.

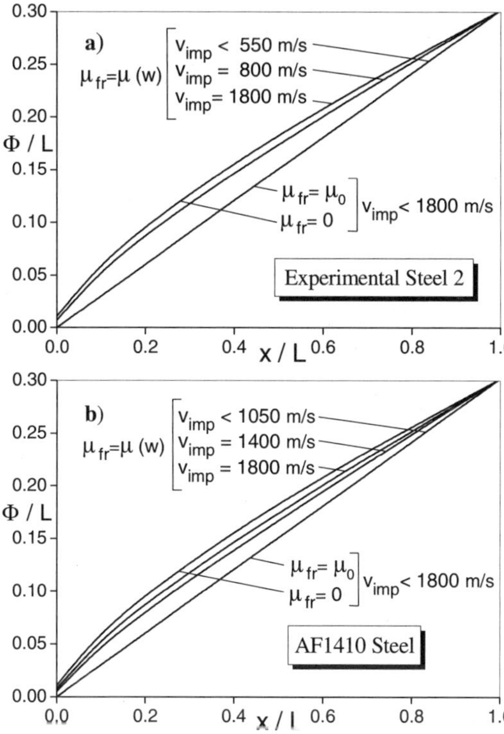

Fig. 14.2.2. Shape of the optimal impactor for various impact velocities.

Figs. 14.2.3a-b allow us to compare the DOP of the optimum impactor with the DOPs of the impactors having other shapes. Inspection of these figures shows that the DOPs of the optimum impactor and of the optimum truncated cone are very close, whereas the DOP of the ogive-nose impactor (the corresponding curve coincides with that from Davis, 2003) is slightly lower than the maximum. Note that the values of the DOP for these shapes are close to the experimental data (open circles) from Davis, 2003 for ogive-nosed impactors. In the case of the model without friction, the DOP is close to the optimum "SV dependent FC" value for Experimental Steel 2 (Fig. 14.2.2a) and is significantly less than for AF1410 Steel. This difference can be explained mainly by the difference in the magnitudes of the parameter w_0 determining the range of velocities where the effect of friction is essential.

Although the results of calculations using real physical parameters of the impactor and the shield are illustrative, more information about the

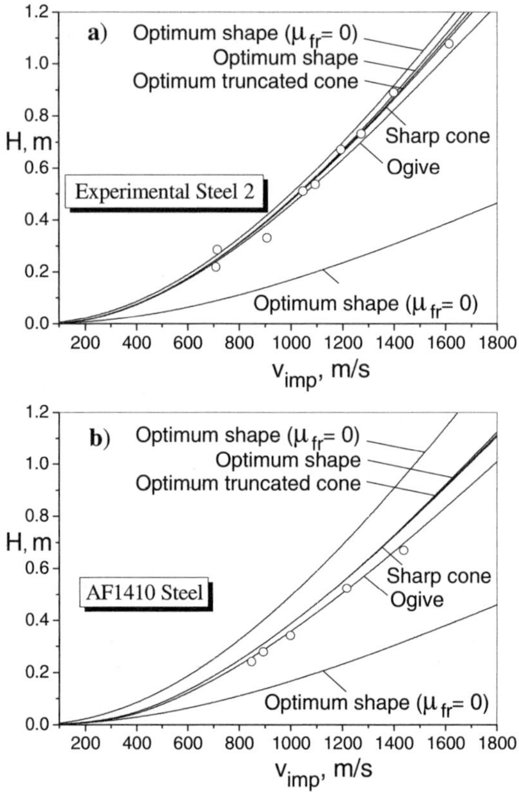

Fig. 14.2.3. Comparison of the DOP for impactors having different shapes; open circles – experimental data for ogive-nosed impactor from Davis, 2003.

optimal solution may be obtained when the problem is investigated using the dimensionless variables. The dimensionless DOP of the optimum impactor (the ratio of the DOP of the optimum impactor to the DOP of a sharp cone with the same maximum size), η, and the function determining the generatrix of the optimum impactor in the dimensionless coordinates, depend on the following dimensionless parameters:

$$t_2 = v_{imp} / w_0 , \quad t_3 = t_0 / t_2^2 = a_2 w_0^2 / a_0 , \tag{14.2.14}$$

μ_0, μ_∞ and τ. Some typical results of numerical calculations are shown in Figs. 14.2.4-14.2.7 for $\mu_0 = 0.5, \mu_\infty = 0.03, \tau = 0.5$, where the

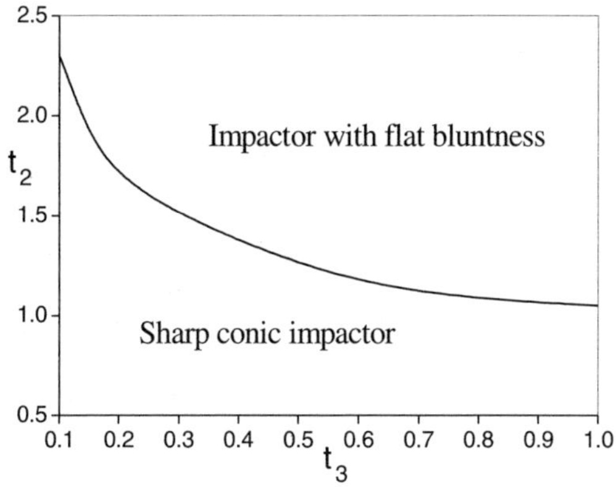

Fig. 14.2.4. Type of the optimum impactor on t_2 and t_3 plane.

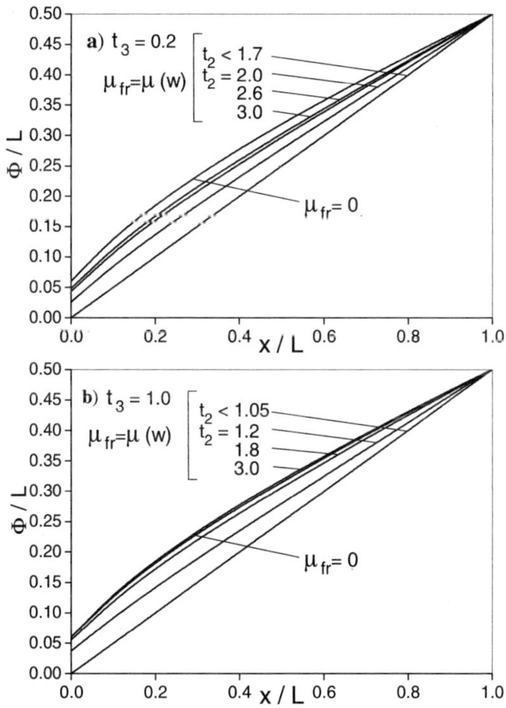

Fig. 14.2.5. Shape of the optimum impactor as a function of the parameter t_2.

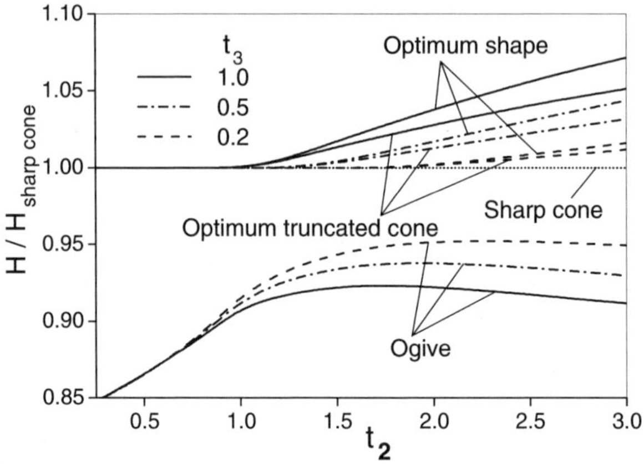

Fig. 14.2.6. Dimensionless DOP (normalized by the DOP of the sharp cone) of the optimum impactor, optimum truncated-conical impactor and ogive-shaped impactor as a function of the parameters t_2 and t_3.

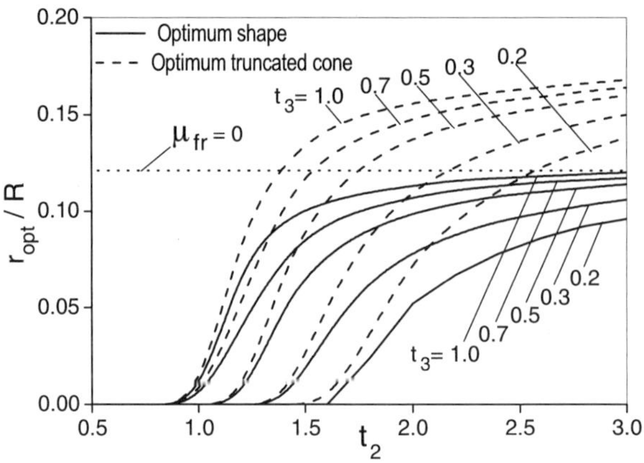

Fig. 14.2.7. Dimensionless radius (normalized by the radius of the shank) of flat bluntness of the optimum impactor and the optimum truncated-conical impactor as a function of parameters t_2 and t_3.

influence of the parameters t_2 and t_3 on the shape of the optimum impactor and its DOP is analyzed.

Analysis shows that the optimum impactor is either a sharp cone or a body with a convex upward generatrix and a flat bluntness. The domains where these two shapes are realized are shown in Fig. 14.2.4. Inspection of Fig. 14.2.4 reveals that for a given material of the shield (given t_3 and w_0), these shapes correspond to relatively low and relatively high impact velocities, v_{imp}, respectively. When v_{imp} increases the radius of the bluntness of the optimum impactor increases, and the curvature of its generatrix is more pronounced (Fig. 14.2.5).

Comparison of the DOP of the optimum impactor with the DOPs of the optimum truncated conical impactor, sharp conical impactor and ogive-nosed impactor is shown in Fig. 14.2.6. Inspection of Fig. 14.2.6 shows that the DOPs of the optimum impactor and of the optimum truncated cone are close, whereas the DOP of the ogive nose impactor is considerably smaller. Radius of the bluntness of the optimum truncated conical impactor is always larger than the radius of the bluntness of the optimum impactor (Fig. 14.2.7).

All numerical calculations for determining the optimum shape of the impactor have been performed with the accuracy that guarantees the error of the magnitude of the optimization criterion less than 0.1%. One calculation requires several Central Processing Unit minutes of IBM PC. Consequently, the suggested adaptation of the method of local variations is applicable and very convenient for solving the problems of shape optimization of penetrating impactors using engineering models. Moreover, this modification of the method of local variations is valid for the more general functionals. In particular, the method is applicable when the optimization criterion can be written as an integral over a parameter with the integrand depending on this parameter and several integrals which in their turn depend on the parameter and on the unknown function.

Chapter 15

Semi-Infinite Concrete Shields

Nomenclature

$a_0,, a_1, a_2$	parameters determining the model, Eq. (4.1.55);
$\tilde{a}_0,, \tilde{a}_1, \tilde{a}_2$	parameters in Table 4.1.1;
DOP	depth of penetration;
E_0, E_1, E_2	parameters in Eq. (15.1.5);
f_c'	unconfined compressive strength;
h_0	depth of crater in the two-stage model;
H	depth of penetration (DOP);
$\overline{H}$	$= H / h_0$;
H^{opt}	DOP of the optimal impactor;
L	length of the nose of the impactor, Fig. 1.1.2;
m	mass of projectile;
r	radius of flat bluntness of the impactor nose;
$\bar{r}$	$= r/L$;
R	shank radius of impactor;
v_{imp}	impact (initial) velocity of projectile;
x	coordinate associated with impactor, Fig .1.1.2;
$\beta_0, \beta_1, \beta_3$	parameters, Eq. (15.1.1);
λ	$= m/h_0$;
μ_{fr}	friction coefficient;
ξ	parameter, Eq. (15.1.1);
ρ	coordinate associated with impactor, Fig .1.1.2;
ρ_{sh}	material density of shield;
τ	$= R/L$;

$\overline{\Phi}(\overline{x})$ function determining the shape of impactor using dimensionless coordinates x/L and ρ/L.

15.1 Penetration Model

We consider normal penetration of a rigid striker (a body of revolution) into a concrete semi-infinite shield and use the $2\,\&\,3$ model (see Section 4.1.2.1) in order to describe impactor-shield interaction, where $h_0 = 4R$, R is shank radius of the impactor and a_i is determined by Eq. (4.1.55).

It is convenient to use the dimensionless variables. Let us introduce parameters β_0, β_1 and β_2:

$$\beta_0 = \frac{a_0}{\xi v_{imp}^2}, \quad \beta_1 = \frac{a_1^2}{a_0 a_2}, \quad \beta_2 = \frac{a_2}{\xi}, \quad \xi = \frac{m}{\pi h_0 L^2}. \tag{15.1.1}$$

Hence

$$a_0 = \xi v_{imp}^2 \beta_0, \quad a_1 = \xi v_{imp}\sqrt{\beta_0\beta_1\beta_2}, \quad a_2 = \xi\beta_2. \tag{15.1.2}$$

Substituting a_0, a_1 and a_2 from Eqs. (15.1.2) into Eqs. (1.6.10), (1.6.15), (1.7.3) for A_i and Eq. (1.7.4) for B_i yields the following formulas:

$$B_0 = \lambda\overline{r}^2 v_{imp}^2 \beta_0, \; B_1 = \lambda\overline{r}^2 v_{imp}\sqrt{\beta_0\beta_1\beta_2}, \; B_2 = \lambda\overline{r}^2\beta_2, \tag{15.1.3}$$

$$A_i = \lambda v_{imp}^{2-i} E_i, \quad i = 0,1,2, \tag{15.1.4}$$

where

$$E_0 = \beta_0 J_0, \; E_1 = \sqrt{\beta_0\beta_1\beta_2}\, J_1, \; E_2 = \beta_2 J_2, \; i = 0,1,2, \tag{15.1.5}$$

$$\overline{r} = \frac{r}{L}, \; \lambda = \frac{m}{h_0}, \; J_i = \overline{r}^2 + 2\int_0^1 \frac{\overline{\Phi}\,\overline{\Phi}_{\overline{x}}^{\,i}(\overline{\Phi}_{\overline{x}} + \mu_{fr})}{(\overline{\Phi}_{\overline{x}}^2 + 1)^{i/2}} d\overline{x}, \; i = 0,1,2, \tag{15.1.6}$$

m is mass of the impactor, r and L are the radius of the flat bluntness and the length of the nose of the impactor, correspondingly, the function

$\overline{\Phi}(\overline{x})$ determines the shape of the impactor in dimensionless coordinates x/L and ρ/L.

Using Eqs. (15.1.4) and (15.1.3) for A_i and B_i, Eq. (4.1.17) yields the following expression for the dimensionless DOP, $\overline{H}$:

$$\overline{H} = \frac{H}{h_0} = 1 + K(\overline{v}_0, E_0, E_1, E_2), \tag{15.1.7}$$

where the function K is determined by Eqs. (1.7.10)-(1.7.13),

$$\begin{aligned}\overline{v}_0 &= \frac{v_0}{v_{imp}} \\ &= \frac{\sqrt{E_1^2 + 4(E_2+1)[1 - E_0 - \overline{r}^2(\beta_2 + \sqrt{\beta_0\beta_1\beta_2} + \beta_0)]} - E_1}{2(E_2+1)}.\end{aligned} \tag{15.1.8}$$

Let us analyze the obtained relationships. Eq. (15.1.8) implies the following constraint:

$$1 - E_0 - \overline{r}^2(\beta_2 + \sqrt{\beta_0\beta_1\beta_2} + \beta_0)] > 0, \tag{15.1.9}$$

that affords positive value of $\overline{v}_0$.

Hereafter we consider the case $\mu_{fr} = 0$. Hence

$$\begin{aligned}J_0 &= \overline{r}^2 + 2\int_0^1 \overline{\Phi}\,\overline{\Phi}_x d\overline{z} = \overline{r}^2 + \int_0^1 \frac{d\overline{\Phi}^2}{d\overline{x}} d\overline{x} = \tau^2, \\ E_0 &= \beta_0\tau^2, \quad \tau = \frac{R}{L},\end{aligned} \tag{15.1.10}$$

where inequality given by Eq. (15.1.9) can be rewritten as follows:

$$\beta_0\tau^2 + \overline{r}^2(\beta_2 + \sqrt{\beta_0\beta_1\beta_2} + \beta_0)] < 1. \tag{15.1.11}$$

In particular, if $\overline{r} = 0$ (sharp impactor) Eq. (15.1.11) yields the following simple condition:

$$\beta_0\tau^2 < 1. \tag{15.1.12}$$

Since our goal is shape optimization, in the general case we require that Eq. (15.1.11) must be satisfied for all $0 \le \bar{r} \le \tau$ in order to allow variation of $\bar{r}$. Because the left hand side of Eq. (15.1.11) increases when $\bar{r}$ increases, after substituting $\bar{r} = \tau$ Eq. (15.1.11) can be rewritten as follows:

$$\tau^2(2\beta_0 + \sqrt{\beta_0\beta_1\beta_2} + \beta_2) < 1. \tag{15.1.13}$$

In the case of the model determined by Eq. (4.1.55), formulas for β_0, β_1 and β_2 in Eq. (4.1.55) read:

$$\beta_0 = \frac{f_c'\tilde{a}_0}{\xi v_{imp}^2}, \quad \beta_1 = e^2, \quad \beta_2 = \frac{\rho_{sh}\tilde{a}_2}{\xi}, \quad e = \frac{\tilde{a}_1}{\sqrt{\tilde{a}_0\tilde{a}_2}}, \tag{15.1.14}$$

where f_c' and ρ_{sh} are unconfined compressive strength and density of the shield material, respectively, and parameters $\tilde{a}_i$ are given in Table 4.1.1. Using Eqs. (15.1.14) we can rewrite Eq. (15.1.13) as follows:

$$(\tau\sqrt{\beta_2})^2 + e(\tau\sqrt{\beta_0})(\tau\sqrt{\beta_2}) + 2(\tau\sqrt{\beta_0})^2 - 1 < 0. \tag{15.1.15}$$

The solution of the inequality (15.1.15) reads:

$$\begin{cases} 0 < \tau^2\beta_0 < 0.5 \\ 0 < \tau^2\beta_2 < 0.25[\sqrt{(e^2-8)\tau^2\beta_0 + 4} - \tau\sqrt{\beta_0}]^2 \end{cases} \tag{15.1.16}$$

For the compressible, elastic-cracked-plastic model (see Table 4.1.1) when

$$\tilde{a}_0 = 3.45, \quad \tilde{a}_1 = 1.6, \quad \tilde{a}_2 = 1.12, \quad e = 0.814, \tag{15.1.17}$$

the domain determined by Eq. (15.1.16) is shown in Fig. 15.1.1.

In the case of the 2&2 model (Section 4.1.2.5) when

$$a_0 = 82.6\left(\frac{10^6}{f_c'}\right)^{0.544} f_c', \quad a_1 = 0, \quad a_2 = \rho_{sh}, \tag{15.1.18}$$

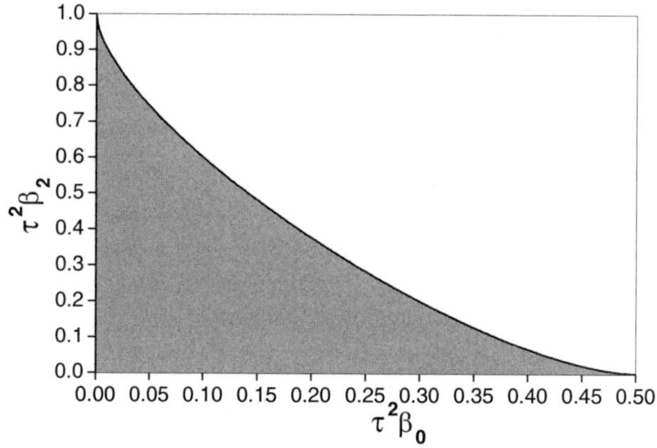

Fig. 15.1.1. Domain of variation of parameters β_0, β_2 and τ.

we obtain that

$$\beta_0 = \frac{82.6Y(10^6/Y)^{0.544}}{\xi v_{imp}^2}, \quad \beta_1 = 0, \quad \beta_2 = \frac{\rho_{sh}}{\xi} \tag{15.1.19}$$

and Eq. (15.1.12) must be satisfied.

15.2 Shape Optimization of Impactor

We consider the problem of determining the shape of the impactor by maximizing the DOP into a semi-infinite concrete shield. Parameters describing mechanical properties of the material of the shield, mass of the impactor and geometry of the nose are assumed to be given. Using the dimensionless variables, the problem reduces to optimization of the functional $\overline{H}$ in Eq. (15.1.7) with given parameters $\beta_0, \beta_1, \beta_2$ and τ. The solution, function $\overline{\Phi}(\overline{x})$, must satisfy the following constraints:

$$\overline{\Phi}(1) = \tau, \quad \overline{\Phi}_x(\overline{x}) \geq 0, \quad \tau\overline{x} \leq \overline{\Phi}(\overline{x}) \leq \tau, \quad 0 \leq \overline{x} \leq 1. \tag{15.2.1}$$

The constraint $\tau\overline{x} \leq \overline{\Phi}$ is introduced for eliminating the shapes that are of no practical significance, e.g. needles.

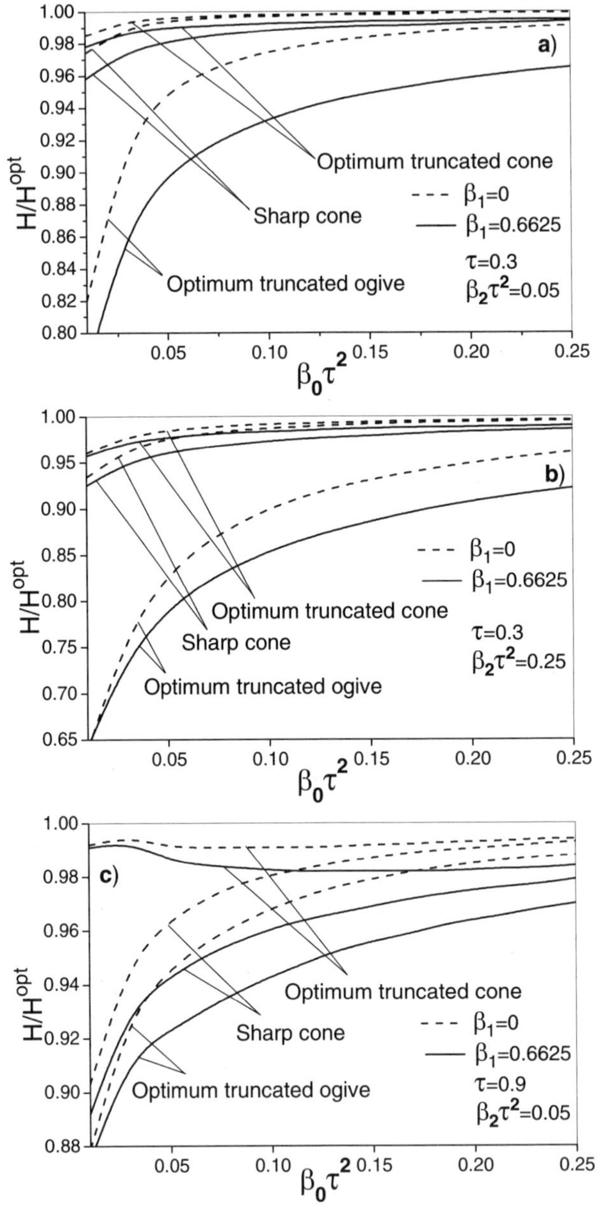

Fig. 15.2.1. Depth of penetration of sharp cone, optimal truncated cone and truncated ogive impactors normalized by the DOP of the optimal impactor vs. $\beta_0\tau^2$.

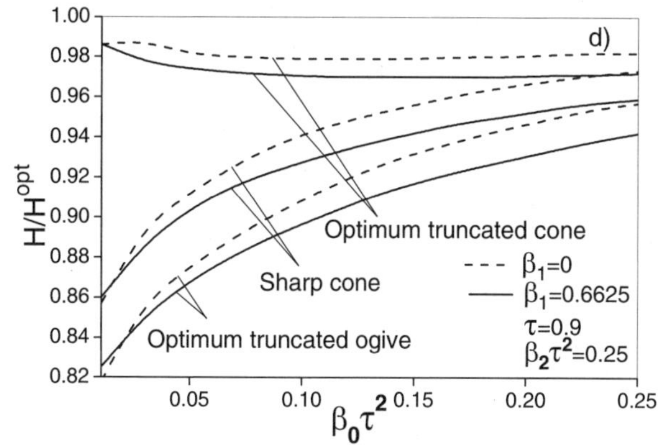

Fig. 15.2.1. Continued.

The solution of the above variational problem can be found using method of local variations (Banichuk *et al.*, 1969). Calculations were conducted for the described above *2&3* model (Eqs. (15.1.14) and (15.1.17)) and *2&2* model of Teland and Sjøl, 2004 rewritten for arbitrary bodies of revolution (Section 4.1.2.5), $\mu_{fr} = 0$. Here the characteristics of *2&3* and *2&2* models are presented in compliance with the classification scheme of the models that is suggested in Section 4.1.1.

Representative results of numerical calculations are shown in Figs. 15.2.1-15.2.3 where solid and dashed curves denote $2\,\&\,3$ model and $2\,\&\,2$ model, correspondingly.

The behavior of the performance parameter, $\eta = H/H^{opt}$ vs. β_0, β_2 and τ, is shown in Figs. 15.2.1a-d, where H is the DOP of the sharp cone, optimal truncated cone or optimal truncated ogival impactor and H^{opt} is the DOP of the optimal impactor. Inspection of these figures shows that η is very close to 1 for optimal truncated cone (OTC). Taking into account the approximate character of the adopted penetration model, it is conceivable to consider the OTC as a quasi-optimal impactor. This is very convenient because the shape of the truncated cone is determined only by two dimensionless parameters, the radius of flat bluntness, $r_{tr\,cone}$, and τ. The effect of parameters β_0 and β_2 on the optimal value of $r_{tr\,cone}$ is shown in Fig.15.2.2 for $\tau = 0.6$ and $\tau = 0.9$; for small τ, OTC is close to a sharp cone.

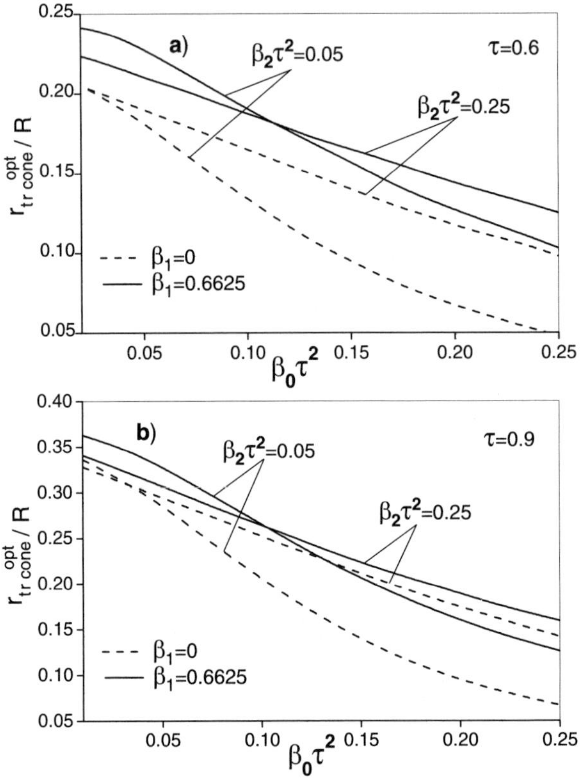

Fig. 15.2.2. Normalized radius of bluntness of the optimal truncated cone vs. $\beta_0\tau^2$.

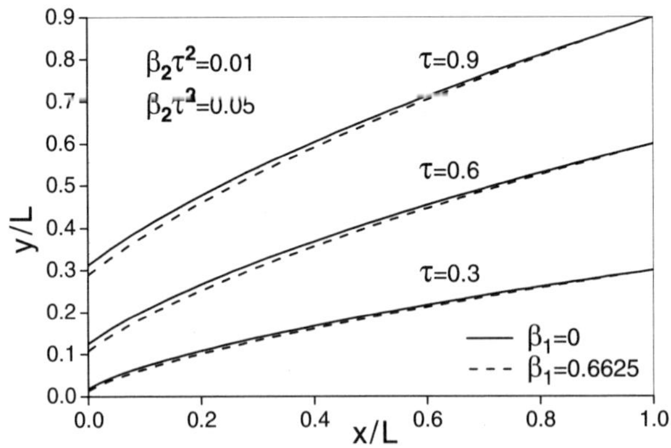

Fig. 15.2.3. Optimum shapes of impactors.

The choice of the particular model does not cause qualitative changes in results of shape optimization. Indeed, variation of the optimal nose radius of OTC with increase of $\beta_0\tau^2$ (see Fig. 15.2.2) is balanced by the reduced sensitivity of the DOP to the deviation of the nose radius from the optimal value. Consider, e.g., the case with $\beta_0\tau^2 = 0.15$, $\beta_2\tau^2 = 0.05$ and $\tau = 0.6$ (see Fig. 15.2.2a) whereby the dimensionless optimal nose radii, calculated using *2&3* model and *2&2* model, are 0.16 and 0.10, respectively. At the same time, variation of the dimensionless nose radius from 0 to 0.40 results in deviation of the DOP from the corresponding value for OTC by less than 3% for both models.

Fig. 15.2.3 illustrates the shapes of the optimal impactors.

Chapter 16

Metal Shields Having a Finite Thickness

Nomenclature

a_0, a_2	parameters determining the model, Eq. (1.4.2);
b	shield thickness;
$\bar{b}$	$= b/L$;
BLV	ballistic limit velocity;
D_{lat}	resistance of the lateral part of the nose of impactor;
f	function, Eqs. (16.1.9);
g	function, Eqs. (16.1.9);
h	instantaneous depth of penetration, Fig. 6.8.1,
$\bar{h}$	$= h/L$;
$\bar{h}_i$	$= i\Delta\bar{h}$;
I_0, I_2	integrals, Eq. (16.1.2);
k_*	parameter, Eq. (16.1.9);
L	length of nose of impactor, Fig. 6.8.1a;
m	mass of projectile;
m_*	parameter, Eq. (16.1.9);
N	number of intervals;
r	radius of flat bluntness of projectile nose;
$\bar{r}$	$= r/L$;
v	instantaneous velocity of projectile;
v_{bl}	BLV of projectile;
$\bar{v}_{bl}$	$= v_{bl}/\sqrt{w_*}$;
v_{bl}^{cone}	BLV of conical nose projectile;

v_{bl}^{cyl}	BLV of cylindrical projectile;
w	$=v^2=\dot{h}^2$;
$\overline{w}$	$=w/w_*$;
$\overline{w}_{bl}^{cone}$	value of $\overline{w}_{bl}$ for sharp cone;
$\overline{w}_{bl}^{cyl}$	value of $\overline{w}_{bl}$ for cylinder;
w_*	$=a_0/\rho_{sh}$;
$\overline{w}_{bl}$	$=\overline{v}_{bl}^2$;
x	coordinate associated with impactor, Fig. 6.8.1a;
$\overline{x}$	$=x/L$;
$\alpha_i^{(N)}$	coefficients depending on the type of the Newton-Cotes formula;
δ	function associated with description drag of projectile nose, Eq. (1.3.9);
$\Delta\overline{h}$	$=(\overline{b}+1)/N$;
θ, Θ	functions determining the impactor-shield interaction contact surface, Eqs. (1.2.2) and (1.2.4);
μ_{fr}	friction coefficient;
ρ	
ρ_{sh}	material density of shield;
τ	$=R/L$;
τ_s	shear strength of shield material;
τ_*	parameter, Eq. (16.1.9);
τ_*^{div}	value of the parameter τ_* when $\overline{w}_{bl}^{cone}=\overline{w}_{bl}^{cyl}$, Eqs. (16.3.2), (16.3.4);
φ	functions in Eq. (16.1.9);
Φ	function determining the shape of impactor, Fig. 6.8.1a;
$\overline{\Phi}$	$=\Phi/L$;
$\overline{\Phi}_{\overline{x}}$	$=d\overline{\Phi}/d\overline{x}$.

16.1 Formulation of the Problem and Mathematical Model

Consider a high speed normal penetration of a rigid body of revolution into a ductile (metal) shield having thickness b. We assume that projectile-shield interaction is described by the model presented in Section 6.8.1 (the notations are shown in Fig. 6.8.1) which must be supplemented by specifying the relationship for the resistance force D_{lat} at the lateral surface of the impactor.

Let us select the two-term model of impactor-shield interaction given by Eq. (1.4.2). Hence $D_{lat}(h,\dot{h})$ is determined by Eq. (1.4.4), where Eqs. (1.4.5) and (1.4.7) are used for calculating f_0 and f_2, the term associated with the resistance of flat bluntness is omitted, and Eqs. (1.2.2) and (1.2.4) are used for calculating Θ and θ.

It is convenient to write the expression for $D_{lat}(h,\dot{h})$ in the following form ($\mu_{fr}=0$):

$$D_{lat}(h,\dot{h}) = 2\pi(a_0 I_0 + a_2 I_2 \dot{h}^2), \qquad (16.1.1)$$

where

$$I_\mu(h) = \int_{\theta(h;b,L)}^{\Theta(h;b,L)} G_\mu(\Phi,\Phi_x)dx = \int_0^L \eta(x,h;b,L)\, G_\mu(\Phi,\Phi_x)dx, \qquad (16.1.2)$$

$$G_\mu(\Phi,\Phi_x) = \Phi\Phi_x[\Phi_x^2/(\Phi_x^2+1)]^{\mu/2}, \quad \mu = 0,2, \qquad (16.1.3)$$

$$\eta(x,h;b,L) = \begin{cases} 0 & if \quad 0 \le x \le \theta(h;b,L) \\ 1 & if \quad \theta(h;b,L) \le x \le \Theta(h;b,L) \\ 0 & if \quad \Theta(h;b,L) \le x \le L \end{cases}, \qquad (16.1.4)$$

h is the instantaneous depth of penetration, function $\Phi(x)$ determines the generatrix of the impactor nose having length L. Parameters b and L are listed as arguments of functions Θ and θ.

Therefore D_{lat} can be written in the form of Eq. (6.8.6) with

$$\chi_0(h) = 2\pi a_0 I_0, \quad \chi_1(h) = 2\pi a_2 I_2, \quad \chi_2(h) = 0. \qquad (16.1.5)$$

Taking into account Eqs.(6.8.1) and (6.8.2), we can rewrite Eq. (6.8.7) as follows:

$$\begin{aligned} &[m + \pi\rho_{sh} r^2 b\delta(h;b)]\frac{dw}{dh} \\ &+ 4\pi[a_2 I_2(h) w + a_0 I_0(h) + \tau_s r(b-h)\delta(h;b)] = 0, \end{aligned} \qquad (16.1.6)$$

where ρ_{sh} and τ_s are density and shear strength of the material of the shield, correspondingly, r is the radius of flat bluntness of a projectile, function δ is determined in Eq. (1.3.9) (b is included in the list of arguments) and a new function $w = v^2 = \dot{h}^2$ of the independent variable h is introduced.

Using the dimensionless variables

$$\bar{h} = \frac{h}{L}, \quad \bar{w} = \frac{w}{w_*}, \quad w_* = \frac{a_0}{\rho_{sh}}, \tag{16.1.7}$$

Eq. (16.1.6) can be rewritten as follows:

$$\frac{d\bar{w}}{d\bar{h}} + f(\bar{h})\bar{w} + g(\bar{h}) = 0, \tag{16.1.8}$$

where

$$f(\bar{h}) = \frac{k_* \bar{I}_1(\bar{h})}{\varphi(\bar{h})}, \; g(\bar{h}) = \frac{\tau_* \bar{r}(\bar{b} - \bar{h})\,\delta(\bar{h};\bar{b}) + \bar{I}_0(\bar{h})}{\varphi(\bar{h})},$$

$$\varphi(\bar{h}) = 0.25[m_* + \bar{b}\bar{r}^2\,\delta(\bar{h};\bar{b})], \quad \bar{r} = \tau - \bar{J},$$

$$\bar{I}_\mu(\bar{h}) = \int_0^1 \eta(\bar{x},\bar{h};\bar{b},1)\,G_\mu(\bar{\Phi},\bar{\Phi}')d\bar{x}, \quad \bar{J} = \int_0^1 \bar{\Phi}' d\bar{x}, \tag{16.1.9}$$

$$\bar{\Phi} = \frac{\Phi}{L}, \bar{\Phi}' = \frac{d\bar{\Phi}}{d\bar{x}}, \bar{x} = \frac{x}{L}, \bar{b} = \frac{b}{L}, \tau = \frac{R}{L}, \bar{r} = \frac{r}{L},$$

$$k_* = \frac{a_1}{\gamma_{sh}}, \tau_* = \frac{\tau_{sh}}{a_0}, m_* = \frac{m}{\pi \gamma_{sh} L^3},$$

m is mass of the impactor.

Eq. (16.1.8) yields the following general expression for the dimensionless ballistic limit velocity (BLV):

$$\bar{v}_{bl}^2 = \bar{w}_{bl} = \int_0^{\bar{b}+1} g(\bar{h})\exp\left(\int_0^{\bar{h}} f(\bar{H})d\bar{H}\right)d\bar{h}, \quad \bar{v}_{bl} = \frac{v_{bl}}{\sqrt{w_*}}. \tag{16.1.10}$$

For the particular cases of the sharp cone nose impactor and a cylindrical impactor with a flat bluntness, Eq. (16.1.10) yields the following formulas:

$$\overline{w}_{bl}^{cone} = \frac{\tau^2+1}{k_*\tau^2}\left[\exp\left(\frac{2k_*\tau^4\overline{b}}{(\tau^2+1)m_*}\right)-1\right], \quad \overline{\Phi}(\overline{x}) = \tau\overline{x}, \qquad (16.1.11)$$

$$\overline{w}_{bl}^{cyl} = \frac{2\tau_*\tau\overline{b}^2}{m_*+\overline{b}\tau^2}, \quad \overline{\Phi}(\overline{x}) = \tau. \qquad (16.1.12)$$

Let us consider the problem of minimizing the BLV. The model describing impactor-shield interaction, parameters determining mechanical properties of the shield, mass of the impactor, the length and the shank radius of the nose of the projectile are assumed to be given. Using the dimensionless variables, the problem reduces to determining the non-decreasing function $\overline{\Phi}(\overline{x})$ that satisfies the following constraints:

$$\overline{\Phi}(1) = \tau,\ \tau\overline{x} \le \overline{\Phi}(\overline{x}) \le \tau,\ \overline{\Phi}'(\overline{x}) \ge \frac{\tau-\overline{\Phi}(\overline{x})}{1-\overline{x}} \ge 0,\ 0 \le \overline{x} < 1, \qquad (16.1.13)$$

and minimizes the functional in Eq. (16.1.10) whereas parameters b, τ, k_*, τ_* and m_* are assumed to be known. The second constraint in Eqs. (16.1.13) stipulates that the optimum shape is sought between the sharp cone and the cylinder. The third inequality represents the necessary (but not sufficient) condition of convexity upward of the generatrix: the tangent at every point of the generatrix must be located not lower than the straight segment between this point and the end point of the generatrix. Hereafter we assume that parameter a_2 has the meaning of material density of the shield, i.e., $k_* = 1$.

16.2 Method of Solution

The functional $\overline{w}_{bl}[\overline{\Phi}(\overline{x})]$ in Eq. (16.1.10) is quite involved and standard methods of minimization cannot be applied directly. We will reduce the problem of minimization of this functional to a variational problem for the functional that is a function of some integral functionals. The latter problem can be solved using an efficient numerical method of local variations (Banichuk *et al.*, 1969; Chernousko and Banichuk, 1973).

The Newton-Cotes formulas for approximate calculation of the external integral in Eq. (16.1.10) can be written as follows (Korn and Korn, 1968):

$$\overline{w}_{bl} \approx \Delta\overline{h}\sum_{i=0}^{N}\alpha_i^{(N)} g(\overline{h}_i)\exp\left(\int_0^{\overline{h}_i} f(\overline{H})d\overline{H}\right), \qquad (16.2.1)$$

where $N+1$ is the number of nodes in which the values of the integrand are calculated,

$$\Delta\overline{h} = (\overline{b}+1)/N, \quad \overline{h}_i = i\Delta\overline{h}, \quad , \quad i = 0,1,...,N. \qquad (16.2.2)$$

Coefficients $\alpha_i^{(N)}$ depend on the type of the Newton-Cotes formula. In particular, $\alpha_0^{(N)} = \alpha_1^{(N)} = ... = \alpha_{N-1}^{(N)} = 1$, $\alpha_N^{(N)} = 0$ for the rectangular integration rule formula, and $\alpha_0^{(N)} = \alpha_N^{(N)} = 0.5$, $\alpha_1^{(N)} = \alpha_2^{(N)} = ... = \alpha_{N-1}^{(N)} = 1$ when the trapezoidal integration rule formula is used.

The integral in Eq. (16.2.1) can be approximated in a similar manner and Eq. (16.2.1) can be rewritten as follows:

$$\overline{w}_{bl} \approx \Delta\overline{h}\sum_{i=0}^{N}\alpha_i^{(N)} g(\overline{h}_i)\exp\left(\Delta\overline{h}\sum_{j=0}^{i}\alpha_j^{(i)} f(\overline{h}_j)\right), \qquad (16.2.3)$$

where for simplicity we used the same nodes in both sums. The appropriate choice of N allows attaining the required accuracy of the approximation.

Inspection of Eq. (16.1.9) and Eq. (16.2.3) shows that the functional $\overline{w}_{bl}[\overline{\Phi}(\overline{x})]$ is, generally, a function of $(2N+1)$ integrals, $\overline{I}_0(\overline{h}_1)$, $\overline{I}_0(\overline{h}_2), ..., \overline{I}_0(\overline{h}_N)$; $\overline{I}_2(\overline{h}_1), \overline{I}_2(\overline{h}_2), ..., \overline{I}_2(\overline{h}_N)$ and $\overline{J}$, taking into account that $\overline{I}_0(0) = \overline{I}_2(0) = 0$.

16.3 Numerical Results. Discussion and Conclusions

Similar to most existing numerical methods, the method of local variations allows determining only the local extrema. In order to overcome this shortcoming, calculations must be performed with different initial functions, and the obtained results must be compared. In our calculations, we use the following initial functions:

$$\overline{\Phi}(\overline{x}) = \xi + (\tau - \xi)\overline{x}^{\varepsilon}, \quad \xi \le \tau \tag{16.3.1}$$

with different ξ and ε.

In most calculations performed using these initial functions, iterations converge to one of the following three functions describing 1) sharp cone, 2) cylinder and 3) a convex downward generatrix that is very close to the generatrix of a sharp cone and yields practically the same BLV. Therefore, the minimum value of the BLV is attained for the sharp cone if $\overline{w}_{bl}^{cone} < \overline{w}_{bl}^{cyl}$ or for the cylinder if $\overline{w}_{bl}^{cone} > \overline{w}_{bl}^{cyl}$. If $\overline{w}_{bl}^{cone} = \overline{w}_{bl}^{cyl}$, then both shapes yield the same BLV. Using Eq. (16.1.11) and Eq. (16.1.12) one can determine the value of the parameter τ_*^{div} such that the condition $\overline{w}_{bl}^{cone} = \overline{w}_{bl}^{cyl}$ is satisfied:

$$\tau_*^{div} = \frac{(\tau^2+1)(m_* + \overline{b}\tau^2)}{2\overline{b}^2\tau^3}\left[\exp\left(\frac{2\tau^4\overline{b}}{(\tau^2+1)m_*}\right) - 1\right], \quad k_* = 1. \tag{16.3.2}$$

Clearly, sharp cone or cylinder is an optimum shape if $\tau_* > \tau_*^{div}$ or $\tau_* < \tau_*^{div}$, respectively. For the relatively large m_*,

$$m_* \gg \overline{b}\tau^2, \quad 2\tau^4\overline{b} \ll (\tau^2+1)m_*, \tag{16.3.3}$$

Eq. (16.3.2) can be simplified by keeping only two terms in Taylor series expansion for the exponent:

$$\tau_*^{div} = \tau/\overline{b}\,. \tag{16.3.4}$$

the variation of τ_*^{div} as a function of $\overline{b}$ and m_* for $\tau = 0.25$ and $\tau = 0.5$. Generally, the shape of the optimum impactor depends on τ_*, $\overline{b}$, m_* and τ but if $\tau_* \le \tau/\overline{b}$, then cylinder is the optimum shape for impactor of any mass. Fig. 16.3.2 compares the BLV of the impactors with sharp conical, flat (cylinder) and ogive-shaped noses. The values of the BLV in these figures, v_{bl}, are normalized by the BLV of the conical-nosed impactor, v_{bl}^{cone}.

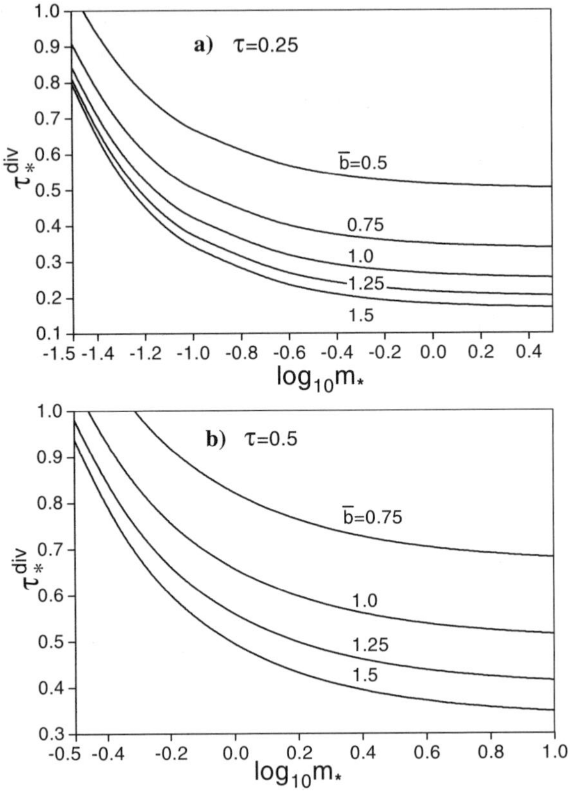

Fig. 16.3.1. Dependencies τ_*^{div} vs. m_* for different $\overline{b}$.

The obtained solution has a clear physical meaning in some limiting cases. If $\tau_* \to 0$, then the BLV of the cylinder $v_{bl}^{cyl} \to 0$ whereas the BLV of any non-cylindrical impactor is positive since energy is required for expanding the cavity in the shield during penetration. Clearly, the advantage of the cylinder is retained for small τ_* . On the other hand, for large τ_*, perforation with plugging becomes not advantageous, and sharp-nosed projectile is an impactor having the optimum shape under these conditions.

In summary, we demonstrated that i) impactor with the optimum sharp nose is a cone and ii) the minimum BLV is attained if the nose of the impactor is conical or the impactor is a cylinder. The choice of the optimum shape is performed by comparing the BLVs of these two impactors.

Verification of these results requires analysis of large sets of data obtained in experiments and numerical simulations. Each cycle of experiments/simulations must be conducted for the same shields varying the shapes of the impactors (including cylindrical and conical impactors) while keeping R, L and m constant. To the best of our knowledge only a few series of such calculations and experiments were reported in the literature.

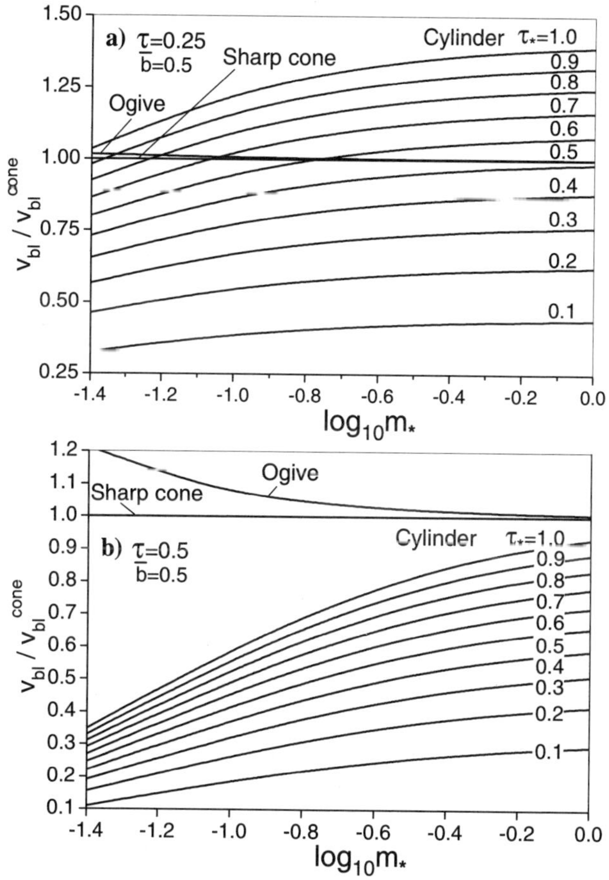

Fig. 16.3.2. Comparison of the BLVs of impactors with conical, flat and ogive-shaped noses.

Analysis of these results allows us to draw a conclusion that they do not contradict our findings. Børvik *et al.*, 2002a, 2002b studied experimentally and numerically perforation of the thick steel plate by the projectiles having the same mass and diameter with flat, semispherical and conical noses. It was found that the BLV of cylinder is considerably smaller than the BLVs of two other impactors. Dey *et al.*, 2004 published the results of experiments and computer simulations on perforation of shields manufactured from three structural steels by projectiles with flat, conical and ogival noses. It appears that the cylinder is the impactor with the minimum BLV. In order to comply with the constraint that all projectiles have the same length of the nose, we considered the cylindrical part of the semispherical-nose impactor as a part of the nose.

Chapter 17

Fiber-Reinforced Plastic Laminates

Nomenclature

A	parameter, Eqs. (17.2.3);
b	thickness of shield;
$\bar{b}$	$=b/R$;
B	parameter, Eqs. (17.2.3)
BLV	ballistic limit velocity;
C	parameter, Eqs. (17.2.7);
D	drag force acting on projectile;
FRP	fiber-reinforced plastic;
h	instantaneous depth of penetration, Fig. 1.1.2;
J	functional in Eq. (17.3.2)
L	length of the nose of impactor, Fig. 1.1.2;
m	mass of projectile;
$\bar{m}$	$=m/(\pi R^2 L\gamma)$;
$\bar{r}$	$=r/R$;
$\bar{r}^{opt}$	value of $\bar{r}$ for optimum impactor;
R	shank radius of impactor;
S	radius of generatrix of ogive-shaped nose of impactor;
u	cosine of the angle between the local inner normal vector to projectile surface and vector in the direction that is opposite to the impactor velocity;
$v_{bl}^{blunt\ cone}$	BLV of blunt cone;
$v_{bl}^{sharp\ cone}$	BLV of sharp cone;
v_{imp}	impact (initial) velocity of projectile;
v	velocity of projectile;

v_{bl}	ballistic limit velocity (BLV) of projectile;
v_{res}	residual velocity of projectile;
β	empirical parameter depending on the shape of impactor;
β_*	value of β in the model by He *et al.*, 2007a;
$\beta_\times, \beta^\times$	range of variation of β;
δ	function in formula for a drag force acting on impactor nose, Eqs. (1.3.6) and (1.3.9);
η	permissible relative deviation of the BLV of impactor from the maximum value;
θ, Θ	functions determining impactor-shield interaction surface area, Eqs. (1.2.2) and (1.2.4);
ϑ	semi-vertex angle of cone-nosed impactor;
λ	$= L/R$;
μ	$= \sqrt{\rho_{sh}/\sigma_e}$;
μ_{fr}	friction coefficient;
ρ_{sh}	material density of shield;
σ_e	quasi-static linear elastic limit of the material of shield;
φ	functions in Eq. (17.2.3);
Φ	function determining the shape of impactor, Fig.1.2.2;
Φ_x	$= d\Phi/dx$;
Ω_n, Ω_τ	functions determining the model, Eq. (17.2.1).

17.1 Introduction

In order to formulate and solve variational problem of shape optimization of impactors a mathematical model that yields the optimized characteristics of penetration is required. Commonly these shape dependent characteristics are either the depth of penetration (DOP) or ballistic limit velocity (BLV). Clearly, analytical models (Zhu *et al.*, 1992b; Reid and Wen, 2000; Wen, 2000, 2001; Babu *et al.*, 2007), which allow determining the resistance force for penetration only for strikers having certain characteristic shapes, cannot be used directly for projectile shape optimization.

For lack of models which are valid for impactors having an arbitrary shape, Ben-Dor *et al.*, 2002a, 2002b, 2002c suggested using the well known in aerodynamics phenomenological approach, the tangent cones method. In the case of normal penetration of strikers having a shape of a body of revolution, this approach can be described as follows. A striker is embedded inside a virtual cone such that in the cross-section, which is normal to the axis of the striker, these two bodies are tangent. It is assumed that in all points of this cross-section the forces acting on both bodies are the same. This approach allows determining the resistance force acting at the impactor having a shape of an arbitrary body of revolution using the model for a conical impactor whereby the parameter of the model is an apex angle of the cone. Ben-Dor *et al.*, 2002a, 2002b, 2002c selected the model suggested in Reid and Wen, 2000; Wen, 2000, 2001 as the model for conical impactors. The latter model is based on the assumption that the local pressure at the surface of the impactor exerted by the shield can be expressed as a sum of a quasi-static resistive pressure due to the elastic–plastic deformations and dynamic resistive pressure depending on the impact velocity.

Using a modification of this model, Ben-Dor with coauthors solved a number of variational problems and determined the optimum shapes impactor (bodies of revolution) for a shield having a finite thickness (analytical solution, Ben-Dor *et al.*, 2002a) and for a semi-infinite shield (numerical solution, Ben-Dor *et al.*, 2002b). It was shown that the optimum impactor having the minimum BLV has a flat bluntness and its BLV is very close to that for the optimum blunt cone impactor, i.e., a quasi-optimum strikers can be selected among the blunt cones. A comparison of impactors with different shapes showed that the optimum impactor has a significant advantage over sharp-cone shaped and ogive-shaped impactors. Extension of the tangent cones method approach to 3-D impactors was suggested by Ben-Dor *et al.*, 2002c.

He *et al.*, 2007a, 2007b, 2008 suggested a modified model that is also valid for cone-shaped and ogive-shaped impactors but, in contrast with Reid and Wen, 2000; Wen, 2000, 2001, accounts for the effect of the instantaneous velocity of penetrating impactor on the resistance force.

Ben-Dor *et al.*, 2009c used this model as a starting point for investigating problems of shape optimization of penetrating impactors. The dependence of the local pressure on shape of impactor is determined by two parameters. The first parameter depends only upon the angle between the unit normal vector to the impactor surface and the impactor velocity while the second parameter, β, depends on the shape of impactor. This is the reason that the model based on tangent cones method yields the resistance force of the ogive-nosed impactor, different from the resistance force which is determined by direct application of the model by He *et al.*, 2007a, 2007b, 2008. Ben-Dor *et al.*, 2009c investigated the choice of the parameter β using theoretical considerations and numerical calculations. Their analysis is based on the available extensive experimental data. Using this analysis allows determining the optimal shapes of impactors in the class of truncated cone-shaped bodies. This chapter is based on the results obtained by Ben-Dor *et al.*, 2009c.

17.2 Mathematical Model

He *et al.*, 2007a, 2007b, 2008 suggested the following localized interaction model for normal penetration of a conical-nosed and ogive-nosed projectiles into fiber-reinforced plastic (FRP) laminate shields:

$$\Omega_n(u,v) = \sigma_e + \sqrt{\rho_{sh}\sigma_e}\,\beta uv, \quad \Omega_\tau(u,v) = 0\,(\mu_{fr} = 0), \qquad (17.2.1)$$

where Ω_n and Ω_τ are functions that determine normal and tangent components of the local force exerted by a shield on a projectile; ρ_{sh} and σ_e are density and quasi-static linear elastic limit of the shield material, respectively, u is cosine of the angle between the inner normal vector to the projectile surface and the vector in the direction opposite to the direction of impactor velocity; v is the instantaneous velocity of projectile; μ_{fr} is friction coefficient; β is an empirical parameter depending on the shape of impactor.

The parameter β is assumed to be $\beta = 4 + 1.5\sin\vartheta$ for a conical-nosed impactor (ϑ is semi-vertex angle of the cone) and $\beta = 2 + 2R/S$

for an ogive-nosed impactor, where S is the radius of the generatrix and R is radius of the shank of impactor. Since this model is applicable only for two shapes of impactors and implies different values of the local resistance pressure for different locations with the same value of the parameter u, it cannot be used directly for solving shape optimization problems.

Let us consider the case of cone-shaped impactors. The relationship between the impact velocity, v_{imp}, and the residual velocity, v_{res}, given by Eq. (1.3.30) for the non-averaged localized interaction model that is determined by Eq. (17.2.1) can be readily obtained after integration:

$$\varphi(v_{imp}, v_{res}, B) = A, \tag{17.2.2}$$

where m is mass of projectile,

$$\varphi(v_{imp}, v_{res}, B) = \int_{v_{res}}^{v_{imp}} \frac{v dv}{1+Bv} = \frac{v_{imp} - v_{res}}{B} - \frac{1}{B^2}\ln\left(\frac{Bv_{imp}+1}{Bv_{res}+1}\right),$$
$$A = \frac{\pi b R^2 \sigma_e}{m}, \quad B = \beta \sin\vartheta\sqrt{\frac{\rho_{sh}}{\sigma_e}}. \tag{17.2.3}$$

Equation (17.2.2) is similar to the corresponding relationship in He *et al.*, 2007a but is derived differently.

Equation (17.2.2) can be used for determining B when v_{imp} and v_{res} (or, in a particular case, the BLV, v_{bl}) are known from the experiments. Since

$$\frac{\partial\varphi}{\partial B} = -\int_{v_{res}}^{v_{imp}} \frac{v^2 dv}{(1+Bv)^2} < 0,$$
$$\varphi(v_{imp}, v_{res}, 0) = \int_{v_{res}}^{v_{imp}} v dv = \frac{v_{imp}^2 - v_{res}^2}{2} > 0, \quad \varphi(v_{imp}, v_{res}, +\infty) = 0, \tag{17.2.4}$$

Eq. (17.2.2) has one root if

$$v_{imp}^2 - v_{res}^2 > 2A, \tag{17.2.5}$$

and does not have roots if Eq. (17.2.5) is not valid.

The results of our calculations are shown in Fig. 17.2.1 where we have processed summarized in He *et al.*, 2007a results of 56 experiments by Zhu *et al.*, 1992a, 1992b; Wen *et al.*,1998; Mines *et al.*,1999; Ulven *et al.*, 2003. Two outliers and one point where the condition in Eq. (17.2.5) is not valid were excluded from the analysis. Inspection of Fig. 17.2.1 and a low value of the correlation coefficient between β and $\sin\alpha$, 0.13, are indicative of the absence of the pronounced linear dependence between these parameters.

The results of calculations performed with different β for conical nose impactors are shown in Fig. 17.2.2 where the asterisk denotes values of $\beta = \beta_*$ that correspond to the model He *et al.*, 2007a and symbols denote the experimental data (references are given in the figures captions). Clearly, the choice $\beta = \beta_*$ does not always provides the best fit between the model and experimental data while most of the experimental data on v_{res} and v_{bl} (definition of the ballistic limit velocity, v_{bl}, can be found in Section 1.3.2) fall in the band which is determined by the model given by Eq. (17.2.1) where β varies in the range from $\beta_\times = 3$ to $\beta^\times = 6$.

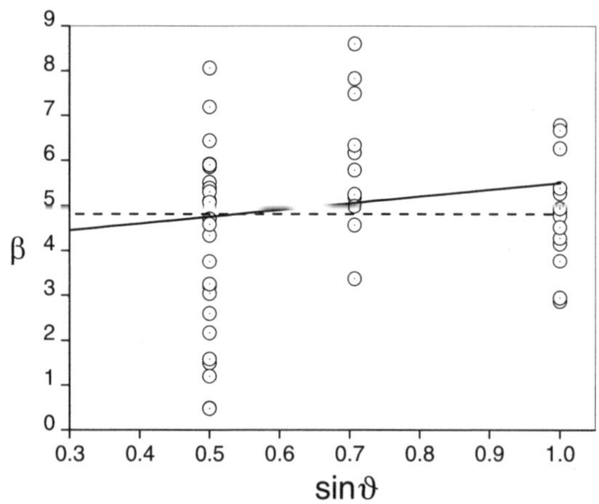

Fig. 17.2.1. Values of β calculated from the experimental data from Zhu *et al.*, 1992a, 1992b; Wen *et al.*, 1998; Mines *et al.*, 1999; Ulven *et al.*, 2003 vs. sinus of the half-angle of the cone apex. Solid line – model by He *et al.*, 2007a, dash line – averaged β.

Consider now an ogive-nosed impactor having the generatrix determined by function $\Phi(x)$ in the coordinates shown in Fig. 1.1.2:

$$\Phi(x) = \sqrt{S^2 - (x-L)^2} - C, \qquad (17.2.6)$$

where

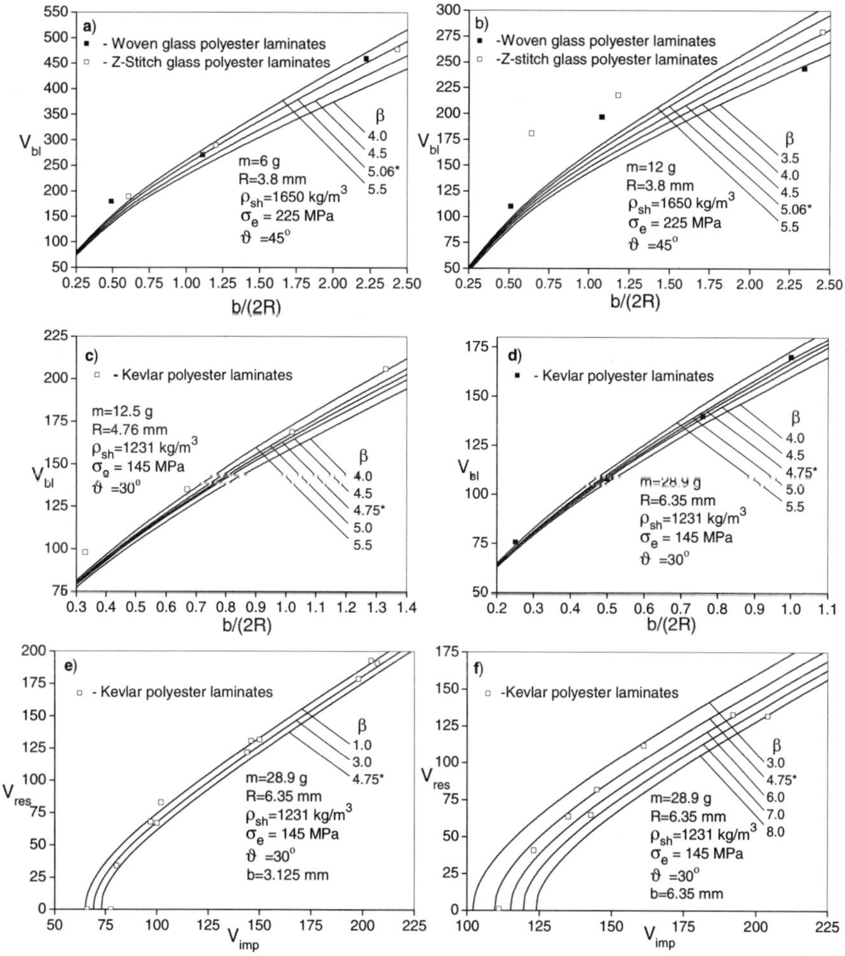

Fig. 17.2.2. Comparison of the calculated ballistic characteristics for sharp cone-shaped impactors with the experimental data from a), d), e), f) Zhu *et al.*,1992b; b), h), i) Mines *et al.*,1999; c) Zhu *et al.*,1992a; g), j) Ulven *et al.*, 2003.

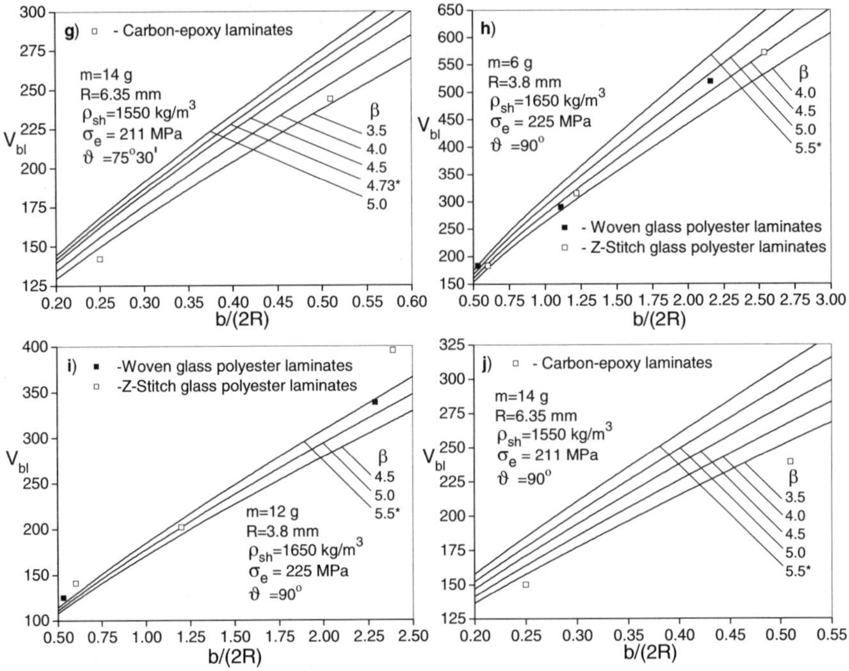

Fig. 17.2.2. Continued.

$$S = 0.5(L^2 + R^2)/R, \quad C = 0.5(L^2 - R^2)/R. \qquad (17.2.7)$$

For the model determined by Eq. (17.2.1) after substituting $\Phi(x)$ from Eq. (17.2.6), the expression for a drag force $D(h,v)$ in Eq. (1.3.25) can be written as follows:

$$D(h,v) = \pi\sigma_e S^2[\Xi(v,\Theta(h;b,L)) - \Xi(v,\theta(h;b,L))]. \qquad (17.2.8)$$

Here Eqs. (1.2.2) and (1.2.4) are used for calculating functions Θ and θ; the thickness of the shield, b, and the length of the nose of impactor, L, are listed as arguments of these functions and

$$\Xi(v,x) = (\sqrt{1-z^2} - \hat{C})^2 + \beta\mu v[(2/3)z^3 - \hat{C}\sin^{-1} z + \hat{C}z\sqrt{1-z^2}],$$
$$z = (x-L)/S, \quad \hat{C} = C/S, \quad \mu = \sqrt{\rho_{sh}/\sigma_e}. \qquad (17.2.9)$$

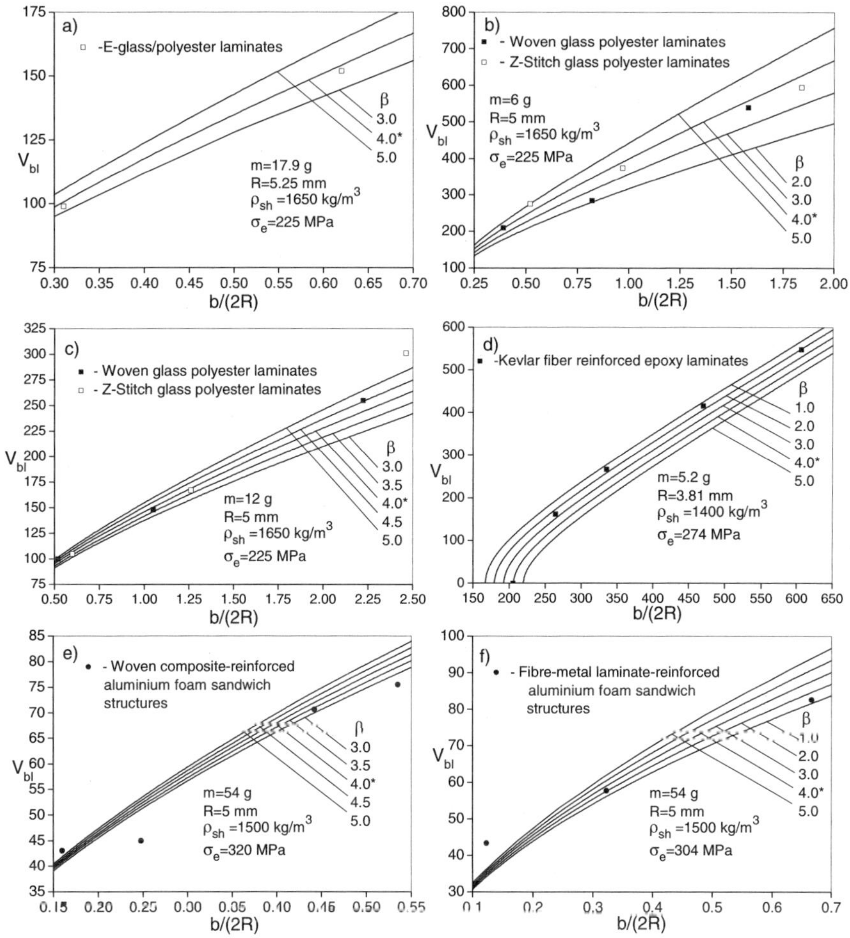

Fig. 17.2.3. Comparison of the calculated ballistic characteristics for hemispherical nosed impactors with the experimental data from a) Reid *et al.*,1999; b), c) Mines *et al.*,1999; d) Wang *et al.*, 2005a; e), f) Reyes-Villanueva and Cantwell, 2004.

The values of BLVs obtained by numerical solution of the equation of impactor motion with the resistance force, $D(h,v)$, determined by Eq. (17.2.8), are shown in Fig. 17.2.3 for different β, where asterisk denotes $\beta = \beta_*$ that corresponds to the model by He *et al.*, 2007b. Experimental data for hemispherical nosed impactors ($C = 0$) obtained by Mines *et al.*, 1999; Reid *et al.*, 1999; Wang *et al.*, 2005a and summarized by He *et al.*, 2007b together with some additional experimental data from Reyes-Villanueva and Cantwell, 2004 are shown in Fig. 17.2.3. Inspection of these figures shows that situation in this case is similar to that for cone-nosed impactors.

Analysis of Fig. 17.2.1, Fig. 17.2.2 and Fig. 17.2.3 shows that without a considerable loss of accuracy it is possible to neglect the dependence of β on the shape of the impactor.

17.3 Shape Optimization of Impactors

17.3.1 *Classical solution on the basis of averaged model*

Consider the problem of minimizing the BLV in a "traditional" formulation on the basis of the averaged model. The model describing impactor-shield interaction, parameters determining the mechanical properties of the shield, the mass of the impactor, the length and the shank radius of the nose of the projectile are assumed to be given.

For $\Omega_n(u,v)$ given by Eq. (17.2.1), Eq. (1.5.11) can be written as follows:

$$\int_0^{v_{bl}} \frac{v dv}{1+\beta\mu J[\overline{\Phi}(\overline{x})]v} = \frac{\pi b \sigma_e R^2}{m}, \tag{17.3.1}$$

where

$$J[\overline{\Phi}(\overline{x})] = \overline{r}^2 + 2\int_0^{\lambda} \frac{\overline{\Phi}\dot{\overline{\Phi}}^2 d\overline{x}}{\sqrt{\dot{\overline{\Phi}}^2+1}}, \tag{17.3.2}$$

$$\overline{r} = r/R, \quad \overline{\Phi} = \Phi/R, \quad \overline{x} = x/R, \quad \lambda = L/R. \tag{17.3.3}$$

Considering Eq. (17.3.1) as the dependence between the optimization criterion, v_{bl}, and the functional J depending on the unknown function $\overline{\Phi}(\overline{x})$ it can readily assessed that the minimum v_{bl} is attained when $J[\overline{\Phi}(\overline{x})]$ is minimum. Consequently, the considered optimization problem is reduced to minimizing the functional *J*. The latter result implies the following three conclusions. 1) One can use directly the results obtained in Ben-Dor *et al.*, 2002a where the variational problem for the functional J had been solved, without studying the problem of shape optimization of impactors in the "averaged" version. 2) Ben-Dor *et al.*, 2002a showed that the optimum impactor with the minimum BLV has a flat bluntness and its shape is very close to that of the optimum blunt cone impactor. In view of these reasons, we consider below the problem of shape optimization of impactors in a non-classical formulation for blunt cones. 3) In the framework of the averaged model, the optimal shape of the impactor is determined only by the given ranges for the sizes of the impactor and does not depend on the properties of the shield material, on its thickness and on the parameter β which is assumed to be the same for all compared impactors.

17.3.2 *Interval optimal solution for truncated cones*

Let us consider now exact formulation of shape optimization problem for impactors having truncated cone noses without averaging the model. Parameters $\sigma_e, \gamma, b, m, R, L, \eta, \beta_\times, \beta^\times$ are assumed to be given. The problem can be formulated as follows: to determine the interval of the bluntness radius, r, $[r_{\inf}^{opt}, r_{\sup}^{opt}]$, such that for an arbitrary r in this interval, the deviation of the BLV of the impactor from the maximum BLV does not exceed $100 \cdot \eta\%$ for any value of β (β is assumed the same for all the considered impactors) from the interval $[\beta_\times, \beta^\times]$. This formulation of the problems allows taking into account the ambiguity in the choice of β. Note in passing that the suggested formulation of shape optimization problem is related with the interval analysis (Moore, 1979).

In the case of a truncated cone, the generatrix of the impactor is given by the following formula:

$$\Phi(x) = r + \tan \vartheta\, x . \tag{17.3.4}$$

For $\Phi(x)$ determined by Eq. (17.3.4) and Ω_n given by Eq. (17.2.1), equation of the impactor motion, Eq. (1.3.13), can be written in the following form:

$$-\bar{m}\lambda\bar{v}(d\bar{v}/d\bar{h}) \\ =\bar{r}^2(1+\beta\bar{v})\bar{\delta}+(1+\beta\bar{v}\sin\vartheta)\tan\alpha[\bar{\Theta}-\bar{\theta}]2\bar{r}+\tan\vartheta(\bar{\Theta}+\bar{\theta}), \tag{17.3.5}$$

where

$$\bar{\theta}=\theta(\bar{h};\bar{b},\lambda), \bar{\Theta}=\Theta(\bar{h};\bar{b},\lambda), \bar{\delta}=\delta(\bar{h};\bar{b}), \\ \bar{h}=h/R, \bar{b}=b/R, \ \bar{v}=\mu v, \bar{m}=m/(\pi R^2 L\rho_{sh}), \tag{17.3.6}$$

$\delta(h;b)$ is determined by Eq. (1.3.9); and b is listed as an argument. The dimensionless BLV, $\bar{v}_{bl}=\mu v_{bl}$, is determined as the solution of Eq. (17.3.5) with the initial condition $\bar{v}(\bar{b}+\bar{L})=0$ for $\bar{h}=0$.

The problem of shape optimization of the truncated cone-nosed impactors reduces to determining the optimal radius of bluntness, $\bar{r}^{opt}$. Generally, $\bar{r}^{opt}$ depends on $\bar{m},\bar{b},\beta$ and λ.

Our calculations show that the BLV of a blunt cone-nose impactor, $v_{bl}^{blunt\ cone}$, only weakly depends on the bluntness radius, $\bar{r}$, in the vicinity of the minimum for every practically relevant set of parameters $\bar{m},\bar{b},\beta$ and λ. Typical result of calculations are shown in Fig. 17.3.1 for $\lambda=1$, where $v_{bl}^{sharp\ cone}$ is the BLV of the corresponding sharp cone. In these figures we showed only those segments of the curves for which the BLV of a blunt cone-nosed impactor deviates by less than 1% from its optimum value. Selecting the optimum radius of the cone bluntness for the averaged model, $\bar{r}_*^{opt}$, instead of the optimum radius, $\bar{r}^{opt}$, implies that the difference in BLV is less than 1% for all considered $\bar{m},\bar{b}$ and β.

From the engineering point of view it is not theoretically optimal value of $\bar{r}^{opt}$ that is of interest but rather the interval for $\bar{r}$ where the magnitude of the deviation of the BLV from the optimal value is small. Mathematical definition of this interval is elucidated in the following.

Denote the left and the right end points of this interval by $\bar{r}_{\min}^{opt}(\bar{m},\bar{b},\lambda,\beta,\eta)$ and $\bar{r}_{\max}^{opt}(\bar{m},\bar{b},\lambda,\beta,\eta)$, where η denotes the deviation of the BLV from the optimal value, $v_{bl}^{opt}(\bar{m},\bar{b},\lambda,\beta)$. Then for all $\bar{r}$ (and only for these values of $\bar{r}$) in the interval

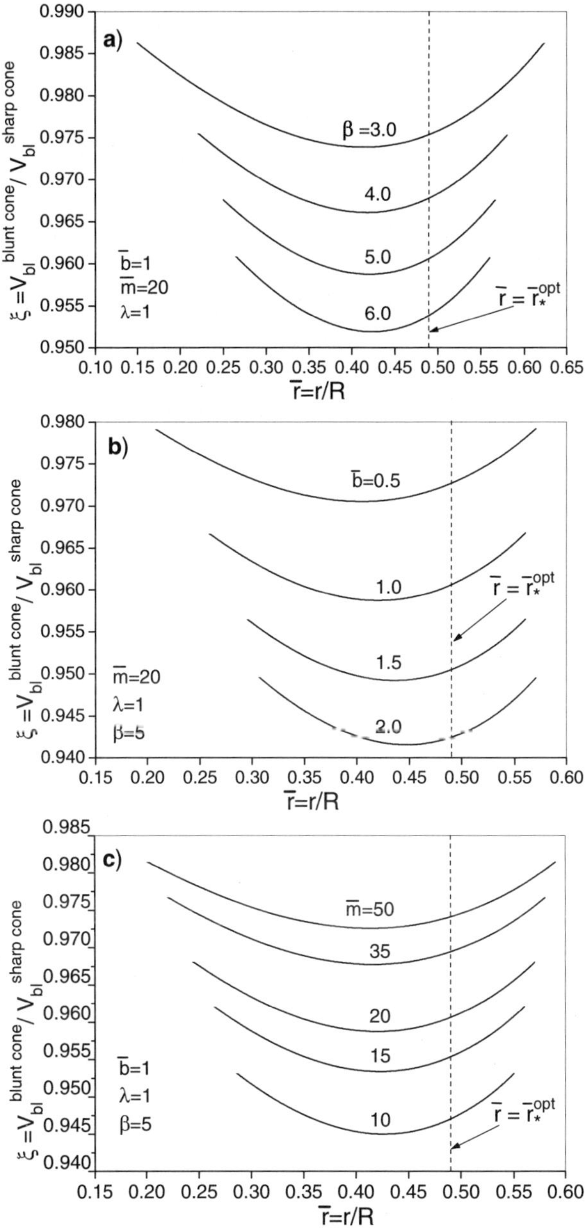

Fig. 17.3.1. Normalized BLV of truncated cone-shape impactors vs. radius of nose bluntness

$$\bar{r}^{opt}_{\min}(\bar{m},\bar{b},\lambda,\beta,\eta) \le \bar{r} \le \bar{r}^{opt}_{\max}(\bar{m},\bar{b},\lambda,\beta,\eta), \tag{17.3.7}$$

the following condition is valid:

$$v_{bl}(\bar{r};\bar{m},\bar{b},\lambda,\beta) \le (1+\eta)v^{opt}_{bl}(\bar{m},\bar{b},\lambda,\beta). \tag{17.3.8}$$

In the considered problem such interval exists since v_{bl} is a convex downward function of $\bar{r}$.

Clearly, Eq. (17.3.8) will be satisfied if the interval of $\bar{r}$ given by Eq. (17.3.7) is replaced by the following shorter interval:

$$\bar{r}^{opt}_{\inf}(\bar{m},\bar{b},\lambda,\eta) \le \bar{r} \le \bar{r}^{opt}_{\sup}(\bar{m},\bar{b},\lambda,\eta), \tag{17.3.9}$$

where

$$\begin{aligned} \bar{r}^{opt}_{\inf}(\bar{m},\bar{b},\lambda,\eta) &= \max_{\beta_\times \le \beta \le \beta^\times} \bar{r}^{opt}_{\min}(\bar{m},\bar{b},\lambda,\beta,\eta) = \bar{r}^{opt}_{\min}(\bar{m},\bar{b},\lambda,\beta^\times,\eta) \\ \bar{r}^{opt}_{\sup}(\bar{m},\bar{b},\lambda,\eta) &= \min_{\beta_\times \le \beta \le \beta^\times} \bar{r}^{opt}_{\max}(\bar{m},\bar{b},\lambda,\beta,\eta) = \bar{r}^{opt}_{\max}(\bar{m},\bar{b},\lambda,\beta_\times,\eta) \end{aligned} \tag{17.3.10}$$

Table 17.3.1. The ranges of the normalized radius of the impactor's nose bluntness that provides the deviation of the BLV from its minimum less than 3%, for $3 \le \beta \le 6$.

b/R	$\lambda = L/R$							
	0.25	0.50	0.75	1.00	1.25	1.50	1.75	2.00
$\bar{m}=5$								
0.25	0.48÷0.93	0.31÷0.82	0.19÷0.72	0.10÷0.65	0.04÷0.59	0.00÷0.54	0.00÷0.51	0.00÷0.49
0.75	0.59÷0.93	0.41÷0.82	0.28÷0.71	0.19÷0.63	0.12÷0.56	0.06÷0.51	0.02÷0.47	0.00÷0.44
2.00	0.63÷0.92	0.47÷0.82	0.35÷0.71	0.25÷0.63	0.18÷0.56	0.12÷0.51	0.08÷0.46	0.04÷0.42
$\bar{m}=15$								
0.25	0.30÷0.95	0.17÷0.85	0.07÷0.76	0.00÷0.70	0.00÷0.65	0.00÷0.60	0.00÷0.57	0.00÷0.55
0.75	0.54÷0.94	0.34÷0.84	0.20÷0.74	0.11÷0.66	0.04÷0.60	0.00÷0.55	0.00÷0.52	0.00÷0.49
2.00	0.61÷0.93	0.44÷0.83	0.30÷0.73	0.20÷0.65	0.12÷0.59	0.06÷0.53	0.02÷0.49	0.00÷0.46
$\bar{m}=50$								
0.25	0.00÷0.99	0.00÷0.91	0.00÷0.83	0.00÷0.78	0.00÷0.74	0.00÷0.70	0.00÷0.67	0.00÷0.66
0.75	0.39÷0.96	0.20÷0.87	0.07÷0.78	0.00÷0.71	0.00÷0.66	0.00÷0.62	0.00÷0.59	0.00÷0.56
2.00	0.55÷0.94	0.36÷0.85	0.22÷0.77	0.11÷0.69	0.04÷0.63	0.00÷0.58	0.00÷0.54	0.00÷0.51

In Eq. (17.3.10) we took into account the results of numerical simulations which show that $\bar{r}^{opt}_{\min}(\bar{m},\bar{b},\lambda,\beta,\eta)$ and $\bar{r}^{opt}_{\max}(\bar{m},\bar{b},\lambda,\beta,\eta)$ are increasing and decreasing functions of β, respectively.

The functions $\bar{r}^{opt}_{\inf}(\bar{m},\bar{b},\lambda,\eta)$ and $\bar{r}^{opt}_{\sup}(\bar{m},\bar{b},\lambda,\eta)$ determine the solution of the optimization problem formulated in the beginning of this Section. The results of calculations for $\eta = 0.03$ are summarized in Table 17.3.1. Clearly there is a wide choice of shapes of impactors with the improved ballistic properties.

17.4 Concluding Remarks

There are two interconnected aspects of the results obtained in this chapter. Direct results are associated with the use of the approximate models for shape optimization of impactors penetrating into FRP laminates. However, we emphasize also the need to apply nontraditional approaches for solving these problems. These approaches must take into account an ambiguity in the choice of the parameters of the model and peculiarities of the behavior of the optimization criterion in the vicinity of the optimum. Although the feasibility of using universal methods and analytical techniques for shape optimization of penetrating impactors is rather restricted, the approaches based on numerical simulation can be successful.

Part 8

Effectiveness of Segmented Impactors

Non-monolithic (segmented, spaced) impactors attract attention because in many instances their performance is superior to that of monolithic impactors having the same mass.

Numerous experimental and numerical investigations of ballistic properties of non-monolithic impactors for hypervelocity penetration into various media were conducted in the past (Anderson *et al.*, 1997; Bell, 1987; Cao *et al.*, 2010; Charters, 1986; Charters *et al.*, 1990; Chou *et al.*, 1975; Chou and Toland, 1977; Cuadros, 1990; De Rosset, 1981; De Rosset and Kimsey, 1986; Eichelberger, 1956; Evstrop'ev-Kudryevatyj, *et al.*, 1992; Fedorov and Veldanov, 2006; Fedorov *et al.*, 2007; Fomin, 1999; Franzen *et al.*, 1994; Han *et al.*, 2005; Hauver and Melani, 1990; Herbette, 1989; Hohler and Stilp, 1990; Holland *et al.*, 1990; Hunkler, 1990; Kivity *et al.*, 1989; Kucher, 1982; Lang *et al.*, 2011; Lee, 2001; Lee and Normandia, 1999; Littlefield, 2001; Littlefield *et al.*, 1999; Lynch *et al.*, 2001; Magier, 2008, 2010; Naz and Lehr, 1990; Normandia and Lee, 1999; Orphal, 2006; Orphal and Franzen, 1990; Orphal and Miller, 1991; Presnell and Rajendran, 2010; Raatschen *et al.*, 1987; Scheffler, 1989; Scheffler, 1990; Scheffler and Zukas, 1990; Sedgwick *et al.*, 1987; Sorensen *et al.*, 1990; Trushkov *et al.*, 2007; Trushkov and Kramshonkov, 2008a, 2008b; Wang *et al.*, 1995; Weihrauch and Wollmann, 1993; Westerling *et al.*, 1997; Zukas, 1990). However, there are only a few studies (Tate, 1990; Zhao *et al.*, 1995; Ben-Dor *et al.*, 2010d) where analytical methods have been applied.

Some interesting effects which were discovered in the case of hypervelocity penetration stimulated investigations for high-speed velocities. These studies have been conducted by Ben-Dor *et al.*, 2007e, 2008b, 2010d.

In discrete formulation of the problem it is assumed that the segmented striker consists of a number of segments which penetrate the shield in a consecutive order. Using engineering models we investigated the effects of segmentation and the size of the segments on ballistic characteristics of projectiles.

In discrete problems, which are considered in Chapter 18, it is assumed that each successive segment impinges on the penetrator composed from the previous segments which have been stopped by the drag force.

In Chapter 19, the discrete problem is analyzed in a more general formulation where the successive segment can impinge upon the composite body consisting of the preceding segments before they have been stopped by drag force. We also suggest a limiting continuous model which affords an arbitrary time-dependent increase of the mass of a penetrator during penetration, and determine analytically the dependence of the mass increase that provides the maximum penetration depth.

Chapter 20 is dedicated to investigation of penetration by segmented impactors for hypervelocity impact.

In Chapters 18–20 we describe the results obtained by Ben-Dor *et al.*, 2007e, 2008b, 2010d.

Chapter 18

High-Speed Impact. Simplified Discrete Model

Nomenclature

a_0, a_2	parameters in impactor-shield interaction model, Eq. (1.6.1);
A_0, A_2	parameters in Eqs. (1.6.3) and (1.6.4);
DOP	depth of penetration;
f	function, Eq. (18.1.2);
h_i	DOP of composite striker consisting of i sub-strikers;
k_v	parameter, Eq. (18.2.3);
L	length of the nose of impactor, Fig. 1.1.2;
m	$= M_N$, mass of projectile;
m_i	mass of i th segment;
M	variable in Eq. (18.1.3), mass of impactor;
M_i	mass of composite striker consisting of i segments;
N	number of segments;
P	total DOP of segmented impactor;
P_0	total DOP of monolithic impactor;
v_i	velocity of composite projectile consisting of i th segments after impact of the ($i+1$)th segment
v_{imp}	impact (initial) velocity of projectile;
V	variable in Eq. (18.1.3), impact velocity;
V_0	$=\sqrt{A_0/A_2}$;
$V_{\min}, V_{\max}$	minimum and maximum velocities of impactor in function $\psi(V)$;

α	parameter, Eq. (18.2.3);
Δh_i	penetration depth of composite striker in the shield;
ε	$= a_2/a_0$;
η	$= P/P_0$;
λ_i	$= m_i/M_i$;
$\lambda_{\min}$	$= V_{\min}/v_{imp}$;
ξ	parameter in Eq. (18.2.5);
$\Phi(x)$	function determining the shape of impactor, Fig.1.2.2;
ψ	function, Eq. (18.4.1).

18.1 Formulation of the Problem

Consider a high speed normal orderly penetration of rigid sharp sub-strikers into a semi-infinite shield. In this chapter we use the following model that describes this multi-step penetration.

The first sub-striker having mass m_1 and velocity v_{imp} (see Fig. 18.1.1a) hits a non-damaged shield and penetrates to the depth h_1 (Fig. 18.1.1b). We assume that this sub-striker forms a cavity in a shield for other sub-strikers, and only its nose interacts with the shield at later stages of penetration. Once the first sub-striker is slowed down by drag to zero velocity, the next sub-striker having mass m_2 and velocity v_{imp} penetrates in the shield (see Fig. 18.1.1c for $i = 2$). After inelastic collision, the composite striker consisting of two sub-strikers and having velocity v_2 (Fig. 18.1.1d) penetrates to the depth h_2 (Fig. 18.1.1e). Once this composite sub-striker is slowed down to zero velocity, the next sub-striker penetrates in the shield and the process is repeated.

When a sub-striker having mass m_i and moving with velocity v_{imp} collides inelastically with a composite striker consisting of $i-1$ sub-strikers having the total mass $m_1 + m_2 + ... + m_{i-1}$, the velocity of the newly formed after collision composite striker, v_i, reads:

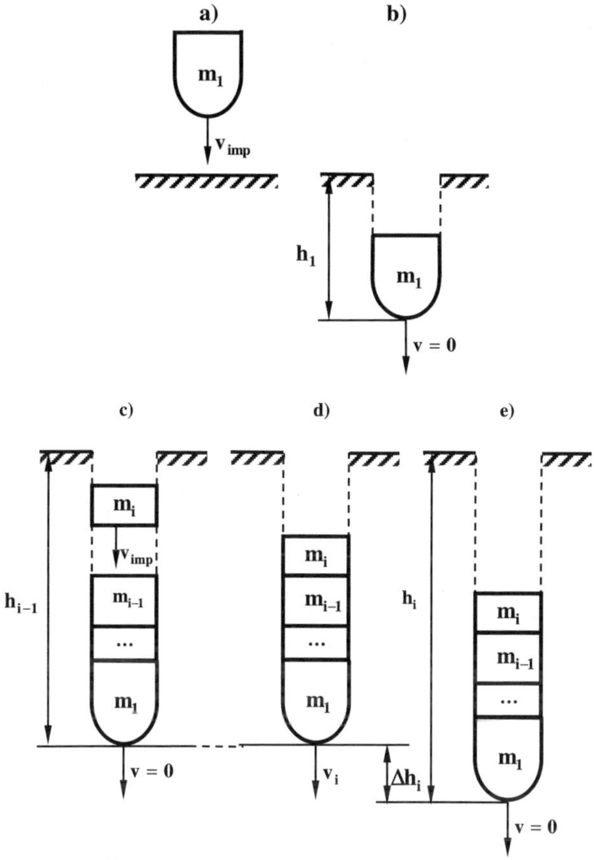

Fig. 18.1.1. Formulation of the problem.

$$v_i = m_i v_{imp} \Big/ \sum_{\nu=1}^{i} m_\nu \,, \quad i = 1,2,...,N \,, \tag{18.1.1}$$

where N is the number of sub-strikers.

Assuming that mechanical properties of the material of the shield are constant and that the composite striker composed of the sub-strikers continues its translational motion in the shield we arrive at the following relation (see Fig. 18.1.1):

$$\Delta h_i = h_i - h_{i-1} = f(M_i, v_i), M_i = \sum_{\nu=1}^{i} m_\nu, h_0 = 0, \quad i = 1,2,...,N \,, \tag{18.1.2}$$

where h_i is the DOP of a composite striker consisting of i sub-strikers having mass M_i, Δh_i is the depth of penetration of a composite striker in the shield until it slows down to the zero velocity. Function f determines the model, i.e., the dependence of the penetration depth, Δh, of the striker having the same nose shape in a homogeneous shield manufactured from the same material, on the mass of the impactor, M, and on the initial (impact) velocity, V:

$$\Delta h = f(M,V), \quad f(M,0)=0. \tag{18.1.3}$$

A number of such models are described in Parts 2-4 of this monograph.

The expression for the total DOP, P, reads:

$$P=\sum_{i=1}^{N}\Delta h_i=\sum_{i=1}^{N} f(M_i,v_i), \tag{18.1.4}$$

while the DOP of the homogeneous impactor having the same mass $m = M_N$ and the same impact velocity, v_{imp}, is determined by the following formula:

$$P_0 = f(m, v_{imp}). \tag{18.1.5}$$

Our goal is to determine when such sequential penetration has an advantage over penetration by a homogeneous impactor having the same m and v_{imp} and to estimate the maximum increase of the DOP (if any). This problem reduces to the maximization of P or $\eta = P/P_0$ when m, v_{imp} and N are given.

18.2 Study of the Problem Using Two-Term Impactor-Shield Interaction Model

18.2.1 *Mathematical model*

Here we employ the model of impactor-shield interaction that is described in details in Section 1.6.3; function $f(M,V)$ in this case is determined by Eqs. (1.6.22):

$$f(M,V)=\frac{M}{2A_2}\ln\left(\frac{A_2}{A_0}V^2+1\right), \tag{18.2.1}$$

where parameters A_0 and A_2 are determined by Eqs. (1.6.3) and (1.6.4); they depend on the chosen model for impactor-shield interaction and on the shape of impactor nose. Substituting Eq. (18.2.1) for the function $f(M,V)$ in Eqs. (18.1.4) and (18.1.5) yields:

$$P=\frac{m}{2A_2}\sum_{i=1}^{N}\left(\sum_{\nu=1}^{i}k_\nu\right)\ln\left(1+\frac{\alpha k_i^2}{(\sum_{\mu=1}^{i}k_\mu)^2}\right),\quad P_0=\frac{m}{2A_2}\ln(1+\alpha), \tag{18.2.2}$$

where

$$\alpha=\frac{A_2}{A_0}v_{imp}^2,\quad k_\nu=\frac{m_\nu}{m},\ \nu=1,2,...,N,\quad \sum_{\nu=1}^{N}k_\nu=1\ . \tag{18.2.3}$$

Equation (18.2.2) implies that

$$\eta=P/P_0=1+\ln\xi/\ln(\alpha+1)\,, \tag{18.2.4}$$

where

$$\xi=\frac{1}{1+\alpha}\prod_{i=1}^{N}\left(1+\alpha k_i^2\Big/(\sum_{\mu=1}^{i}k_\mu)^2\right)^{\sum_{\nu=1}^{i}k_\nu}\ . \tag{18.2.5}$$

Clearly, for a given value of the parameter α, the conditions $P>P_0$, $\eta>1$ and $\xi>1$ are interchangeable, and the problems of maximization for P, η and ξ are equivalent.

18.2.2 *Analytical study of case of two segments*

For $N=2$, Eq. (18.2.5) can be rewritten as follows:

$$\xi(k) = (\alpha k^2 + 1)/(\alpha + 1)^k, \quad k = k_2, \quad 0 \le k \le 1. \qquad (18.2.6)$$

Let us investigate the behavior of the function $\xi(k)$. Formula for the first derivative of $\xi(k)$ reads:

$$\xi'(k) = \xi_1(k)/(\alpha+1)^k, \ \xi_1(k) = -\alpha \ln(\alpha+1)k^2 + 2\alpha k - \ln(\alpha+1). \quad (18.2.7)$$

Since $\xi(k)$ is smooth and $\xi(0) = \xi(1) = 1$, Rolle theorem guarantees the existence of such $k = k_a$ in the interval $[0,1)$ that $\xi'(k_a) = \xi_1(k_a) = 0$. Consequently, the quadratic equation $\xi_1(k) = 0$ has two real roots, k_a and k_b (we assume here that $k_a \ne k_b$, i.e., the discriminant of this equation is not equal to zero). According to Vieta theorem, $k_a k_b = 1/\alpha > 0$ and $k_a + k_b = 2/\ln(\alpha+1) > 0$; these inequalities imply that both roots are positive. Assume that $k_a < k_b$ and taking into account that $\xi'(0) < 0$ let us consider two possible cases.

If $k_b \ge 1$ then $\xi'(k) < 0$ for $0 \le k < k_a$, $\xi'(k_a) = 0$ and the derivative is positive, $\xi'(k) > 0$ for $k_a < k < 1$. Hence, $\xi(k)$ decreases from 1 to $\xi(k_a)$ when k increases from 0 to k_a, and increases further up to 1 when k increases up to 1. Therefore, $\xi \le 1$, i.e., the sequential penetration has no advantage over penetration by the homogeneous impactor.

If $0 < k_a < k_b < 1$ then the behavior of the function $\xi(k)$ is as follows: $\xi(k)$ decreases from $\xi(0) = 1$ to $\xi(k_a) < 1$ in the interval $k \in [0, k_a)$, increases from $\xi(k_a)$ to $\xi(k_b) > 1$ in the interval $k \in (k_a, k_b)$ and decreases from $\xi(k_b)$ to $\xi(1) = 1$ in the interval $k \in (0, k_a]$. Therefore the equation $\xi(k) = 1$ has a single root, k_*, which is located between $k = k_a$ and $k = k_b$, and the sequential penetration has an advantage over penetration by the monolithic impactor if $k_* < k < 1$.

Let us investigate the condition, $k_b < 1$. Since

$$k_b = \frac{\alpha + \sqrt{\alpha^2 - \alpha \ln^2(\alpha+1)}}{\alpha \ln(\alpha+1)}, \qquad (18.2.8)$$

this condition can be written in the following form:

$$\sqrt{\alpha^2 - \alpha \ln^2(\alpha+1)} < \alpha[\ln(\alpha+1)-1]\,. \tag{18.2.9}$$

Equation (18.2.9) shows that the solution of the inequality is located in the interval $\alpha > \alpha_0 = e - 1 \approx 1.72$. After some algebra Eq. (18.2.9) yields:

$$\varphi(\alpha) > 0, \quad \varphi(\alpha) = (\alpha+1)\ln(\alpha+1) - 2\alpha\,. \tag{18.2.10}$$

Since $\varphi'(\alpha) = \ln(\alpha+1) - 1 > 0$ for $\alpha > \alpha_0$, $\varphi(\alpha_0) = 2 - e < 0$ and $\varphi(e^2 - 1) = 2 > 0$, the inequality $\varphi(\alpha) > 0$ is valid for

$$\alpha > \alpha_* \approx 3.92\,, \tag{18.2.11}$$

where α_* is the solution of the equation $\varphi(\alpha) = 0$.

Therefore, sequential penetration can have an advantage over the penetration by the homogeneous impactor only when $\alpha > \alpha_*$. This advantage is realized for $k_* < k_2 < 1$, where k_* is the solution of the equation $\xi(k) = 1$ in the interval $[0, 1]$. The maximum advantage occurs for $k_2 = k_b$ with k_b given by Eq. (18.2.8).

18.2.3 *Results of numerical calculations*

The above theoretical results for $N = 2$ are confirmed by numerical

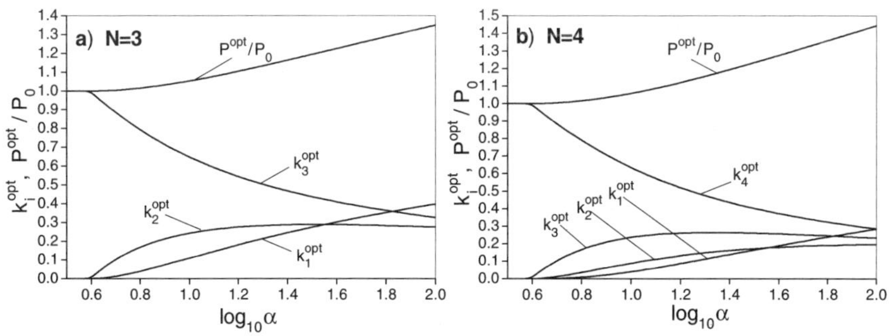

Fig. 18.2.2. Optimum solutions for N - segment penetration; P^{opt} is the maximum DOP for sequential impacts, P_0 is the DOP of the monolithic impactor; k_i^{opt} determine the optimum distribution of masses between the segments. Two term model.

calculations shown in Fig. 18.2.1. Clearly, significant advantage of the sequential penetration is attained for the relatively large values of the coefficient α, i.e., when the dynamic component of the drag is much larger than the static component.

The results of calculations for $N=3$ and $N=4$ are shown in Fig. 18.2.2, where P^{opt} is the maximum DOP in the case of sequential penetration which is attained for $k_v = k_v^{opt}$. Naturally, the ratio $\eta^{opt} = P^{opt}/P_0$ increases with the increase of N for the same α but the increase is small.

If all sub-strikers have the same mass, then $k_i = 1/N$ and Eq. (18.2.5) can be written as follows:

$$\xi = \frac{1}{(\alpha+1)} \prod_{i=1}^{N} \left(1+\alpha/i^2\right)^{i/N} . \tag{18.2.12}$$

Comparison of the optimum DOPs for the sequential penetration with the DOP for penetration by the homogeneous impactor is shown in Fig. 18.2.3 for different N. Inspection of Fig. 18.2.3 reveals that $\eta^{opt} = P^{opt}/P_0$ can exceed 1.5 for large N and α.

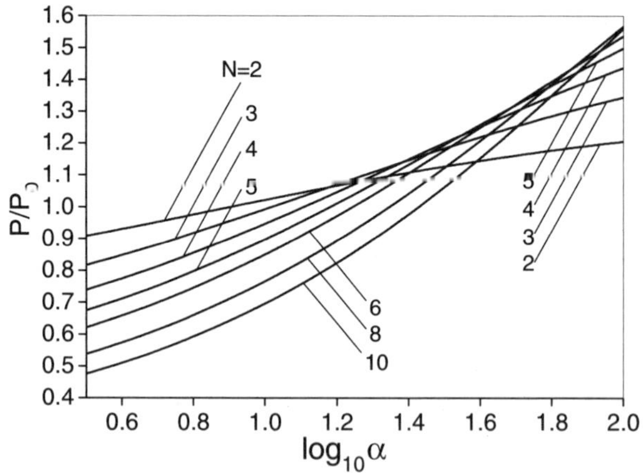

Fig. 18.2.3. Segments having the same mass. Effect of the number of segments, N, on the DOP, P, after sequential impacts. Two term model.

Table 18.2.1. Values $\varepsilon = a_2/a_0$ for different materials.

Shield material	$\varepsilon \cdot 10^6$, s^2/m^2	*Reference*	*Remark*
Aluminum alloy 6061-T651	1.9	Forrestal *et al.*, 1988	Compressible elastic-plastic model
Aluminum alloy 6061-T651	2.5	Forrestal *et al.*, 1991	Incompressible strain-hardening model
Aluminum alloy 7075-T651	2.0	Forrestal *et al.*, 1992	Incompressible strain-hardening model
Aluminum alloy 7075-T651	1.5	Forrestal *et al.*, 1992	Compressible strain-hardening model
Aluminum alloy 6061-T651	3.0	Warren and Forrestal, 1998	Incompressible model without strain rate effects
Duraluminum	2.1	Vitman and Ioffe, 1948; Vitman and Stepanov, 1959	
Soft steel	4.2	Vitman and Ioffe, 1948; Vitman and Stepanov, 1959	
Concrete	3.4–6.9	Forrestal *et al.*, 1994	Tunnel region of penetration
Concrete	12.4	Forrestal *et al.*, 2003	Tunnel region of penetration
Concrete	13.0	Forrestal and Tzou, 1997	Incompressible elastic-plastic model
Concrete	1.7– 2.3	Jones *et al.*, 2003	
Limestone	2.2–3.2	Frew *et al.*, 2000	
Soil	100	Forrestal and Luk, 1992	
Dry sand	200–300	Sierakowski *et al.*, 1977	

Expression for α for the most widely used sub-model described in Section 1.6.2.3. Eqs. (1.6.3) and (1.6.15) reads:

$$\alpha = \frac{A_2}{A_0} v_{imp}^2 = \frac{r^2 + 2\int\limits_0^L \frac{\Phi'^2}{\Phi'^2+1}(\mu_{fr}+\Phi')\Phi dx}{r^2 + 2\int\limits_0^L (\mu_{fr}+\Phi')\Phi dx} \varepsilon v_{imp}^2 \le \varepsilon v_{imp}^2 , \qquad (18.2.13)$$

where L is the length of the nose of the first segment that interacts with the shield, function $\Phi(x)$ determines the generatrix, $r = \Phi(0)$ is radius of the bluntness (if any), μ_{fr} is friction coefficient between the impactor and the shield,

$$\varepsilon = a_2 / a_0 , \tag{18.2.14}$$

and the equality in Eq. (18.2.13) occurs for the cylindrical projectile. Typical values of ε are presented in Table 18.2.1. Inspection of this table allows to conclude that for the impact velocities less than $1000\ m/s$, the pronounced effect of segmentation to increase the DOP can be attained for soils. For instance, the results of experiments with the cylindrical projectile penetrating into dry sand (Sierakowski *et al.*, 1977) showed that $\varepsilon = 0.0002 \div 0.0003\ s^2/m^2$, and the calculated increase of the DOP for the sequential penetration is about $20 \div 25\%$ for impact velocity $v_{imp} \approx 700\ m/s$.

18.3 Study of Problem on the Basis of Young Model

Young equations for penetration into concrete and geological media are described in Sections 3.23, 7.4.1, 7.4.2.

Typical results of numerical calculations are shown in Fig. 18.3.1 for sub-strikers with different masses and in Fig. 18.3.2 for sub-strikers having the same mass. Inspection of these figures shows that an advantage of a sequential penetration can be expected for high-speed impactors having a relatively large mass.

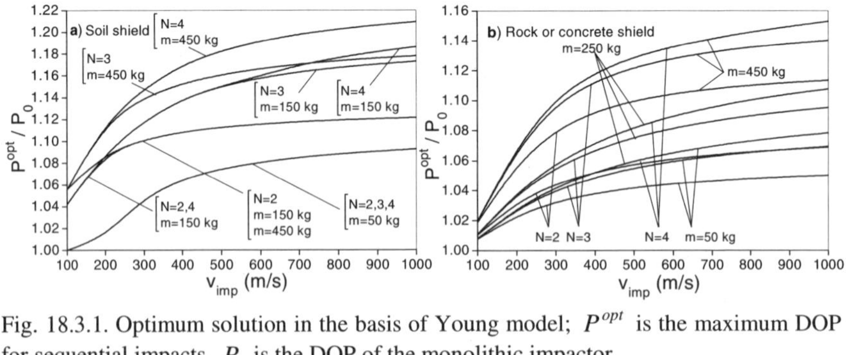

Fig. 18.3.1. Optimum solution in the basis of Young model; P^{opt} is the maximum DOP for sequential impacts, P_0 is the DOP of the monolithic impactor.

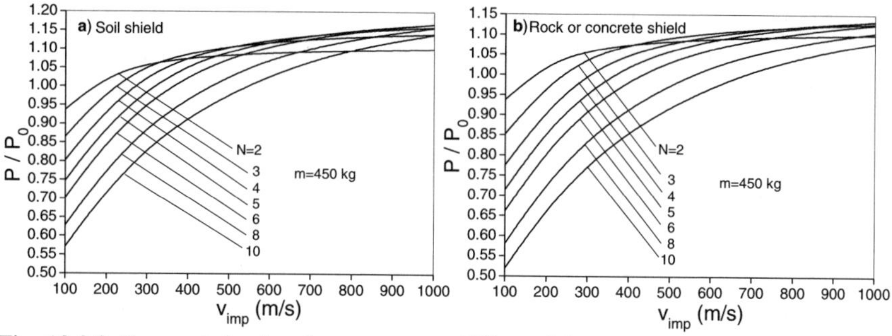

Fig. 18.3.2. Segments having the same mass. Effect of the number of segments, N, on the DOP, P, after sequential impacts. Young model.

18.4 General Penetration Model

18.4.1 *Formulation of problem*

In this section, we consider the case when

$$f(V) = M\psi(V) \tag{18.4.1}$$

and assume that $\psi(V)$ is a positive increasing function in the interval

$$V_{\min} < V \leq V_{\max} \tag{18.4.2}$$

that satisfies some additional conditions discussed below.

In this case Eqs. (18.1.2), (18.1.4), (18.1.5) can be written as follows:

$$\Delta h_i = h_i - h_{i-1} = M_i\psi(v_i), \quad h_0 = 0, \quad i = 1,2,...,N\,, \tag{18.4.3}$$

$$M_i = \sum_{\nu=1}^{i} m_\nu, \quad i = 1,2,...,N \tag{18.4.4}$$

$$P = \sum_{i=1}^{N} \Delta h_i = \sum_{i=1}^{N} M_i\psi(v_i)\,. \tag{18.4.5}$$

$$P_0 = m\psi(v_{imp})\,. \tag{18.4.6}$$

Our goal of is to determine when sequential penetration has an advantage over penetration by a homogeneous impactor for the same m and v_{imp}, and vice versa.

18.4.2 *Analysis of problem*

Here we do not specify the expression for the function $\psi(V)$. In this case meaningful results can be obtained assuming only that $\psi(V)$ is a convex downward, convex upward or a linear function that satisfies some additional conditions. In order to avoid cumbersome mathematical details we do not consider situations when the domain of the function $\psi(V)$ includes intervals where this function is convex downward (convex upward) as well as intervals where this function is linear.

Let $\psi(V)$ be a convex downward positive increasing function in the interval $0 \le V_{\min} < V < V_{\max}$ that satisfies the condition,

$$V_{\min}\psi'(V_{\min}) - \psi(V_{\min}) \ge 0. \tag{18.4.7}$$

Substituting v_i from Eq. (18.1.1) into Eq. (18.4.5) and taking into account Eq. (18.4.4) we obtain the following relation:

$$P = \sum_{i=1}^{N} M_i \psi(\frac{m_i}{M_i} v_{imp}) = \sum_{i=1}^{N} \frac{m_i}{\lambda_i} \psi(\lambda_i v_{imp}),\ \lambda_i = \frac{m_i}{M_i},\ i = 1,2,...,N \tag{18.4.8}$$

The values of the impact velocity and of masses of the segments are subjected to the constraint that the argument of the function $\psi(V)$ belongs to the interval where this function is convex downward, positive and increasing. Consequently, the following constraints must be taken into account: $V_{\min} < \lambda_i v_{imp} < V_{\max}$, , $m_i > 0$, $m_1 + m_1 + ... + m_N = m$ for $i = 1,2,...,N$. These constraints can be rewritten using the dimensionless masses of the segments:

$$V_{\min} < v_{imp} < V_{\max},\quad \overline{m}_i > 0,\quad \sum_{\nu=1}^{N} \overline{m}_\nu = 1,\ \overline{m}_i = \frac{m_i}{m},\ i = 1,2,...,N, \tag{18.4.9}$$

$$\overline{m}_i > \frac{\lambda_{\min}}{1-\lambda_{\min}} \sum_{\nu=1}^{i-1} \overline{m}_\nu,\quad \lambda_{\min} = V_{\min}/v_{imp},\quad i = 2,...,N, \tag{18.4.10}$$

Excluding

$$\overline{m}_N = 1 - \sum_{\nu=1}^{N-1} \overline{m}_\nu \tag{18.4.11}$$

from Eq. (18.4.9) we obtain the following constraints:

$$\overline{m}_i > \frac{\lambda_{\min}}{1-\lambda_{\min}} \sum_{\nu=1}^{i-1} \overline{m}_\nu, \quad i = 2,\dots,N-1, \quad N > 2,$$
$$\sum_{\nu=1}^{N-1} \overline{m}_\nu < 1-\lambda_{\min}, \quad \overline{m}_j > 0, \quad j = 1,2,\dots,N-1 \tag{18.4.12}$$

It is important to note that only "natural" constraints given by Eq. (18.4.9) remain if $V_{\min} = 0$.

Equations (18.4.11) and (18.4.12) imply that

$$\overline{m}_1 < 1-\lambda_{\min}, \quad \overline{m}_1 > 0, \quad \overline{m}_2 = 1-\overline{m}_1 \tag{18.4.13}$$

for $N = 2$ and

$$\overline{m}_2 > \frac{\lambda_{\min}}{1-\lambda_{\min}} \overline{m}_1, \quad \overline{m}_1 + \overline{m}_2 < 1-\lambda_{\min},$$
$$\overline{m}_1 > 0, \quad \overline{m}_2 > 0, \overline{m}_3 = 1-\overline{m}_1 - \overline{m}_2 \tag{18.4.14}$$

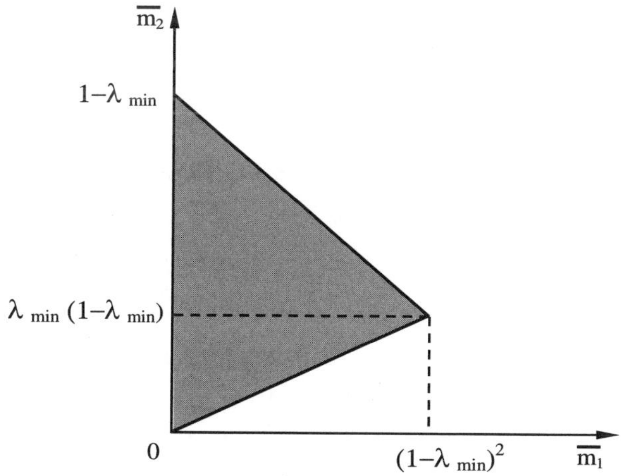

Fig. 18.4.1. Domain of admissible values $\overline{m}_1$ and $\overline{m}_2$ for a case of three segments.

for $N = 3$. The domain that satisfied inequalities in Eq. (18.4.14) is shown in Fig. 18.4.1.

Since function $\psi(V)$ is a convex downward positive increasing function in the domain $0 \le V_{\min} < V < V_{\max}$ that satisfies the condition given by Eq. (18.4.7), then the following inequality is valid (see Property 1 of convex functions in Appendix A):

$$\psi(\lambda_i v_{imp}) < \lambda_i \psi(v_{imp}), \quad i = 1,2,...,N. \tag{18.4.15}$$

Consequently,

$$\begin{aligned} P &= \sum_{i=1}^{N} \frac{m_i}{\lambda_i} \psi(\lambda_i v_{imp}) < \sum_{i=1}^{N} \frac{m_i}{\lambda_i} \lambda_i \psi(v_{imp}) \\ &= \psi(v_{imp}) \sum_{i=1}^{N} m_i = m\psi(v_{imp}) = P_0. \end{aligned} \tag{18.4.16}$$

It can be shown that $P = P_0$ only in a degenerate case when a segmented impactor is replaced by a monolithic impactor.

In summary, we proved the following claim. Let $\psi(V)$ be a convex downward, positive, increasing function in the interval $0 \le V_{\min} < V < V_{\max}$ that satisfies the condition given in Eq. (18.4.7). Then, for any impact velocity, $V_{\min} < v_{imp} < V_{\max}$, the monolithic impactor having mass m has an advantage over the segmented strikers with the same total mass provided a distribution of masses of the segments satisfies the constraints given by Eqs. (18.4.11) and (18.4.12). If $V_{\min} = 0$, then monolithic impactor has the advantage over any segmented striker having the same mass.

Let us now assume that $\psi(V)$ is a convex upward positive increasing function in the domain $0 \le V_{\min} < V < V_{\max}$ that satisfies the condition:

$$V_{\min} \psi'(V_{\min}) - \psi(V_{\min}) \le 0. \tag{18.4.17}$$

Similar analysis, as in the case of convex downward function $\psi(V)$, shows that segmented strikers with the total mass m have an advantage over the monolithic impactor having the same mass provided the distribution of masses of the segments satisfies the constraints given by

Eqs. (18.4.11) and (18.4.12). If $V_{\min} = 0$ then all segmented strikers have an advantage over the monolithic impactor having the same mass.

Let $\psi(V)$ be a positive linear function, $\psi(V) = k_1 V + k_0$ in the interval $0 \le V_{\min} < V < V_{\max}$, where k_0 and $k_1 > 0$ are parameters. Then

$$P_0 = k_1 m v_{imp} + k_0 m, \quad P = P_0 + k_0 s, \tag{18.4.18}$$

where

$$s = (N-1)m_1 + (N-2)m_2 + \ldots + (N-i)m_i + \ldots + m_{N-1} > 0. \tag{18.4.19}$$

Consequently, the above results for convex downward and convex upward $\psi(V)$ can be extended to the linear model when $k_0 < 0$ and $k_0 > 0$, correspondingly. If $k_0 = 0$ then all segmented impactors and the monolithic impactor penetrate to the same depth.

18.4.3 *Velocity-dependent resistance*

Consider widely used engineering penetration models based on the assumption that the resistance force D is a function of the instantaneous velocity of the impactor v. Let us demonstrate that the above analysis in Section 18.4.2 is valid for these models.

Indeed, in this case the equation describing motion of the impactor having mass M can be written as $Mv\,dv/dh = -D(v)$, where h is the instantaneous depth of penetration. Assume that the impactor that penetrated to the depth $h = h_{start}$, begins its motion with the impact velocity $v = V$ and is slowed down to zero velocity at the depth $h = h_{stop}$. Hence equation of impactor motion implies that

$$M\int_V^0 \frac{vdv}{D(v)} = -\int_{h_{start}}^{h_{stop}} dh. \tag{18.4.20}$$

Therefore

$$\Delta h = h_{stop} - h_{start} = M\psi(V), \quad \psi(V) = \int_0^V \frac{vdv}{D(v)} \tag{18.4.21}$$

and we recover the model given by Eq. (18.4.1) where the function $\psi(V)$ is determined by Eq. (18.4.21).

Now let us consider a problem which was analyzed in Section 18.2. Here we pursue two goals: (i) to illustrate the proposed above approach and (ii) to generalize the results which were obtained in Section 18.2.2 for $N = 2$ using analytical methods.

The model given by Eq. (18.2.1) belongs to the considered class of models for which $D = D(v)$ with $D(v) = A_2 v^2 + A_0$. Introducing function $\psi(V)$, Eq. (18.2.1) can be rewritten as follows:

$$\psi(V) = \frac{1}{2A_2} \ln\left(1 + \frac{A_2}{A_0} V^2\right). \tag{18.4.22}$$

Formulas for the first and second order derivatives of $\psi(V)$ read:

$$\psi'(V) = \frac{V}{A_2 V^2 + A_0} > 0, \quad V > 0, \quad \psi''(V) = \frac{A_0 - A_2 V^2}{(A_2 V^2 + A_0)^2}. \tag{18.4.23}$$

Since $\psi''(V) > 0$ when $0 < V < V_0 = \sqrt{A_0/A_2}$, positive function $\psi(V)$ is convex downward in this interval. Consequently, the monolithic impactor has an advantage over any segmented striker if $0 < v_{imp} < V_0$.

Function $\psi(V)$ is convex upward if $V > V_0$. Let us determine $V_{\min}$ such that Eq. (18.4.17) is valid. Using Eq. (18.4.23), the inequality in Eq. (18.4.17) is transformed to the inequality given by Eq. (18.2.10) where

$$\alpha = (V_{\min}/V_0)^2. \tag{18.4.24}$$

The solution of the inequality given by Eq. (18.2.10) is $\alpha > \alpha_* \approx 3.92 > 1$. Hence, one can select $V_{\min} = \sqrt{\alpha_*}\, V_0 \approx 2.0 V_0$.

Therefore for any impact velocity, $v_{imp} > V_{\min}$, segmented strikers having the total mass m have an advantage over the monolithic impactor having the same mass provided the distribution of masses of the segments satisfies the constraints given by Eqs. (18.4.11) and (18.4.12).

Chapter 19

High-Speed Impact. Generalized Discrete and Continuous Models

Nomenclature

A_0, A_2	parameters in Eqs. (1.6.3) and (1.6.4);
D	resistance force acting on projectile;
DOP	depth of penetration;
h_i	instantaneous DOP of the composite striker consisting of i sub-strikers when impacted by ($i+1$)-th segment;
h	instantaneous DOP, Fig. 1.1.2;
H	DOP;
k_i	parameter, Eq. (19.1.6);
K_i	parameter, Eq. (19.1.6);
m	$= \tilde{M}_N$, mass of projectile;
m_i	mass of i th segment;
M	instantaneous mass of impactor;
$\overline{M}$	$= M/m$;
$\tilde{M}_i$	mass of the composite striker consisting of i segments;
N	number of segments;
$\overline{P}^{opt}$	dimensionless optimum (maximum) value of DOP;
P	total DOP of segmented impactor;
$\overline{P}$	dimensionless total DOP of segmented impactor, Eq. (19.2.16);
$\overline{P}_0$	dimensionless total DOP of monolithic impactor, Eq. (19.2.16);

P_0	total DOP of monolithic impactor;
t	time;
$\bar{t}$	dimensionless time, Eq. (19.2.16);
v	instantaneous velocity of impactor;
$\bar{v}$	$= v/v_{imp}$;
v_i	velocity of the composite projectile consisting of i segments before impact by ($i+1$)-th segment;
v_{imp}	impact (initial) velocity of projectile;
v'	$= dv/dM$;
$\bar{v}_{\times}$	$= 1/\sqrt{\alpha}$;
$\bar{v}'$	$= d\bar{v}/d\bar{M}$;
w_i	velocity of the composite projectile consisting of i segments after impact by i th segment;
$\bar{w}_i$	$= w_i/v_{imp}$;
Δh_i	depth of penetration of composite striker;
Δh_{YZ}	distance on the plane (M, v) that a penetrator traverses from the point $Y(M_Y, v_Y)$ to the point $Z(M_Z, v_Z)$;
$\Delta\bar{h}_{YZ}$	dimensionless Δh_{YZ}, Eq. (19.2.16);
η	$= P/P_0$;
θ_v	$= v_v/w_v$;
μ	lower bound for the mass of impactor;
$\bar{\mu}$	$= \mu/m$;
ξ	parameter, Eq. (19.1.13);
$\bar{\varphi}$	function, Eq. (19.2.11);
$\bar{\psi}$	function, Eq. (19.2.11);
ω	function, Eq. (19.2.23);
Ω_0	function, Eq. (1.1.6).

19.1 Investigation of Problem Using Generalized Discrete Model

19.1.1 *Formulation of problem*

Consider a high-speed normal impact of a cluster of sub-strikers which penetrate sequentially into a semi-infinite shield. The first sub-striker having mass m_1 and velocity v_{imp} approaches a non-damaged shield (see Fig. 19.1.1a) and penetrates into the shield. We assume that this sub-striker forms a cavity for other sub-strikers and that only its nose interacts with the shield during the later stages of penetration. When a striker reaches the depth h_1 and has the velocity v_1, it is overtaken by the second sub-striker having mass m_2 and moving with the velocity v_{imp} (see Fig. 19.1.1b). After inelastic collision, the striker composed of two sub-strikers acquires velocity w_2 and continues penetration (see Fig. 19.1.1c). The same occurs with the subsequent sub-strikers (see Fig. 19.1.1d-f).

Hence, in the general case it is assumed that each subsequent segment collides with a moving composite striker that is composed of the preceding segments. This is in contrast to the simplified model from the previous chapter where it is assumed that at the moment of collision the composite striker has ceased its motion.

Let a sub-striker having mass m_i and moving with velocity v_{imp} collides inelastically with the striker composed of $i-1$ sub-strikers having a total mass $\tilde{M}_{i-1}$ and moving with velocity v_{i-1}, where

$$\tilde{M}_1 = m_1, \quad \tilde{M}_\nu = m_1 + ... + m_\nu, \quad \nu > 1. \tag{19.1.1}$$

The velocity of the newly formed sub-striker after the collision, w_i, can be found from the momentum conservation equation:

$$m_i v_{imp} + \tilde{M}_{i-1} v_{i-1} = \tilde{M}_i w_i, \quad i = 2,...,N, \tag{19.1.2}$$

where N is the number of sub-strikers. Introducing variables θ_ν instead of v_ν,

$$v_\nu = \theta_\nu w_\nu, \quad 0 \le \theta_\nu \le 1, \tag{19.1.3}$$

we obtain the following recursive relation:

$$w_i = \frac{m_i v_{imp}}{\tilde{M}_i} + \frac{\tilde{M}_{i-1}\theta_{i-1}}{\tilde{M}_i} w_{i-1}, \quad w_1 = v_{imp}, \quad i = 2,...,N. \quad (19.1.4)$$

Writing Eq. (19.1.4) for $i = 2,..., j$ and eliminating the terms with $w_2, w_3,..., w_{j-1}$ in these equations we obtain the following relation:

$$\overline{w}_j = \frac{1}{K_j}\left(k_j + \sum_{\vartheta=1}^{j-1} k_\vartheta \prod_{\nu=\vartheta}^{j-1} \theta_\nu\right), \quad \overline{w}_1 = 1, \quad j = 2,..., N \quad (19.1.5)$$

where

$$k_i = \frac{m_i}{m}, \; K_i = \frac{\tilde{M}_i}{m}, \; \overline{w}_i = \frac{w_i}{v_{imp}}, \; m = \sum_{\nu=1}^{N} m_\nu, \; \sum_{\nu=1}^{N} k_\nu = 1, \; K_N = 1. \quad (19.1.6)$$

Assume that at the depth $h = h_{i-1}$ the velocity of the composite impactor $v = w_i$ and it continues motion so that at the depth $h = h_i$ the velocity of the impactor $v = v_i$ (see Fig. 19.1.1). The model for impactor-shield interaction that is described in details in Section 1.6.3 yields the following equations of motion:

$$\tilde{M}_i v\, dv/dh = -D(v), \quad D(v) = A_2 v^2 + A_0, \quad (19.1.7)$$

where D is the resistance force, parameters A_0 and A_2 are determined by Eqs. (1.6.3) and (1.6.4) and depend on the chosen particular model of impactor-shield interaction and on the shape of the nose of impactor. The solution of this equation that satisfies the conditions at the boundaries reads:

$$\tilde{M}_i \int_{w_i}^{v_i} \frac{v dv}{A_2 v^2 + A_0} = -\int_{h_{i-1}}^{h_i} dh. \quad (19.1.8)$$

After integration we obtain:

$$\Delta h_i = h_i - h_{i-1} = \frac{\widetilde{M}_i}{2A_2}\ln\left(\frac{A_2 w_i^2 + A_0}{A_2 v_i^2 + A_0}\right), \qquad (19.1.9)$$
$$h_0 = 0, \quad v_N = 0, \quad i = 1,2,\ldots,N.$$

Consequently, the DOP of the segmented impactor, P, and the DOP of the monolithic impactor having the same total mass, P_0, are given by the following formulas:

$$P = \sum_{i=1}^{N}(h_i - h_{i-1}) = \frac{m}{2A_2}\sum_{i=1}^{N} K_i \ln\left(\frac{\alpha \overline{w}_i^2 + 1}{\alpha\theta_i^2 \overline{w}_i^2 + 1}\right),$$
$$P_0 = \frac{m}{2A_2}\ln(1+\alpha), \qquad (19.1.10)$$

where $\overline{w}_i$ are determined by Eq. (19.1.5) and

$$K_i = \sum_{\nu=1}^{i} k_\nu, \quad \alpha = \frac{A_2}{A_0} v_{imp}^2, \quad \theta_N = 0. \qquad (19.1.11)$$

Here we took into account that the DOP is defined as the depth when the velocity of impactor vanishes.

Eq. (19.1.10) implies that

$$\eta = P/P_0 = 1 + \ln\xi/\ln(\alpha+1), \qquad (19.1.12)$$

where

$$\xi = \frac{1}{1+\alpha}\prod_{i=1}^{N}\left(\frac{\alpha\overline{w}_i^2 + 1}{\alpha\theta_i^2\overline{w}_i^2 + 1}\right)^{K_i}. \qquad (19.1.13)$$

Clearly, for a given α, the conditions $P > P_0$, $\eta > 1$ and $\xi > 1$ are interchangeable.

Our goal is to determine when a segmented penetrator has an advantage over the monolithic penetrator having the same m, v_{imp} and the shape of the nose, and to estimate the maximum increase of the DOP (if any) due to segmentation. Taking into account Eqs. (19.1.5) and (19.1.11), this problem can be reduced to the problem of determining

$k_1, k_2, ..., k_N$ and $\theta_1, \theta_2, ..., \theta_{N-1}$ which satisfy the following constraints

$$0 \le k_i \le 1, \quad k_1 + k_2 + ... + k_N = 1, \quad 0 \le \theta_i \le 1 \qquad (19.1.14)$$

and maximize η or ξ when α and N are given.

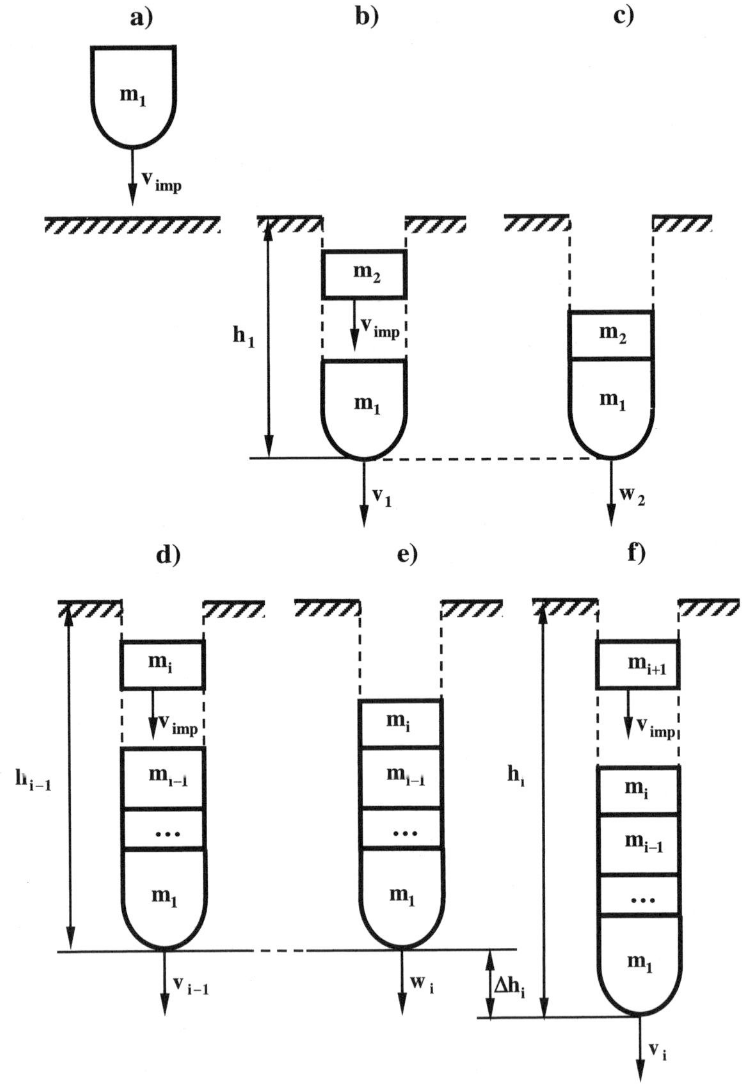

Fig. 19.1.1. Formulation of the problem for discrete model.

19.1.2 Analysis of problem

The case of $N = 2$ can be investigated analytically. Using Eqs. (19.1.5) and (19.1.11) we can write Eq. (19.1.13) as follows:

$$\xi = \frac{\alpha[k_2 + (1-k_2)\theta_1]^2 + 1}{(\alpha+1)^{k_2}(\alpha\theta_1^2 + 1)^{1-k_2}}. \qquad (19.1.15)$$

Let us investigate the behavior of ξ as a function of θ_1. Formula for the first derivative of ξ reads:

$$\xi'(\theta_1) = \frac{2\alpha k_2(1-k_2)}{(\alpha+1)^{k_2}(\alpha\theta_1^2+1)^{2-k_2}}(\theta_1 - 1)\chi(\theta_1), \qquad (19.1.16)$$

where

$$\chi(\theta_1) = \alpha(1-k_2)\theta_1^2 + \alpha k_2\theta_1 - 1. \qquad (19.1.17)$$

Since the discriminant of the quadratic equation $\chi(\theta_1) = 0$,

$$\Delta = \alpha^2 k_2^2 + 4\alpha(1-k_2), \qquad (19.1.18)$$

is positive, this equation has two real roots,

$$\theta_1^{(a)} = -\frac{\sqrt{\Delta} + \alpha k_2}{2\alpha(1-k_2)} < 0, \quad \theta_1^{(b)} = \frac{\sqrt{\Delta} - \alpha k_2}{2\alpha(1-k_2)} > 0. \qquad (19.1.19)$$

It is easy to prove that $\theta_1^{(b)}<1$ when $\alpha > 1$, $\theta_1^{(b)}>1$ when $\alpha < 1$ and $\theta_1^{(b)}=1$ when $\alpha = 1$.

Consider the case of $\alpha \le 1$ (see Fig. 19.1.2). Then $\xi'(\theta_1) > 0$ for $0 \le \theta_1 < 1$ and, consequently, $\xi(\theta_1)$ increases in this interval. Since $\xi(1) = 1$, we conclude that $\xi \le 1$ for all permissible θ_1 and k_2.

In the case of $\alpha > 1$ (see Fig. 19.1.2), $\xi'(\theta_1) > 0$ for $0 \le \theta_1 < \theta_1^{(b)}$, $\xi'(\theta_1^{(b)}) = 0$ and $\xi'(\theta_1) < 0$ for $\theta_1^{(b)} < \theta_1 < 1$. Consequently, $\theta_1 = \theta_1^{(b)}$ is a point of maximum of the function $\xi(\theta_1)$ and $\xi(\theta_1) > \xi(1) = 1$. If $\xi(0) \ge 1$ then $\xi(\theta_1) > 1$ for $0 < \theta_1 < 1$. If $\xi(0) < 1$ then $\xi(\theta_1) > 1$ for $\tilde{\theta}_1 < \theta_1 < 1$ where $\tilde{\theta}_1 < 1$ is the root of the equation $\xi(\theta_1) = 1$.

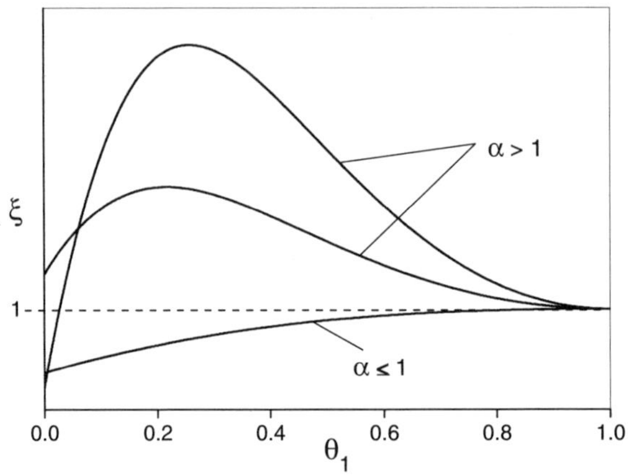

Fig. 19.1.2 . Different types of dependence ξ vs. θ_1 .

Therefore there is no advantage of a two-segment impactor over a monolithic impactor when $\alpha \le 1$. On the contrary, if $\alpha > 1$ then, for every pair of k_1 and k_2 $(0 < k_i < 1, k_1 + k_2 = 1)$, there exists an interval, $\tilde{\theta}_1 < \theta_1 < 1$, where a two-segment impactor is advantageous over the monolithic one, and this advantage is maximal when $\theta_1 = \theta_1^{(b)}$.

Generally, distribution of the mass between segments can also vary and ξ must be considered as a function of two variables, k_2 and θ_1. The character of the dependence $\xi = \xi(k_2, \theta_1)$ is illustrated in Fig. 19.1.3. Analysis shows that this function has a single maximum, at $k_2 = k_2^*$, $\theta_1 = \theta_1^*$. The value k_2^* can be found as a solution of the minimization problem for $\xi = \xi(k_2, \theta_1^{(b)}) = \psi(k_2)$; θ_1^* is determined by substituting $k_2 = k_2^*$ into the expression for $\theta_1^{(b)}$ in Eq. (19.1.19).

Maximum η that can be attained for a two-segment penetrator, and the corresponding optimum values of the parameters k_1, k_2 and θ_1 are shown in Fig. 19.1.4a. For the relatively large α, this maximum approaches 1.2-1.3. Data presented in Table 18.2.1 show that such situation is possible for soils where the ratio $\varepsilon = a_2/a_0$ can be in the range $0.0001 \div 0.0003\ s^2/m^2$ and for very high impact velocities.

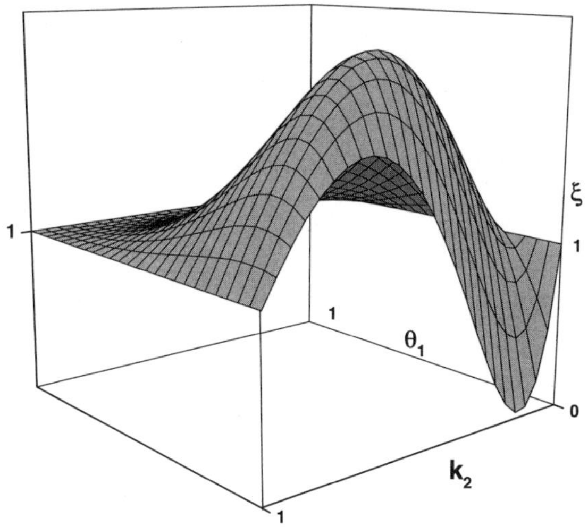

Fig. 19.1.3. Example of the dependence $\xi = \xi(k_2, \theta_1)$.

The values $\tilde{\eta}^{opt}$ correspond to the particular case $v_1 = 0$ which has been studied in Chapter 18. The values η^{opt} and $\tilde{\eta}^{opt}$ are close and practically coincide for large α. Similar situation occurs for three-segment penetrators (Fig. 19.1.4b).

An important remark is in place. Let Ξ_ν be a set of different penetrators that can be assembled from ν segments. Clearly, $\Xi_N \subset \Xi_{N+1}$, since Ξ_{N+1} reduces to Ξ_N, if in the first case we assume that a pair of adjacent segments is considered as one segment. Therefore increasing the number of segments provides more choices for selecting an optimum configuration and cannot decrease (and in practice increases) η^{opt} for given α. In particular, the possibility to increase the DOP using segmented impactors ($\alpha > 1$) that was demonstrated above for the case of $N = 2$ certainly exists in the case of $N > 2$.

Consider the case when all segments have the same mass. Then $k_1 = k_2 = ... = k_N = 1/N$ and Eq. (19.1.5) reads:

$$\overline{w}_j = \frac{1}{j}\left(1 + \sum_{\vartheta=1}^{j-1} \prod_{\nu=\vartheta}^{j-1} \theta_\nu\right), \quad \overline{w}_1 = 1, \quad j = 2, ..., N. \qquad (19.1.20)$$

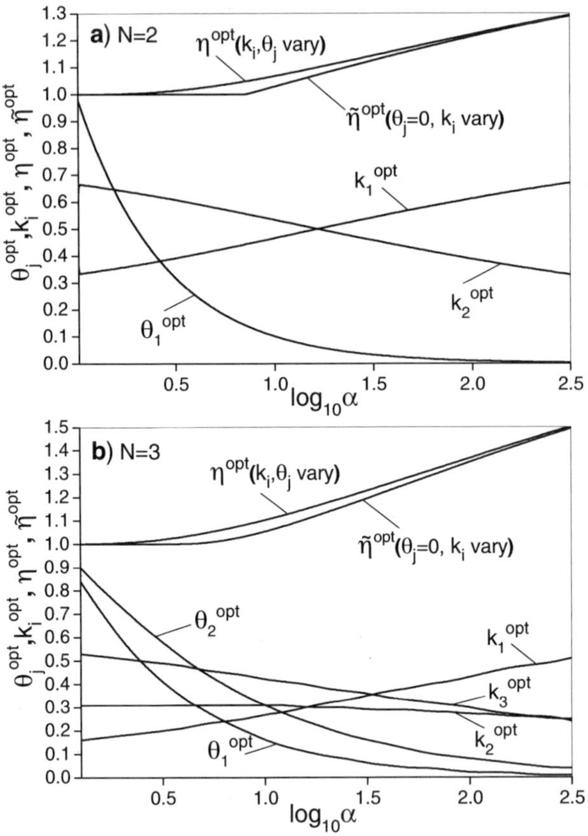

Fig. 19.1.4. Optimum solutions for two- and three-segment impactors.

The optimal values of the parameter η for this type of penetrators are denoted as $\hat{\eta}^{opt}$ and are shown in Fig. 19.1.5 for $N = 2$ and $N = 3$, respectively. For comparison, in these plots we presented also the results of impactor optimization for two other formulations of the problem. There is no advantage of the segmented impactor with the same mass of the segments over a monolithic impactor for relatively small values of α, whereas the effectiveness of such segmentation is close to the optimal (for given N) for large values of α.

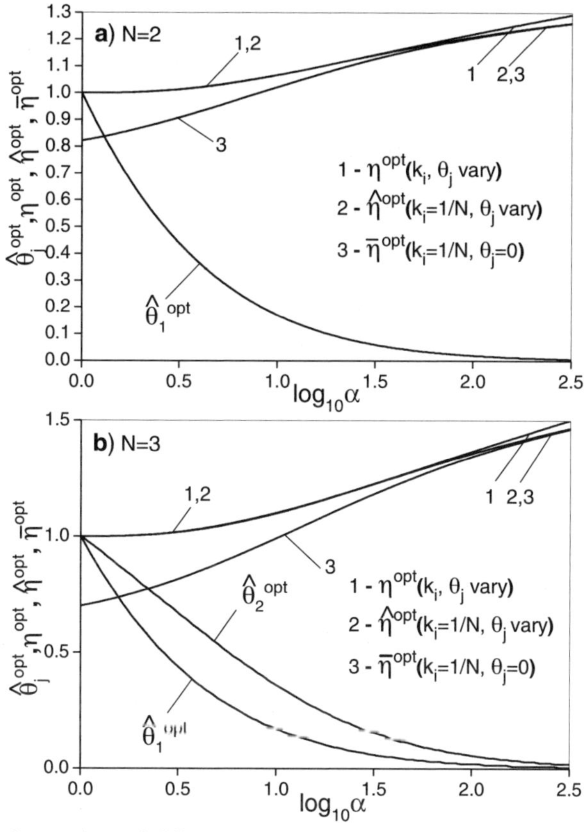

Fig. 19.1.5. Comparison of different solutions for two- and three-segment impactors.

19.2 Investigation of Problem Using Continuous Model

19.2.1 *Formulation of problem*

In the limit $N \to \infty$ a mass of penetrator varies continuously. It is convenient to introduce directly the limiting continuous model. Clearly, the maximum DOP obtained for each value of α using the continuous model is the upper bound (for the same magnitude of α) for the DOP of segmented impactors when k_j, θ_j, and N are varied.

Penetrator can be considered to be a shell having a constant nose shape which is filled with material during penetration. Assume that initially the mass of the shell $\mu < m$ and velocity v_{imp} are small, the acquired material has velocity v_{imp}, and the resistance force is determined by Eq. (19.1.7). Motion of the impactor is governed by the equation of motion of a projectile with a variable mass (see, e.g., Sommerfeld, 1952) that in the considered case reads:

$$M\frac{dv}{dt} + (v - v_{imp})\frac{dM}{dt} = -(A_2 v^2 + A_0), \tag{19.2.1}$$

where v is the velocity of the impactor, t is time, M is the instantaneous mass of the impactor that varies in the range between μ and the final value m, i.e.,

$$\mu \le M \le m. \tag{19.2.2}$$

Since the mass of the penetrator cannot decrease, the following constraint is valid:

$$\frac{dM}{dt} \ge 0. \tag{19.2.3}$$

Taking into account that $dt = dh/v$, Eq. (19.2.1) can be rewritten as follows:

$$dh = \psi dv + \varphi dM; \quad \psi = -\frac{Mv}{A_2 v^2 + A_0}, \quad \varphi = -\frac{v(v - v_{imp})}{A_2 v^2 + A_0}. \tag{19.2.4}$$

Eq. (19.2.4) implies the following expression for the distance on the plane (M, v) that the penetrator traverses from the point $Y(M_Y, v_Y)$ to the point $Z(M_Z, v_Z)$:

$$\Delta h_{YZ} = \int_{(M_Y, v_Y)}^{(M_Z, v_Z)} \varphi\, dM + \psi\, dv, \tag{19.2.5}$$

where integration is performed along the trajectory of the projectile. Hence the DOP can be expressed as follows:

$$P = \int_{(0,v_{imp})}^{(m,0)} \varphi dM + \psi dv . \tag{19.2.6}$$

Using Eq. (19.2.4) we obtain that

$$\frac{dM}{dt} = v\frac{dM}{dh} = -\frac{A_2v^2 + A_0}{Mv' + v - v_{imp}}, \quad v' = \frac{dv}{dM}, \tag{19.2.7}$$

where the function $v = v(M)$ describes the trajectory of the projectile except for the segments where $M = const$ (passive motion of penetrator) and $dM = 0$, i.e., the inequality in Eq. (19.2.3) is valid. Taking into account that the expression $A_2v^2 + A_0$ has an upper bound and is positive, inequality in Eq. (19.2.3) can be rewritten as

$$Mv' + v - v_{imp} \le 0 . \tag{19.2.8}$$

Clearly, the following inequality must be also valid:

$$0 \le v \le v_{imp} . \tag{19.2.9}$$

The problem is reduced now to determining the path between the initial points (μ, v_{imp}) and the final point $(m, 0)$ that provides the maximum DOP, P, in Eq. (19.2.6) taking into account the constraints given by Eqs. (19.2.1), (19.2.2), (19.2.8) and (19.2.9). The parameters A_0, A_2, m and v_{imp} are assumed to be given. The solution of this problem for $\mu \to 0$ gives an upper bound of the DOP which can be achieved by segmented penetrator when the number of segments, their masses and velocity of the immersed in the shield portion of the projectile at the moment of the impact by a successive segment vary.

Equations (19.2.1)–(19.2.2) and (19.2.6)–(19.2.9) imply the following expressions that can be written using for convenience the dimensionless variables that are defined below:

$$\overline{M}\frac{d\overline{v}}{d\overline{t}} + (\overline{v} - 1)\frac{d\overline{M}}{d\overline{t}} = -(\alpha\overline{v}^2 + 1), \tag{19.2.10}$$

$$\Delta\overline{h}_{YZ} = \int_{(M_{\overline{Y}},v_{\overline{Y}})}^{(M_{\overline{Z}},v_{\overline{Z}})} \overline{\varphi} d\overline{M} + \overline{\psi} d\overline{v}, \quad \overline{\psi} = -\frac{\overline{M}\overline{v}}{\alpha\overline{v}^2+1}, \quad \overline{\varphi} = -\frac{\overline{v}(\overline{v}-1)}{\alpha\overline{v}^2+1}, \tag{19.2.11}$$

$$\overline{P} = \int_{(\overline{\mu},1)}^{(1,0)} \overline{\varphi} d\overline{M} + \overline{\psi} d\overline{v}, \tag{19.2.12}$$

$$\overline{M}\overline{v}' + \overline{v} - 1 \le 0, \tag{19.2.13}$$

$$\overline{\mu} \le \overline{M} \le 1, \quad 0 \le \overline{v} \le 1, \tag{19.2.14}$$

$$\frac{d\overline{M}}{d\overline{t}} = -\frac{\alpha\overline{v}^2+1}{\overline{M}\overline{v}' + \overline{v} - 1}, \tag{19.2.15}$$

where masses and velocities are normalized using m and v_{imp}, correspondingly, and

$$\overline{P}_0 = \frac{A_0}{mV_{imp}^2} P_0, \overline{P} = \frac{A_0}{mV_{imp}^2} P, \Delta\overline{h}_{YZ} = \frac{A_0}{mV_{imp}^2} \Delta h_{YZ}, \overline{t} = \frac{A_0}{mV_{imp}} t. \tag{19.2.16}$$

The problem is reduced to the maximization of the functional $\overline{P}$ in Eq. (19.2.12), whereas the solution must satisfy the constraints given by Eqs. (19.2.13)–(19.2.15).

19.2.2 *Connection between discrete and continuous problems*

The above considered discrete version of the problem can be described as a particular case of the continuous model. This assertion for a two-segment projectile is illustrated in Fig. 19.2.1 where the path *ABCDE* describes penetration.

The interval *AB* of the trajectory has the equation $\overline{v} = 1$ and corresponds to formation of the first segment having the dimensionless mass $k_1 = m_1/m$ and the dimensionless velocity equal to 1. Eq. (19.2.11) shows that $\Delta\overline{h}_{AB} = 0$, i.e., the segment is formed before beginning of motion.

The interval BC has the equation $\overline{M} = k_1$ and describes passive penetration of the first segment whereby the dimensionless velocity decreases from 1 to $\overline{v}_1$ and the traveled distance is

$$\Delta\overline{h}_{BC} = -k_1 \int_1^{\overline{v}_1} \frac{\overline{v}d\overline{v}}{\alpha\overline{v}^2 + 1} = \frac{k_1}{2\alpha} \ln\left(\frac{\alpha + 1}{\alpha\overline{v}_1^2 + 1}\right). \qquad (19.2.17)$$

The interval CD describes the collision between the first and the second segments. In this case, $d\overline{M}/d\overline{t} = +\infty$ and Eq. (19.2.15) shows that the equation of the curve CD is $\overline{M}\overline{v}' + \overline{v} - 1 = 0$. The solution of this equation that passes through the point $C(k_1, \overline{v}_1)$ is $\overline{v} = 1 - k_1(1 - \overline{v}_1)/\overline{M}$. Substituting $\overline{M} = 1$ in this solution we obtain the ordinate of the point D, $\overline{w}_2 = 1 - k_1(1 - \overline{v}_1)$. Clearly,

$$\Delta\overline{h}_{CD} = -\int_{k_1}^{1} \frac{(\overline{M}\overline{v}' + \overline{v} - 1)\overline{v}}{\alpha\overline{v}^2 + 1} = 0. \qquad (19.2.18)$$

The interval DE has equation $\overline{M} = 1$ and describes passive penetration of composite striker consisting of two segments in the shield whereby the dimensionless velocity decreases from $\overline{w}_2$ to 0 and the traveled distance is

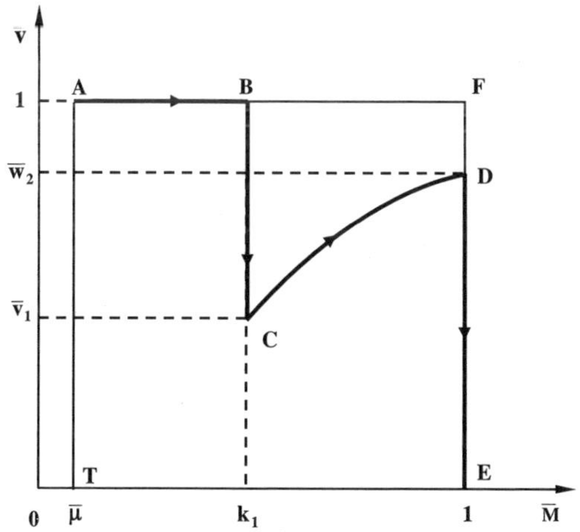

Fig. 19.2.1. Discrete model for a two-segment impactor as a particular case of the continuous model.

$$\Delta\bar{h}_{DE} = -\int_{\bar{w}_2}^{0} \frac{\bar{v}d\bar{v}}{\alpha\bar{v}^2+1} = \frac{1}{2\alpha}\ln(\alpha\bar{w}_2^2+1)\,. \tag{19.2.19}$$

Consequently, formula for the DOP reads:

$$\bar{P} = \Delta\bar{h}_{BC} + \Delta\bar{h}_{DE} = \frac{1}{2\alpha}\ln\left[\frac{(\alpha+1)^{k_1}(\alpha\bar{w}_2^2+1)}{(\alpha\bar{v}_1^2+1)^{k_1}}\right]. \tag{19.2.20}$$

Penetration of the homogeneous impactor is described by the trajectory *AFE* (see Fig. 19.2.1). Since $\Delta\bar{h}_{AF}=0$, the DOP can be calculated as follows:

$$\bar{P}_0 = -\int_{1}^{0} \frac{\bar{v}d\bar{v}}{\alpha\bar{v}^2+1} = \frac{1}{2\alpha}\ln(\alpha+1)\,. \tag{19.2.21}$$

Since $P/P_0 = \bar{P}/\bar{P}_0$, and taking into account Eqs. (19.2.20) and (19.2.21), Eq. (19.1.12) yields:

$$\xi = \exp\left[\left(\frac{\bar{P}}{\bar{P}_0}-1\right)\ln(\alpha+1)\right] = \frac{(\alpha+1)^{k_1-1}(\alpha\bar{w}_2^2+1)}{(\alpha\bar{v}_1^2+1)^{k_1}}. \tag{19.2.22}$$

Substituting $k_1 = 1-k_2, \bar{v}_1 = \theta_1, \bar{w}_2 = k_2+(1-k_2)\theta_1$ into Eq. (19.2.22) we recover Eq. (19.1.15).

19.2.3 *Solution of continuous problem*

Optimization problem which has been formulated in Section 19.2.1 has some features which complicate application of the classical calculus of variations (CV). Firstly, being transformed into the conventional form, the integral in Eq. (19.2.12) has the integrand that depends linearly on the derivative of the unknown function. Methods of classical CV are not effective in the case of such degenerative problem. Secondly, the possibility of the intervals where $\bar{M} = const$ and $\bar{v} = const$ must be also taken into account. Because of these reasons for solving this problem we use a special method (Miele, 1962c) that was developed exactly for such problems.

Let us calculate the function

$$\omega(\bar{v}) = \frac{\partial \psi}{\partial \bar{M}} - \frac{\partial \varphi}{\partial \bar{v}} = \frac{(1-\bar{v})(\alpha\bar{v}^2 - 1)}{(\alpha\bar{v}^2 + 1)^2} \tag{19.2.23}$$

and temporarily ignore the constraint given by Eq. (19.2.13).

Consider the case when $\alpha \le 1$. Hence $\omega < 0$ inside the domain $AFET$ (see Fig. 19.2.2a). According to the method suggested in Miele, 1962c, AFE is the optimum trajectory among the trajectories located inside the domain and connecting the points A and E. The criterion for selecting the optimal trajectory is the following: the optimum trajectory must be a segment of the boundary of the domain and the domain must be located on the right when the point moves from A to E.

If $\alpha > 1$ then the domain $AFET$ (see Fig. 19.2.2b) includes two sub-domains: $AFSG$ where $\bar{v}_\times < \bar{v} < 1$ and $\omega > 0$, $GSET$ where $0 < \bar{v} < \bar{v}_\times$ and $\omega < 0$ where $\bar{v}_\times = 1/\sqrt{\alpha}$. According to the method described in Miele, 1962a, the optimum trajectory is $AGSE$.

Since the constraint given by Eq. (19.2.13) is valid for the trajectories that we called "optimal", the obtained solution is indeed optimal. In the

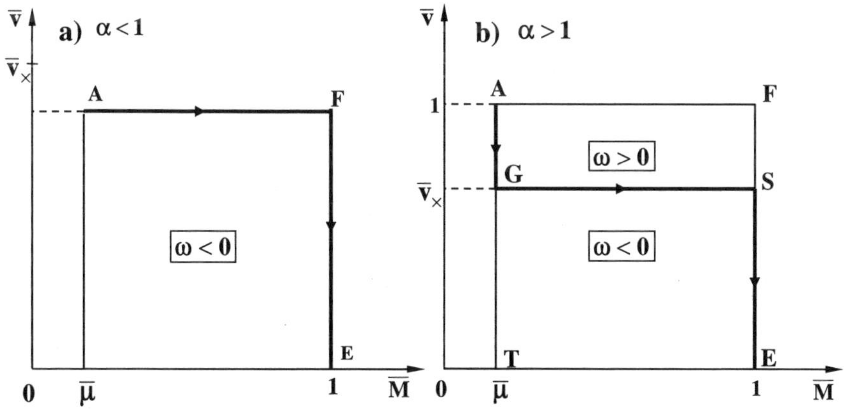

Fig. 19.2.2. Optimum solutions for the continuous model.

case $\alpha \le 1$, the trajectory AFM describes the motion of the monolithic penetrator. Therefore segmentation can not increase the DOP. Clearly, $\bar{\mu}$ does not affect the DOP as shows Eq. (19.2.21).

In the case of $\alpha > 1$, formula for the maximum DOP reads:

$$\begin{aligned}\bar{P}^{opt} &= \Delta\bar{h}_{BC} + \Delta\bar{h}_{GS} + \Delta\bar{h}_{SE} \\ &= -\bar{\mu}\int_1^{\bar{v}_\times} \frac{\bar{v}d\bar{v}}{\alpha\bar{v}^2+1} - \frac{\bar{v}_\times(\bar{v}_\times - 1)}{\alpha\bar{v}_\times^2+1}\int_{\bar{\mu}}^1 d\bar{M} - \int_{\bar{v}_\times}^0 \frac{\bar{v}d\bar{v}}{\alpha\bar{v}^2+1} \\ &= \frac{\bar{\mu}}{2\alpha}\ln\left(\frac{\alpha+1}{\alpha\bar{v}_\times^2+1}\right) + \frac{\bar{v}_\times(1-\bar{v}_\times)}{\alpha\bar{v}_\times^2+1}(1-\bar{\mu}) + \frac{1}{2\alpha}\ln(\alpha\bar{v}_\times^2+1).\end{aligned} \tag{19.2.24}$$

After substituting $\bar{v}_\times = 1/\sqrt{\alpha}$ we obtain:

$$\eta^{opt}(\alpha,\bar{\mu}) = \frac{\bar{P}^{opt}}{\bar{P}_0} = \Theta - \bar{\mu}(\Theta - 1) = \bar{\mu} + (1-\bar{\mu})\Theta, \tag{19.2.25}$$

where

$$\Theta = \frac{\Theta_1}{\Theta_2}, \Theta_1(\alpha) = \sqrt{\alpha} + \ln 2 - 1, \Theta_2(\alpha) = \ln(\alpha+1). \tag{19.2.26}$$

Denote

$$\chi(\alpha) = \Theta_1(\alpha) - \Theta_2(\alpha) = \sqrt{\alpha} + \ln 2 - 1 - \ln(\alpha+1). \tag{19.2.27}$$

Since

$$\chi(1) = 0, \quad \chi'(\alpha) = \frac{(\sqrt{\alpha}-1)^2}{2\sqrt{\alpha}(\alpha+1)} > 0, \tag{19.2.28}$$

it can be concluded that $\chi(\alpha) > 0$ for $\alpha > 1$. Consequently, $\Theta_1(\alpha) > \Theta_2(\alpha)$ and $\Theta(\alpha) > 1$ for $\alpha > 1$. Eq. (19.2.25) implies that $\eta^{opt} > 1$ for all $0 < \bar{\mu} < 1$ and $\alpha > 1$, i.e., as expected, the optimum solution is better than the solution associated with the monolithic impactor. Decreasing of η^{opt} with increase of $\bar{\mu}$ can be also easily explained: increasing $\bar{\mu}$ results in shrinking of the domain where we search for the optimal solution.

The results of calculations presented in Fig. 19.2.3 show that non-monolithic impactors can have a significant advantage over monolithic impactors having the same mass but the ratio of the optimum DOP to the DOP of the monolithic impactor can not exceed $\Theta(\alpha)$ (see the curve for $\bar{\mu}=0$ in Fig. 19.2.3).

Let us determine the optimum control, function $\overline{M}(\bar{t})$, that provides the maximum DOP.

On the interval AG (Fig. 19.2.2 b) function $\overline{M}(\bar{t})$ is known, $\overline{M}(\bar{t})=\bar{\mu}$. In this case Eq. (19.2.10) reads:

$$\bar{\mu}\frac{d\bar{v}}{d\bar{t}}=-(\alpha\bar{v}^2+1)\,. \tag{19.2.29}$$

Hence

$$\bar{\mu}\int_1^{\bar{v}_\times}\frac{d\bar{v}}{\alpha\bar{v}^2+1}=-\int_{\bar{t}_A}^{\bar{t}_G}d\bar{t}\,,\quad \bar{v}_\times=1/\sqrt{\alpha}\,, \tag{19.2.30}$$

where $\bar{t}_A$ and $\bar{t}_G$ are the moments of time when the penetrator is located in the corresponding points; hereafter we assume $\bar{t}_A=0$. Eq. (19.2.30) yields:

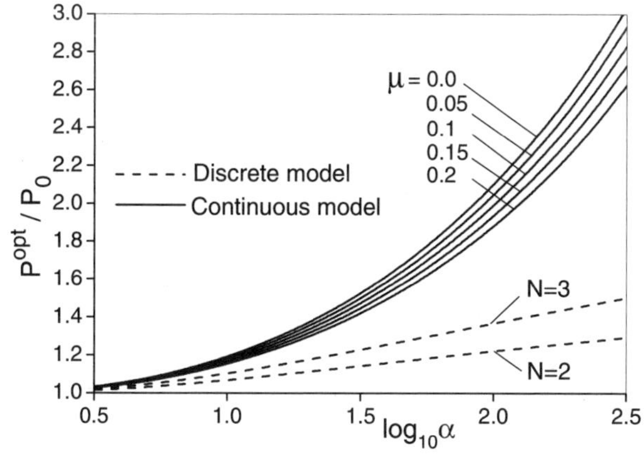

Fig. 19.2.3. Normalized optimum DOP for the considered models

$$\bar{t}_G = \frac{\bar{\mu}}{\sqrt{\alpha}}\left(\tan^{-1}\sqrt{\alpha} - \frac{\pi}{4}\right). \tag{19.2.31}$$

Motion from point G to point S is described by Eq. (19.2.10) with $\bar{v} = \bar{v}_\times$:

$$(\bar{v}_\times - 1)\frac{d\overline{M}}{d\bar{t}} = -(\alpha\bar{v}_\times^2 + 1). \tag{19.2.32}$$

Therefore the instantaneous mass of the impactor, $\overline{M}$, and time, $\bar{t}$, are related by the following formula:

$$(\bar{v}_\times - 1)\int_{\bar{\mu}}^{\overline{M}} d\hat{M} = -(\alpha\bar{v}_\times^2 + 1)\int_{\bar{t}_G}^{\bar{t}} d\hat{t}, \tag{19.2.33}$$

which implies that

$$\begin{gathered}\overline{M}(\bar{t}) = \bar{\mu} + \frac{2\sqrt{\alpha}}{\sqrt{\alpha}-1}(\bar{t} - \bar{t}_G), \\ \bar{t}_G \le \bar{t} \le \bar{t}_S, \quad \bar{t}_S = \bar{t}_G + \frac{\sqrt{\alpha}-1}{2\sqrt{\alpha}}(1-\bar{\mu}).\end{gathered} \tag{19.2.34}$$

The interval SE corresponds to the motion of the impactor without change of its mass until it stops.

Therefore the active optimal control is to increase linearly the mass of penetrator in order to provide constant velocity of the penetrator $\bar{v} = \bar{v}_\times$ during time interval $\bar{t}_G \le \bar{t} \le \bar{t}_S$.

It must be noted that in the limiting case $\bar{\mu} = 0$ the interval AG describes instantaneous reduction of the velocity of a zero mass impactor by a finite increment. In order to prevent from this hardly interpretable situation we consider small but finite values of $\bar{\mu}$. This approximation is quite legitimate since Eq. (19.2.25) and results presented in Fig. 19.2.3 show that small increase of $\bar{\mu}$ above zero results in a small reduction of the optimal DOP in comparison with the DOP for the limiting ideal case with $\bar{\mu} = 0$.

19.3 Concluding Remarks

The main results obtained in this chapter can be summarized as follows. On the basis of the widely used two-term approximate model of impactor-shield interaction we developed two models of high-speed penetration pertinent to discrete and continuous increase of the mass of impactor. We solved analytically the problem of minimizing the DOP using the continuous model. The obtained solution is of interest by itself. However it allows us also to determine the upper bound for the increase of the DOP by discrete segmentation of the penetrator and to evaluate the efficiency of different variants of segmentation. We found a parameter α which determines the feasibility to increase the DOP by gradual increase of the mass of penetrator during penetration, and, in particular, by segmentation. We demonstrated that for $\alpha \leq 1$ these methods do not allow increasing the DOP, irrespectively of the used algorithm for discrete or continuous increase of the mass of penetrator. However, for $\alpha > 1$ the DOP can be increased by using appropriate algorithm for control of the mass of the impactor. We showed analytically that in the case of discrete increase of the mass of penetrator the DOP can be increased for any number of segments. The results of analytical investigation are illustrated and extended using numerical simulations. Clearly, the obtained theoretical predictions require experimental validation.

Chapter 20

Hypervelocity Impact

Nomenclature

d	diameter of segments;
DOP	depth of penetration;
f	function that determines the model, Eq. (20.1.1);
L_i	length of i th segment;
$\overline{L}_i$	$= L_i/d$;
$\overline{L}_{\min}, \overline{L}_{\max}$	minimum and maximum values of $\overline{L}_i$ ($i = 1,2,...,N$);
N	number of segments;
P	DOP of segmented impactor;
$\overline{P}$	$= P/d$;
P_0	DOP of monolithic impactor;
$\overline{P}_0$	$= P_0/d$;
P_i	DOP of the i th segment;
$\overline{P}_i$	$= P_i/d$;
Q	function, Eq. (20.3.2);
v_{imp}	impact velocity of projectile.

20.1 Formulation of Problem

Consider normal sequential penetration of cylindrical segments (rods) having the same very high impact velocity v_{imp} into a semi-infinite shield. The impact velocity is sufficiently high so that penetration is accompanied by the decrease of the length of the projectiles because of their erosion. We assume that segments do not interact with each other during penetration, the materials of the segments and of the shield retain their mechanical properties, and penetrating segments erode completely, i.e., "equal segments penetrate equal amounts" (Charters *et al.*, 1990), independent of the history of the penetration process. A sequence of such penetrating segments can be considered as a model of penetration by a segmented rod.

Assume that the DOP of the i th segment, P_i, depends on its length, L_i, as follows:

$$\overline{P}_i = f(\overline{L}_i), \quad \overline{P}_i = P_i/d, \quad \overline{L}_i = L_i/d, \quad i = 1,2,...,N, \qquad (20.1.1)$$

where the function f determines the model, d is the diameter of the segments, N is the number of segments. In order to simplify notations, the impact velocity and parameters characterizing the mechanical properties of the materials of the segments and of the shield are not listed as arguments of the function f.

We assume that Eq. (20.1.1) is considered in the following range of L_i:

$$\overline{L}_{\min} \le \overline{L}_i \le \overline{L}_{\max}, \quad i = 1,2,...,N, \qquad (20.1.2)$$

where $\overline{L}_{\min}$ and $\overline{L}_{\max}$ are given.

Let us consider also penetration by a monolithic rod with the length that is equal to the sum of the lengths of the segments and having the same impact velocity. Clearly, the impact energy of the monolithic rod and of the segments is the same.

Formulas for the total DOP after impacts by N segments $\overline{P}$ and the DOP of the rod $\overline{P}_0$ read:

$$\overline{P} = \frac{P}{d} = \sum_{i=1}^{n} f(\overline{L}_i), \quad \overline{P}_0 = \frac{P_0}{d} = f(\overline{L}), \quad \overline{L} = \sum_{i=1}^{N} \overline{L}_i. \qquad (20.1.3)$$

Let us investigate the behavior of $\overline{P}$ as a function of $\overline{L}_i$ and compare the magnitudes of $\overline{P}$ and $\overline{P}_0$ for the same $\overline{L}$.

20.2 Comparison of DOP of Monolithic Impactor and Segmented Impactors

Let $f(\tilde{L})$ be a convex downward positive increasing function for $0 \le \bar{L}_{\min} < \tilde{L} < \bar{L}_{\max}$ that satisfies the following condition:

$$\bar{L}_{\min} f'(\bar{L}_{\min}) - f(\bar{L}_{\min}) \ge 0 . \qquad (20.2.1)$$

Using Property 1 of convex downward positive increasing functions (see Appendix A), we obtain the following estimate for $\bar{P}$:

$$\bar{P} = \sum_{i=1}^{N} f(\bar{L}_i) = \sum_{i=1}^{N} f(\frac{\bar{L}_i}{\bar{L}}\bar{L}) < \sum_{i=1}^{N} \frac{\bar{L}_i}{\bar{L}} f(\bar{L}) = f(\bar{L}) = \bar{P}_0 . \qquad (20.2.2)$$

It can be shown that $\bar{P} = \bar{P}_0$ only in the degenerate case when the segmented impactor is replaced by the monolithic impactor.

Therefore for any impact velocity, the monolithic impactor with the length $\bar{L}$ has an advantage over the segmented strikers having the same total length whereby the distribution of the lengths of the segments satisfies the constraints given by Eq. (20.1.2).

Let now $f(\tilde{L})$ be a convex upward, positive, increasing function for $0 \le \bar{L}_{\min} < \tilde{L} < \bar{L}_{\max}$ that satisfies the following condition:

$$\bar{L}_{\min} f'(\bar{L}_{\min}) - f(\bar{L}_{\min}) \le 0 . \qquad (20.2.3)$$

Employing analysis that is similar to the analysis in the case of a convex downward function $f(\bar{L})$ and using Property 2 of convex upward positive increasing functions (see Appendix A), yields that $\bar{P} > \bar{P}_0$, and $\bar{P} = \bar{P}_0$ only in the degenerate case when the segmented impactor is replaced by the monolithic impactor.

Therefore for any impact velocity the segmented strikers with the total length $\bar{L}$ whereby the distribution of the lengths of the segments satisfies the constraints given by Eq. (20.1.2), have the advantage over the monolithic impactor having the same length.

Let $f(\tilde{L})$ be a positive linear function, $f(\tilde{L}) = k_1\tilde{L} + k_0$ for $0 \le \bar{L}_{\min} < \tilde{L} < \bar{L}_{\max}$, where k_0 and $k_1 > 0$ are parameters. Hence

$$\overline{P} = \sum_{i=1}^{N} (k_1 \overline{L}_i + k_0) = k_1 \overline{L} + k_0 N = \overline{P}_0 + k_0 (N-1) . \qquad (20.2.4)$$

Consequently, the above results for the convex downward and convex upward function $f(\overline{L})$ can be extended to the linear model when $k_0 < 0$ and $k_0 > 0$, correspondingly. If $k_0 = 0$ then all segmented impactors and the monolithic impactor penetrate to the same depth.

20.3 Optimum Segmentation

Consider a rod having a length $\overline{L}$ and assume that the following inequality is satisfied:

$$N \overline{L}_{\min} < \overline{L} < N \overline{L}_{\max} . \qquad (20.3.1)$$

Inequality (20.3.1) implies that the set of segments with the same length, $\overline{L}/N$, is permissible. Let us prove that the optimal (minimum/maximum) DOP of the segmented impactor is attained for this partition into segments if f is a positive, increasing, convex downward/convex upward function.

Let f be a convex downward function. Since the monolithic impactor has an advantage over the segmented strikers, one can consider the worst possible segmentation.

It is convenient to exclude the variable $\overline{L}_N$ from the set of the arguments and to consider the following optimization problem:

$$\overline{P} = Q(\overline{L}_1, \overline{L}_2, ..., \overline{L}_{N-1}) = \sum_{j=1}^{N-1} f(\overline{L}_j) + f\left(\overline{L} - \sum_{j=1}^{N-1} \overline{L}_j \right) \rightarrow \min \qquad (20.3.2)$$

subjected to the following constraints:

$$\overline{L} - \overline{L}_{\max} \le \sum_{j=1}^{N-1} \overline{L}_j < \overline{L} - \overline{L}_{\min} , \ \overline{L}_{\min} \le \overline{L}_i \le \overline{L}_{\max} , \quad i = 1,2,..., N-1 . \qquad (20.3.3)$$

Let us investigate the behavior of the function Q using methods of classical mathematical analysis (Korn and Korn, 1968) and of convex analysis (Rockafellar, 1970).

Taking into account that

$$\frac{\partial Q}{\partial \bar{L}_\nu} = f'(\bar{L}_\nu) - f'\left(\bar{L} - \sum_{j=1}^{N-1} \bar{L}_j\right),$$

$$\frac{\partial^2 Q}{\partial \bar{L}_\nu \partial \bar{L}_\mu} = \kappa_{\nu\mu} f''(\bar{L}_\nu) + f''\left(\bar{L} - \sum_{j=1}^{N-1} \bar{L}_j\right), \tag{20.3.4}$$

$$\nu = 1,\ldots, N-1,\ \mu = 1,\ldots, N-1, \quad \kappa_{\nu\mu} = \begin{cases} 1 & if \quad \nu = \mu \\ 0 & if \quad \nu \neq \mu \end{cases}$$

the characteristic equation of the Hessian (matrix with the second derivatives of the function Q) reads:

$$\prod_{j=1}^{N-1} \left[f''(\bar{L}_\nu) - \eta\right] = 0\,. \tag{20.3.5}$$

Since all roots of Eq. (20.3.5) are positive, the Hessian is positive definite in the domain determined by Eq. (20.3.3) and, therefore, $Q(\bar{L}_1, \bar{L}_2, \ldots, \bar{L}_{N-1})$ is a convex downward function. The latter assertion implies that (i) the point $\bar{L}_1 = \bar{L}_2 = \ldots = \bar{L}_{N-1} = \bar{L}/N$ where all first partial derivatives of function Q equal zero (see Eq. (20.3.4)) is a point of the local minimum and (ii) this local minimum is also a global minimum because the domain determined by linear inequalities in Eq. (20.3.3) is convex. Therefore, splitting the rod into the segments having equal lengths is the worst possible segmentation.

The case of convex upward f can be considered similarly. Therefore the best segmentation is achieved when segments have the same length.

If f is a linear function the distribution of the length of segment does not effect the DOP of the segmented impactor.

20.4 Effect of Number of Segments on DOP

In this Section, we investigate the effect of the number of segments on the DOP assuming that the lengths of the segments are the same.

Let $f(\tilde{L})$ be a convex downward positive increasing function for $0 \le \bar{L}_{\min} < \tilde{L} < \bar{L}_{\max}$ that satisfies the constraint given by Eq. (20.2.1) which can be rewritten as follows:

$$\xi(\bar{L}_{\min}) \ge 0, \quad \xi(x) = xf'(x) - f(x). \tag{20.4.1}$$

Since $\xi'(\tilde{L}) = \tilde{L}f''(\tilde{L}) > 0$ for $\bar{L}_{\min} < \tilde{L} < \bar{L}_{\max}$, $\xi(\tilde{L})$ is an increasing function and, consequently, $\xi(\tilde{L}) > 0$ in this interval.

The DOP of the set of segments having the lengths $\bar{L}/N$ is $\bar{P}(N) = N f(\bar{L}/N)$. Assuming that N varies continuously we find that $\bar{P}'(N) = -\xi(\bar{L}/N) < 0$. Therefore the DOP decreases with the increase of the number of segments ($\bar{L}/\bar{L}_{\max} < N < \bar{L}/\bar{L}_{\min}$).

The case when $f(\tilde{L})$ is a convex upward positive increasing function for $0 \le \bar{L}_{\min} < \tilde{L} < \bar{L}_{\max}$ that satisfies the condition, $\xi(\bar{L}_{\min}) \le 0$, can be considered similarly. Consequently we can conclude that the DOP increases with the increase of the number of segments ($\bar{L}/\bar{L}_{\max} < N < \bar{L}/\bar{L}_{\min}$).

20.5 Verification of Approach

Data on the dependence of the DOP on the length of the rods are very scarce. However, the available information indicates that $f(\tilde{L})$ is a convex upward, positive, increasing function for $0 \le \bar{L}_{\min} < \tilde{L} < \bar{L}_{\max}$ that satisfies the constraint given by Eq. (20.2.3).

Anderson *et al.*, 1996 summarized the results of experiments and numerical simulations on ordnance impact velocity range. These results in the rearranged form are shown in Fig. 20.5.1a (references corresponding to different symbols in these plots can be found in Anderson *et al.*, 1996). Anderson *et al.*, 1996 proposed the regression model

$$f(\bar{L}) = (\chi_1 - \chi_2 \ln \bar{L})\bar{L}, \tag{20.5.1}$$

where

$$\chi_1 = -0.209 + 1.044(v_{imp}/\tilde{v}) > 0, \; \chi_2 = 0.194, \; \tilde{v} = 1\,km/s \tag{20.5.2}$$

in the case of penetration of tungsten projectiles into rolled homogeneous armor. The curves determined by Eq. (20.5.1) for several values of v_{imp} are shown in Fig. 20.5.1a.

Since

$$f'(\overline{L}) = \chi_1 - \chi_2(\ln\overline{L}+1) = 1.044(v_{imp}/\tilde{v}) - 0.194\ln\overline{L} - 0.403 > 0 \qquad (20.5.3)$$

for $v_{imp} \geq 1.1\,km/s$ and $\overline{L} \leq 30$, $f''(\overline{L}) = -\chi_2/\overline{L} < 0$ and $\lim_{\overline{L}\to 0} f(\overline{L}) = 0$, the function, $f(\tilde{L})$, satisfies the above requirements.

Similar conclusion can be drawn on the basis of the results of computer simulations performed by Anderson *et al.*, 1995. These data are shown in Fig. 20.5.1b (symbols) where curves correspond to the approximation $f(\overline{L}) = \chi_3 + \chi_4\overline{L} + \chi_5\overline{L}^2$, where coefficients χ_3, χ_4 and χ_5 depend on the impact velocity. Values of these coefficients for different impact velocities are presented in Table 20.5.1.

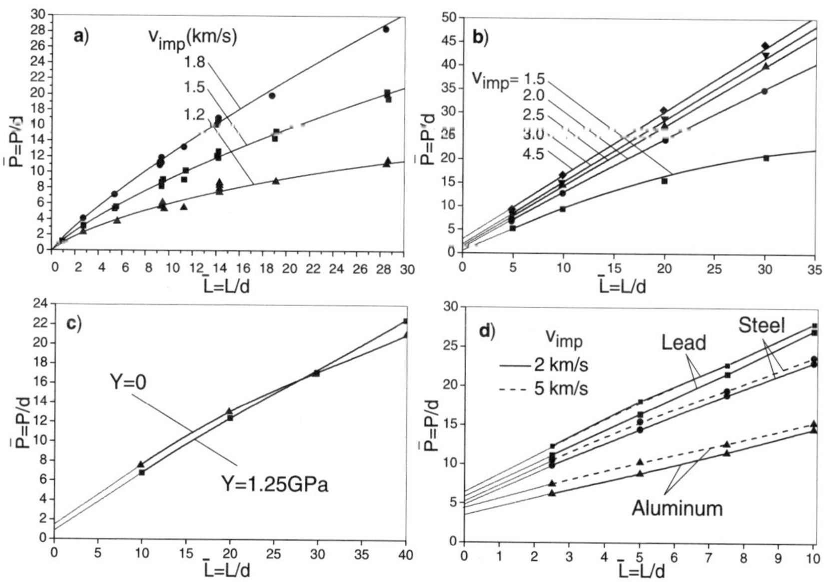

Fig. 20.5.1. Dimensionless DOP $\overline{P}$ vs. dimensionless length of the rod $\overline{L}$: data from a) Anderson *et al.*, 1996, b) Anderson *et al.*, 1995, c) Rosenberg and Dekel, 1994, d) Fedorov *et al.*, 2007.

Table 20.5.1. Parameters of the model of Anderson *et al.*, 1995.

Velocity (*km/s*)	χ_3	χ_4	χ_5
1.5	0.1362	1.0727	-0.012658
2.0	1.5398	1.2041	-0.002358
2.5	1.5186	1.2803	0
3.0	1.9116	1.3279	0
4.5	2.9859	1.3638	0

The model proposed by Gerasimov *et al.*, 2008 implies $f(\bar{L}) \sim \bar{L}^{2/3}$ and, consequently, it satisfies the above requirements.

Results of computer simulations on impact of hardened steel shield by impactor with different yield strengths Y (Rosenberg and Dekel, 1994) shown in Fig. 20.5.1c (symbols) as well as the numerical results obtained by Fedorov *et al.*, 2007 (symbols in Fig. 20.5.1d) on penetration of rods manufactured from different materials into soil, support the above conclusion.

Theoretical approach developed in this chapter predicts an advantage of the segmented impactors over the monolithic impactor and increase of the effectiveness of segmentation with the increase of the number of segments, provided that the specified above requirements are satisfied. These theoretical predictions are supported by experiments and numerical simulations mentioned in the introduction to Part 8.

Without dwelling on the details we just note that the approach suggested by Zhao *et al.*, 1995 can be considered as a particular case of the theory presented in this chapter when $f(\bar{L}) = \zeta_1 \bar{L} - \zeta_2 \bar{L}^2$, where ζ_1 and ζ_2 are positive constants.

Part 9

Modeling and Optimal Control of Impactors with Jet Thruster

Optimization of jet propulsion in media with drag was considered mainly with applications to planes and missiles. Survey of the obtained results and references can be found in the studies by Leitmann, 1962; Kosmodemiansky, 1966; Tertychny-Dauri, 2004. Using jet thrusters for increasing the depth of penetration of a striker into solid media was analyzed only in a few publications. Sagomonyan, 1988 formulated and analyzed two problems associated with penetration by strikers equipped with jet thruster: (i) determining the moment when the jet thruster must be switched on in order to provide the maximum depth of penetration (DOP) into soil provided that jet thruster can operate only during a fixed time interval; (ii) determining the depth where the jet thruster must be switched on in order to provide the maximum DOP provided that jet thruster was allowed to operate only along the given length of the trajectory of a striker. The second problem was solved for a penetrator with a conical nose assuming that the mass of the penetrator remains constant. Gould, 1997 suggested engineering designs whereby a rocket motor is attached to the penetrator and operates during penetration.

Ben-Dor *et al.*, 2007d considered maximization of the DOP as an optimization problem for a penetrator with a variable mass. They noticed a similarity between this problem and maximization of the distance of a horizontal flight of a projectile in the atmosphere. Various formulations of the latter problem were considered in the past by Hibbs, 1952; Cicala and Miele, 1956; Miele, 1957, 1962b; Krotov, 1995. However it transpired that this similarity had limited applications because of different drag laws in the atmosphere and soil. Only general properties of the solutions which are obtained for arbitrary dependence of the resistance force upon the instantaneous mass and velocity, can be used for solving optimization problems of penetration mechanics. Consequently, optimization of a penetrator equipped with a jet thruster must be analyzed separately. In the study of Ben-Dor *et al.*, 2007d, the authors employed the two-term averaged penetration model whereby the resistance force is a linear function of a squared velocity. Combining analytical and numerical methods they determined the optimum burning programs and compared the obtained results with more simple burning programs for controlling motion of a penetrator.

In the study by Ben-Dor *et al.*, 2008a, the authors developed further the approach suggested in Ben-Dor *et al.* 2007d for a penetration model with a resistance force depending upon the instantaneous velocity and the instantaneous mass of the impactor. In particular, results were obtained for the model suggested by Young, 1997 which is widely used for calculating the DOP in soil, rock, concrete, ice and frozen soil.

The performed analysis demonstrates that appropriate choice of the parameters of the jet thruster and of the burning program allows achieving a considerable increase of the DOP.

The results obtained by Ben-Dor *et al.*, 2007d, 2008a are used in Part 7 of this monograph.

Chapter 21

Application of Two-Term Model of Penetration

Nomenclature

a_0, a_2	parameters determining the model, Eq. (1.6.14);
$\overline{A}_0$, $\overline{A}_2$	parameters in Eqs. (21.2.4);
BP	burning program;
c	relative exit velocity of gases from the nozzle of jet thruster;
$\overline{c}$	Dimensionless c, Eq. (21.1.12);
D	resistance force acting on projectile;
$\overline{D}$	Dimensionless d, Eqs. (21.1.12) and (21.2.3);
DOP	depth of penetration;
h	instantaneous DOP, Fig.1.1.2;
$\overline{h}$	dimensionless h, Eq. (21.1.12);
H	DOP;
$\overline{H}$	dimensionless H, Eq. (21.1.12);
k	parameter, Eq. (21.2.10);
L	length of the nose of impactor, Fig. 1.1.2;
$\overline{L}$	$= L/R$;
m	instantaneous mass of projectile;
$\overline{m}$	dimensionless m, Eq. (21.1.12);
$\overline{m}^{(j)}$	mesh points of interpolation for $\overline{m}$, Eq. (21.3.4);
m_{imp}	initial value of m;
m_{res}	final value of m;
N_v	number of mesh intervals for $\overline{v}$;
N_m	number of mesh intervals for $\overline{m}$;

$\overline{P}$	dimensionless DOP, Eq. (21.2.9);
$\overline{P}_0$	dimensionless DOP for "passive" penetration, Eq. (21.2.51);
$\overline{P}_1$	dimensionless DOP corresponding to the BP given by Eq. (21.2.49), see Eq. (21.2.50);
$\overline{P}_\beta$	dimensionless DOP corresponding to the BP given by Eq. (21.2.40), see Eq. (21.2.42);
$\overline{P}_{\max}$	maximum dimensionless DOP;
r	radius of the flat bluntness of the nose of impactor;
$\overline{r}$	dimensionless r, Eq. (21.2.6);
R	radius of the shank of impactor;
t	time;
$\overline{t}$	dimensionless time, Eq. (21.1.12);
v_{imp}	impact velocity of projectile;
v	instantaneous velocity of projectile;
$\overline{v}'$	$=d\overline{v}/d\overline{m}$;
$\overline{v}$	dimensionless v, Eq. (21.1.12);
$\overline{v}^{(j)}$, $\overline{w}^{(j)}$	interpolation mesh points for $\overline{v}$, Eqs. (21.3.1) and (21.3.2);
$\overline{v}_*$	parameter, Eq. (21.2.14);
$\overline{V}$	dimensionless v, Eq. (21.2.12);
x	coordinate associated with impactor, Fig. 1.1.2;
$\overline{x}$	dimensionless x, Eq. (21.2.6);
α	parameter, Eq. (21.2.12);
$\overline{\beta}$	parameter, Eq. (21.2.33);
$\overline{\beta}_{\min}$	parameter, Eq. (21.2.10);
γ_0, γ_1	parameters, Eq. (21.2.45);
δ	$=\overline{P}_{\max}/\overline{P}_0$;
$\Delta\overline{m}$	mesh width for $\overline{m}$, Eq. (21.3.4);
$\Delta\overline{P}_m$	change of the DOP for penetration with a constant velocity;
$\Delta\overline{P}_v$	change of the DOP for penetration with a constant mass;
$\Delta\overline{P}_{vm}$	change of the DOP when mass and velocity vary;
$\Delta\overline{v}$	mesh width for $\overline{v}$, Eq. (21.3.5);
θ_0, θ_1	parameters in Eq. (21.2.36);

$\lambda_0^{(j)}, \lambda_1^{(j)}, \lambda_2^{(j)}$	parameters in Eq. (21.3.15);
$\bar{\mu}$	dimensionless constant consumption of fuel, Eq. (21.2.33);
$\mu_{\max}$	maximum value of mass flux of a thruster;
$\bar{\mu}_{\max}$	dimensionless $\mu_{\max}$, Eq. (21.1.12);
μ_{fr}	friction coefficient;
Φ	function determining shape of impactor, Fig. 1.2.2;
$\bar{\Phi}$	dimensionless Φ, Eq. (21.2.6);
$\bar{\Phi}_{\bar{x}}$	$= d\bar{\Phi}/d\bar{x}$;
$\bar{\omega}$	function determined by Eq. (21.2.1);
$\bar{\omega}_0$	function determined by Eq. (21.2.12).

21.1 Formulation of Problem

We consider a high speed normal penetration of impactor equipped with a jet thruster into a semi-infinite dense medium. Since the effect of gravity during penetration can be neglected, motion of the impactor is governed by the equation of motion of a projectile with a variable mass (see, for instance, Banach, 1951):

$$m\frac{dv}{dt} + c\frac{dm}{dt} = -D(v), \tag{21.1.1}$$

where v is the instantaneous velocity of the projectile, D is the resistance force, t is time, c is the relative exit velocity of gases from the nozzle of a jet thruster, m is a variable mass of the impactor that changes from the initial value of m_{imp} to the final value m_{res}. It is assumed that the thruster is capable to deliver a mass flux in the range between zero and the maximum value, $\mu_{\max}$:

$$-\mu_{\max} \le \frac{dm}{dt} \le 0. \tag{21.1.2}$$

Since $d/dt = v\,d/dh$, Eq. (21.1.1) can be rewritten as follows:

$$v\left(m\frac{dv}{dh}+c\frac{dm}{dh}\right)=-D(v) \tag{21.1.3}$$

Following Hibbs, 1952 we consider v as a function of m. This is convenient because penetration is associated with the decrease of m, and v is a single-valued function of m, except for the case when $m = const$ (free motion of the penetrator). Hence Eq. (21.1.3) can be rewritten as

$$v\frac{dm}{dh}\left(m\frac{dv}{dm}+c\right)=-D(v)\,, \tag{21.1.4}$$

and the expression for the DOP, H, reads:

$$H=\int_{m_{res}}^{m_{imp}}\frac{(m\,dv/dm+c)v}{D(v)}dm\,. \tag{21.1.5}$$

Using Eq. (21.1.4) we obtain that

$$\frac{dm}{dt}=v\frac{dm}{dh}=-\frac{D(v)}{m\,dv/dm+c}\,, \tag{21.1.6}$$

and Eq. (21.1.2) can be rewritten as

$$0\le\frac{D(v)}{m\,dv/dm+c}\le\mu_{\max}\,. \tag{21.1.7}$$

The problem is to determine a function $v(m)$ that provides the maximum DOP, H, subjected to the constraint given by Eq. (21.1.7). The model describing impactor/shield interaction and shape of the impactor which determine the function $D(v)$), parameters $\mu_{\max}, m_{imp}, m_{res}$ and v_{imp} are assumed to be given.

Equations (21.1.1), (21.1.5) and (21.1.7) imply the following expressions that can be written using for convenience the defined below dimensionless variables:

$$\bar{m}\frac{d\bar{v}}{d\bar{t}}+\bar{c}\frac{d\bar{m}}{d\bar{t}}=-\bar{D}(\bar{v})\,, \tag{21.1.8}$$

$$\overline{H} = \int_{\overline{m}_{res}}^{1} \frac{(\overline{m}\overline{v}' + \overline{c})\overline{v}}{\overline{D}(\overline{v})} d\overline{m} , \tag{21.1.9}$$

$$\frac{\overline{D}(\overline{v})}{\overline{m}\overline{v}' + \overline{c}} \geq 0 , \tag{21.1.10}$$

$$\frac{\overline{D}(\overline{v})}{\overline{m}\overline{v}' + \overline{c}} \leq \overline{\mu}_{\max} , \tag{21.1.11}$$

where L is the length of the nose of the impactor,

$$\begin{gathered}\overline{v} = \frac{v}{v_{imp}}, \overline{c} = \frac{c}{v_{imp}}, \overline{m}_{res} = \frac{m_{res}}{m_{imp}}, \overline{m} = \frac{m}{m_{imp}} \ \overline{\mu}_{\max} = \frac{L\mu_{\max}}{v_{imp}m_{imp}}, \\ \overline{t} = \frac{v_{imp}t}{L}, \overline{h} = \frac{h}{L}, \overline{H} = \frac{H}{L}, \overline{D}(\overline{v}) = \frac{L}{v_{imp}^2 m_{imp}} D(v_{imp}\overline{v}), \overline{v}' = \frac{d\overline{v}}{d\overline{m}},\end{gathered} \tag{21.1.12}$$

The problem reduces to the optimization of the functional $\overline{H}$ in Eq. (21.1.9), whereas the solution must satisfy the constraints given by Eqs. (21.1.10)–(21.1.11) and

$$\overline{v}(1) = 1, \quad \overline{v}(\overline{m}_{res}) = 0, \quad \overline{v}(\overline{m}) \geq 0, \quad \overline{m}_{res} \leq \overline{m} \leq 1 . \tag{21.1.13}$$

It must be noted that this optimization problem is degenerate (see, e.g, Van Brunt, 2003) since the integrand in the functional $\overline{H}$ is a linear function of $\overline{v}'$. Consequently, solutions of the Euler-Lagrange equation in this case do not satisfy the boundary conditions, i.e., they do not yield the solution of the optimization problem.

Eqs. (21.1.10)–(21.1.11) afford two situations: the possibility of "vertical sub-arcs"

$$\overline{m} = const , \tag{21.1.14}$$

when $\overline{v}' = \infty$, and the constraints

$$\overline{m}\overline{v}' + \overline{c} \geq \overline{D}(\overline{v}) / \overline{\mu}_{\max} , \tag{21.1.15}$$

$$\overline{m}\overline{v}' + \overline{c} \geq 0 , \tag{21.1.16}$$

when $\bar{v}' < \infty$. Clearly, Eq. (21.1.16) is implied by Eq. (21.1.15) and can be omitted. It is important to note that Eq. (21.1.16) or, formally, Eq. (21.1.15) with $\bar{\mu} = \infty$ must be taken into account even if the upper bound on the absolute value of $d\bar{m}/d\bar{t}$ is removed.

21.2 Analytical Study of Problem

21.2.1 *General properties of extremal*

Using the results obtained by Cicala and Miele, 1956 (see also Miele, 1962b) which are adapted to the case $\bar{D} = \bar{D}(\bar{v})$, the optimum curve must consist of the following sub-arcs:

(1) sub-arc $\bar{\omega}(\bar{v}) = 0$ where

$$\bar{\omega}(\bar{v}) = (\bar{v} - \bar{c})\bar{D}(\bar{v}) + \bar{c}\bar{v}\, d\bar{D}/d\bar{v} ; \tag{21.2.1}$$

(2) sub-arc $\bar{m} = const$, and

(3) sub-arc $\dot{\bar{m}} = -\bar{\mu}_{\max}$ where $\dot{\bar{m}} = d\bar{m}/d\bar{t}$.

The sub-arc 1 represents the solution of the Euler-Lagrange equation, while the sub-arc 2 and the sub-arc 3 are determined by Eqs. (21.1.10) and (21.1.11), respectively, where the inequality sign is replaced with equal sign.

These sub-arcs can join in the following sequences: a) if $\bar{\omega} > 0$ in some point then only the sequence $\bar{m} = const \Rightarrow \dot{\bar{m}} = -\bar{\mu}_{\max}$ is possible in this point, b) if $\bar{\omega} < 0$ then the inverse sequence $\dot{\bar{m}} = -\bar{\mu}_{\max} \Rightarrow \bar{m} = const$ is permissible, and c) if $\bar{\omega} = 0$ then the sequences $\bar{\omega} = 0 \Leftrightarrow \bar{m} = const$ and $\bar{\omega} = 0 \Leftrightarrow \dot{\bar{m}} = -\bar{\mu}_{\max}$ are possible. The arrows indicate the direction of motion of the penetrator along the sub-arcs in the $\bar{m}, \bar{v}$ plane.

21.2.2 *Formulation of problem for two-term model*

Hereafter we use a two-term model of impactor-shield interaction with a constant friction which is described in Section 1.6.2.3. This model implies the following relationship for the resistance force:

$$\frac{D(v)}{\pi} = r^2(a_2 v^2 + a_0) + 2\int_0^L \left(a_2 \frac{\Phi_x^2}{\Phi_x^2 + 1} v^2 + a_0 \right)(\Phi_x + \mu_{fr})\Phi\, dx\,, \quad (21.2.2)$$

where μ_{fr} is friction coefficients, coefficients a_0 and a_2 depend on the mechanical properties of the material of the shield, function $\Phi(x)$ determines the shape of impactor, r is radius of the flat bluntness of the nose of impactor.

Using the dimensionless variables defined by Eqs. (21.1.12), we can rewrite Eq. (21.2.2) as follows:

$$\overline{D}(\overline{v}) = \frac{L}{v_{imp}^2 m_{imp}} D(v_{imp}\overline{v}) = \overline{A}_2 \overline{v}^2 + \overline{A}_0\,, \quad (21.2.3)$$

where

$$\overline{A}_0 = \frac{\pi L^3 a_0}{m_{imp} v_{imp}^2} \overline{I}_0\,, \quad \overline{A}_2 = \frac{\pi L^3 a_2}{m_{imp}} \overline{I}_2\,, \quad (21.2.4)$$

$$\begin{aligned} \overline{I}_0 &= \overline{r}^2 + 2\int_0^1 (\overline{\Phi}_{\overline{x}} + \mu_{fr})\overline{\Phi} d\overline{x} > 0, \\ \overline{I}_2 &= \overline{r}^2 + 2\int_0^1 \frac{(\overline{\Phi}_{\overline{x}} + \mu_{fr})\overline{\Phi}\,\overline{\Phi}_{\overline{x}}^2}{\overline{\Phi}_{\overline{x}}^2 + 1} d\overline{x} > 0, \end{aligned} \quad (21.2.5)$$

$$\overline{x} = \frac{x}{L}, \quad \overline{r} = \frac{r}{L}, \quad \overline{\Phi} = \frac{\Phi}{L}, \quad \overline{\Phi}_{\overline{x}} = \frac{d\overline{\Phi}}{d\overline{x}}\,. \quad (21.2.6)$$

It is convenient also to rewrite Eqs. (21.1.9) and (21.1.15) as follows:

$$\overline{P} = \int_{\overline{m}_{res}}^1 \frac{(\overline{m}\overline{v}' + \overline{c})\overline{v}}{k\overline{v}^2 + 1} d\overline{m}\,, \quad (21.2.7)$$

$$\overline{m}\overline{v}' + \overline{c} \ge \overline{\beta}_{\min}(k\overline{v}^2 + 1)\,, \quad (21.2.8)$$

where

$$\overline{P} = k_0 \overline{H} = \frac{\pi L^2 a_0 \overline{I}_0}{m_{imp} v_{imp}^2} H , \tag{21.2.9}$$

$$k = \frac{\overline{A}_2}{\overline{A}_0} = \frac{a_2 v_{imp}^2 \overline{I}_2}{a_0 \overline{I}_0}, \quad \beta_{\min} = \frac{\overline{A}_0}{\overline{\mu}_{\max}} = \frac{\pi L^2 a_0 \overline{I}_0}{v_{imp} \mu_{\max}}. \tag{21.2.10}$$

The problem reduces to maximizing the functional $\overline{P}$ in Eq. (21.2.7), whereas the solution must satisfy the constraints given by Eqs. (21.1.13) and (21.2.8).

21.2.3 *Optimum sub-arcs*

21.2.3.1 *Sub-arc 1*

Substituting expression for $\overline{D}(\overline{v})$ from Eq. (21.2.3) into Eq. (21.2.1) we obtain:

$$\overline{\omega}(\overline{v}) = \overline{c}^3 \overline{A}_2 \overline{\omega}_0(\overline{V}), \tag{21.2.11}$$

where

$$\overline{\omega}_0(\overline{V}) = \overline{V}^3 + \overline{V}^2 + \alpha(\overline{V} - 1), \quad \alpha = \frac{1}{k\overline{c}^2}, \quad \overline{V} = \frac{\overline{v}}{\overline{c}}. \tag{21.2.12}$$

The sub-arc *1* is described by the equation $\overline{\omega}(\overline{v}) = 0$. Since

$$\begin{gathered} \overline{\omega}_0(0) = -\alpha < 0, \quad \overline{\omega}_0(1) = 2 > 0, \\ \overline{\omega}_0'(\overline{V}) = 3\overline{V}^2 + 2\overline{V} + \alpha > 0, \quad \overline{V} \geq 0, \end{gathered} \tag{21.2.13}$$

the equation $\overline{\omega}(\overline{v}) = 0$ has one positive root, $\overline{v} = \overline{v}_*$, where

$$\frac{\overline{v}_*}{\overline{c}} = \begin{cases} \dfrac{1}{3}\left(\sqrt[3]{\sqrt{\zeta} + 18\alpha - 1} - \sqrt[3]{\sqrt{\zeta} - 18\alpha + 1} - 1\right) & \text{if} \quad \zeta \geq 0 \\ \dfrac{1}{3}\left(2\sqrt{1 - 3\alpha}\cos\dfrac{\varphi}{3} - 1\right) & \text{if} \quad \zeta < 0 \end{cases} \tag{21.2.14}$$

and

$$\zeta = (3\alpha - 1)^3 + (1 - 18\alpha)^2, \quad \varphi = \cos^{-1}\left(\frac{18\alpha - 1}{(1 - 3\alpha)^{3/2}}\right). \quad (21.2.15)$$

Therefore, the constraints $\overline{\omega}(\overline{v}) < 0, \overline{\omega}(\overline{v}) = 0, \overline{\omega}(\overline{v}) > 0$ correspond to the conditions $\overline{v} < \overline{v}_*, \overline{v} = \overline{v}_*, \overline{v} > \overline{v}_*$, respectively. The root of the equation $\overline{\omega}(\overline{v}) = 0$ as a function of parameter α is shown in Fig. 21.2.1a.

The equation of the arc 1 reads:

$$\overline{v} = \overline{v}_*. \quad (21.2.16)$$

Consider a point moving along the curve $\overline{v} = \overline{v}_*$ from the location with the coordinates $(\overline{m}_b, \overline{v}_*)$ to the location with the coordinates $(\overline{m}_a, \overline{v}_*)$, where $\overline{m}_{res} \le \overline{m}_a < \overline{m}_b \le 1$. Hence the increase of the functional given by Eq. (21.1.9) along this path reads:

$$\Delta\overline{P}_m(\overline{v}_*, \overline{m}_a, \overline{m}_b), = \int_{\overline{m}_a}^{\overline{m}_b} \frac{\overline{v}_* d\overline{m}}{k\overline{v}_*^2 + 1} = \frac{\overline{v}_*(\overline{m}_b - \overline{m}_a)}{k\overline{v}_*^2 + 1}. \quad (21.2.17)$$

21.2.3.2 *Sub-arc 2*

The sub-arc 2 corresponds to the motion with a constant mass. Assume that the velocity of the impactor decreases from $\overline{v} = \overline{v}_b$ to $\overline{v} = \overline{v}_a$ while its mass remains constant, $\overline{m} = \overline{m}_0$. Then the increase of the functional given by Eq. (21.1.9) is as follows:

$$\Delta\overline{P}_v(\overline{m}_0, \overline{v}_a, \overline{v}_b) = \overline{m}_0 \int_{\overline{v}_a}^{\overline{v}_b} \frac{\overline{v} d\overline{v}}{k\overline{v}^2 + 1} = \frac{\overline{m}_0}{2k} \ln\left(\frac{k\overline{v}_b^2 + 1}{k\overline{v}_a^2 + 1}\right). \quad (21.2.18)$$

21.2.3.3 *Sub-arc 3*

Along the *sub-arc 3*, $\dot{\overline{m}} = -\overline{\mu}_{\max}$. The equation of this sub-arc in the coordinates $(\overline{m}, \overline{v})$ is determined by Eq. (21.2.8) where the inequality sign is replaced with equal sign:

$$\overline{m}\overline{v}' + \overline{c} = \overline{\beta}_{\min}(k\overline{v}^2 + 1). \tag{21.2.19}$$

The solution of the latter ODE with separable variables passing through the point $(1, \overline{v}_0)$ can be written as follows:

$$\overline{m} = \exp\left(\int_{\overline{v}_0}^{\overline{v}} \xi(\overline{\beta}_{\min}, \overline{V})\, d\overline{V}\right), \xi(\overline{\beta}, \overline{V}) = \frac{1}{\overline{\beta}kV^2 + \overline{\beta} - \overline{c}}. \tag{21.2.20}$$

Consider a point moving along this curve from the point $(\overline{m}_b, \overline{v}_b)$ to the point $(\overline{m}_a, \overline{v}_a)$, where $\overline{m}_{res} \le \overline{m}_a < \overline{m}_b \le 1$. In this case Eqs. (21.2.7) and (21.2.19) yield the following expression for the trajectory of the impactor:

$$\begin{aligned} \Delta\overline{P}_{mv}(\overline{\beta}_{\min}, \overline{v}_0, \overline{v}_a, \overline{v}_b) &= \overline{\beta}_{\min} \int_{\overline{v}_a}^{\overline{v}_b} \overline{v} \frac{d\overline{m}}{d\overline{v}} d\overline{v} \\ &= \overline{\beta}_{\min} \int_{\overline{v}_a}^{\overline{v}_b} \overline{v}\, \xi(\overline{\beta}_{\min}, \overline{v}) \exp\left(\int_{\overline{v}_0}^{\overline{v}} \xi(\overline{\beta}_{\min}, \overline{V})\, d\overline{V}\right) d\overline{v}. \end{aligned} \tag{21.2.21}$$

If the function $\overline{v}(\overline{m})$ can be written in the explicit form, then the expression

$$\Delta\overline{P}_{vm}(\overline{\beta}_{\min}, \overline{v}_0, \overline{m}_a, \overline{m}_b) = \overline{\beta}_{\min} \int_{\overline{m}_a}^{\overline{m}_b} \overline{v}\, d\overline{m} \tag{21.2.22}$$

is more convenient.

21.2.4 *Analytical solution for particular case*

The condition $\overline{\beta}_{\min} = 0$ ($\overline{\mu}_{\max} = \infty$) implies that the step change of the mass of the penetrator (pulse burning) is allowed. Eq. (21.2.21) shows that this change of mass and of the velocity of penetrator occurs without change of its location. Equation of the sub-arc 3 follows from Eq. (21.2.20):

$$\overline{v} = \overline{v}_0 - \overline{c} \ln \overline{m}. \tag{21.2.23}$$

Analysis of the conditions presented in Section 21.2.3 shows that three version of burning program (BP) are possible, depending on the values of the parameters $k, \bar{c}$ and $\bar{m}_{res}$.

Consider the case when $\bar{v}_* < 1$. Let us first determine the range of the coordinates $\bar{c}$ and k where this inequality is valid. Since the root of the equation $\bar{\omega}_0(\bar{v}_*/\bar{c}) = 0$ is less than 1, $\bar{v}_* < 1$ for every $\bar{c} \le 1$. For the analysis of the case $\bar{c} > 1$, let us express k as a function of $\bar{c}$ and $\bar{v}_*$ using equation $\bar{\omega}_0(\bar{v}_*/\bar{c}) = 0$:

$$k = \frac{\bar{c} - \bar{v}_*}{\bar{v}_*^2(\bar{v}_* + \bar{c})}. \tag{21.2.24}$$

Clearly, increasing $\bar{v}_*$ from 0 to 1 is associated with decreasing k from $+\infty$ to $(\bar{c}-1)/(\bar{c}+1)$. Therefore, $\bar{v}_* < 1$ when $k \ge \Theta(\bar{c})$, where

$$\Theta(\bar{c}) = \begin{cases} 0 & if \quad 0 < \bar{c} \le 1 \\ (\bar{c}-1)/(\bar{c}+1) & if \quad \bar{c} \ge 1 \end{cases}. \tag{21.2.25}$$

Therefore $\bar{v}_* \ge 1$ only if $\bar{c} > 1$ and $0 < k \le (\bar{c}-1)/(\bar{c}+1) < 1$. This situation is illustrated in Fig. 21.2.1b.

In the considered case, $\bar{v}_* < 1$, the optimal BP includes three sub-arcs (Fig. 21.2.2a) in the coordinates ($\bar{m}, \bar{v}$):

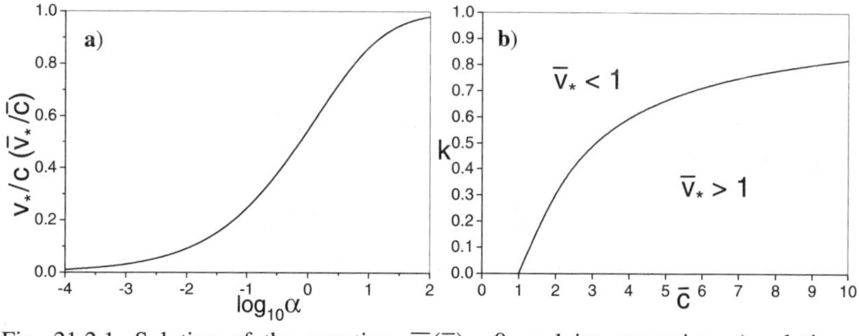

Fig. 21.2.1. Solution of the equation $\bar{\omega}(\bar{v}) = 0$ and its properties; a) solution as a function of the parameter α; b) $\bar{v}_*$ vs. k and $\bar{c}$.

$$\begin{cases} AC: & \overline{m}=1, \overline{v}:1 \to \overline{v}_* \\ CD: & \overline{v}=\overline{v}_*, \overline{m}:1 \to \overline{m}_{res} \\ DB: & \overline{m}=\overline{m}_{res}, \overline{v}:\overline{v}_* \to 0 \end{cases} \quad (21.2.26)$$

The corresponding expression for the DOP reads:

$$\begin{aligned} \overline{P}_{\max} &= \Delta\overline{P}_v(1,\overline{v}_*,1) + \Delta\overline{P}_m(\overline{v}_*,\overline{m}_{res},1) + \Delta\overline{P}_v(\overline{m}_{res},0,\overline{v}_*) \\ &= \frac{\overline{v}_*(1-\overline{m}_{res})}{k\overline{v}_*^2+1} + \frac{\overline{m}_{res}+1}{2k}\ln(k\overline{v}_*^2+1) - \frac{\ln(k+1)}{2k} \end{aligned} \quad (21.2.27)$$

In the case $\overline{v}_*=1$ the segment AC vanishes.

Consider now the case when $\overline{v}_*>1$ and the curve $\overline{v}=1-\overline{c}\ln\overline{m}$ intersects with the line $\overline{v}=\overline{v}_*$ in the point $(\overline{m}_*,\overline{v}_*)$, whereas $\overline{m}_*>\overline{m}_{res}$. Hence the optimal BP also includes three sub-arcs (see Fig. 21.2.2b):

$$\begin{cases} AC: & \overline{v}=1-\overline{c}\ln\overline{m}, \overline{m}:1 \to \overline{m}_*; \\ CD: & \overline{v}=\overline{v}_*, \overline{m}:\overline{m}_* \to \overline{m}_{res}; \\ DB: & \overline{m}=\overline{m}_{res}, \overline{v}:\overline{v}_* \to 0, \end{cases} \quad (21.2.28)$$

where $\overline{m}_*=\exp\left((1-\overline{v}_*)/\overline{c}\right)$. The corresponding expression for the DOP is as follows:

$$\begin{aligned} \overline{P}_{\max} &= \Delta\overline{P}_m(\overline{v}_*,\overline{m}_{res},\overline{m}_*) + \Delta\overline{P}_v(\overline{m}_{res},0,\overline{v}_*) \\ &= \frac{\overline{v}_*(\overline{m}_*-\overline{m}_{res})}{k\overline{v}_*^2+1} + \frac{\overline{m}_{res}}{2k}\ln(k\overline{v}_*^2+1). \end{aligned} \quad (21.2.29)$$

If $v_*>1$ and $\overline{m}_*\le\overline{m}_{res}$ then the optimal BP includes two sub-arcs (see Fig. 21.2.2c):

$$\begin{cases} AC: & \overline{v}=1-\overline{c}\ln\overline{m}, \overline{m}:1 \to \overline{m}_{res}; \\ CB: & \overline{m}=\overline{m}_{res}, \overline{v}:\overline{v}_1 \to 0, \end{cases} \quad (21.2.30)$$

where $\overline{v}_1=1-\overline{c}\ln\overline{m}_{res}$. The DOP is determined using only the sub-arc CB:

$$\overline{P}_{\max} = \Delta\overline{P}_v(\overline{m}_{res},0,\overline{v}_1) = \frac{\overline{m}_{res}}{2k}\ln(k\overline{v}_1^2+1). \quad (21.2.31)$$

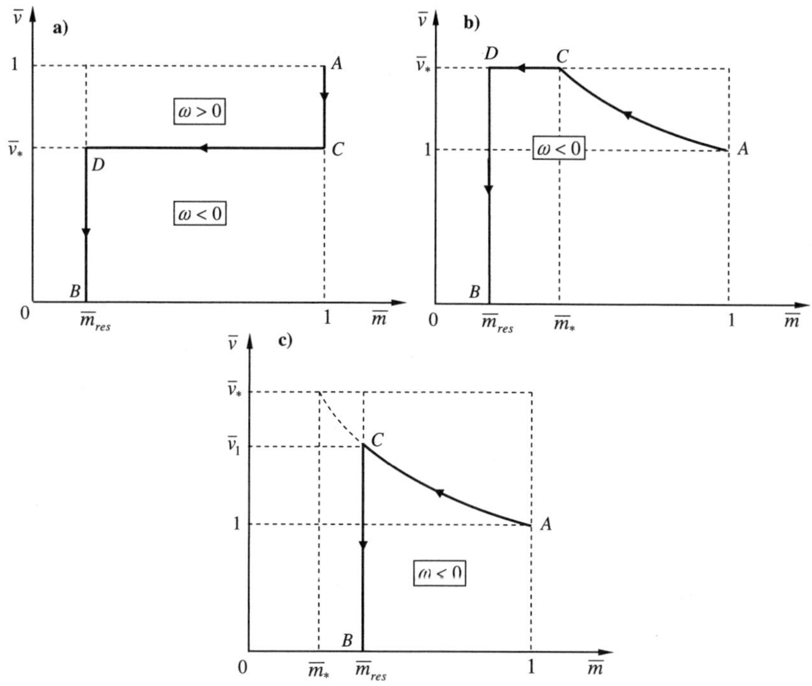

Fig. 21.2.2. Modes of the optimum BPs for $\overline{\beta}_{\min} = 0$.

21.2.5 *Some simple burning programs*

Let us consider BPs which include two regimes. The first regime is penetration with a constant consumption of fuel, μ, until fuel is available. The following relationship is valid in this case:

$$\frac{\overline{m}\overline{v}' + \overline{c}}{k\overline{v}^2 + 1} = \overline{\beta}, \qquad (21.2.32)$$

where

$$\overline{\beta} = \frac{k_0}{\overline{\mu}}, \quad \overline{\mu} = \frac{L\mu}{v_{imp} m_{imp}}. \qquad (21.2.33)$$

When fuel supply is exhausted, penetrator continues free motion (the second regime) with $\overline{m} = \overline{m}_{res}$ until it stops. We assume that the regime when penetrator stops before the fuel supply is completely exhausted does not occur. This regime occurs only when consumption of fuel is too small.

The above regime with a constant consumption of fuel is described by Eq. (21.2.20) with $\overline{v}_0 = 1$ after replacing $\overline{\beta}_{\min}$ with $\overline{\beta}$. The integral in Eq. (21.2.20) can be calculated in a closed form. Let us consider some particular cases.

Let

$$\overline{\beta} > \overline{c}\,. \tag{21.2.34}$$

Hence Eq. (21.2.20) yields:

$$\tan^{-1}(\overline{v}/\theta_0) = \tan^{-1}(1/\theta_0) + \theta_1 \ln \overline{m}\,, \tag{21.2.35}$$

where

$$\theta_0 = \sqrt{\frac{\overline{\beta} - \overline{c}}{k\overline{\beta}}}, \quad \theta_1 = \sqrt{k\overline{\beta}(\overline{\beta} - \overline{c})}\,. \tag{21.2.36}$$

Equation (21.2.35) shows that $\overline{v}$ increases from 1 to 0 when $\overline{m}$ increases from 1 to $\overline{m}_*$ (see Fig. 21.2.3a) where

$$\overline{m}_* = \exp\left(-\frac{\tan^{-1}(1/\theta_0)}{\theta_1} \right). \tag{21.2.37}$$

Penetrator must not stop before fuel supply is completely exhausted, i.e., the condition $\overline{m}_* \le \overline{m}_{res}$ must be satisfied. It is easy to show that the right hand size of the Eq. (21.2.37) increases from 0 to 1 when $\overline{\beta}$ increases from $\overline{c}$ to $+\infty$. Consequently, Eq. (21.2.37) has one root, $\overline{\beta}_1$, for arbitrary $0 < \overline{m}_* < 1$, and the inequality $\overline{m}_* \le \overline{m}_{res}$ is satisfied for all $\overline{c} < \overline{\beta} \le \overline{\beta}_1$.

Equation (21.2.35) allows determining the dependence $\overline{v}(\overline{m})$ in the explicit form:

$$\overline{v} = \Xi(\overline{m}), \quad \overline{m}_{res} \le \overline{m} \le 1, \tag{21.2.38}$$

where

$$\Xi(\overline{m}) = \frac{\theta_0 \tan(\theta_1 \ln \overline{m}) + 1}{\theta_0 - \tan(\theta_1 \ln \overline{m})} \theta_0, \quad \overline{m}_{res} \le \overline{m} \le 1. \tag{21.2.39}$$

The BP is described as follows:

$$\begin{cases} AC: & \overline{v} = \Xi(\overline{m}), \overline{m}: 1 \to \overline{m}_{res}; \\ CB: & \overline{m} = \overline{m}_{res}, \overline{v}: \overline{v}_1 \to 0, \end{cases} \tag{21.2.40}$$

where

$$\overline{v}_1 = \Xi(\overline{m}_{res}). \tag{21.2.41}$$

The expression for the DOP in this case reads:

$$\begin{aligned} \overline{P}_\beta &= \Delta\overline{P}_{vm}(\overline{\beta}, 1, \overline{m}_{res}, 1) + \Delta\overline{P}_v(\overline{m}_{res}, 0, \overline{v}_1) \\ &= \overline{\beta} \int_{\overline{m}_{res}}^{1} \Xi(\overline{m}) d\overline{m} + \frac{\overline{m}_{res}}{2k} \ln(k\overline{v}_1^2 + 1). \end{aligned} \tag{21.2.42}$$

Assume that

$$0 < \overline{\beta} < \overline{c}. \tag{21.2.43}$$

Hence Eq. (21.2.20) yields:

$$\overline{m}^{\gamma_1} = \left| \frac{(1 + \gamma_0)(\overline{v} - \gamma_0)}{(1 - \gamma_0)(\overline{v} + \gamma_0)} \right|, \tag{21.2.44}$$

where

$$\gamma_0 = \sqrt{\frac{\overline{c} - \overline{\beta}}{k\overline{\beta}}}, \quad \gamma_1 = 2\sqrt{k\overline{\beta}(\overline{c} - \overline{\beta})}. \tag{21.2.45}$$

After some algebra, using Eq. (21.2.44) we can obtain the dependence $\overline{v}$ versus $\overline{m}$ in the explicit form given by Eq. (21.2.38) with

$$\Xi(\overline{m}) = \frac{\gamma_0 + 1 - (\gamma_0 - 1)\overline{m}^{\gamma_1}}{\gamma_0 + 1 + (\gamma_0 - 1)\overline{m}^{\gamma_1}} \gamma_0. \tag{21.2.46}$$

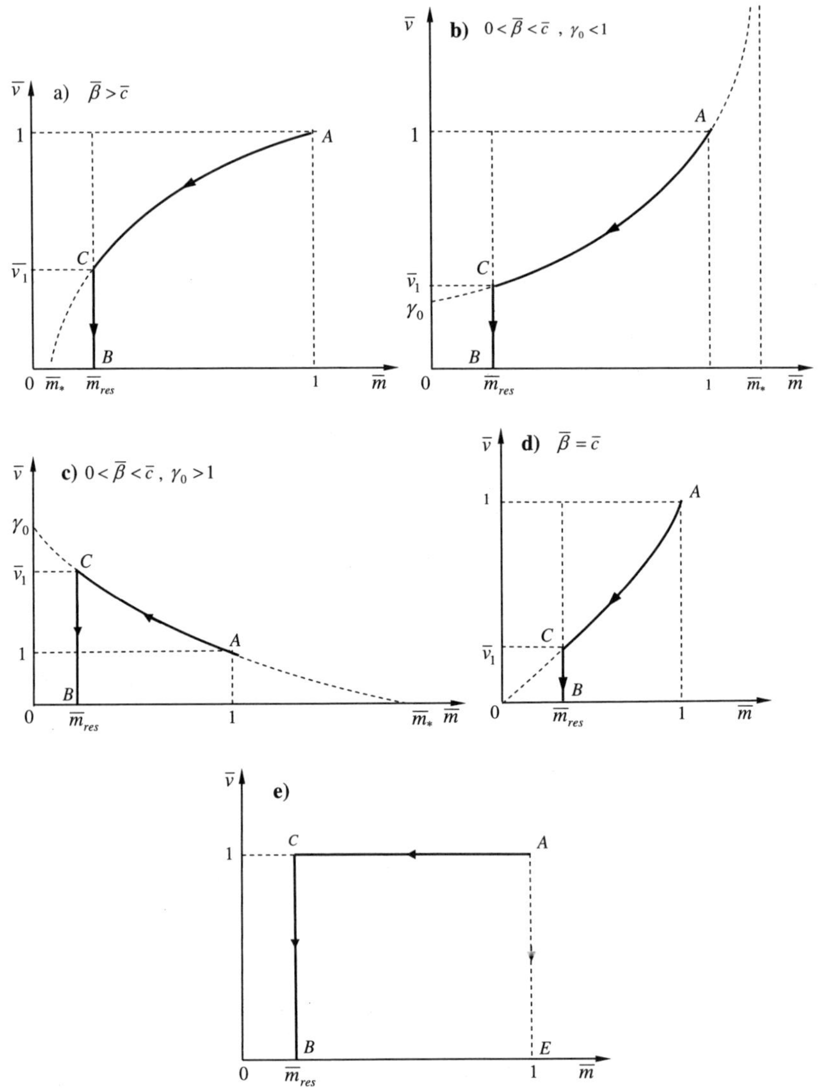

Fig. 21.2.3. Modes of some simple BPs; a-d) $\bar{\beta}_{\min} = const.$ e) $\bar{\beta}_{\min} \neq const.$

These burning programs are shown in Figs. 21.2.3b-c where $\bar{m}_* = \left|(\gamma_0 + 1)/(\gamma_0 - 1)\right|^{1/\gamma_1} > 1$ for $\gamma_0 > 0$ and $\gamma_0 \neq 1$. The BP and the DOP are determined by Eqs. (21.2.40)–(21.2.42).

If

$$\overline{\beta} = \overline{c}\,, \tag{21.2.47}$$

Eq. (21.2.20) implies Eq. (21.2.38) with

$$\varXi(\overline{m}) = \frac{1}{1-\overline{c}k\ln\overline{m}}\,. \tag{21.2.48}$$

The corresponding curve is shown in Fig. 21.2.3d, and the BP and the DOP are determined by Eqs. (21.2.40)–(21.2.42).

Let us consider two additional simple regimes of penetration (Fig. 21.2.3e).

The BP with a constant velocity at the first stage of penetration (the permissibility of this regime can be guaranteed only if $\overline{\beta}_{\max} = 0$) is described as follows:

$$\begin{cases} AC: & \overline{v} = 1, \overline{m}: 1 \to \overline{m}_{res}; \\ CB: & \overline{m} = \overline{m}_{res}, \overline{v}: 1 \to 0. \end{cases} \tag{21.2.49}$$

The corresponding formula for the DOP reads:

$$\overline{P}_1 = \varDelta\overline{P}_v(\overline{m}_{res}, 0, 1) + \varDelta\overline{P}_m(1, \overline{m}_{res}, 1) = \frac{\overline{m}_{res}}{2k}\ln(k+1) + \frac{1-\overline{m}_{res}}{k+1}\,. \tag{21.2.50}$$

In the case of "passive" penetration with $\overline{m} = 1, \overline{v}: 1 \to 0$ (the trajectory AE in Fig. 21.2.3e), the DOP is determined by the following formula:

$$\overline{P}_0 = \varDelta\overline{P}_v(1,1,0) = \frac{1}{2k}\ln(k+1)\,. \tag{21.2.51}$$

This regime with a constant mass of the impactor is used as a reference regime when different BPs are compared.

21.2.6 *Results of numerical calculations*

The results of calculations described in this section are shown in Fig. 21.2.4 and in Table 21.2.1.

Comparison of the DOPs for the optimal BP in the case $\overline{\beta}_{\min} = 0$, $\overline{P}_{\max}$, with some simple BPs is shown in Table 21.2.1. All values of DOP are normalized by the DOP for the "passive" penetration, $\overline{P}_0$, that is determined by Eq. (21.2.51), i.e., the performance parameter $\delta = \overline{P}_{\max}/\overline{P}_0$ is used. The burning regimes with DOPs $\overline{P}_1$ and $\overline{P}_\beta$ are described above. Empty cells denote that the regime with corresponding $\overline{\beta}$ is impossible. Table 21.2.1 demonstrates the advantage of using a jet thruster.

Curves in Fig. 21.2.4 correspond to three values of $\overline{c}$. Three curves are plotted in each figure. Curve 1 is the optimal solution for $\overline{\beta}_{\min} = 0$; curve 2 is the best solution among BPs with $\overline{\beta} = const$; curve 3 is the BP determined by Eq. (21.2.49). Clearly, varying the magnitude of $\overline{\beta}$ allows obtaining the DOP close to the optimal. On the other hand ,the BP determined by Eq. (21.2.49) is commonly inefficient.

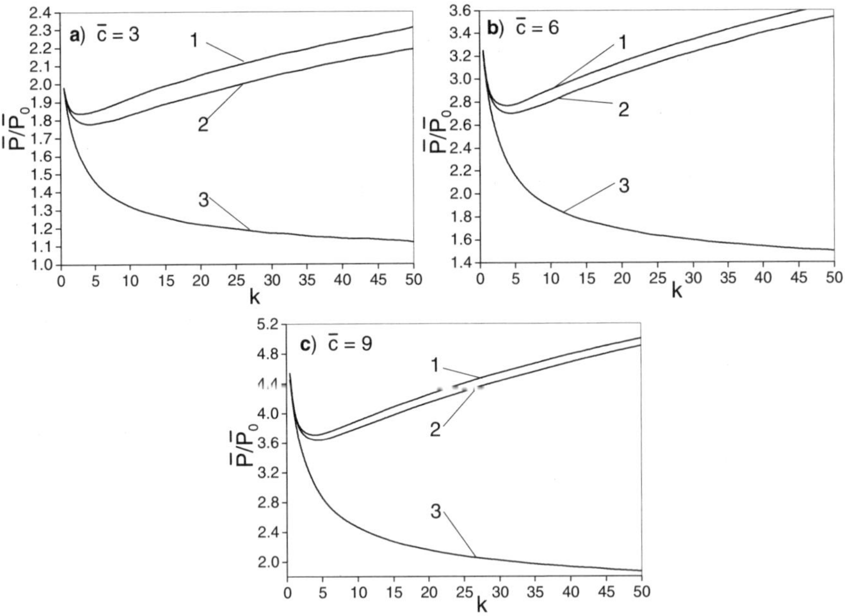

Fig. 21.2.4. Comparison of different BPs; 1 – maximum depth of penetration for $\overline{\beta}_{\min} = 0$; 2 – the best among the BPs with $\overline{\beta}_{\min} = const.$; 3 – two-stage BP (see the text for the details).

Table 21.2.1. Comparison of DOP for different BPs: $\overline{P}_{\max}$ - optima for $\overline{\beta}_{nim}=0$; $\overline{P}_0$ - $\overline{m}=1, \overline{v}:1\rightarrow 0$; $\overline{P}_1$ - $\overline{v}=1$, $\overline{m}:1\rightarrow \overline{m}_{res}$; $\overline{P}_\beta$ - $\overline{v}:1\rightarrow \overline{v}_1$, $\overline{m}:1\rightarrow \overline{m}_{res}$ with $\overline{\beta}=const.$; $\overline{m}=\overline{m}_{res}$; $\overline{v}:\overline{v}_1\rightarrow 0$.

k	$\frac{\overline{P}_{\max}}{\overline{P}_0}$	$\frac{\overline{P}_1}{\overline{P}_0}$	$\frac{\overline{P}_\beta}{\overline{P}_0}$											
	$\beta \Rightarrow$		0	0.5	1	2	3	4	5	6	7	8	9	9.5
							$\overline{c}=3$							
0.5	1.98	1.98	1.86	1.92	1.95	1.98	1.95	1.87	1.74					
1.0	1.88	1.83	1.62	1.71	1.78	1.86	1.86	1.78	1.64					
2.0	1.83	1.66	1.41	1.56	1.66	1.78	1.79	1.69						
8.0	1.89	1.36	1.15	1.44	1.63	1.80	1.75							
14.0	1.98	1.27	1.08	1.47	1.70	1.88	1.77							
20.0	2.05	1.22	1.05	1.51	1.76	1.94	1.78							
26.0	2.11	1.19	1.03	1.56	1.83	2.00	1.79							
32.0	2.17	1.17	1.01	1.60	1.88	2.06	1.80							
38.0	2.22	1.15	1.00	1.63	1.93	2.11	1.80							
44.0	2.26	1.14	0.99	1.67	1.98	2.15	1.81							
50.0	2.31	1.12	0.98	1.70	2.02	2.19	1.81							
							$\overline{c}=6$							
0.5	3.25	3.22	2.87	2.98	3.07	3.19	3.24	3.22	3.12	2.96	2.72	2.42		
1.0	2.94	2.91	2.31	2.47	2.61	2.81	2.91	2.93	2.86	2.71	2.47			
2.0	2.79	2.57	1.89	2.12	2.31	2.57	2.71	2.75	2.68	2.51	2.22			
8.0	2.86	1.96	1.40	1.85	2.16	2.55	2.73	2.75	2.61	2.27				
14.0	3.01	1.78	1.29	1.88	2.24	2.68	2.88	2.88	2.69	2.22				
20.0	3.15	1.69	1.23	1.93	2.34	2.81	3.02	3.01	2.77	2.20				
26.0	3.27	1.63	1.20	1.99	2.43	2.93	3.14	3.12	2.85	2.18				
32.0	3.38	1.58	1.17	2.04	2.51	3.04	3.25	3.22	2.93	2.18				
38.0	3.48	1.55	1.16	2.09	2.58	3.13	3.35	3.32	3.00	2.17				
44.0	3.57	1.52	1.14	2.14	2.65	3.22	3.45	3.40	3.06	2.17				
50.0	3.65	1.50	1.13	2.19	2.72	3.30	3.53	3.48	3.12	2.17				
							$\overline{c}=9$							
0.5	4.54	4.45	3.71	3.89	4.04	4.27	4.43	4.51	4.52	4.45	4.31	4.09	3.80	3.62
1.0	4.01	4.00	2.85	3.08	3.29	3.61	3.83	3.96	4.01	3.98	3.87	3.67	3.38	3.20
2.0	3.75	3.48	2.24	2.56	2.82	3.21	3.48	3.64	3.71	3.69	3.58	3.37	3.04	2.82
8.0	3.82	2.57	1.59	2.17	2.57	3.12	3.46	3.66	3.73	3.68	3.50	3.17	2.60	
14.0	4.05	2.30	1.44	2.19	2.67	3.29	3.67	3.87	3.94	3.87	3.66	3.25	2.50	
20.0	4.25	2.16	1.37	2.25	2.78	3.46	3.86	4.08	4.14	4.06	3.81	3.35	2.46	
26.0	4.43	2.06	1.32	2.32	2.89	3.61	4.03	4.26	4.32	4.23	3.95	3.44	2.43	
32.0	4.59	2.00	1.29	2.38	2.99	3.74	4.19	4.42	4.48	4.38	4.08	3.53	2.41	
38.0	4.74	1.95	1.27	2.44	3.08	3.87	4.33	4.57	4.63	4.52	4.21	3.62	2.40	
44.0	4.88	1.91	1.25	2.50	3.16	3.98	4.46	4.71	4.77	4.65	4.32	3.71	2.40	
50.0	5.00	1.87	1.23	2.56	3.24	4.09	4.58	4.84	4.90	4.77	4.43	3.78	2.40	

21.3 Numerical Study of Problem

Determining formulas for the optimal BP in a general case when $\overline{\beta}_{\min} \neq 0$ requires a thorough analysis of different combinations of the sub-arcs. The derived formulas are quite involved and are not presented here. It must be emphasized that a high-speed penetration is a very fast process. Consequently, realization of the complicated BP with several switching in this case is not feasible. Therefore, we are interested not in the BP but rather in the maximum DOP that can be used for comparing the efficiencies of simple and technically feasible BPs at the design stage. We must also take into account the uncertainties in estimating the characteristics of the materials. Under these conditions such advantages of the analytical approach as accuracy and speed of calculations are not that important. It should be emphasized also that analytical approach is feasible only for very simple formulations of the problem. On the other hand, numerical techniques are very efficient when applied to the above considered formulations of the problem or for more general formulations of the problem.

21.3.1 *Application of dynamic programming*

Dynamical programming (Bellman *et al.*, 1958; Roberts, 1964; Pedregal, 2003) is an appropriate method for determining the optimal BPs. This method allows determining a global extremum taking into account constraints. In the following we describe application of this method for the above formulated optimization problem, although it can be applied for analyzing advanced formulations of this problem by introducing additional constraints.

In order to solve the variational problem for the functional given by Eq. (21.2.7), the function $\overline{v}(\overline{m})$ can be approximated as a piecewise linear function determined by the values (see Fig. 21.3.1)

$$\overline{v}^{(0)} = 0, \overline{v}^{(1)}, \ldots, \overline{v}^{(j)}, \ldots, \overline{v}^{(N_m-1)}, \overline{v}^{(N_m)} \tag{21.3.1}$$

and

$$\overline{w}^{(0)}, \overline{w}^{(1)}, \ldots, \overline{w}^{(j)}, \ldots, \overline{w}^{(N_m-1)}, \overline{w}^{(N_m)} = 1 \tag{21.3.2}$$

in $N_m + 1$ equally spaced mesh points of interpolation,

$$\overline{m}^{(0)} = \overline{m}_{res}, \overline{m}^{(1)},...,\overline{m}^{(j)},..., \overline{m}^{(N_m-1)}, \overline{m}^{(N_m)} = 1, \tag{21.3.3}$$

where

$$\overline{m}^{(j)} = \overline{m}_{res} + j\Delta\overline{m}, \ \ \Delta\overline{m} = (1 - \overline{m}_{res})/N_m, \ \ j = 0,1,2,...,N_m. \tag{21.3.4}$$

The unknown values of the function, $\overline{v}^{(j)}$ and $\overline{w}^{(j)}$ are chosen among the finite set of values $0, \Delta\overline{v}, 2\Delta\overline{v},..., N_v\,\Delta\overline{v}$, where

$$\Delta\overline{v} = \overline{v}_{\max}/N_v \tag{21.3.5}$$

and $v_{\max}$ is a given upper bound for the velocity of the penetrator.

Penetration of the impactor is associated with the trajectory

$$\begin{aligned}&(\overline{m}^{(N_m)}, \overline{w}^{(N_m)}) \to (\overline{m}^{(N_m)}, \overline{v}^{(N_m)}) \to (\overline{m}^{(N_m-1)}, \overline{w}^{(N_m-1)})\\ &\to (\overline{m}^{(N_m-1)}, \overline{v}^{(N_m-1)}) \to ... \to (\overline{m}^{(1)}, \overline{w}^{(1)}) \to (\overline{m}^{(1)}, \overline{v}^{(1)})\\ &\to (\overline{m}^{(0)}, \overline{w}^{(0)}) \to (\overline{m}^{(0)}, \overline{v}^{(0)}).\end{aligned} \tag{21.3.6}$$

Since the velocity of the impactor can not increase without burning of fuel, $\overline{v}^{(j)} \le \overline{w}^{(j)}$ for all j.

Note that possibility of "vertical" segments in the solution is not usually taken into account when describing discrete versions of the dynamic programming for solving variational problems.

The equation of the line between the points $(\overline{m}^{(j-1)}, \overline{w}^{(j-1)})$ and $(\overline{m}^{(j)}, \overline{v}^{(j)})$ can be written as follows:

$$\overline{v} = e^{(j)}\overline{m} + f^{(j)}, \ \ j = 1,2,...,N_m, \tag{21.3.7}$$

where

$$e^{(j)} = \frac{\overline{v}^{(j)} - \overline{w}^{(j-1)}}{\overline{m}^{(j)} - \overline{m}^{(j-1)}}, \ \ f^{(j)} = \frac{\overline{m}^{(j)}\overline{w}^{(j-1)} - \overline{m}^{(j-1)}\overline{v}^{(j)}}{\overline{m}^{(j)} - \overline{m}^{(j-1)}}. \tag{21.3.8}$$

Then the integral in Eq. (21.2.7) along the piecewise linear contour can be written as follows:

$$\overline{P} = Q^{(0)} + \sum_{j=1}^{N_m} (F^{(j)} + Q^{(j)}), \tag{21.3.9}$$

where

$$Q^{(j)} = \overline{m}^{(j)} \int_{\overline{v}^{(j)}}^{\overline{w}^{(j)}} \frac{\overline{v}\, d\overline{v}}{k\overline{v}^2 + 1} = \frac{\overline{m}^{(j)}}{2k} \ln\left(\frac{k\overline{w}^{(j)2} + 1}{k\overline{v}^{(j)2} + 1} \right), \; j = 0,1,2,\ldots, N_m, \tag{21.3.10}$$

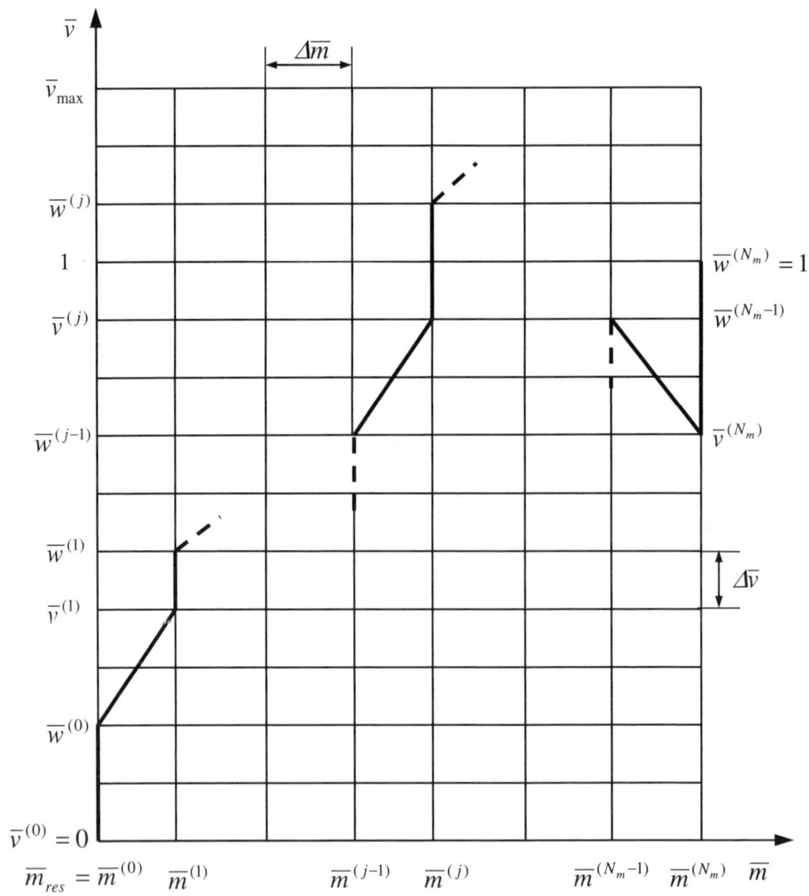

Fig. 21.3.1. Discretization of the problem.

$$F^{(j)} = \begin{cases} F_1^{(j)} & if \quad \bar{v}^{(j)} > \bar{w}^{(j-1)} \\ F_0^{(j)} & if \quad \bar{v}^{(j)} = \bar{w}^{(j-1)} \end{cases}, \quad j = 1,2,...,N_m. \tag{21.3.11}$$

$$F_0^{(j)} = \int_{\bar{m}^{(j-1)}}^{\bar{m}^{(j)}} \frac{\bar{c}\,\bar{v}^{(j)} d\bar{m}}{k\bar{v}^{(j)2} + 1} = \frac{\bar{c}\,\bar{v}^{(j)} \Delta\bar{m}}{k\bar{v}^{(j)2} + 1}, \quad j = 1,2,...,N_m, \tag{21.3.12}$$

$$F_1^{(j)} = \int_{\bar{w}^{(j-1)}}^{\bar{v}^{(j)}} \frac{(\bar{m} + \bar{c}\bar{v}')\bar{v}d\bar{v}}{k\bar{v}^2 + 1} = \frac{1}{e^{(j)}} \int_{\bar{w}^{(j-1)}}^{\bar{v}^{(j)}} \frac{(\bar{v} + \bar{c} - f^{(j)})\bar{v}d\bar{v}}{k\bar{v}^2 + 1}$$

$$= \frac{1}{ke^{(j)}} \left[\bar{v}^{(j)} - \bar{w}^{(j)} + 0.5(\bar{c} - f^{(j)}) \ln\left(\frac{k\bar{w}^{(j)2} + 1}{k\bar{v}^{(j)2} + 1} \right) \right] \tag{21.3.13}$$

$$- \frac{1}{k^{3/2} e^{(j)}} \left[\tan^{-1}(\sqrt{k}\bar{v}^{(j)}) - \tan^{-1}(\sqrt{k}\bar{w}^{(j)}) \right], \quad j = 1,2,...,N_m.$$

The constraint given by Eq. (21.2.8) is always satisfied on the "vertical" segments of the solution. Consider now the "non-vertical" segments of the solution. Substituting $\bar{v}$ from Eq. (21.3.7) into Eq. (21.2.8) yields the following inequality:

$$\lambda_2^{(j)} \bar{m}^2 + \lambda_1^{(j)} \bar{m} + \lambda_0^{(j)} \le 0, \quad j = 1,2,...,N_m, \tag{21.3.14}$$

where

$$\begin{aligned} \lambda_0^{(j)} &= \bar{\beta}_{\min} (kf^{(j)2} + 1) - \bar{c}, \\ \lambda_1^{(j)} &= (2k\bar{\beta}_{\min} f^{(j)} - 1)e^{(j)}, \\ \lambda_2^{(j)} &= k\bar{\beta}_{\min} e^{(j)2}. \end{aligned} \tag{21.3.15}$$

After some algebra, it can be shown that Eq. (21.3.14) is satisfied for $\bar{m}^{(j-1)} \le \bar{m} \le \bar{m}^{(j)}$ when one of the following sets of conditions is satisfied:

$$\bar{\beta}_{\min} = 0, \quad e^{(j)} = 0 \; (\bar{v}^{(j)} = \bar{w}^{(j-1)}), \tag{21.3.16}$$

$$\bar{\beta}_{\min} > 0, e^{(j)} = 0, \lambda_0^{(j)} \le 0, \tag{21.3.17}$$

$$\overline{\beta}_{\min} = 0, e^{(j)} \neq 0 \quad e^{(j)}\overline{m}^{(j-1)} + \overline{c} \geq 0, \quad e^{(j)}\overline{m}^{(j)} + \overline{c} \geq 0, \tag{21.3.18}$$

$$\overline{\beta}_{\min} \neq 0, e^{(j)} \neq 0 \quad \varepsilon^{(j)} \geq 0, \quad \overline{m}^{(j-1)} \geq \overline{m}_a^{(j)}, \overline{m}^{(j)} \leq \overline{m}_b^{(j)}, \tag{21.3.19}$$

where

$$\begin{aligned} &\overline{m}_a^{(j)} = -0.5(\lambda_1^{(j)} + \sqrt{\varepsilon^{(j)}})/\lambda_2^{(j)} \; \overline{m}_b^{(j)} = 0.5(-\lambda_1^{(j)} + \sqrt{\varepsilon^{(j)}})/\lambda_2^{(j)}, \\ &\varepsilon^{(j)} = \lambda_1^{(j)2} - 4\lambda_0^{(j)}\lambda_2^{(j)}. \end{aligned} \tag{21.3.20}$$

We do not allow regimes whereby fuel is exhausted while the velocity of the penetrator remains zero, i.e., the following condition must be satisfied:

$$\overline{v}^{(j)} + \overline{w}^{(j-1)} > 0, \quad j = 1,2,\ldots,N_m. \tag{21.3.21}$$

Unlike the standard approach (see, e.g., Roberts, 1964; Pedregal, 2003), we do not use approximations for the argument and the unknown function on the subintervals. This modification of the method allows reducing the original problem to the exact piecewise linear discrete problem and decreases the error of approximation in the case of sharp changes of the function. Clearly, this approach is expedient if the functions $Q^{(j)}$ and $F^{(j)}$ in Eq. (21.3.9) can be calculated.

21.3.2 *Results of numerical optimization*

In Fig. 21.3.2 we showed the dependencies of the maximum DOP as a function of the parameter k for different values of $\overline{\beta}_{\min}$. These figures differ by the values of the dimensionless relative exit velocity of gases from the nozzle of the jet thruster. In order to simplify the analysis, we considered the cases when the upper bound on $\overline{v}$ is irrelevant, i.e., $\overline{v}_{\max} = +\infty$. The maximum effect of using a jet thruster is observed for small k, although the effect is quite significant also for larger k. Increase of $\overline{\beta}_{\min}$ implies stronger constraints, namely, reducing the upper bound for the admissible fuel consumption rate of the thruster. Consequently, the curves $\delta = \delta(k)$, where $\delta = \overline{P}_{\max}/\overline{P}_0$, are located lower with the increase of $\overline{\beta}_{\min}$.

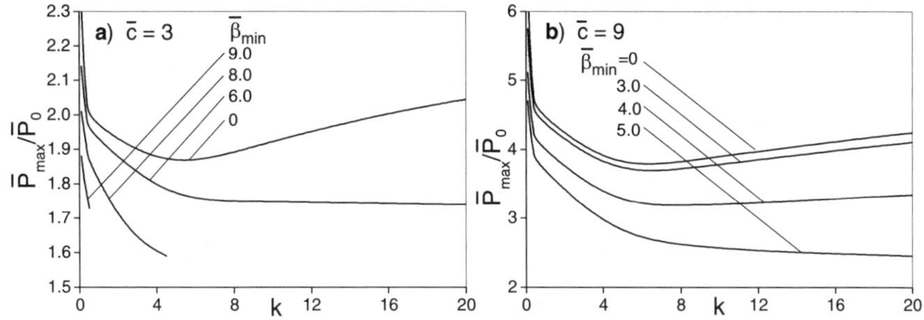

Fig. 21.3.2. Maximum depth of penetration.

Moreover, inspection of Fig. 21.3.2a, shows that some curves of the dependencies $\delta = \delta(k)$ are terminated. The latter implies that the capabilities of the thruster are inadequate, and penetration terminates before exhausting fuel supply.

Figs. 21.3.3a-d show typical variants of the optimal BPs. Fig. 21.2.1b shows that for $\bar{c} > 1$ and relatively small values of k, $\bar{v}_* > 1$, i.e., at some segments of the trajectory a penetrator moves faster than the impact velocity. Such situation is shown in Figs. 21.3.3a,c; regimes with $\bar{v} > 1$ correspond to small values of $\overline{\beta}_{\min}$ (Figs. 21.3.3a,c show such regime for $\overline{\beta}_{min} = 0$). The BP includes motion with the maximum fuel flow rate until the velocity $\bar{v}_*$ is attained, further motion with a constant velocity $\bar{v}_*$ until the fuel supply is exhausted, and inertial motion until the state of rest. When $\overline{\beta}_{\min}$ increases and reaches some threshold, the second segment of the trajectory vanishes, and depending on the magnitude of $\overline{\beta}_{\min}$ penetration proceeds with velocities larger or smaller than the impact velocity.

Typical situations corresponding to $\bar{v}_* < 1$ are shown in Figs. 21.3.3b,d. The regime of motion for relatively small $\overline{\beta}_{\min}$ is as follows: inertial motion until velocity $\bar{v}_*$ is reached; motion with a constant velocity $\bar{v}_*$ until exhausting fuel supply, and, finally, inertial motion until the state of rest. When $\overline{\beta}_{\min}$ increases the optimal regime of motion changes. Inertial motion is replaced by penetration with a maximum fuel flow rate until velocity $\bar{v}_*$ is attained, and this motion is executed until exhausting fuel supply. The final stage of penetration is inertial motion until the state of rest.

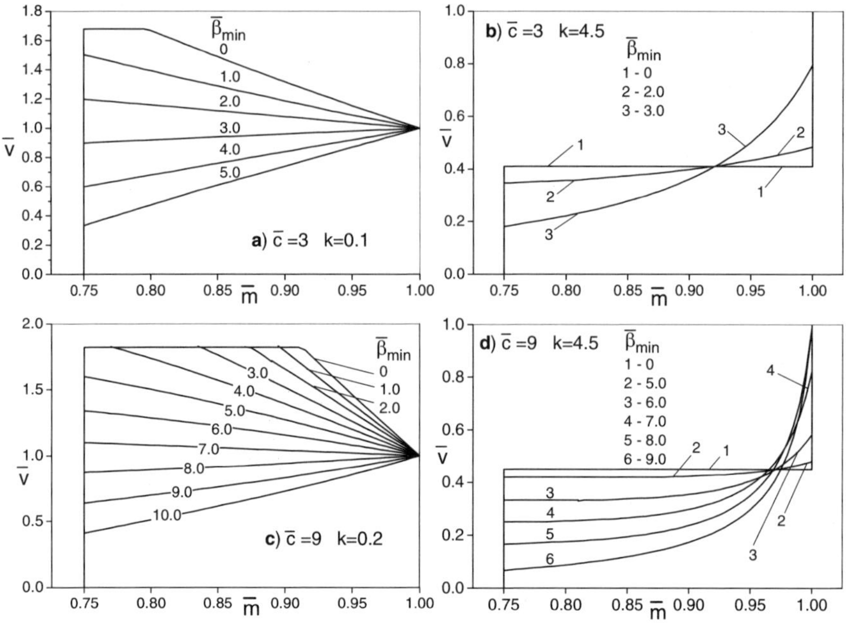

Fig. 21.3.3. Modes of optimum BPs.

21.4 Concluding Remark

Nelson, 2002 emphasized that there is a limitation on increasing impact velocity in order to attain a higher penetration depth because of the constraints on the impact velocity and acceleration required for the survival of a penetrator on impact. Using a penetrator with a jet thruster in order to increase the depth of penetration can be considered, among other approaches, as an alternative to the increase of impact velocity. It must be noted that constraints on impact velocity and acceleration can be readily incorporated in the model without complicating the procedure of dynamic programming. Indeed, the constraint on the impact velocity is taken into account on the stage of discretization of the problem and mesh generation. The constraint on the acceleration is as follows:

$$|\, dv/dt \,| \le \chi_{\max}\, g\,, \tag{21.4.1}$$

where g is the acceleration of gravity, $\chi_{\max}$ is a given upper bound for the overload of the impactor. Using the dimensionless variables determined by Eq. (21.1.12) we can rewrite Eq. (21.4.1) as follows:

$$|\, d\bar{v}/d\bar{t}\, | \leq \bar{k}_0 \bar{\chi}_{\max}, \quad \bar{\chi}_{\max} = \chi_{\max}\, Lg/(\bar{k}_0 v_{imp}^2). \tag{21.4.2}$$

In the case of the two-term model, $\overline{D}(\bar{v}) = k_0(k\bar{v}^2 + 1)$, and Eq. (21.1.8) implies that $d\bar{t} = -(1/\bar{k}_0)(\overline{m}d\bar{v} + \bar{c}d\overline{m})/(k\bar{v}^2 + 1)$. Substituting this expression into Eq. (21.2.51) allows rewriting the constraint:

$$-\bar{\chi}_{\max} \leq \frac{(k\bar{v}^2 + 1)\bar{v}'}{\overline{m}\bar{v}' + \bar{c}} \leq \bar{\chi}_{\max}\,. \tag{21.4.3}$$

Discretization of the above inequalities is performed similarly to the disretization of the constraints on the mass flux of the thruster.

Chapter 22

Application of the Modified Young Model

Nomenclature

BP	burning program;
c	relative exit velocity of gases from the nozzle of jet thruster;
$\hat{c}$	parameters, Eq. (22.2.7) or (22.2.32);
$\overline{c}$	dimensionless c, Eq. (22.3.6);
D	resistance force acting on projectile;
$\overline{D}$	dimensionless D, Eq. (22.3.6);
DOP	depth of penetration;
$e^{(j)}, f^{(j)}$	parameters, Eq. (21.3.8);
g	acceleration of gravity;
h	instantaneous depth of penetration;
$\overline{h}$	dimensionless h, Eq. (22.3.6);
H	depth of penetration;
H_0	DOP for "passive" penetration, Eq. (22.2.26);
$\overline{H}$	dimensionless H , Eq. (22.3.6);
IFS	ice and frozen soil;
k , k_0	parameters, Eq. (22.1.5);
k_1	parameter in Eq. (22.1.4), specified in Tab. 22.1.1;
L	reference length;
m	mass of projectile;
$\overline{m}$	dimensionless m, Eq. (22.3.6);
$\overline{m}^{(j)}$	mesh points of interpolation for $\overline{m}$, Eq. (21.3.4);
$\Delta\overline{m}$	mesh width for $\overline{m}$, Eq. (21.3.4);
m_{imp}	initial value of m ;

m_{res}	final value of m;
$\overline{m}_{res}$	dimensionless m_{res}, Eq. (22.3.6);
N_v	number of mesh intervals for $\overline{v}$;
N_m	number of mesh intervals for $\overline{m}$;
P	DOP;
P_a	function, Eq. (22.1.3) or Eq. (22.1.8);
P_b	function, Eq. (22.1.4);
SRC	soil, rock and concrete;
t	time;
v_{imp}	impact velocity of projectile;
$v_{\max}$	maximum velocity of impactor;
$\overline{v}_{\max}$	dimensionless $v_{\max}$, Eq. (22.3.6);
v	instantaneous velocity of projectile;
$\overline{v}'$	$=d\overline{v}/d\overline{m}$;
$\overline{v}$	dimensionless v, Eq. (22.3.6);
$\overline{v}^{(j)}$, $\overline{w}^{(j)}$	mesh points of interpolation for $\overline{v}$, Eqs. (21.3.1) and (21.3.2);
v_*	parameter of the model in Eq. (22.1.2); see Eqs. (22.1.7) or Eq. (22.1.13) ;
$\overline{v}_*$	$=v_*/v_{imp}$;
v_0	parameter in Eq. (22.1.4); see Eqs. (22.1.7);
$v_\times$	root of the equation $\Omega_a(v)=0$, Eq. (22.2.9);
α_1	parameter in Eq. (22.1.3), specified in Tab. 22.1.1;
β_0, β_2	parameters, Eq. (22.1.8);
$\overline{\beta}_{\min}$	parameter, Eq. (22.3.6);
γ_0, γ_2	parameters, Eq. (22.1.13);
$\hat{\gamma}_0$	$=\gamma_0/\gamma_2$;
δ	$=H_{\max}/H_0$;
ΔH	change of the depth of penetration;
$\Delta\overline{v}$	mesh size for $\overline{v}$, Eq. (21.3.5);
η	$=H/H_0$;
θ	function, Eq. (22.1.6);
κ	parameter in Eq. (22.2.2);

κ_1, κ_2	parameters in Eqs. (22.1.5) and (22.1.6), specified in Tab. 22.1.1.;
λ	function, Eq. (22.2.32);
$\mu_{\max}$	maximum mass flux of thruster;
σ	parameter in Eq. (22.1.6), specified in Tab. 22.1.1;
τ	parameter in Eq. (22.2.13);
φ	function, Eq. (22.3.21);
$\chi_{\max}$	upper bound for the overload of impactor;
$\bar{\chi}_{\max}$	dimensionless $\chi_{\max}$, Eq. (22.3.6);
Ψ	function, Eq. (22.1.2);
ψ	function, Eq. (22.3.20);
Ω	function determined by Eq. (22.2.4); see also Eq. (22.2.5) and (22.2.31);
Ω_a, Ω_b	functions, Eq. (22.2.6).

22.1 Formulation of Problem

22.1.1 *Equation for resistance force*

Young penetration equations (Young, 1997) for the depth of penetration (DOP) into soil, rock and concrete (SRC) shields as well as into ice and frozen soil (IFS) which are considered in Sections 3.23.1 and 7.4 can be written in SI Units in the following unified form:

$$P = k\Psi(v_{imp}), \tag{22.1.1}$$

where

$$\Psi(v_{imp}) = \begin{cases} P_a(v_{imp}) & if \quad v_{imp} < v_* \\ P_b(v_{imp}) & if \quad v_{imp} \geq v_* \end{cases} \tag{22.1.2}$$

$$P_a(v_{imp}) = \alpha_1 \ln(1 + \alpha_2 v_{imp}^2), \tag{22.1.3}$$

$$P_b(v_{imp}) = k_1(v_{imp} - v_0), \tag{22.1.4}$$

$$k = k_0\theta(m), \quad k_0 = K_s K_{shape}\left(\frac{4}{\pi d^2}\right)^{\kappa_1}, \tag{22.1.5}$$

$$\theta(m) = K_m = \begin{cases} \sigma m^{\kappa_1+\kappa_2} & if \quad SRC \\ m^{\kappa_1}\ln(50+0.29m^2) & if \quad IFS \end{cases} \tag{22.1.6}$$

P is the DOP; K_s K_{shape} and K_m, are coefficients depending on mechanical properties of the shield, shape of the impactor and mass of impactor, correspondingly (see Sections 3.23.1 and 7.4); v_{imp} is impact velocity, d is diameter of impactor,

$$v_0 = 30.5, \ \alpha_2 = 0.000215, \ v_* = 2v_0 = 61, \tag{22.1.7}$$

and other coefficients are presented in Tab. 22.1.1.

Hereafter we use smooth approximation of the function $\Psi(v_{imp})$ that provides the continuous second derivative for all values of impact velocity, $v_{imp} > 0$. Such approximation is described in Section 3.23.2.2 and yields the following relationship for $P_a(v_{imp})$ instead of the relationship in Eqs. (22.1.3):

$$P_a(v_{imp}) = (\beta_2 v_{imp}^2 + \beta_0)v_{imp}^2, \tag{22.1.8}$$

where

$$\beta_0 = 0.00922k_1, \quad \beta_2 = 0.2323295\cdot 10^{-6}k_1, \tag{22.1.9}$$

and a new value $v_* = 81.33$.

Table 22.1.1. Parameters of the models.

Shield's material	m	κ_1	κ_2	σ	α_1	k_1
Soil	$2 \le m < 27$	0.7	0.4	0.27	0.0008	0.000018
	$m \ge 27$	0.7	0	1	0.0008	0.000018
Rock, Concrete	$5 \le m < 182$	0.7	0.15	0.46	0.0008	0.000018
	$m \ge 182$	0.7	0	1	0.0008	0.000018
Ice, Frozen Soil		0.6			0.00024	0.0000046

In order to determine the resistance force, D, that is assumed to be a function of the instantaneous velocity of impactor, v, and yields the relationship for the DOP given by Eqs. (22.1.1), (22.1.2), (22.1.4) and (22.1.8), we use the method described in Chapter 9. The expression for the resistance force reads:

$$D(m,v) = \frac{m}{k} \times \begin{cases} D_a(v) & if \quad v < v_* \\ D_b(v) & if \quad v \geq v_* \end{cases} \tag{22.1.10}$$

where m is a mass of impactor that is assumed to be a constant and

$$D_a(v) = v / P_a'(v) = (4\beta_2 v^2 + 2\beta_0)^{-1}, D_b(v) = v / P_b'(v) = v / k_1. \tag{22.1.11}$$

Substituting β_0 and β_2 given by Eq. (22.1.9) into Eq. (22.1.11) we obtain:

$$D(m,v) = \frac{m}{kk_1} \times \begin{cases} 1/(\gamma_0 - \gamma_2 v^2) & if \quad v < v_* \\ v & if \quad v \geq v_* \end{cases}, \tag{22.1.12}$$

where

$$\gamma_0 = \frac{2\beta_0}{k_1} = 0.01844, \gamma_2 = -\frac{4\beta_2}{k_1} = 0.929318 \cdot 10^{-6}, v_* = 81.33. \tag{22.1.13}$$

We assume also that Eq. (22.1.12) for the drag force is valid for a penetrator having a variable mass and that the mass of the impactor does not exceed $m = 27\,kg$ or $m = 182\,kg$ in the case of penetration into soil or rock and concrete, respectively. The latter restriction can be relaxed using a more smooth approximation for the function $\theta(m)$.

22.1.2 *Mathematical model and formulation of problem*

Consider a high speed normal penetration of a rigid sharp striker (a body of revolution) equipped with a jet thruster into a semi-infinite shield along the axis h, where the coordinate h, the instantaneous depth of penetration, is defined as the distance between the nose of the impactor and the front surface of the shield.

Equation of motion of a projectile with a variable mass reads:

$$m\frac{dv}{dt}+c\frac{dm}{dt}=-D(m,v), \quad D(m,v)>0\,, \tag{22.1.14}$$

where t is time, c is the relative exit velocity of gases from the nozzle of a jet thruster, m is the instantaneous mass of the impactor that varies in the range between the initial value, m_{imp}, and the final value, m_{res}, i.e.,

$$m_{res} \le m \le m_{imp}\,. \tag{22.1.15}$$

It is assumed that the thruster is capable to deliver all mass fluxes in the range between zero and the maximum value $\mu_{\max}$:

$$-\mu_{\max} \le \frac{dm}{dt} \le 0\,. \tag{22.1.16}$$

The constraint imposed on the acceleration of a projectile is as follows:

$$|\,dv/dt\,| \le \chi_{\max}\, g\,, \tag{22.1.17}$$

where g is the acceleration of gravity, $\chi_{\max}$ is a given upper bound for the overload of the impactor.

It is convenient to rewrite Eq. (22.1.14) as follows:

$$v\frac{dm}{dh}\left(mv'+c\right)=-D(m,v), \quad v'=\frac{dv}{dm}\,, \tag{22.1.18}$$

where v is considered as a function of m.

Hence the expression for the DOP, H, for a given impact velocity, v_{imp}, reads:

$$H=\int_{m_{res}}^{m_{imp}} \frac{(mv'+c)v}{D(m,v)}dm\,. \tag{22.1.19}$$

The following conditions are valid for the initial and terminal points at the trajectory of impactor:

$$v(m_{imp})=v_{imp}, \quad v(m_{res})=0\,. \tag{22.1.20}$$

Using Eq. (22.1.18) we obtain that

$$\frac{dm}{dt} = v\frac{dm}{dh} = -\frac{D(m,v)}{mv' + c}, \qquad (22.1.21)$$

and Eq. (22.1.16) can be rewritten as

$$0 \le \frac{D(m,v)}{mv' + c} \le \mu_{\max}. \qquad (22.1.22)$$

Using the relationship $dt = -(c + mv')dm/D(m,v)$ obtained from Eq. (22.1.21) we can rewrite Eq. (22.1.17) in the following form:

$$\left|\frac{D(m,v)v'}{mv' + c}\right| \le \chi_{\max}\, g, \qquad (22.1.23)$$

where g is acceleration due to gravity.

We also assume that

$$v \ge 0, \quad v \le v_{\max}, \qquad (22.1.24)$$

where $v_{\max}$ is the maximum velocity of impactor.

The problem is to determine the function $v(m)$ that provides the maximum DOP, H, in Eq. (22.1.19) taking into account the constraints given by Eqs. (22.1.20), (22.1.22)–(22.1.24). Function $D(m,v)$ and the parameters $\mu_{\max}, \chi_{\max}, v_{\max}, m_{imp}, m_{res}$ and v_{imp} are assumed to be given.

Inequalities in Eq. (22.1.22) imply two situations: (i) possibility of the "vertical sub-arcs", $m = const$ when $v' = \infty$, and (ii) constraints $mv' + c \ge D(m,,v)/\mu_{\max}$ and $mv' + c \ge 0$, when $v' < \infty$. Clearly, the latter inequality can be omitted and we can write this constraint as follows:

$$mv' + c \ge D(m,v)/\mu_{\max} \quad if \quad v' < \infty. \qquad (22.1.25)$$

For $m = const$ ($v' = \infty$), the constraint in Eq. (22.1.23) can be simplified as follows:

$$D(m,v)/m \le \chi_{\max}\, g \quad if \quad m = const. \qquad (22.1.26)$$

22.2 Analytical Investigation of Limiting Case

In order to estimate the upper bound for increasing the DOP by using the jet thruster let us consider the problem with the minimum number of constraints. In this version of the formulation of the problem, Eq. (22.1.19) remains valid, and the constraints given by Eq. (22.1.20), the first inequality in Eqs. (22.1.24) and Eq. (22.1.25) with $\mu_{\max} = \infty$, i.e.,

$$mv' + c \geq 0 \tag{22.2.1}$$

are taken into account.

22.2.1 *Penetration into non-frozen shield*

We consider first penetration into non-frozen soil, rock or concrete. In this case formula for $\theta(m)$ in Eq. (22.1.6) reads:

$$\theta(m) = \sigma m^{\kappa}, \kappa = \kappa_1 + \kappa_2, \tag{22.2.2}$$

where the values κ_1, κ_2 and σ are presented in Tab. 22.1.1. We assume that m varies in the range where σ is constant.

Using Eqs. (22.1.5) and (22.2.2), Eq. (22.1.12) can be written as follows:

$$D = \frac{m^{1-\kappa}}{\sigma k_0 k_1} \times \begin{cases} 1/(\gamma_0 - \gamma_2 v^2) & if \quad v < v_* \\ v & if \quad v \geq v_* \end{cases}. \tag{22.2.3}$$

Cicala and Miele, 1956 showed that the optimum curve must consist of the following sub-arcs:

(1) the sub-arc $\Omega(m, v) = 0$, where

$$\Omega(m, v) = (v - c)D(m, v) + v(c\, dD/dv - m\, dD/dm); \tag{22.2.4}$$

(2) the sub-arc $m = const$, and
(3) the sub-arc $\dot{m} = -\mu_{\max}$ where $\dot{m} = dm/dt$.

The sub-arc 1 represents the solution of the Euler-Lagrange equation, while the sub-arc 3 is described by Eq. (22.2.1) where a larger or equal sign is replaced by equal sign.

These sub-arcs can be joined in the following sequences (the arrows indicate the direction of motion of penetrator along the sub-arcs in the m, v plane):

a) if $\Omega > 0$ in some point, then only the sequence $m = const \Rightarrow \dot{m} = -\mu_{\max}$ is possible in this point,

b) if $\Omega < 0$, then the inverse sequence $\dot{m} = -\mu_{\max} \Rightarrow m = const$ is permissible, and

c) if $\Omega = 0$ then the sequences $\Omega = 0 \Leftrightarrow \overline{m} = const$ and $\Omega = 0 \Leftrightarrow \dot{m} = -\mu_{\max}$ are possible in this point.

Substituting D from Eq. (22.2.3) into Eq. (22.2.4) we obtain:

$$\Omega = \frac{\kappa m^{1-\kappa}}{\sigma k_0 k_1} \times \begin{cases} -\gamma_2(\gamma_0 - \gamma_2 v^2)^{-2}\,\Omega_a(v) & if \quad v < v_* \\ \Omega_b(v) & if \quad v \geq v_* \end{cases}, \qquad (22.2.5)$$

where

$$\Omega_a(v) = v^3 - 3\hat{c}v^2 - \hat{\gamma}_0 v + \hat{\gamma}_0\hat{c}, \quad \Omega_b(v) = v^2, \qquad (22.2.6)$$

$$\hat{c} = c/\kappa, \quad \hat{\gamma}_0 = \gamma_0/\gamma_2\,. \qquad (22.2.7)$$

Since

$$\begin{aligned} &\Omega_a(-\infty) < 0, \quad \Omega_a(0) > 0, \\ &\Omega_a(v_*) = -\Omega_b(v_*)(\gamma_0 - \gamma_2 v_*^2)^2/\gamma_2 < 0, \quad \Omega_a(+\infty) > 0, \end{aligned} \qquad (22.2.8)$$

the cubic equation $\Omega_a(v) = 0$ has 3 real roots but only one root, $v_\times$, is located between 0 and v_*:

$$v_\times = \hat{c} + 2\left(\frac{\hat{\gamma}_0 + 3\hat{c}^2}{3}\right)^{1/2} \cos\left(\frac{\zeta}{3} + \frac{4\pi}{3}\right), \quad \zeta = \cos^{-1}\left(\frac{3\hat{c}^2}{\hat{\gamma}_0 + 3\hat{c}^2}\right)^{3/2}. \qquad (22.2.9)$$

The dependence $v_\times$ vs. $\hat{c} = c/\kappa$ is shown in Fig. 22.2.1.

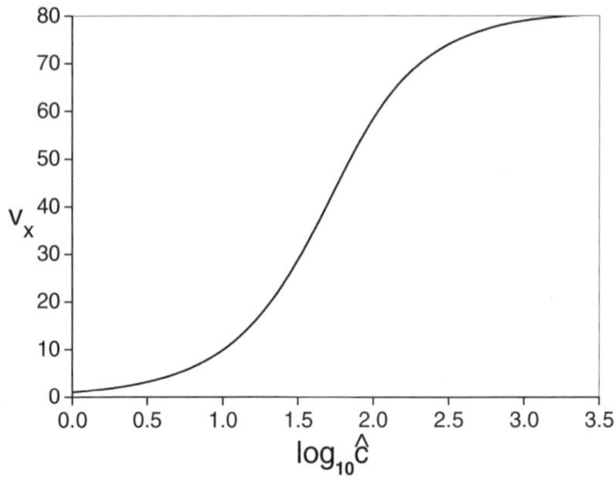

Fig. 22.2.1. Solution of the equation $\Omega(v)=0$ as a function of the parameter $\hat{c}=c/\kappa$.

Taking into account this result and that $\Omega_b(v)>0$ for $v \geq v_*$ we conclude that $\Omega(v)<0$ when $0 \leq v < v_\times$, $\Omega(v_\times)=0$ and $\Omega(v)>0$ when $v > v_\times$. Therefore, the conditions $\Omega(v)<0, \Omega(v)=0, \Omega(v)>0$ correspond to the conditions, $\bar{v}<\bar{v}_\times, \bar{v}=\bar{v}_\times, \bar{v}>\bar{v}_\times$, respectively.

The equation of the sub-arc 1 reads:

$$v = v_\times. \tag{22.2.10}$$

Consider a point moving along the curve $v = v_\times$ from the location with the coordinates $(m_b, v_\times)$ to the location with the coordinates $(m_a, v_\times)$, where $m_{res} \leq m_a < m_b \leq m_{imp}$. Hence the increase of the functional given by Eq. (22.1.19), where D is determined by Eq. (22.2.3), along this path reads:

$$\begin{aligned} \Delta H/(k_0 k_1) &= \Gamma_m(m_a, m_b) \\ &= c v_\times(\gamma_0 - \gamma_2 v_\times^2)\int_{m_a}^{m_b} \frac{dm}{m^{1-\kappa}} = c\hat{v}_\times(\gamma_0 - \gamma_2 v_\times^2)(m_b^\kappa - m_a^\kappa). \end{aligned} \tag{22.2.11}$$

Along the curve $v = v_\times < v_*$, Eq. (22.1.14) can be written as follows:

$$c\frac{dm}{dt} = -\frac{m^{1-\kappa}}{k_0 k_1(\gamma_0 - \gamma_2 v_\times^2)}. \qquad (22.2.12)$$

The solution of Eq. (22.2.12) with the initial conditions $m(t_b) = m_b$ reads:

$$m = [m_b^\kappa - \tau(t - t_b)]^{1/\kappa}, \quad \tau = \frac{\kappa}{ck_0 k_1(\gamma_0 - \gamma_2 v_\times^2)}. \qquad (22.2.13)$$

Mass of penetrator increases from m_b to m_a until the time $t = t_a$, where $t_a = t_b + (m_b^\kappa - m_a^\kappa)/\tau$.

The sub-arc 2 corresponds to penetration with a constant mass. Assume that the velocity of the impactor decreases from $v = v_b$ to $v = v_a$ while its mass remains constant, $m = m_0$. Hence the increase of the functional given by Eq. (22.1.19) along this path reads:

$$\begin{aligned} \Delta H/(k_0 k_1) &= \Gamma_v^{(1)}(m_0, v_a, v_b) = m_0^\kappa \int_{v_a}^{v_b} v(\gamma_0 - \gamma_2 v^2)dv \\ &= 0.25 m_0^\kappa (v_b^2 - v_a^2)[2\gamma_0 - \gamma_2(v_b^2 + v_a^2)] \quad \text{if} \quad v_b \le v_\times, \end{aligned} \qquad (22.2.14)$$

and

$$\begin{aligned} \Delta H/(k_0 k_1) &= \Gamma_v^{(2)}(m_0, v_a, v_b) \\ &= m_0^\kappa \int_{v_a}^{v_b} dv = m_0^\kappa (v_b - v_a) \quad \text{if} \quad v_b \ge v_\times. \end{aligned} \qquad (22.2.15)$$

Along the sub-arc 3, $\dot{m} = -\mu_{\max} = -\infty$. The equation of this sub-arc in the coordinates (m, v) is determined by Eq. (22.2.1) where a larger or equal sign is replaced by equal sign:

$$mv + c = 0. \qquad (22.2.16)$$

The solution of this ODE with separable variables passing through the point (m_{imp}, V_0) can be written as follows:

$$v = V_0 - c\ln(m/m_{imp}). \qquad (22.2.17)$$

This trajectory corresponds to the step change of mass of penetrator (the pulse burning). This change of mass and velocity of penetrator occurs without changing its location.

Analysis of the constraints presented at the beginning of this section shows that three version of burning program (BP) are possible, depending on the values of the parameters $\hat{c}$ and v_{imp}.

Consider the case when $v_{imp} > v_*$. Taking into account that $v_* > v_\times$ the optimal BP includes the following sub-arcs (Fig. 22.2.2a) in the coordinates (m, v):

$$\begin{cases} AE: & m = m_{imp}, \quad v: v_{imp} \to v_* \\ EC: & m = m_{imp}, \quad v: v_* \to v_\times \\ CD: & v = v_\times, \quad m: m_{imp} \to m_{res} \\ DB: & m = m_{res}, \quad v: v_\times \to 0 \end{cases} \tag{22.2.18}$$

and corresponding expression for the DOP reads:

$$H/(k_0 k_1 m_{imp}^{\kappa}) = \Gamma_v^{(2)}(1, v_*, v_{imp}) + \Gamma_v^{(1)}(1, v_\times, v_*) + \Gamma_m(\overline{m}_{res}, 1) + \Gamma_v^{(1)}(\overline{m}_{res}, 0, v_\times), \quad \overline{m}_{res} = m_{res}/m_{imp}. \tag{22.2.19}$$

In the case $v_{imp} = v_*$ the segment AE vanishes.

If $v_\times < v_{imp} < v_*$ then the optimal BP includes the following sub-arcs (Fig. 22.2.2b):

$$\begin{cases} AC: & m = m_{imp}, \quad v: v_{imp} \to v_\times \\ CD: & v = v_\times, \quad m: m_{imp} \to m_{res} \\ DB: & m = m_{res}, \quad v: v_\times \to 0 \end{cases} \tag{22.2.20}$$

and formula for the DOP reads:

$$H/(k_0 k_1 m_{imp}^{\kappa}) = \Gamma_v^{(1)}(1, v_\times, v_{imp}) + \Gamma_m(\overline{m}_{res}, 1) + \Gamma_v^{(1)}(\overline{m}_{res}, 0, v_\times). \tag{22.2.21}$$

In the case when $v_{imp} = v_\times$, the segment AC vanishes.

Motion of penetrator that is described by Eqs. (22.2.18) and (22.2.20), includes (in a general case) the following stages: “passive” penetration when a jet thruster is turned off until the velocity slows down to the magnitude $v_\times$ (segment AC of the trajectory); penetration with a constant velocity $v_\times$ with operating jet thruster until complete exhaustion of fuel supply (segment CD of the trajectory); and passive penetration until a penetrator slows down to zero velocity (segment DB of the trajectory).

Consider now the case when $v_{imp} < v_\times$ and the curve determined by Eq. (22.2.17) with $V_0 = v_{imp}$ intersects with the line $v = v_\times$ in the point $(m_\times, v_\times)$, whereas $m_\times > m_{res}$ (Fig. 22.2.2c). Hence the optimal BP includes three sub-arcs:

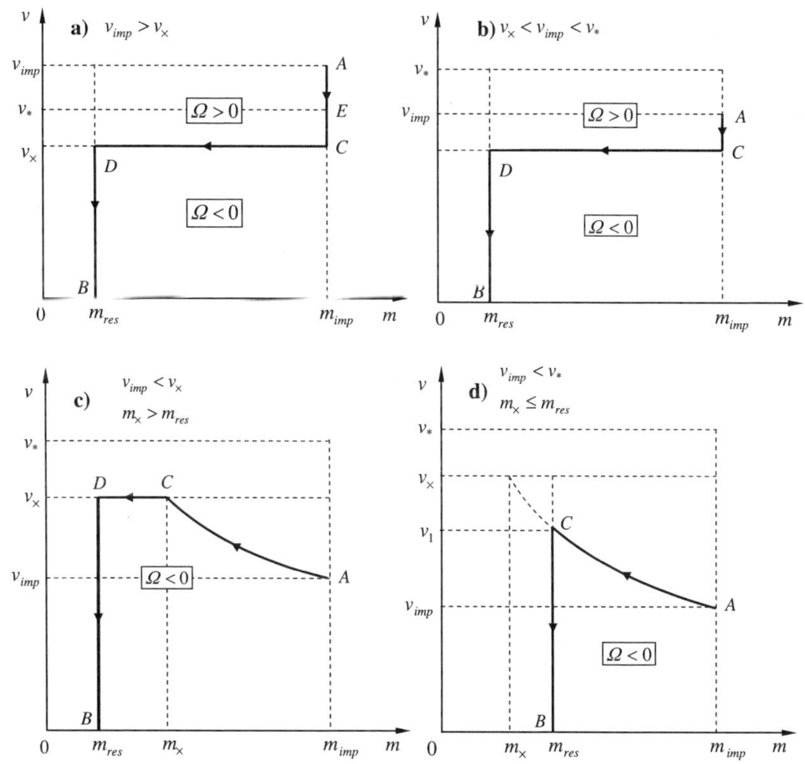

Fig. 22.2.2. Stages of optimal BPs for penetration into non-frozen soil, rock and concrete shields, $\mu_{\max} = \infty$.

$$\begin{cases} AC: & v = v_{imp} - c\ln(m/m_{res}), & m: m_{imp} \to m_{\times} \\ CD: & v = v_{\times}, & m: m_{\times} \to m_{res} \\ DB: & m = m_{res}, & v: v_{\times} \to 0 \end{cases} \quad (22.2.22)$$

where $m_{\times} = m_{imp} \exp((1 - v_{\times})/c)$.

Eq. (22.2.22) describes the following stages of the controlled motion of a penetrator: operation of a jet thruster in the impulse regime whereby the velocity of penetrator instantaneously changes from the initial value, v_{imp}, to $v_{\times}$ by spending mass of fuel equal to $m_{imp} - m_{\times}$ without motion of a penetrator (sub-arc AC); operation of jet thruster for providing the constant velocity of a penetrator, $v_{\times}$, until exhaustion of the fuel supply (segment CD) and passive penetration until penetrator slows down to zero velocity (segment DB of the trajectory).

The expression for the DOP in this case reads:

$$H/(k_0 k_1 m_{imp}^{\kappa}) = \Gamma_m(\overline{m}_{res}, \overline{m}_{\times}) + \Gamma_v^{(1)}(\overline{m}_{res}, 0, v_{\times}), \overline{m}_{\times} = m_{\times}/m_{imp} . \quad (22.2.23)$$

If $v_{imp} < v_{\times}$ and $m_{\times} \le m_{res}$ then the optimal BP includes two sub-arcs (Fig. 22.2.2d):

$$\begin{cases} AC: & v = v_{imp} - c\ln(m/m_{res}), & m: m_{imp} \to m_{res} \\ CB: & m = m_{res}, & v: v_1 \to 0 \end{cases}, \quad (22.2.24)$$

where $v_1 = v_{imp} - c\ln\overline{m}_{res}$.

Here a jet thruster operates in the impulse regime until exhausting the whole fuel supply. In this case the velocity of the penetrator instantaneously changes from the initial velocity to v_1 (segment AC of the trajectory). After this penetrator continues its motion until it slows down to zero velocity (segment CB of the trajectory).

The DOP is determined using only the sub-arc CB:

$$H/(k_0 k_1 m_{imp}^{\kappa}) = \Gamma_v^{(1)}(\overline{m}_{res}, 0, v_1) . \quad (22.2.25)$$

The depth of passive penetration of impactor having a mass m_{imp} is as follows:

$$\frac{H_0}{k_0 k_1 m_{imp}^{\kappa}} = \begin{cases} \Gamma_v^{(1)}(1,0,v_{imp}) & if \quad v_{imp} \le v_* \\ \Gamma_v^{(2)}(1,v_*,v_{imp}) + \Gamma_v^{(1)}(1,0,v_*) & if \quad v_{imp} > v_* \end{cases}. \qquad (22.2.26)$$

Some of the above relationships can be simplified taking into account that $v_\times \approx v_*$ for $c > 2000\, m/s$ (Fig. 22.2.1).

We use the efficiency ratio

$$\eta = H/H_0, \qquad (22.2.27)$$

that depends on $v_{imp}, c, \overline{m}_{res}$ and κ, for comparing the DOP with the optimal BP and the depth of passive penetration (Fig. 22.2.3). Although η can assume very large values for the relatively small impact velocities, the absolute values of the DOP, H, are reasonable. The results for the relatively high impact velocities are the most interesting because of the feasibility of increasing the absolute DOP by using the jet thruster.

The optimum BP for the impact velocities, $v_{imp} > v_* \approx 81\, m/s$, is simple: passive penetration until slowing down to velocity $v_\times$, penetration with the constant velocity $v_\times$ while the propellant is available (the consumption of the propellant is described by Eq. (22.2.13) with $m_a = m_{res}$ and $m_b = m_{imp}$, $m = [m_{res}^{\kappa} - \tau(t - t_{res})]^{1/\kappa}$), and passive penetration until the state of rest. This BP is valid if the constraints on mass fluxes and acceleration of penetrator are not taken into account. Let us determine the conditions when these constraints do not affect this optimum BP.

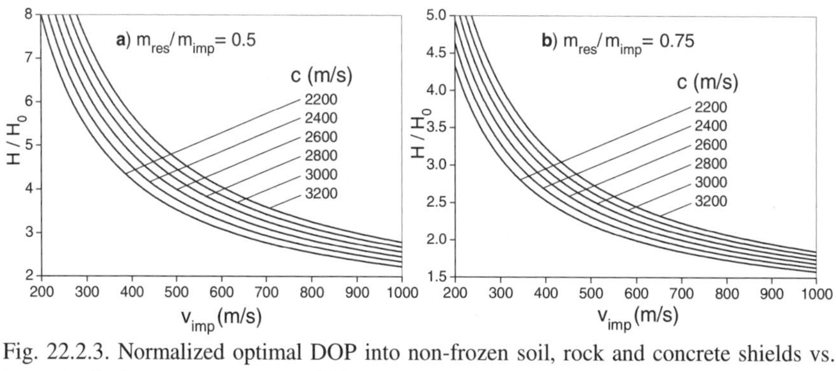

Fig. 22.2.3. Normalized optimal DOP into non-frozen soil, rock and concrete shields vs. impact velocity; $\mu_{\max} = \infty$, $\kappa = 0.7$.

The constraints given by Eq. (22.1.16) are relevant only on the segment of the trajectory where $v = v_{\times}$. Equation (22.1.14) implies that $dm/dt = -D(m, v_{\times})/c = -(\tau/\kappa)m^{1-\kappa}$. Substituting this relationship into Eq. (22.1.16) we obtain the inequality $m \le (\kappa\mu_{\max}/\tau)^{1/(1-\kappa)}$ that must be satisfied for $m_{res} \le m \le m_{imp}$. Since mass decreases on this interval, the latter inequality can be replaces by the following one:

$$m_{imp} \le (\kappa\mu_{\max}/\tau)^{1/(1-\kappa)}. \tag{22.2.28}$$

The constraints given by Eq. (22.1.16) are relevant only on the segment of the trajectory with a constant mass, $m = \tilde{m}$. Equation (22.1.14) implies that $dv/dt = -D(\tilde{m}, v)/\tilde{m}$. Substituting this relationship into Eq. (22.1.16) we find that $D(m_{imp}, v) \le m_{imp} g\chi_{\max}$ if $v_{\times} \le v \le v_{imp}$ and $D(m_{res}, v) \le m_{res} g\chi_{\max}$ if $0 \le v \le v_{\times}$. Since D is an increasing function of v, these constraints are valid if $D(m_{imp}, v_{imp}) \le m_{imp} g\chi_{\max}$ and $D(m_{res}, v_{\times}) \le m_{res} g\chi_{\max}$. The last two inequalities imply that

$$m_{imp} \ge \frac{v_{imp}}{k_0 k_1 g\chi_{\max}}, \quad m_{res} \ge \frac{1}{k_0 k_1 (\gamma_0 - \gamma_2 v_{\times}^2) g\chi_{\max}}. \tag{22.2.29}$$

22.2.2 *Penetration into frozen shield*

Let us consider the case of penetration into ice and frozen soil shields. In this case $\theta(m) = m^{0.6}\ln(50 + 0.29m^2)$ and

$$D(m, v) = \frac{m^{0.4}}{k_0 k_1 \ln(50 + 0.29m^2)} \times \begin{cases} 1/(\gamma_0 - \gamma_2 v^2) & if \quad v < v_* \\ v & if \quad v \ge v_* \end{cases}. \tag{22.2.30}$$

Substituting expression for a drag force D from Eq. (22.2.30) into Eq. (22.2.4) we obtain:

$$\Omega = \frac{m^{0.4}[0.6 + \lambda(m)]}{k_0 k_1 \ln(50 + 0.29m^2)} \times \begin{cases} -\gamma_2(\gamma_0 - \gamma_2 v^2)^{-2}\,\Omega_a(v) & if \quad v < v_* \\ \Omega_b(v) & if \quad v \ge v_* \end{cases}, \tag{22.2.31}$$

where

$$\Omega_a(m,v) = v^3 - 3\hat{c}v^2 - \hat{\gamma}_0 v + \hat{\gamma}_0\hat{c}, \quad \Omega_b(v) = v^2,$$
$$\hat{c} = \frac{c}{0.6+\lambda(m)}, \quad \lambda(m) = \frac{0.58m^2}{(50+0.29m^2)\ln(50+0.29m^2)}. \tag{22.2.32}$$

The equation $\Omega(m,v)=0$ determines the sub-arc $v = v_\times(m)$ where $v_\times(m)$ is determined by Eq. (22.2.9) with $\hat{c} = \hat{c}(m)$ from Eq. (22.2.32). In contrast to the case of SRC shields, the sub-arc is curvilinear in this case.

Let us investigate now the behavior of the function $\lambda(m)$. Changing the variable $50+0.29m^2 = x$ we reduce the problem to analyzing the function

$$\Theta(x) = \lambda\left(\sqrt{\frac{x-50}{0.29}}\right) = \frac{2(x-50)}{x\ln x}. \tag{22.2.33}$$

Let us calculate the derivative:

$$\Theta'(x) = \frac{2}{x\ln x}\Theta_0(x), \Theta_0(x) = 50(\ln x + 1) - x, \Theta_0'(x) = \frac{50}{x} - 1. \tag{22.2.34}$$

For $x > 50$, $\Theta_0'(x) < 0$. Hence, $\Theta_0(x)$ decreases in this semi-infinite interval. Since $\Theta_0(50) > 0$ and $\Theta_0(350) < 0$, we conclude that the equation $\Theta_0(x) = 0$ has a single root which can be easy determined, $x = x_0 \approx 341.7$. This root is a point of the maximum of the function $\Theta(x)$ that increases from $\Theta(50) = 0$ to $\Theta(x_0) \approx 0.293$ and then decreases to 0. Therefore, $0 < \lambda(m) \le 0.293$, and the maximum is attained for $m = m_0 \approx 31.7\ kg$. Taking into account the behavior of function $\lambda(m)$ in Fig. 22.2.4, Eq. (22.2.32) shows that $1.12c < \hat{c} \le 1.67c$ and, for $c > 2000\ m/s$, the value $v_\times$ is very close to v_* and practically does not vary. Consequently, the sub-arc which is determined by equation $\Omega(m,v) = 0$ is only slightly different from a straight line. Therefore, the trajectory of penetrator which is close to the optimum for all considered shield materials is as follows ($v_{imp} > v_* \approx 81\ m/s$): passive

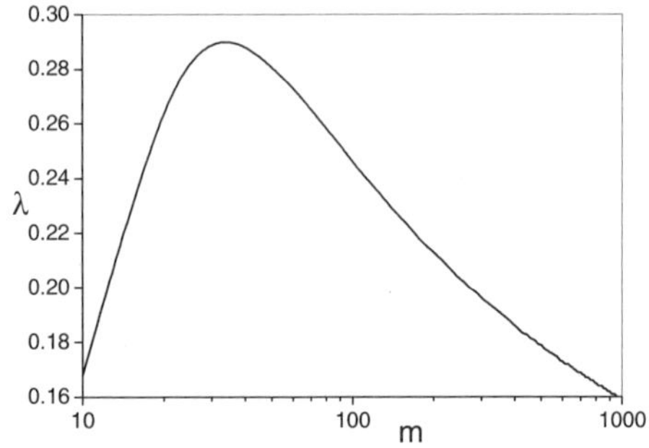

Fig. 22.2.4. Dependence $\lambda = \lambda(m)$ in the model of penetration into ice and frozen soil

penetration until slowing down to velocity v_*, penetration with a constant velocity v_* while the propellant is available, and passive penetration until a state of rest. Analysis shows that the effect of the material of the shield on the efficiency coefficient η is insignificant.

22.3 Numerical Investigation of the Problem Using Dynamic Programming

22.3.1 *Formulation of problem in dimensionless variables*

Equations (22.1.19), (22.1.20), (22.1.23)–(22.1.26), describing the considered variational problem in a general case, imply the following expressions which can be written using for convenience the defined below dimensionless variables:

$$H = \int_{\overline{m}_{res}}^{1} \frac{(\overline{m}\overline{v}' + \overline{c})\overline{v}}{\overline{D}(\overline{m},\overline{v})} d\overline{m}, \tag{22.3.1}$$

$$\overline{v}(1) = 1, \quad \overline{v}(\overline{m}_{res}) = 0, \quad 0 \le \overline{v}(\overline{m}) \le v_{\max}, \quad \overline{m}_{res} \le \overline{m} \le 1, \tag{22.3.2}$$

$$\overline{m}\overline{v}' + \overline{c} \ge \overline{\beta}_{\min} \overline{D}(\overline{m},\overline{v}) \quad if \quad \overline{v}' < \infty, \tag{22.3.3}$$

$$\left| \frac{\overline{D}(\overline{m},\overline{v})\overline{v}'}{\overline{m}\overline{v}' + \overline{c}} \right| \le \overline{\chi}_{\max} \quad if \quad \overline{v}' < \infty. \tag{22.3.4}$$

$$D(\overline{m},\overline{v})/\overline{m} \le \overline{\chi}_{\max} \quad if \quad \overline{m} = const\,, \tag{22.3.5}$$

where L is some reference length and

$$\overline{v} = \frac{v}{v_{imp}}, \overline{c} = \frac{c}{v_{imp}}, \overline{v}_{\max} = \frac{v_{\max}}{v_{imp}}, \overline{m}_{res} = \frac{m_{res}}{m_{imp}}, \overline{m} = \frac{m}{m_{imp}},$$

$$\overline{\beta}_{\min} = \frac{v_{imp} m_{imp}}{L\mu_{\max}}, \overline{\chi}_{\max} = \frac{Lg\chi_{\max}}{v_{imp}^2}, \overline{h} = \frac{h}{L}, \overline{H} = \frac{H}{L}, \overline{v}' = \frac{d\overline{v}}{d\overline{m}}, \tag{22.3.6}$$

$$\overline{D}(\overline{m},\overline{v}) = \frac{L}{v_{imp}^2 m_{imp}} D(m_{imp}\overline{m}, v_{imp}\overline{v}).$$

The problem is reduced to the optimization of the functional $\overline{H}$ in Eq. (22.3.1), whereas the solution must satisfy the constraints given by Eqs. (22.3.2)–(22.3.5).

22.3.2 *Case of general model for resistance force*

In this section we describe application of the dynamic programming (DP) for solving the problem numerically without specification of the function that determines the resistance force.

Discretization of the problem is described by Eqs. (21.3.1)–(21.3.8), (see also Fig. 21.3.1 and accompanying explanations in Section 21.3.1).

The integral in Eq. (22.3.1) along the piecewise linear contour can be written as follows:

$$\overline{H} = Q_v^{(0)} + \sum_{j=1}^{N_m} (Q_m^{(j)} + Q_v^{(j)})\,, \tag{22.3.7}$$

where $Q_m^{(j)}$ and $Q_v^{(j)}$ are the components of the integral over "vertical" and "non-vertical" segments of the trajectory, respectively (Fig. 21.3.1),

$$Q_v^{(j)} = \int_{\overline{v}^{(j)}}^{\overline{w}^{(j)}} G_v^{(j)}(\overline{v})d\overline{v}, \quad j = 0,1,2,\ldots,N_m\,, \tag{22.3.8}$$

$$Q_m^{(j)} = \int_{\overline{m}^{(j-1)}}^{\overline{m}^{(j)}} G_m^{(j)}(\overline{m})d\overline{m}, \quad j = 1,2,\ldots,N_m\,, \tag{22.3.9}$$

where

$$G_v^{(j)}(\overline{v}) = \frac{\overline{m}^{(j)}\overline{v}}{\overline{D}(\overline{m}^{(j)},\overline{v})}, \quad G_m^{(j)}(\overline{m}) = \frac{(e^{(j)}\overline{m}+\overline{c})(e^{(j)}\overline{m}+f^{(j)})}{\overline{D}(\overline{m},e^{(j)}\overline{m}+f^{(j)})}. \quad (22.3.10)$$

Integrals in Eq. (22.3.8) can be often calculated in explicit form; otherwise, one can use numerical integration, e.g. the trapezoidal rule formula with points of interpolation, $\overline{v}_\xi^{(j)}$, in the nodes of the (m,v) mesh. Hence

$$Q_v^{(j)}/\Delta\overline{v} = 0.5[G_v^{(j)}(\overline{v}_0^{(j)}) + G_v^{(j)}(\overline{v}_{n_v^{(j)}}^{(j)})] + \sum_{\substack{\xi=1 \\ n_v^{(j)}>1}}^{n_v^{(j)}-1} G_v^{(j)}(\overline{v}_\xi^{(j)}), \quad (22.3.11)$$

where $n_v^{(j)}$ is the integer part of $(\overline{w}^{(j)} - \overline{v}^{(j)})/\Delta\overline{v}$ and

$$\overline{v}_\xi^{(j)} = \overline{v}^{(j)} + \xi\Delta\overline{v}, \quad \xi = 0,1,\ldots,n_v^{(j)}. \quad (22.3.12)$$

In a similar manner, integrals in Eq. (22.3.9) can be represented as follows:

$$Q_m^{(j)}/\Delta\overline{m}^{(j)} = 0.5[G_m^{(j)}(\overline{m}_0^{(j)}) + G_m^{(j)}(\overline{v}_{n_m^{(j)}}^{(j)})] + \sum_{\substack{\xi=1 \\ n_m^{(j)}>1}}^{n_m^{(j)}-1} G_m^{(j)}(\overline{v}_\xi^{(j)}), \quad (22.3.13)$$

where $n_m^{(j)}+1$ is the selected number of points of interpolations in the interval $[\overline{m}^{(j-1)},\overline{m}^{(j)}]$,

$$\overline{m}_\xi^{(j)} = m^{(j)} + \xi\Delta\overline{m}^{(j)}, \xi = 0,1,\ldots,n_m^{(j)}, \Delta\overline{m}^{(j)} = \Delta m/n_m^{(j)}. \quad (22.3.14)$$

Inequalities in Eqs. (22.3.3) and (22.3.4) can be written in the points of interpolations in the following form:

$$A_\xi^{(j)} \geq \overline{\beta}_{\min}, \quad \left|e^{(j)}\right| \leq \overline{\chi}_{\max}\left|A_\xi^{(j)}\right|, \quad (22.3.15)$$

where

$$A_\xi^{(j)} = \frac{e^{(j)}\overline{m}_\xi^{(j)} + \overline{c}}{\overline{D}(\overline{m}_\xi^{(j)}, e^{(j)}\overline{m}_\xi^{(j)} + f^{(j)})}, \; j = 1,2,\ldots,N_m, \xi = 0,1,\ldots,n_m^{(j)}. \quad (22.3.16)$$

Inequality (22.3.5) implies the following constraints:

$$\overline{D}(\overline{m}^{(j)},\overline{v}_\xi^{(j)})/\overline{m}^{(j)} \le \overline{\chi}_{\max}, \quad j=1,2,...,N_m, \quad \xi=0,1,...,n_v^{(j)} \tag{22.3.17}$$

Since we do not allow regimes whereby fuel is exhausted while the velocity of penetrator remains zero, the following constraint must be imposed:

$$\overline{v}^{(j)}+\overline{w}^{(j-1)}>0, \quad j=1,2,...,N_m. \tag{22.3.18}$$

22.3.3 *Case of modified Young model*

In the case of the generalized Young model (Section 22.1.1), we obtain:

$$\overline{D}(\overline{m},\overline{v})=\frac{L\psi(\overline{v})}{k_0k_1v_{imp}^2\varphi(\overline{m})}, \tag{22.3.19}$$

where

$$\psi(\overline{v})=\begin{cases}1/(\gamma_0-\gamma_2v_{imp}^2\overline{v}^2) & if \quad \overline{v}<\overline{v}_*\\ v_{imp}\overline{v} & if \quad \overline{v}\ge\overline{v}_*\end{cases} \tag{22.3.20}$$

$$\varphi(\overline{m})=\begin{cases}\sigma m_{imp}^{\gamma_1+\gamma_2}\overline{m}^{\gamma_1+\gamma_2-1} & if \quad SRC\\ m_{imp}^{\gamma_1}\overline{m}^{\gamma_1-1}\ln(50+0.29m_{imp}^2\overline{m}^2) & if \quad IFS\end{cases} \tag{22.3.21}$$

$\overline{v}_*=v_*/v_{imp}$ and it is assumed that relation $m=m_{imp}\overline{m}$ is used in Tab. 22.1.1.

For the "vertical" segments of the trajectory, the analysis can be performed in the exact analytical form. Clearly, the integral in Eq. (22.3.8) can be calculated:

$$Q_v^{(j)}=\frac{k_0k_1v_{imp}^2\overline{m}^{(j)}\varphi(\overline{m}^{(j)})}{L}\int_{\overline{v}^{(j)}}^{\overline{w}^{(j)}}\frac{\overline{v}d\overline{v}}{\psi(\overline{v})}, \quad j=0,1,2,...,N_m \tag{22.3.22}$$

where

$$\int_{\bar{v}^{(j)}}^{\bar{w}^{(j)}} \frac{\bar{v}d\bar{v}}{\psi(\bar{v})} = \begin{cases} \Phi(\bar{v}^{(j)}, \bar{w}^{(j)}) & if \quad \bar{w}^{(j)} < \bar{v}_* \\ \Phi(\bar{v}^{(j)}, \bar{v}_*) + \Psi(\bar{v}_*, \bar{w}^{(j)}) & if \quad \bar{v}^{(j)} \le \bar{v}_* \le \bar{w}^{(j)} \\ \Psi(\bar{v}^{(j)}, \bar{w}^{(j)}) & if \quad \bar{v}_* < \bar{v}^{(j)} \end{cases} \tag{22.3.23}$$

$$\begin{aligned} \Phi(\bar{V}_a, \bar{V}_b,) &= \int_{\bar{V}_a}^{\bar{V}_b} (\gamma_0 - \gamma_2 v_{imp}^2 \bar{v}^2)\bar{v}d\bar{v} \\ &= 0.25(\bar{V}_b^2 - \bar{V}_a^2)[2\gamma_0 - \gamma_2 v_{imp}^2(\bar{V}_b^2 + \bar{V}_a^2)] \\ \Psi(\bar{V}_a, \bar{V}_b,) &= \frac{1}{v_{imp}} \int_{\bar{V}_a}^{\bar{V}_b} d\bar{v} = (\bar{V}_b - \bar{V}_a)/v_{imp} \end{aligned} \tag{22.3.24}$$

Instead of Eq. (22.3.17), the constraints in Eq. (22.3.5) can be written as follows:

$$\bar{w}^{(j)} \le \psi^{-1}\left(\frac{k_0 k_1 v_{imp}^2 \bar{m}^{(j)} \varphi(\bar{m}^{(j)})\bar{\chi}_{\max}}{L}\right), \quad j = 0,1,2,\ldots,N_m, \tag{22.3.25}$$

where

$$\psi^{-1}(z) = \begin{cases} \sqrt{\dfrac{\gamma_0 z - 1}{\gamma_2 v_{imp}^2 z}} & if \quad z \le v_{imp}\bar{v}_* \\ z/v_{imp} & if \quad z > v_{imp}\bar{v}_* \end{cases}. \tag{22.3.26}$$

22.4 Results of Numerical Optimization

Figs. 22.4.1a-c show typical results of calculations of penetration into non-frozen soil, rock and concrete shields, $v_{imp} > v_*$, $\chi_{\max} = \infty$.

In Figs. 22.4.1a,b we demonstrate the optimal BPs. For small values of $\bar{\beta}_{\min}$ when the constraint on the fuel flow rate is irrelevant, the regime of penetration is as follows: inertial penetration until velocity $\bar{v}_\times$ is attained; penetration with a constant velocity $\bar{v}_\times$ until exhausting fuel supply, and, finally, inertial penetration until a state of rest. Beginning from some value, the effect of $\bar{\beta}_{\min}$ on the optimal BP becomes pronounced: the regime with a maximum fuel flow rate appears between the initial passive penetration regime and penetration with a constant velocity, $\bar{v}_{\times}$. Further increase of $\bar{\beta}_{\min}$ results in the following changes: transition from a passive penetration to penetration with the maximum fuel flow rate at larger velocities, the regime of penetration with a constant velocity $\bar{v}_{\times}$ gradually vanishes and penetration regime with the maximum fuel flow rate switches to the regime of inertial penetration with mass $\bar{m}_{res}$. For values of $\bar{\beta}_{\min}$ larger than some threshold value, optimum solution does not exist. This means that penetrator slows down to zero velocity before exhausting the fuel supply for any BP.

In Fig. 22.4.1c, we showed the dependencies of the maximum normalized DOP, $\delta = H_{\max}/H_0$, as a function of parameter $\bar{\beta}_{\min}$ for different values of the dimensionless relative exit velocity of gases from the nozzle of the thruster, c; clearly, H_0 is the same when the variants

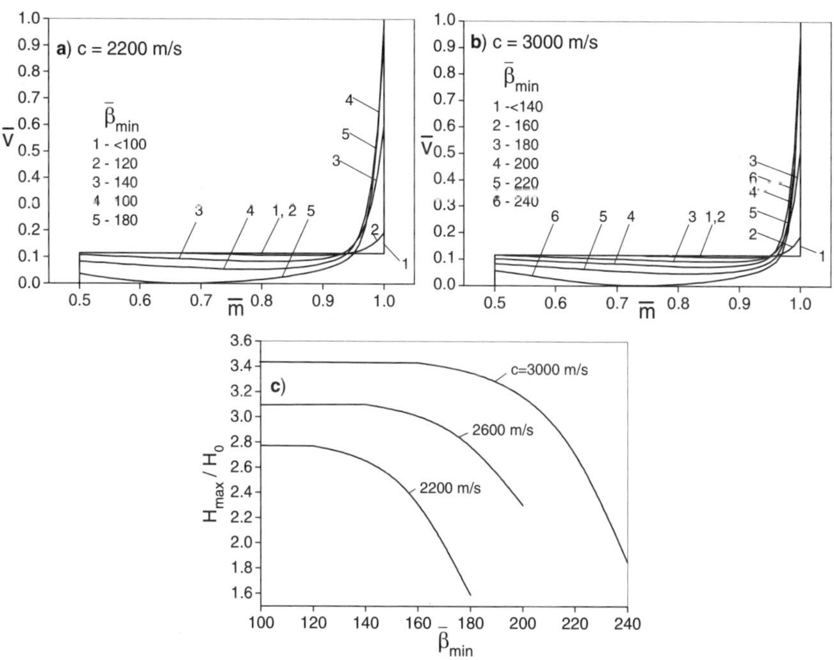

Fig. 22.4.1. Typical optimal solutions for penetration into non-frozen soil, rock and concrete shields with relatively high impact velocities ($m_{imp} = 400\ kg$, $m_{res} = 200\ kg$, $v_{imp} = 700\ m/s$); a-b) optimal BPs; c) normalized optimal DOP vs. $\bar{\beta}_{\min}$.

with different $\bar{\beta}_{\min}$ and c are compared. The maximum effect of using jet thruster is observed for small $\bar{\beta}_{\min}$. Increase of $\bar{\beta}_{\min}$ implies stronger constraints, namely, reducing the upper bound for the admissible fuel consumption rate of the thruster. The curves $\delta = \delta(\bar{\beta}_{\min})$ are located higher with the increase of c, i.e., a negative effect of $\bar{\beta}_{\min}$ can be compensated by increasing c. The fact that some curves of the dependencies $\delta = \delta(\bar{\beta}_{\min})$ are terminated implies that the capabilities of the thruster are inadequate, and penetration terminates before exhausting fuel supply.

Part 10

Effect of Order of Plates, Layering and Spacing on Protective Properties of Ductile Shields

Part 10 includes a survey (Chapter 23) and Chapters 24–27 devoted to particular problems.

The survey includes overviews of analytical, numerical and experimental studies of multi-layer shields which can be useful for determining general tendencies, e.g. effects of layering, spacing and change of the order of plates on the ballistic performance of shields, and also analytical investigations dedicated to optimization of multi-layer shields. In the survey we included characteristics of shields for most of the experimental or numerical studies.

Chapter 24 is concerned with investigation of the effect of spacing on ballistic properties of shields. Ben-Dor *et al.*, 1998a proved that localized interaction model (LIM) for shield-impactor interaction implies independence of the ballistic limit velocity (BLV) of spaced shield on air gaps and their widths for a conical striker. In Chapter 24 the effect of deviation from a conical shape on ballistic properties of spaced shield is investigated numerically using a two-term LIM.

The goal of Chapter 25 is similar to that of Chapter 24. Using a two-term LIM Ben-Dor *et al.*, 1999b determined the order of the plates with different mechanical properties in a multi-layer shield that affords a maximum BLV against conical impactors. In Chapter 25 the validity of this solution for non-conical impactors is investigated numerically.

In Chapter 26 we investigate, on the basis of general assumptions that are pertinent to penetration modeling, the effect of layering of metal shields on their protective characteristics against sharp-nosed rigid strikers, and determine or explain a number of general features of the penetration phenomena. The results of experiments and numerical calculations available in the literature are used for validating the obtained theoretical conclusions.

In Chapter 27, the problem of optimization of a multi-layer metal shield is investigated analytically. The goal is to determine the sequence of the materials and the thicknesses of the layers that provide the maximum BLV of the shield. Optimization is performed under the following constraints: fixed areal density of the shield, the upper bound on the total thickness of the shield and the bounds on the thicknesses of the plates manufactured from every material. For the most interesting in practice cases of two-layer and three-layer shields the solution is obtained in the explicit analytical form.

Chapters 23-27 are based on the results obtained by Ben-Dor *et al.*, 2006b, 2006d, 2010b, 2011a.

Chapter 23

Survey

Some aspects of layering and spacing of shields have been extensively investigated in the literature on high speed penetration mechanics. Many studies compared ballistic characteristics of monolithic shields with those of layered shields, i.e. shields composed of several plates having the same total thickness and manufactured from the same material. The plates may be in contact or there may be air gaps between the plates. Hereafter the latter case is referred to as "spacing". Hence, as alternatives to the monolithic shield, many types of shields having different numbers of plates and different thicknesses of the plates and of the air gaps are feasible. Analysis of the effect of the order of plates manufactured from different materials on ballistic characteristics of the shield attracted particular interest. The simplest case of the latter problem is interchanging the plates in a two-layer shield. In the general case, the number of plates may vary and they may be manufactured from different materials. The combined effects, e.g., changing the order of plates and using air gaps, on the ballistic performance of the shield and various problems of optimization of the structure of the shield were also studied in a number of investigations.

This survey includes analytical, numerical and experimental investigations which can be useful for elucidating general properties of multi-layer metal shields, e.g. effect of layering, spacing and change of the order of plates on the ballistic performance of shields, and also studies which suggested analytical methods for optimization of multi-layer shields. Studies devoted to other aspects of the problem, e.g. analysis and comparisons of ballistic characteristics of various particular shields, are not included in the survey.

A survey of the state-of-the-art in the field of metal shields presented below supports the assessment of Radin and Goldsmith, 1988 that "only limited results for multiple target materials exist in the literature…, and the results obtained cannot easily be correlated since different target and projectile materials, nose shapes, impact geometries and striker speeds were used". Clearly, the latter assessment is not related to the problem of selecting the best shield among the given set of shields against the impactor having a given shape. This problem can be often solved experimentally, and the obtained results can be explained using relatively simple physical reasoning. The problem is to determine a more-or-less general law that will enable predicting the change of the ballistic characteristics of the shield by varying the structure of the shield. This problem has not been solved as yet, although a number of experimental and theoretical studies have been performed in this direction.

It is convenient to demonstrate the notations used below in describing the structure of the shield with examples. The notation 2 mm + 6 mm denotes the shield consisting of two plates (in contact, if not indicated otherwise) with the thicknesses 2 mm and 6 mm, and listed in the direction of projectile penetration. The notation 2x4 mm substitutes the notation 4 mm + 4 mm. In the case of the same air gaps between all adjacent plates, the value of the air gap is written in parentheses, e.g., 6 mm + 8 mm (6 mm) or 2x5 mm + 3 mm (4 mm).

If the particular model takes into account the possibility of simultaneous interaction of the impactor with several layers, this feature of the model hereafter is called the "dependent layers approach", in contrast to the "independent layers approach" when it is assumed that the residual velocity after perforation of some layer serves as an impact velocity of the next layer. Strictly speaking, the latter assumption is valid in the case of large air gaps between layers whereby penetration into a successive plate begins after interaction with the previous plate is terminated. Hereafter using the latter assumption in the model is not indicated. The "dependent layers approach" enhances the adequacy of the projectile-shield interaction model but usually considerable complicates investigation of the problem.

Honda *et al.*, 1930 (see also Goldsmith, 1960) investigated experimentally penetration of steel plates by conical-nosed projectiles having impact velocity of about 500 m/s. Monolithic plates with the thickness of 5 mm to 12 mm and 1 mm layers were considered. It was found that a shield composed of thin, 1 mm thick plates had a lower ballistic resistance than a monolithic shield with the same thickness. However, a spaced shield with the thicknesses of the plates equal to the half-thickness of a monolithic shield performed better than a monolithic shield.

Ilyushin, 1935 proposed an analytical method for calculating ballistic characteristics of monolithic and multi-layered relatively thin shields and arrived at the conclusions that (i) monolithic shields are less effective than multi-layer shields having equivalent thickness and (ii) air gaps improve ballistic properties of the shield.

Hurlich, 1950 noted that the earliest study on the modern use of spaced armor that he had found, was performed in 1913 for the armor of naval vessels. He presented some qualitative arguments in favor of the spaced armor (mostly for tanks), a number of tables with experimental results, some references and curious historical information.

Zaid and Travis, 1974 investigated experimentally single and two-layer in contact steel plates having equivalent total thickness from 1.1 mm to 9.5 mm impacted by cylindrical flat-nosed projectile at velocities up to 500 m/s. They concluded that thin monolithic shields are more effective than multi-layer shields having equivalent thickness, while increasing the thickness of the shields results in the increase of the resistance of layered shields in comparison with monolithic shields.

Marom and Bodner, 1979 conducted a combined analytical and experimental comparative study of monolithic, layered and spaced thin aluminum shields. Experimental investigation were conducted with aluminum alloy shields impacted by round-nosed lead projectiles at average impact velocity 375 m/s. Layered shields consisted of plates in contact having the thicknesses from 1 mm to 3.6 mm, and 13 mm air gaps. The authors found that the ballistic resistance of a monolithic shield is higher than that of a multi-layer shield consisting of plates in contact and lower than the ballistic resistance of a spaced shield. The advantage of the spaced shield over the monolithic shield was revealed by Fomin, 1999 on the basis of numerical calculations using the data from Marom and Bodner, 1979.

Corran *et al.*, 1983a, 1983b, using experimental results on penetration of mild steel plates by impactors having "increasingly rounded nose shape", plotted a curve of perforation energy vs. plate thickness. Tests were performed on combinations of 1.18 mm, 2.00 mm and 3.00 mm plates when the plates in the shields were either in contact or separated by a 12 mm air gap, as well as on the shields 3.0 mm + 1.2 mm and 1.2 mm + 3.0 mm. For all considered shields the authors found a "kink" in the curve "at about 3.5 mm total thickness". The occurrence of the kink was explained by the change of the character of energy absorption. The authors arrived at the conclusions that the order of the plates having unequal thicknesses is important. No advantage was found in using multi-layer shields below the kink. Above the kink the best combinations may approach the best-fit curve to the single layer tests conducted below the kink. The performed experiments demonstrated an advantage in placing the layers in contact.

Nixdorff, 1984a, 1984b, 1987b, 1987a compared ballistic performance of a monolithic metal shield with that of a shield manufactured from the same material, having the same total thickness, and consisting of several plates in contact. Using the theory developed by Awerbuch and Bodner, 1974, Nixdorff showed that separation of a homogeneous shield into several layers results in a reduction of the ballistic limit velocity (BLV) of the shield.

Aptukov, 1985, 1991a and Aptukov *et al.*, 1985, suggested numerical procedures based on the Pontrjagin maximum principle to determine the optimum distribution of mechanical characteristics of a non-homogeneous plate assuming that these characteristics can vary continuously along the direction of penetration. Using conventional penetration mechanics terminology the problem can be formulated as minimization of the areal density of the shield for a given BLV. They employed the two-term impactor-shield interaction model wherein the assumption about the universal dependence on the coefficients of the model (dynamic hardness and density of the material of the plate) was used. Aptukov and Khasanov, 2011 generalized this approach for the case when friction is taken into account. Using a cylindrical cavity expansion model, Aptukov *et al.*, 1986; Aptukov and Belousov, 1991 solved similar discrete optimization problems when the shield consisted of several layers and the materials for the layers could be chosen from a given set of materials. Many of the above results are presented also in the monograph by Aptukov *et al.*, 1992.

The study by Radin and Goldsmith, 1988 was based on semi-empirical models and experimental investigations. Experimental investigations were conducted with 2024-0 aluminum shields impacted with the velocity 50-210 m/s. The following shields were impacted by 12.7 mm diameter blunt and 60° conical-nosed projectiles: 1x3.2 mm, 2x1.6 mm; 1x4.8 mm, 3x1.6 mm; 1x6.4 mm, 4x1.6 mm. The following shields, 2x1.6 mm, 1.6 mm + 3.2 mm, 2x3.2 mm and 2x3.2 mm (6.4 mm), were impacted only by conical-nosed projectile. It was found that (i) a monolithic aluminum shield is superior to a layered shield with the same total thickness for conical-nose and blunt projectiles, (ii) the spaced shields are less effective than layered shield in contact.

Nurick and Walters, 1990 used an energy approach to model ballistic resistance of two-layer shield and compared experimentally monolithic, layered and spaced steel shields with the same total thickness perforated by conical (with semi-apex angles of 30° and 60°) and cylindrical projectiles. Theoretical analysis implies that resistance of a layered shield is less than homogeneous shield with the same thickness. In order to compare effectiveness of layering and spacing, experiments were performed with the following shields: 1x2 mm, 2x0.5 mm, 2x0.5 mm (2 mm), 2x0.5 mm (5 mm), 2x0.5 mm (10 mm), 2x0.5 mm (15 mm), 2x0.5 mm (25 mm). It was found that (i) the BLV of monolithic plate was by 4-8% larger than that of the layered plates in-contact shield having an equivalent thickness; (ii) the spaced shield was less efficient compared to the layered shield consisting of plates in-contact; (iii) BLV of spaced shield decreased when the width of air gap between the plates increased.

Weidemaier *et al.*, 1993 conducted experiments and numerical simulations on the perforation of steel barriers by spherical impactors with a diameter of 17 mm. They studied a monolithic shield with a thickness of 43 mm and shields composed of plates in contact having the same total thickness: 3x14.3 mm, 31 mm + 2x6 mm, 6 mm + 31 mm + 6 mm and 2x6 mm+31 mm. It was found that the ballistic characteristics of layered shields strongly depended on the order of the plates having different thicknesses and that layering could improve or impair the ballistic performance of the shield.

Almohandes *et al.*, 1996 conducted a comprehensive experimental study on the perforation of mild steel barriers by standard 7.62 mm bullets. They investigated monolithic steel shields, as well as layered in contact and spaced shields with the same air gaps between each pair of adjacent plates. The first set of configurations included 6 mm thick monolithic shield, 2 mm + 6 mm, 6 mm + 2 mm, and 2x4 mm shields in contact and with 6 mm air gaps. The second set included 6 mm + 8 mm and 8 mm + 6 mm shields with plates in contact and the following spaced shields: 2x4 mm (24 mm), 4 mm + 6 mm + 4 mm (9 mm), 4 mm + 8 mm + 4 mm (8 mm), 2x2 mm + 4 mm (12 mm), 2 mm + 3 mm + 2 mm + 3 mm +4 mm (4.5 mm), 2 mm + 3x(4 mm) + 2 mm (4 mm), 2 mm + 4 mm + 2 mm (12 mm), 2 mm + 3 mm + 4 mm + 3 mm + 2 mm (4.5 mm), 2 mm + 4 mm + 2 mm + 4 mm + 4 mm (4.5 mm), 2 mm + 6 mm (8 mm). The third set of shields included spaced shields consisting of plates manufactured from steel with other mechanical properties (compared with the sets 1 and 2): 4 mm + 4 mm (24 mm), 4 mm + 5 mm + 4 mm (9.5 mm), 3x4 mm (10 mm), 2 mm + 5 mm + 6 mm (9.5 mm), 2 mm + 4 mm + 6 mm (10 mm), 2x2 mm + 4 mm (12 mm), 2 mm + 2.5 mm + 2 mm + 2.5 mm + 4 mm (4.75 mm), 4x2 mm + 4 mm (5 mm). It was found that monolithic shields were more effective than laminated shields with the same total thickness, regardless of the configuration or impact velocity, and that the difference in the ballistic performance diminished as the impact velocity increased. Moreover, the effectiveness of laminated shields – either in contact or spaced – increased as the number of plates comprising each shield decreased. Ballistic performance of laminated shields is further enhanced by using the thickest lamina as the rear one. The authors also studied shields with different structures where the fiberglass reinforced polyester was used as the filler material.

Gogolewski *et al.*, 1996 conducted experiments on perforation of 6061-T6 aluminum alloy barriers by a blunt 7.62 mm diameter projectiles manufactured from hard steel at speeds varying in the range from 150 m/s to 500 m/s. They examined the following shields: (1) 1x6.72 mm, (2) 3x2.24 mm with layers in contact, (3) 3x2.24 mm with 1 mm Teflon strips between plates and lubricated interplate surfaces (in order to exclude the effect of interplate friction); (4) 3x2.24 mm (2.54 mm), and (5) 3x2.24 mm (7.62 mm) and obtained the following values of the BLV: (1) 233 m/s, (2) 287 m/s, (3) 268 m/s, (4) 228 m/s, (5) 217 m/s. The authors arrived at the following conclusions: (i) a three-layer shield with layers in contact is the best among the considered shields; (ii) eliminating friction between plates impairs ballistic properties of the shield although its protective properties are still superior to those of the monolithic shield; (iii) the BLV of shields with air gaps is lower then the BLV of the monolithic shield.

Zukas, 1996 and Zukas and Scheffler, 2001 found, on the basis of numerical simulations of penetration of metallic shields, that layering dramatically weakens thin $(b/d<1)$ and intermediate thickness $(3<b/d<10)$ shields, while thick shields $(b/d>10)$ show small changes in residual properties of projectile (residual mass and residual velocity) when compared to their monolithic equivalents. Here b and d are the thickness of the shield and the shank diameter of the impactor, respectively.

Gupta and Madhu, 1997 and Madhu *et al.*, 2003 conducted experiments with mild steel plates impacted normally by ogive-nosed projectiles. Steel shields, including the two-layer shields 10 mm + 16 mm, 10 mm + 20 mm, 16 mm + 20 mm and shields consisting of the same pairs of plates arranged in the inverse order, as well as the 3x4.7 mm configurations with the plates in contact and with air gaps of 30 mm and 40 mm, were tested. Aluminum shields, 3x6.1 mm with plates in contact and with air gaps of 20 mm and 30 mm, were also examined. Using some experimental results on monolithic shields, the authors concluded that layering of plates having the intermediate thickness did not result in a significant change of the ballistic performance. For thin shields, the layered combinations in contact yielded higher residual velocity as compared with a monolithic shield manufactured from either aluminum or steel. It was also found that for a spaced shield the residual velocity was higher than in the case of plates in contact and for the same impact velocity.

Woodward and Cimpoeru, 1998 studied experimentally perforation of shields manufactured from 2024-T351 aluminum alloy by cylindrical and conical-nosed projectiles: monolithic shield with a thickness 9.53 mm and several layered shields having the same total thickness. The following configurations were tested: 1x9.53 mm, 3.18 mm+ 6.35 mm,

2x4.76 mm, 6.35 mm + 3.18 mm, 3x3.18 mm, 6x159 mm. In the case of conical-nose projectile the maximum reduction of the BLV, 9%, in comparison with monolithic shield, was observed for 3.18 mm+ 6.35 mm shield, while the increase of the BLV for some layered shields was negligible. For cylindrical projectile the BLV of layered shields increased by up to 13%.

In the series of the paper by Ben-Dor and coauthors (Ben-Dor, et, al., 1998a, 1998d, 1998e, 1998f, 1999b, 1999c, 1999d, 2000c, 2006a, 2006b, 2006d, 2010b, 2011a) using mostly analytical methods the authors analyzed and optimized protective properties of multilayer ductile barriers with or without air gaps. The main results of these studies are summarized below.

Ben-Dor *et al.*, 1998a studied penetration of a 3D rigid sharp conical nosed impactor into a ductile layered shield with air gaps between the plates (the plates are manufactured from the same material) on the basis of the general localized interaction model (LIM) with the "dependent layers approach" The model used may be considered as the model described in Chapter 24, extended to the general LIM and adapted for 3D conical impactors. It was proved analytically that the ballistic performance of the layered shield is independent of the air gap widths between the layers and on the sequence of the plates in the shield. It must be emphasized that the latter result is obtained using a very general model that describes the projectile-shield interaction. Since the above results were obtained for cone-nosed impactors, Ben-Dor *et al.*, 2006b investigated numerically the effect of deviation of the shape of the impactor from a conical shape on the ballistic properties of the spaced shield using a two-term LIM. It was found that the effect of deviation from a conical shape is insignificant and results only in small changes (of order a few percent) in the magnitude of the BLV and energy absorbed by a shield. These theoretical predictions are in agreement with the available experimental results.

Ben-Dor *et al.*, 1998d, 2006a also investigated the effect of air gaps on ballistic properties of multi-layered shields. Unlike Ben-Dor *et al.*, 1998a, here the authors considered cone-nosed bodies of revolution as impactors, used three-term dynamical cylindrical cavity expansion model of the kind determined by Eq. (6.6.5) and along with analytical methods (for the case of large air gaps) applied also numerical simulations (for the case of air gaps when the impactor interacts at the same time with adjacent plates). It was found that a spaced shield performs better than a monolithic shield with the same total thickness. Ben-Dor *et al.*, 2006a proved that the shield with the same thicknesses of the plates has the best defense property if the number of layers is given, and the BLV of the shield increases when the number of plates with the same thickness increases.

Ben-Dor *et al.*, 1999b considered penetration of a 3D rigid sharp conical-nosed impactor into a ductile layered shield which consists of plates manufactured from different materials. In the analysis they used a two-term LIM described in Section 1.6.2.2 and adapted for layered shields:

$$\Omega_n = a_2^{(i)}\omega(u)v^2 + a_0^{(i)}, \quad u = -\vec{v}^{\,0}\cdot\vec{n}^{\,0}, \tag{23.1}$$

where $a_0^{(i)}$ and $a_2^{(i)}$ are parameters describing mechanical properties of the materials from which the ith plate is manufactured. Note that ballistic properties of layered shields against the conical-nosed impactors which were determined in this study for the particular case, $\omega(u) = u^2$, i.e., for the model

$$\Omega_n = a_2^{(i)}v_n^2 + a_0^{(i)}, \tag{23.2}$$

where v_n is the normal component of the impactor velocity at a given location on its surface, are also valid for the general model given by Eq. (23.1). In most models, $a_2^{(i)}$ is proportional to the density of the material of the layer while $a_0^{(i)}$ depends on the mechanical properties of the material.

Ben-Dor *et al.*, 1999b found the following important property of the layered shields: the maximum BLV is attained when the plates are arranged in the order of increasing values of the parameter χ, and the minimum BLV is attained when the plates are arranged in the inverse order , where formula for parameter χ reads:

$$\chi^{(i)} = a_0^{(i)} / a_2^{(i)}. \tag{23.3}$$

For two-layered shields with plates in contact, this property was proved when the dependent layers approach was used in impactor-shield interaction modeling. The established property is also valid if three-term dynamical cylindrical cavity expansion model of impactor-shield interaction is used (Ben-Dor *et al.*, 1998e).

For the particular case of two-layer shield, when the layers are perforated independently, the above criterion for the arrangement of the plates was determined by Baranov *et al.*, 1991. Seliverstov, 2001 attempted to determine a general law for the optimal order of plates in a multi-layered shield using the same assumptions.

Using numerical simulations for two-layered shields penetrated by ogive-shaped impactors, Ben-Dor *et al.*, 2006d showed that the found by Ben-Dor *et al.*, 1999b law of optimal arrangement of the plates can be extended to non-conical sharp penetrators.

Using the same assumptions Ben-Dor *et al.*, 1999c (see also Ben-Dor *et al.*, 2006a) generalized the results of Ben-Dor *et al.*, 1999b for two-layer shields in contact to two-layer shields having arbitrary width air gaps. The following claim was proved analytically: (i) if $\chi^{(1)} < \chi^{(2)}$, then the BLV increases with increasing the air gap width up to the value when the striker still interacts with a frontal plate; after this the BLV remains constant; (ii) if $\chi^{(1)} > \chi^{(2)}$, then instead of increasing the BLV decreases; (iii) if $\chi^{(1)} = \chi^{(2)}$ then the BLV does not depend on the width of the air gap.

Using the same approximations in the model Ben-Dor *et al.*, 2000c generalized the results of Ben-Dor *et al.*, 1999b for two-layer shields in contact to the case when the shield consists of adjacent plates manufactured from one of two possible materials and the total thickness of the plates manufactured from every material is given. It has been found that the maximal BLV is obtained for the shield that consists of two blocks whereas each block includes the plates manufactured from the same material. The order of the blocks is governed by the rule established in Ben-Dor *et al.*, 1999b.

Ben-Dor *et al.*, 1998f investigated analytically the following optimization problem for layered shields. Assume that there are several materials with different properties that can be used for manufacturing the plates in a multi-layer shield. The goal is to determine the structure of the shield, i.e., the order and the thicknesses of the plates manufactured from different materials that provide the maximum BLV of the shield with a given areal density against a normal impact by a conical-nosed 3-D impactor. The employed impactor-shield interaction model was similar to that used by Ben-Dor *et al.*, 1999b, where $a_2^{(i)}$ in Eq. (23.1) was the density of the materials of i th plate. It has been proved analytically that the shield with the maximum BLV consists of one plate or several adjacent plates (these cases are equivalent in the model) manufactured from the material with the maximum value of the parameter χ.

Using a simplified impactor-shield interaction model, Ben-Dor *et al.*, 2011a investigated a problem of optimization of the multi-layered shield when a set of materials with the known mechanical properties is given and the plates manufactured from different materials can be arranged in different order in the shield. The goal was to determine the sequence of the materials in the layers and the thicknesses of the layers that provide the maximum BLV of the shield when the areal density, upper bound on the total thickness of the shield and constraints on the thicknesses of the plates for every material are given. This problem was reduced to the problem of geometric programming which can be solved numerically using the existing methods. For the practically most interesting cases of two layers and three layers, the solutions were obtained analytically in the explicit form.

Ben-Dor *et al.*, 2010b investigated the effect of layering, namely, the thicknesses and the number of the layers, on the protective property of a shield against sharp-nosed rigid impactors. The special feature of the used approach is that it does not require a closed mathematical model for calculating BLV of monolithic and layered shields. It was demonstrated that many characteristics of layered shields can be explained and predicted on the basis of general properties of the functions which determine penetration models without invoking particular expressions for these functions. The latter result is quite important since the qualitative behavior of the characteristics in penetration mechanics is generally determined using simple models which include approximations of the experimental data. The main results of this study can be formulated as follows: (i) any layering has an adverse effect on protective properties of the shield against perforation, (ii) increasing the number of layers with he same thicknesses impairs protective properties of the shield. These assertions were validated by analyzing results of experiments and numerical calculations available in the literature that allowed comparing protective properties of monolithic and layered shields. Theoretical investigation by Masri *et al.*, 2010, based on different assumptions, yielded the same conclusions. Amde *et al.*, 1997 showed the inefficiency of layering using empirical models for consecutive calculation of perforation of layers and monolithic shield.

Systematic exposition of the results obtained in this field up to 2006 in the Ben-Gurion University of the Negev may be found in the monograph by Ben-Dor *et al.*, 2006a.

On the basis of numerical calculations, Partom *et al.*, 2001 concludes that the ballistic performance of 2x3.3 mm aluminum shield is lower than that of the 6.6 mm shield.

Buchar *et al.*, 2001, Buchar *et al.*, 2002 studied numerically and experimentally the effect of 0.5 mm thick carbon steel interlayer on the ballistic resistance of the shield in which the front plate is made from the steel with a higher hardness as compared with the rear steel plate. The following combinations of thicknesses of front and rear plates with and without interlayer were tested against 7.62 mm AP projectile: 2 mm + 4 mm, 3 mm + 3 mm, 4 mm + 2 mm, and 5 mm + 1 mm. It was found that the 0.5 mm thick carbon steel interlayer improved the ballistic performance of the shield.

Yossifon *et al.*, 2002 generalize the penetration model by Yarin *et al.*, 1995 to the case of normal penetration of a rigid projectile into a multi-layered metal shield and made two small series of calculations for two-layer shields with 60 mm layers and ovoid-shaped projectile using the suggested model and hydrocode Autodyn2D. The first series of calculations was performed for the shield with plates manufactured from aluminum alloys and differing only by their hardnesses. In the second series of calculations, plates made of different materials (Rolled Homogenous Armor and Aluminum) were considered. The authors concluded: "The results show that in the cases considered there is no significant difference in the residual velocities when the order of the two layers of different materials is changed. However, this result is most likely limited to the case of a rigid projectile which was considered here".

In their numerical study Glazyrin and Orlov, 2003 compared the ballistic resistance of monolithic 6 mm thick metallic shield and two-layer shields with the thicknesses of the layers 2 mm and 4 mm arranged in two possible configurations, with different conditions at the contact surface between the layers. Cylindrical, spherical, conical-nosed, and ogive-nosed impactors were considered. Glazyrin *et al.*, 2006 presented the results of a numerical investigation of the ballistic performance of the following shields having the same total thickness against an ogive-nosed impactor: 1 mm + 5 mm, 2 mm + 4 mm, 3x2 mm, 2x3 mm, 4 mm + 2 mm, and 1x6 mm. Simulations showed that only 2 mm + 4 mm shield had the advantage over the monolithic shield while the performance of the configuration 4 mm + 2 mm was the worst. Orlov, 2008 conducted similar calculations for the following shields: 3x2 mm, 2x1 mm + 2x2 mm, 2 mm + 2x1 mm + 2 mm, and 2x2 mm + 2x1 mm. The results of numerical simulations revealed that 3x2 mm shield had the highest resistance while the 2x1 mm + 2x2 mm shield had the smallest resistance. It should be noted that the differences between all resistance indexes were very small. Summary of the studies in this field until 2007 conducted in the Institute of Applied Mathematic and Mechanics of the Tomsk State University can be found in the book edited by Gerasimov, 2007.

The experimental results of Almohandes *et al.*, 1996 were used by Liang *et al.*, 2005 for validating their approximate penetration model. This model was used for comparative analysis of shields having different structure. It was concluded that the ballistic performance was the best for the two-layer shield when the ratio of the first-layer thickness to the total thickness was about 0.75, and the worst performance was observed when this ratio was equal to 0.5. An air gap only slightly influenced the resistance to perforation in multi-layer shields.

Elek *et al.*, 2005 developed a simple model to describe perforation of monolithic and multi-layer thin metal plates by a flat-ended cylindrical impactor and used their model for the analysis of the ballistic properties of multi-layer spaced shields. The main results of this study may be summarized as follows. The suggested model predicts that the monolithic shield will have larger resistance than any other multi-layer shield with standoff distance between the layers and equivalent total mass. Analysis of penetration of a two-layer shield shows that the maximum resistance can be obtained for very low (<20% of the total thickness) or very high (>80% of the total thickness) front-layer thickness. The increase of the number of the spaced layers in a multi-layered shield having a constant total mass, results in further decrease of the ballistic resistance. Deterioration of the ballistic performance of thin steel shields against flat-ended cylindrical impactors caused by layering had been noticed earlier by Zaid *et al.*, 1973.

Park *et al.*, 2005 suggested a multistage procedure for optimization of a two-layer shield. At the first stage, using numerical simulations to describe penetration into shields with different thicknesses of the layers, b_1 and b_2, they determined the average temperature of a shield, T_{ave}, the average equivalent plastic strain ε_{ave} and the maximum equivalent plastic strain in a critical element of the shield $\varepsilon_{\max}$. At the second stage, the approximate functions describing the dependencies of T_{ave}, ε_{ave} and $\varepsilon_{\max}$ vs. b_1 and b_2, were determined. At the third stage, using a reduction to a single-criterion problem by a linear combination of criteria, a multi-objective optimization problem was solved. The authors considered two pairs of the optimization criteria: the first criterion in a pair is either T_{ave} or ε_{ave}, while the second is the weight of a shield. The constraints included an upper bounds on $\varepsilon_{\max}$ and constraints on the thicknesses of the plates and on the total thickness of a shield.

Liaghat *et al.*, 2005 developed an analytical model for oblique and normal penetration of conical projectiles into multi-layer thin metal shields. Results of numerical simulations of normal impact are given for the following shields: 1x10 mm, 2x5 mm, 3x3.33 mm, 4x2.5 mm, 5x2 mm. It was found that layered shields in contact have smaller penetration resistance than monolithic shields with the same total thicknesses, and the BLV decreases with increase of the number of layers.

Dey *et al.*, 2007 (see also Dey and Børvik, 2007; Børvik *et al.*, 2009c) compared ballistic perforation resistance of monolithic 12 mm thick shield, two-layer (2x6 mm) shield consisting of plates in contact and spaced shield with 24 mm air gap between the 6 mm thick plates. Penetration of these shields manufactured from steel plates (Weldox 700 E) by blunt (cylinder-shaped) and ogival projectiles was investigated experimentally and numerically. The findings in these studies are summarized in the following: (i) increase of the BLV by almost 50% for plates in contact and by almost 40% for spaced shields as compared to the monolithic shield in case of penetration by blunt projectile; (ii) decrease of the BLV by about 10% for plates in contact and spaced shields in comparison with the monolithic shield in case of penetration by ogival projectile.

Teng *et al.*, 2007 compared the effectiveness of metal monolithic 12 mm thick shield, two-layer shield consisting of 6 mm plates in contact, and the shield consisting of the same plates with 3 mm air gap between them. The authors examined "heavy" ($v_{bl} = 186 m/s$ for the monolithic shield) and "light" ($v_{bl} = 306 m/s$) cylinder-shaped projectiles and "heavy" ($v_{bl} = 487 m/s$) and "light" ($v_{bl} = 526 m/s$) conical-nosed projectiles using numerical simulations. It was found that (i) there is no a essential difference between the values of the BLV for multi-layer plates in contact and spaced shields; (ii) in the case of "heavy" impactors, layering improves the BLV by 25% for cylinder-shaped projectile and impairs the BLV by 8% for conical-shaped projectile; (iii) in the case of the "light" impactors, layering improves the BLV by 7% for cylinder-shaped projectile and does not influence the BLV for conical-shaped projectile.

Gupta *et al.*, 2007, 2008; Iqbal and Gupta, 2008, 2011 presented the results of the experimental and numerical investigations of protective properties of monolithic and layered thin aluminum plates having different thicknesses under the impact of flat, ogive and hemispherical nosed steel projectiles. In order to compare monolithic and layered shields with the same thickness, the data on the following sets of shields was used (the total thickness of the shield is given in square brackets): 1x1.0 mm, 2x0.5 mm [1.0 mm]; 1x1.5 mm, 3x0.5 mm [1.5 mm]; 1x2.0 mm, 2x1.0 mm, 4x0.5 mm [2.0]; 1x2.5 mm, 5x0.5 mm [2.5 mm]; 1x3.0 mm, 2x1.5 mm, 3x1.0 mm [3 mm]. The main findings in these studies are as follows: (i) monolithic shield has the highest resistance against perforation as compared with layered shields with the same total thicknesses and (ii) ballistic resistance of the layered shields decreases with the increase of the number of layers. Iqbal *et al.*, 2012 studied ballistic properties of 1 mm thick 1100-H12 aluminum shields impacted normally by blunt and ogive nosed projectiles. They studied monolithic shield, two-layer in-contact shield and two-layer spaced shields with different widths of air gap between the layers so that shields having the thicknesses 2 mm, 5 mm, 10 mm, 20 mm and 30 mm were examined. The highest BLV was observed for the monolithic shield followed by the layered in-contact and spaced shields, respectively. Variation of widths of the air gap did not have significant influence on the BLV for ogive-nosed projectile while some effect for the blunt projectile was observed. Raguraman *et al.*, 2009 discussed particular details of numerical simulations of penetration into layered shields.

Using numerical simulations Teng *et al.*, 2008 examined the ballistic resistance of two-layer steel shields against projectile impact at velocities 400 m/s and 800 m/s. "Heavy" and "light" conical-nosed impactors and "heavy" and "light" cylindrical impactors and armor shields consisting of two 6 mm plates manufactured from different materials, Domex Protect 500 and Weldox 460E, were considered. It was found that the best shield configuration is when the front layer is made of high ductility and low strength material and the rear layer is manufactured from low ductility and high strength material.

Zhou and Stronge, 2008 investigated experimentally and numerically ballistic properties of monolithic and two-layer steel sheets under normal and oblique impact by cylindrical and hemispherical-nosed projectiles. The main results obtained for the angle of obliquity in the range between 0° to 45° are the following: (i) layered shields cause more energy loss than monolithic shield having the same total thickness when perforated by flat-nosed projectile; (ii) there was no significant difference in the measured BLV between the monolithic shields and layered shield in the case of perforation by a hemispherical-nosed projectile.

Børvik *et al.*, 2009b compared experimentally the BLV of monolithic steel 12 mm thick shield, two-layer shield consisting of the two 6 mm thick plates in contact and with 30 mm air gap against 7.62 mm bullets. They found that there is no difference between the BLVs for the monolithic shield and the two-layer shield while spacing clearly reduces the BLV.

Rosenberg and Dekel, 2010b performed numerical simulations of penetration of several shields by ogive-nosed projectile: 1x40 mm and 2x20 mm for the impact velocities 900 m/s and 1000 m/s, 1x60 mm, 2x30 mm and 3x20 mm for the impact velocities 1150 m/s and 1250 m/s, and 3x20 mm (30 mm) for the impact velocities 1250 m/s. On the basis of these results, analysis of the available data in the literature and physical considerations the authors arrived at the conclusion that layering reduces ballistic resistance of the shield against sharp-nosed projectiles, and using spacing further impairs the ballistic performance of the shield.

Dey *et al.*, 2011 presented and compared the results of three series of numerical calculations of the BLVs for monolithic and layered metal shields with the total thickness of 12 mm against cylindrical and ogive-nosed projectiles: 1x12 mm Weldox 700E, 2x6 mm Weldox 700E, 2x6 mm Weldox 700E with air gaps (series 1); 6 mm Weldox 700E + 6 mm Armox 560T, 6 mm Armox 560T + 6 mm Weldox 700E, 6 mm Weldox 460E + 6 mm Armox 560T, 2x6 mm Armox 560T (series 2); 12x1 mm Armox 560T, 8 mm Weldox 700E + 4 mm Armox 560T, 4 mm Weldox 700E + 8 mm Armox 560T, 2 mm Weldox 700E + 10 mm Armox 560T, 4 mm Armox 560T + 8 mm Armox 560T (series 3). For some of these shields large discrepancies between the numerical simulations and experimental data were found, especially, for the blunt impactor. The authors explain these discrepancies by fragmentation of the hard plate and fragmentation of the projectile that were not taken into account in the simulations.

Flores-Johnson *et al.*, 2011 presented the results of a numerical investigation of the ballistic performance of monolithic, two- and three-layer metal shields consisting of plates and manufactured from steel (Weldox 700E), aluminum (Al 7075-T651) or combination of these materials, impacted by a 7.62-mm APM2 projectile in the velocity range between 775 m/s and 950 m/s. It was assumed that layered shields consist of plates having the same thickness. It is convenient to classify the examined configurations as follows. Series 1 includes monolithic, two- and three-layer steel shields with the total thickness 12 mm, 14 mm, 16 mm, 18 mm, 20 mm. Series 2 includes monolithic, two- and three-layer aluminum shields with the total thickness 12 mm, 20 mm, 30 mm, 36 mm, 40 mm. This classification allows analyzing the effect of layering on the ballistic resistance of a shield. Series 3 includes three three-layer 20–mm thick configurations with single aluminum plate and two steel plates arranged in different order. In series 1-3, the layered shields consist of plates with the same thickness. Series 4 includes two two-layer configurations with single 6.66-mm aluminum plate and 13.33-mm steel plates arranged in two possible configurations. On the basis of Series 3-4, the authors analyzed the effect of the order of the plates on the ballistic resistance of a shield. The obtained results allowed to arrive at the following conclusions: (i) monolithic plate manufactured from Weldox 700E performs better than three-layer shields made from the same material; and the difference becomes more pronounced with the decreasing of the impact velocity while the difference in the performance between monolithic and two-layer shields is not significant; (ii) for the shields manufactured from Al 7075-T651 the difference in performance between monolithic and multi-layer shields is not significant if the shield's thickness is less than 20 mm while there is a large difference in the ballistic performance if the thickness of the shield is larger than 30 mm; (iii) order of the plates manufactured from different materials may strongly affect ballistic resistance of the shield.

Nia and Hoseini, 2011 studied experimentally perforation of monolithic 3 mm thick shield and three-layer shield manufactured from Al 1100 and having the same total thickness, by hemispherical-nosed cylindrical projectiles. They studied effects of the order of layers and the width of air gaps between the layers on the BLV. In the first series of experiments, monolithic shield and shields consisting of 0.5-mm, 1.0-mm and 1.5-mm plates in contact were examined, with all possible arrangements of the plates in the shield (study of layering). In the second experimental series, spaced 1 mm + 0.5 mm + 1.5 mm layered shields were tested with the following combinations of air gaps between the first and the second plates and the second and the third plates: 1 mm + 5 mm, 2 mm + 4 mm, 3 mm + 3 mm, 4 mm + 2 mm, and 5 mm + 1 mm. The main conclusions obtained in this study are summarized in the following: (i) ballistic resistance of monolithic shields is larger than ballistic resistance of plates in-contact and spaced layered shields; (ii) ballistic resistance of plates in-contact layered shields is larger than ballistic resistance of the spaced shields; (iii) order of layers affects the ballistic performance of the shield; (iv) increasing the air gap between the first and the second plate decreases the BLV of the shield.

Babaei *et al.*, 2011 investigated experimentally and numerically ballistic properties of two-layer thin shields with the layer thickness of 1 mm under normal impact by cylindrical projectile at impact velocity in the range from 50 to 400 m/s. Comparison of the shields with plates manufactured from steel and aluminum and arranged in two possible configurations showed that the order of the plates is of importance since the shield with a front steel plate absorbs more energy than a shield with aluminum front plate.

Durmuş *et al.*, 2011 compared experimentally ballistic performances of monolithic 1x2 mm and two-layer 2x1 mm shields consisting of cold rolled sheet metal plates against 9 mm standard NATO projectile. It was found that layering reduced the BLV of the shield.

Zhang *et al.*, 2011b conducted experiments with monolithic A3 steel 2 mm plate, 2x1 mm two-layer shield, and spaced 2x1 mm (6 mm), 2x1 mm (100 mm) shields perforated by hemispherical-nosed projectile and obtained the initial-residual velocity curves. The results of experiments showed that (i) ballistic resistance of a monolithic shield is higher than that of a two-layer shield with or without gap; (ii) air gap width has slight effect on the BLV of the spaced shield; (iii) BLV of layers-in-contact shield is close to or larger than that of the spaced layered shield.

Deng *et al.*, 2012 performed experiments with the following shields: 1x2 mm, 2x1 mm, 0.5 mm + 1.5 mm, 1.5 mm+ 0.5 mm, 4x0.5 mm, 2x1 mm (6 mm), 3 mm, 6x0.5 mm, 1x6 mm, 3x2 mm impacted by hemispherical-nosed projectiles. The BLV was between 120 m/s and 185 m/s. The results showed that (i) monolithic thin shields were more effective than multi-layer shields when the dominant response of monolithic shields were dishing and bulging which involved bending and stretching; and the BLVs of multi-layer plates increased with the increase of the number of layers; (ii) monolithic shields had lower ballistic resistance than multi-layered shields if the total thickness exceeds a certain threshold value, and the dominant shield response changed from shearing to bulging and dishing when using multi-layer shields instead of monolithic shields; (iii) air gap decreased ballistic resistance of multi-layer shields, but the air gap width had a slight effect on the ballistic resistance; (iv) order of layers affected the ballistic resistance of multi-layer shields, the multi-layer shields consisting of two equal thick plates had the highest ballistic resistance followed by multi-layer shields having a thicker front plate and a thinner back plate, and multi-layer shields having a thinner front plate and a thicker back plate. Wei *et al.*, 2012 performed experiments with the following shields: 1x2 mm, 2x1 mm, 0.5 mm + 1.5 mm, 1.5 mm+ 0.5 mm, 4x0.5 mm, 1 mm + 1 mm (6 mm), 1 mm + 1 mm (100 mm), 1x6 mm, 3x2 mm impacted by blunt-nosed cylindrical projectiles. The BLV varied in the range between 100 m/s and 220 m/s. The results of the analysis are similar to those obtained by Deng *et al.*, 2012.

Presented above analysis allows us to draw some cautious conclusions: (i) in most cases layering impairs ballistic properties of the shields and spacing causes further degradation of the ballistic characteristics; (ii) ballistic characteristics of spaced shields having identical layers are adversely affected by the increase of the number of layers while in the case of layers having different thicknesses the ballistic performance depends on the order of the layers; (iii) changing the order of plates manufactured from different materials in multi-layer shields may strongly affect the ballistic performance of the shield; (iv) effect of layering diminishes with the increase of the impact velocity; (v) the character and the magnitude of the effect of layering and spacing on ballistic characteristics of the shield depend upon the shape of the striker (primarily on whether the striker is blunt or sharp) and on the widths of the plates; situation when the failure mode (plugging, petaling, ductile hole growth) changes during switching from monolithic shields to layered and/or spaced shields is particularly involved.

Consequently, the effect of layering warrants further systematic investigations using the existing theoretical, numerical and experimental approaches.

Chapter 24

Effect of Spacing for Non-Conical Impactors. Numerical Simulation

Nomenclature

a_0, a_2	parameters determining the model, Eq. (24.1.4);
b	total thickness of a shield (sum of the thicknesses of all layers including air gaps);
$\bar{b}$	$=b/L$;
b_{sum}	sum of the thicknesses of all plates;
$\bar{b}_{sum}$	$=b_{sum}/L$;
$b^{(2i+1)}$	thicknesses of plates;
$\bar{b}^{(2i+1)}$	$=b^{(2i+1)}/L$;
$b^{(2i)}$	thicknesses of air gaps;
$\bar{b}^{(2i}$	$=b^{(2i)}/L$;
BLV	ballistic limit velocity;
$d\vec{F}$	force acting on the surface element of projectile;
dS	surface element of projectile (area of surface element);
D	resistance force acting on projectile;
e_{abs}	normalized energy absorbed by spaced shield, Eq. (24.3.3);
e_{abs}^0	normalized energy absorbed by non-spaced shield, Eq. (24.3.4);
E_{imp}	impact energy of impactor;
E_{res}	residual energy of impactor (spaced shield);
E_{res}^0	residual energy of impactor (non-spaced shield);
f_0, f_2	functions, Eqs. (24.1.10);
$\vec{F}$	total force acting on impactor;

g	function, Eq. (24.1.13);
h	instantaneous depth of penetration, Fig. 24.1.1;
$\bar{h}$	$=h/L$;
k	parameter, Eq. (24.3.7);
L	length of the nose of impactor, Fig. 24.1.1;
m	mass of projectile;
$\vec{n}^0$	inner normal unit vector at a given location on the impactor surface;
q	function, Eq. (24.1.13);
q_*	parameter, Eq. (24.3.2);
Q	function, Eq. (24.1.20);
v	instantaneous velocity of projectile;
$\vec{v}^0$	unit vector of impactor velocity;
v_{bl}	BLV of projectile (spaced shield);
v_{bl}^0	BLV of projectile (non-spaced shield);
v_{imp}	impact velocity of projectile;
v_{res}	residual velocity of projectile (spaced shield);
x	coordinate associated with the impactor, Fig. 24.1.1;
$\bar{x}$	$=x/L$;
α	parameter, Eq. (24.1.19);
δ	$=v_{bl}/v_{bl}^0$;
ε	function, Eq. (24.1.5);
θ, Θ	functions determining impactor-shield interaction surface, Fig. 24.1.1, Eq. (24.1.1);
λ	$=h^{(1)}/h_{sum}$;
$\mu_{\max}$	number of plates in shield;
ξ	coordinate associated with shield, Fig. 24.1.1;
$\bar{\rho}$	$=\rho/L$;
ρ	coordinate associated with impactor, Fig. 24.1.1;
ϕ	coordinate associated with impactor, Fig. 24.1.1;
Φ	function determining the shape of impactor;
$\bar{\Phi}$	$=\Phi/L$;
χ	parameter, Eq. (24.1.19);
Ψ	function, Eq. (24.1.18).

24.1 Formulation of Problem

Consider high speed normal penetration of a rigid sharp striker (a body of revolution) into a ductile spaced shield having a finite thickness. The main notations are shown in Fig. 24.1.1, and it is assumed that only the nose of the cylindrical impactor can interact with the shield. The coordinate h, the instantaneous depth of penetration, is defined as the distance between the leading edge of the nose of the impactor and the rear surface of the shield. The coordinate ξ is associated with the shield. In cylindrical coordinates, x, ρ, ϕ, associated with the impactor, the surface of the nose is described by the following equation:

$$\rho = \Phi(x), \quad 0 \le x \le L, \quad 0 \le \phi \le 2\pi\,, \tag{24.1.1}$$

where L is the length of the nose of impactor, $\Phi(x)$ is an increasing convex upward function. Assume that the shield consists of $2\mu_{\max} - 1$ layers including $\mu_{\max}$ plates having the thicknesses $b^{(1)}, b^{(3)}, ..., b^{(2\mu_{\max}-1)}$ and air gaps between the plates having the widths $b^{(2)}, b^{(4)}, ..., b^{(2\mu_{\max}-2)}$. The plate with number $2\mu - 1$ is located between the cross-sections $\xi = \xi^{(2\mu-2)}$ and $\xi - \xi^{(2\mu-1)}$, where $\mu - 1,2,..., \mu_{\max}$ and $\xi^{(0)} = 0$. The total thickness of the shield (the sum of the thicknesses of all layers including the widths of air gaps) and the sum of the thicknesses of all plates are denoted by b and b_{sum}, respectively. It is assumed that the plates are manufactures from the same material.

In order to make presentation self-contained we have included several formulas from Chapter 1 instead of simply referring to the results obtained there.

The part of the lateral surface of the impactor between the cross-sections $x = \theta(h)$ and $x = \Theta(h)$ (see Fig. 24.1.1) interacts with some layers of the shield or is in contact with some air gaps (see Fig. 24.1.2a):

$$\theta(h) = \begin{cases} 0 & if \quad 0 \le h \le b \\ h - b & if \quad b \le h \le b + L \end{cases}, \quad \Theta(h) = \begin{cases} h & if\ 0 \le h \le L \\ L & if\ h \ge L \end{cases}. \tag{24.1.2}$$

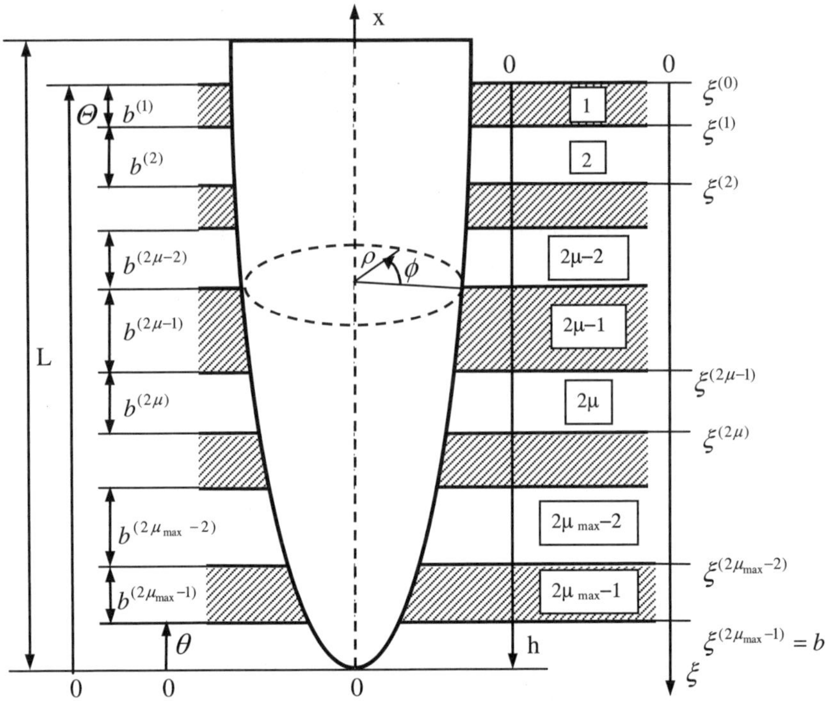

Fig. 24.1.1. Notations.

The equation of motion of the impactor can be written as follows:

$$mv(dv/dh) = -D, \tag{24.1.3}$$

where the velocity of the impactor v is considered to be a function of h, m is the mass of the impactor, and D is the resistance force. We consider such range of impact velocities v_{imp} whereby projectile perforates the shield. Perforation occurs when the position of the striker is $h = b + L$ and the residual velocity is v_{res}. As before, the ballistic limit velocity (BLV), v_{bl}, is defined as the impact velocity of the impactor required to emerge from the shield with zero residual velocity.

We assume that the impactor-shield interaction at a given location on the surface of the impactor that is in contact with a plate is described by a two-term model:

$$d\vec{F} = [a_2(-\vec{v}^0 \cdot \vec{n}^0)^2 v^2 + a_0]\vec{n}^0 dS, \qquad (24.1.4)$$

where $d\vec{F}$ is the force acting on the surface element dS of the impactor, $\vec{n}^0$ is the inner normal unit vector at a given location on the impactor surface, $\vec{v}^0$ the unit vector of the impactor velocity, parameters a_0 and a_2 depend on the properties of shield material.

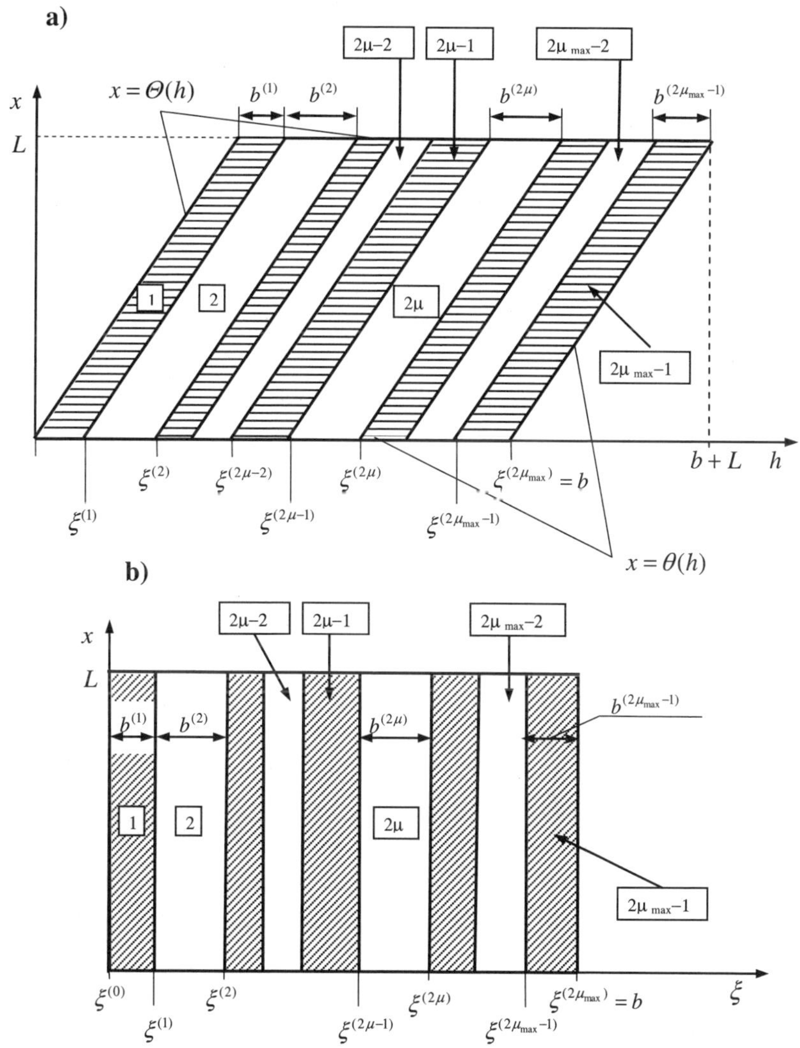

Fig. 24.1.2. Model of spaced shield.

In order to adapt Eq. (24.1.4) for a spaced shield let us introduce a function $\varepsilon(\xi)$, which is equals 1 if the point with the coordinate ξ is located inside any plate and equals 0 if this point is located inside an air gap (Fig. 24.1.1):

$$\begin{aligned} &\varepsilon(\xi) = \varepsilon^{(j)} \quad if \quad \xi^{(j-1)} \le \xi \le \xi^{(j)}, \quad j = 1,2,\dots,\mu_{\max} - 1; \\ &\varepsilon^{(1)} = \dots = \varepsilon^{(2\mu-1)} = \dots = \varepsilon^{(2\mu_{\max}-1)} = 1; \\ &\varepsilon^{(2)} = \dots = \varepsilon^{(2\mu)} = \dots = \varepsilon^{(2\mu_{\max}-2)} = 0. \end{aligned} \tag{24.1.5}$$

Hence Eq. (24.1.4) can be rewritten as follows:

$$d\vec{F} = \varepsilon(\xi)[a_2(-\vec{v}^0 \cdot \vec{n}^0)^2 v^2 + a_0]\vec{n}^0 dS\,. \tag{24.1.6}$$

The total force $\vec{F}$ acting on the impactor surface inside the shield is found by integrating the local force over the impactor-shield contact area that formally includes air gaps. Hence integration is performed over a portion of the impactor surface S that is determined by inequalities $0 \le \phi \le 2\pi$ and $\theta(h) \le x \le \Theta(h)$. Taking into account the identity

$$\xi = h - x, \tag{24.1.7}$$

and using the following formulas of differential geometry:

$$-\vec{v}^0 \cdot \vec{n}^0 = \Phi' / \sqrt{\Phi'^2 + 1}, \quad dS = \sqrt{\Phi'^2 + 1}\, dx\, d\phi, \quad \Phi' = d\Phi/dx\,, \tag{24.1.8}$$

we arrive at the following expression for the resistance force D:

$$\begin{aligned} D &= \vec{F} \cdot (-\vec{v}^0) = \iint_S \varepsilon(\xi)[a_2(-\vec{v}^0 \cdot \vec{n}^0)^2 v^2 + a_0](-\vec{v}^0 \cdot \vec{n}^0) dS \\ &= \frac{m}{2}[f_2(h)v^2 + f_0(h)], \end{aligned} \tag{24.1.9}$$

where

$$\begin{aligned} f_2(h) &= \frac{4\pi a_2}{m} \int_{\theta(h)}^{\Theta(h)} \frac{\varepsilon(h-x)\Phi\Phi'^3}{\Phi'^2 + 1} dx, \\ f_0(h) &= \frac{4\pi a_0}{m} \int_{\theta(h)}^{\Theta(h)} \varepsilon(h-x)\Phi\Phi' dx \end{aligned} \tag{24.1.10}$$

Substituting D from Eq. (24.1.9) into Eq. (24.1.3), after some algebra, we obtain an ordinary linear differential equation with respect to v^2:

$$dv^2/dh + f_2(h)v^2 + f_0(h) = 0 . \tag{24.1.11}$$

The solution of Eq. (24.1.11) with the initial condition $v(0) = v_{imp}$, which corresponds to the beginning of the motion of the impactor with the impact velocity v_{imp}, reads:

$$v^2(h) = \frac{1}{q(h)}\left[v_{imp}^2 - g(h)\right], \tag{24.1.12}$$

where

$$q(h) = \exp\left(\int_0^h f_2(\eta)d\eta\right), \quad g(h) = \int_0^h f_0(H)q(H)dH . \tag{24.1.13}$$

Eq. (24.1.12) yields formulas for residual velocity, $v_{res} = v(b+L)$, and ballistic limit velocity, v_{bl}:

$$v_{res}^2 = \frac{1}{q(b+L)}\left[v_{imp}^2 - g(b+L)\right], \quad v_{bl}^2 = g(b+L) . \tag{24.1.14}$$

Let us prove that $q(b+L)$ does not depend on the widths of air gaps. To this end, let us change the variables (see Fig. 24.1.2b), $x \to x,\ h \to x + \xi$, in the integral in the expression for $q(b+L)$:

$$\begin{aligned} \frac{m}{4\pi a_2}\ln\left(q(b+L)\right) &= \int_0^{b+L} dh \int_{\theta(h)}^{\Theta(h)} \varepsilon(h-x)\psi(x)dx \\ &= \int_0^L \psi(x)dx \int_0^b \varepsilon(\xi)d\xi = b_{sum}\int_0^L \psi(x)dx, \end{aligned} \tag{24.1.15}$$

where

$$\psi(x) = \Phi\Phi'^3/(\Phi'^2 + 1) . \tag{24.1.16}$$

For further analysis it is convenient to rewrite the expression for v_{bl} using dimensionless variables, where L is chosen as a characteristic length:

$$v_{bl} = \sqrt{\chi}\,\Psi(\alpha)\,, \tag{24.1.17}$$

function Ψ depends also on structure of the shield and the shape of impactor, and

$$\Psi(\alpha) = \sqrt{\alpha \int_0^{\bar{b}+1} Q(\bar{h})d\bar{h} \int_{\bar{\theta}(\bar{h})}^{\bar{\Theta}(\bar{h})} \bar{\varepsilon}(\bar{h}-\bar{x})\overline{\Phi}\,\overline{\Phi}'d\bar{x}}\,, \tag{24.1.18}$$

$$\alpha = \frac{4\pi L^3}{m}a_2,\ \chi = \frac{4\pi L^3}{m}a_0,$$
$$\bar{x} = \frac{x}{L},\ \bar{\rho} = \frac{\rho}{L},\ \overline{\Phi} = \frac{\Phi}{L},\ \overline{\Phi}' = \frac{d\overline{\Phi}}{d\bar{x}},\ \bar{h} = \frac{h}{L}, \tag{24.1.19}$$

$$Q(\bar{h}) = \exp\left(\alpha \int_0^{\bar{h}} d\overline{H} \int_{\bar{\theta}(\overline{H})}^{\bar{\Theta}(\overline{H})} \frac{\bar{\varepsilon}(\overline{H}-\bar{x})\overline{\Phi}\,\overline{\Phi}'^3}{\overline{\Phi}'^2+1}d\bar{x}\right),\ \bar{\varepsilon}(\bar{\xi}) = \varepsilon(L\bar{\xi})\,, \tag{24.1.20}$$

$$\bar{\theta}(\bar{h}) = \begin{cases} 0 & if \quad 0 \le \bar{h} \le \bar{b} \\ \bar{h}-\bar{b} & if \quad \bar{b} \le \bar{h} \le \bar{b}+1 \end{cases},\quad \bar{\Theta}(\bar{h}) = \begin{cases} \bar{h} & if\ 0 \le \bar{h} \le 1 \\ 1 & if\ \bar{h} \ge 1 \end{cases}. \tag{24.1.21}$$

Let v_{bl}^0 be the BLV of the shield that consists of plates in contact having the same total thickness, b_{sum}. Clearly, in the framework of our model, this shield is equivalent to the monolithic shield with the total thickness b_{sum}. Eq. (24.1.17) implies that the ratio $\delta = v_{bl}/v_{bl}^0$ depends on the dimensionless parameter α as well as on the shape of the impactor and on the dimensionless thicknesses of the layers (plates and air gaps).

Ben-Dor *et al.*, 1998a analytically found that spacing does not affect ballistic properties of shields against conical impactors. In the next section, we study numerically the effect of spacing on the BLV of shields for non-conical impactors.

24.2 Result of Numerical Calculations and Discussion

We performed calculations for two impactors (cylindrical bodies of revolution with different nose shapes). The first impactor (Impactor A) is a cylinder with a semi-spherical nose. The generatrix of a nose of the second impactor (Impactor B) is the broken line consisting of segments between three points with the dimensionless coordinates $(\bar{x},\bar{\rho}):(0,0)$; $(0.3,0.7);(1,1)$. Validity of this model for blunt, thick impactors against shield having a finite thickness is questionable, and, consequently, Impactor A can be considered only as a model.

In following analysis, we use the dimensionless variables. In Fig. 24.2.1 we showed plots of the function Ψ for a monolithic plate. Using Eq. (24.1.17), one can determine the ranges of variation of parameters $\bar{b}_{sum}$ and α that correspond to a given range of BLVs for a shield manufactured from a given material. The plots shown in Fig. 24.2.1 correspond to the Impactor A, and they differ only insignificantly from the plots obtained for the Impactor B.

In each pair of Figs. 24.2.2-24.2.8, the letters 'a" and "b" correspond to the Impactor A and B, respectively.

Fig. 24.2.2 shows the dependence of the parameter $\delta = v_{bl}/v_{bl}^{0}$ vs. α for different widths of a gap (a shield consisting of two plates).

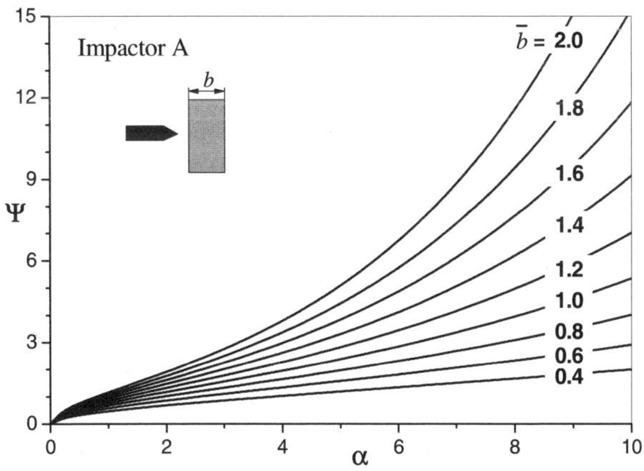

Fig 24.2.1. Function $\Psi(\alpha)$ for different $\bar{b}$.

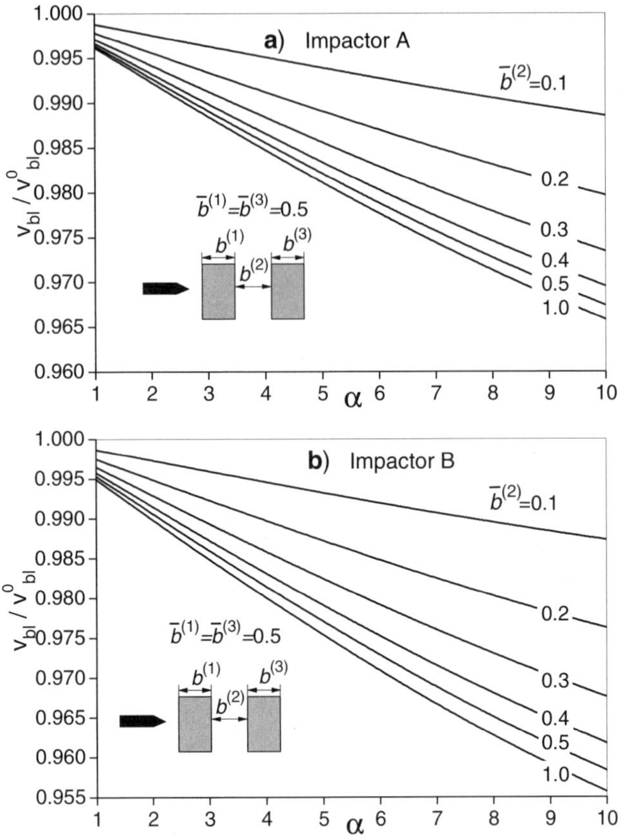

Fig. 24.2.2. Effect of thickness of air gap on ballistic properties of two-layer shield.

Inspection of these figures shows that ballistic efficiency of the shield decreases when the width of the air gap between the plates increases (the larger is parameter $\bar{b}^{(2)}$, the lower is the curve). The rate of change of parameter δ that characterizes ballistic efficiency, sharply decreaes with increase of $\bar{b}^{(2)}$. Hence inspection of Fig. 24.2.2 shows that increase of $\bar{b}^{(2)}$ from 0.4 to 0.5 and from 0.5 to 1.0 is accompanied by approximately the same change of δ. The effect of air gaps on δ becomes more pronounced with the increase of parameter α. Inspection of Fig. 24.2.3 allows us to arrive at similar conclusions. In this figure we showed the results obtained for a shield consisting of three plates whereby the widths of air gaps between the first and the second plate and between the second and the third plate are varied but kept equal. Consequently, the maximum negative effect of spacing (which is the most interesting) occurs for large air gaps equal to the length of the nose of impactor since in the framework of the used model, further increase of the air gap width does not change the BLV of the shield. Our further analysis is performed exactly for this width of the air gap ($\bar{b}^{(2)} \geq 1$).

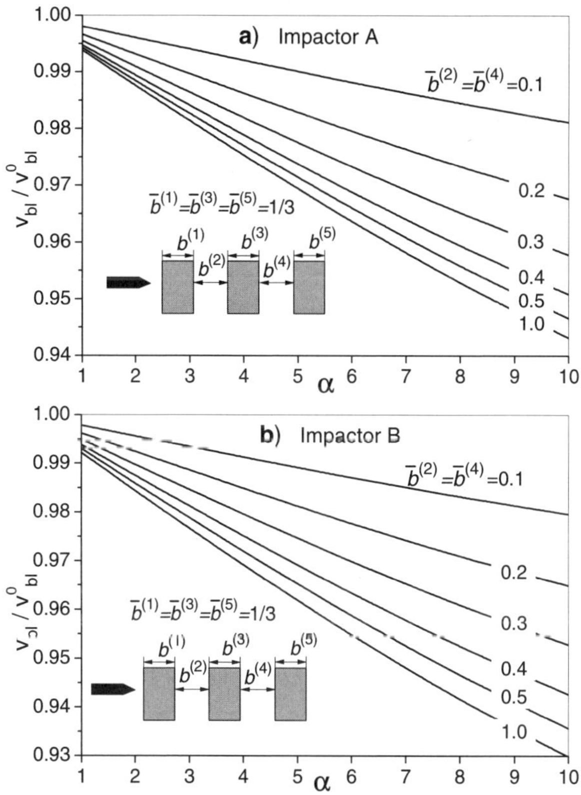

Fig. 24.2.3. Effect of thickness of air gap on ballistic properties of three-layer shield.

In Fig. 24.2.4, we showed the plots of function $\delta(\alpha)$ for the case of two identical plates and for different values of the total width of the shield, $\bar{b}_{sum}$.

Inspection of these plots shows that the effect of air gaps is more pronounced for large α. It must be noted that some curves in Fig. 24.2.4 intersect, i.e., for two shields having different total thicknesses parameter δ can be larger for the first shield than for a second one for the same value of the parameter α, and it can become smaller than for the second shield for different values of parameter α. The same dependencies are observed in the case when a plate is separated into three identical plates with equal widths air gaps between the plates (Fig. 24.2.5).

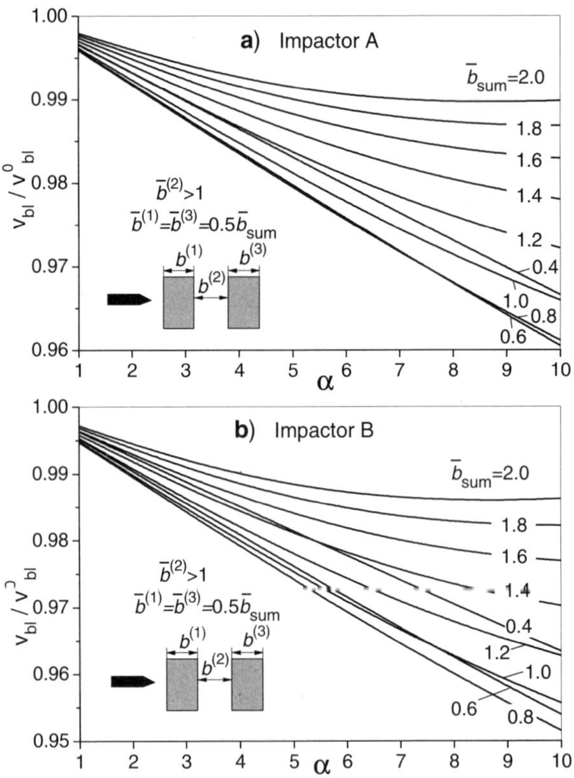

Fig. 24.2.4. Effect of parameters α and $\bar{b}_{sum}$ on ballistic properties of two-layer shield with large air gap.

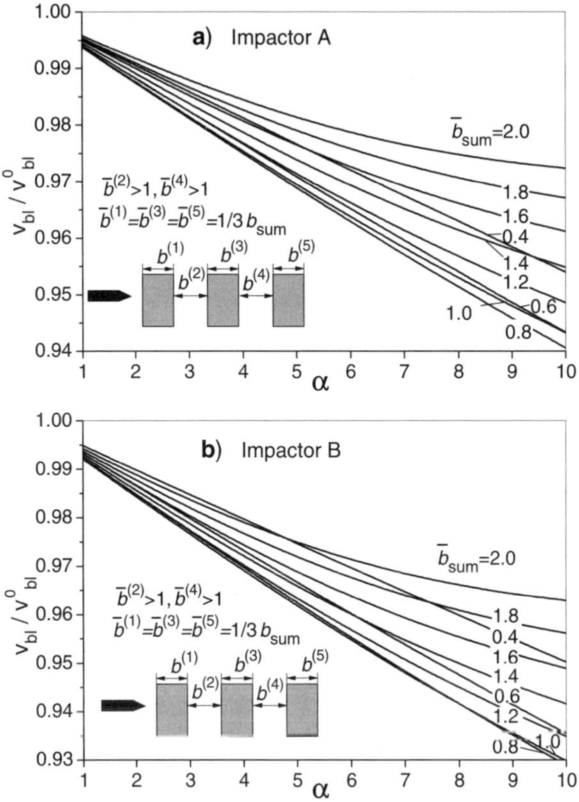

Fig. 24.2.5. Effect of parameters α and $\bar{b}_{sum}$ on ballistic properties of three-layer shield with large air gaps.

In Figs. 24.2.6 and 24.2.7, we showed the plots of $\delta(\lambda)$ for $\bar{b}_{sum} = 1$ and $\bar{b}_{sum} = 2$, respectively, and for different values of parameter α (λ is the ratio of the width of the first plate to the total width, $\lambda = b^{(1)}/b_{sum}$). Inspection of these figures allows to conclude that (i) shape of the impactor only weakly affects the plots; (ii) for every value of parameter α, initially the increase of the dimensionless thickness of the first plate causes the increase of the negative effect of spacing until some maximum value is attained, and then the negative effect of the increase of the dimensionless thickness of the first plate decreases, (iii) the magnitude of λ whereby the negative effect of spacing is maximum

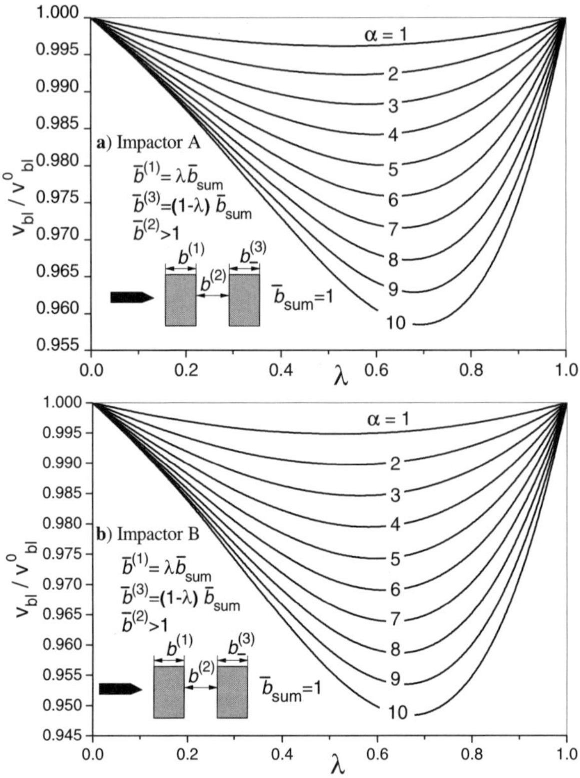

Fig. 24.2.6. Effect of thicknesses of plates on ballistic properties of two-layer shield with large air gap and $\bar{b}_{sum} = 1$.

depends upon the total thickness, and increases with the increase of parameter α.

Fig. 24.2.8 compares the results obtained for a spaced shield consisting of three plates with different dimensionless thicknesses of the plates subjected to a constraint that the total thickness, $\bar{b}_{sum}$, is kept constant for $\bar{b}_{sum} = 1$ and $\bar{b}_{sum} = 2$. Calculations were performed for three sets of the dimensionless thicknesses of the plates, $b^{(1)} : b^{(2)} : b^{(3)}$, namely, $1:2:3$, $1:1:1$ and $3:2:1$. The results show that the negative effect of spacing is minimum for the configuration $1:2:3$, and depends on parameter α for two other configurations.

Therefore, in the framework of the employed model, the effect of spacing on the BLV for non-conical impactors is of the order of several

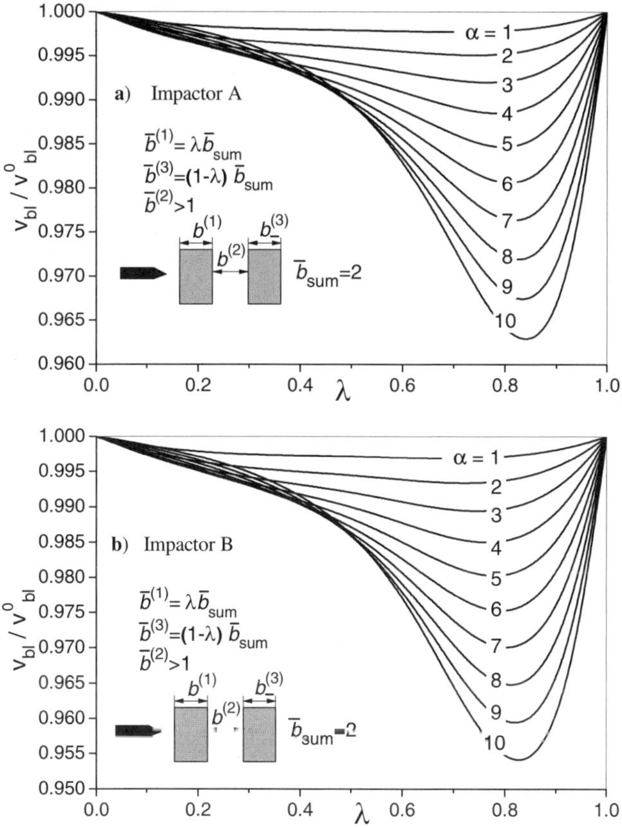

Fig. 24.2.7. Effect of the thicknesses of plates on ballistic properties of two-layer shield with large air gap and $\bar{b}_{sum} = 2$.

percent. The results of calculations showed that for slender projectiles this effect becomes even smaller.

24.3 Experimental Validation

In our analysis, we use the results of the experiments performed with spaced and non-spaced shields for the same impact velocity, v_{imp} that were published by Almohandes *et al.*, 1996 for 7.62 mm bullet perforating mild steel shield.

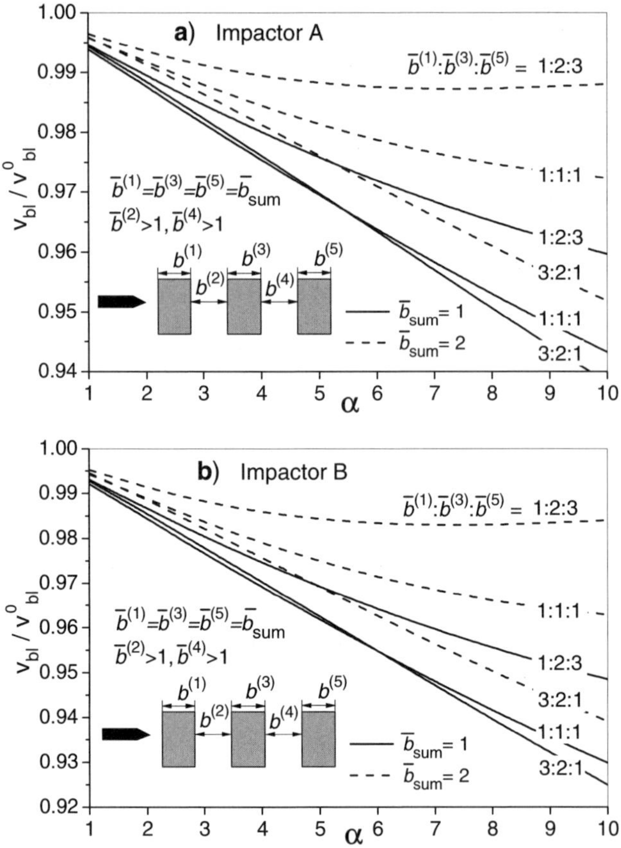

Fig. 24.2.8. Effect of thicknesses of plates on ballistic properties of three-layer shield with large air gaps.

First, let us analyze, using the employed model, the connection between the amount of energy absorbed and the BLV of spaced and non-spaced shields for the same v_{imp}.

Using the relationships for the impact energy and residual energy of the impactor (E_{imp} and E_{res}, respectively),

$$E_{imp} = 0.5mv_{imp}^2 \ , \ E_{res} = 0.5mv_{res}^2 \ , \qquad (24.3.1)$$

and Eq. (24.1.14) which is rewritten as

$$v_{res}^2 = \frac{1}{q_*}\left[v_{imp}^2 - v_{bl}^2\right], \quad q_* = q(b+L) = Q(\bar{b}+1), \tag{24.3.2}$$

we obtain the following formula for the normalized energy absorbed by a spaced shield:

$$e_{abs} = \frac{E_{imp} - E_{res}}{E_{imp}} = \frac{q_* - 1}{q_*} + \frac{1}{q_*}\left(\frac{v_{bl}}{v_{imp}}\right)^2. \tag{24.3.3}$$

Since q_* is the same for spaced and non-spaced shields (see below), an equation similar to Eq. (24.3.3) is also valid for the spaced shield:

$$e_{abs}^0 = \frac{E_{imp} - E_{res}^0}{E_{imp}} = \frac{q_* - 1}{q_*} + \frac{1}{q_*}\left(\frac{v_{bl}^0}{v_{imp}}\right)^2, \tag{24.3.4}$$

where a superscript 0 denotes the corresponding parameter for non-spaced shield with the same total thickness.

Hence

$$e_{abs} - e_{abs}^0 = \frac{1}{q_*}\left(\frac{v_{bl}^0}{v_{imp}}\right)^2\left[\left(\frac{v_{bl}}{v_{bl}^0}\right)^2 - 1\right]. \tag{24.3.5}$$

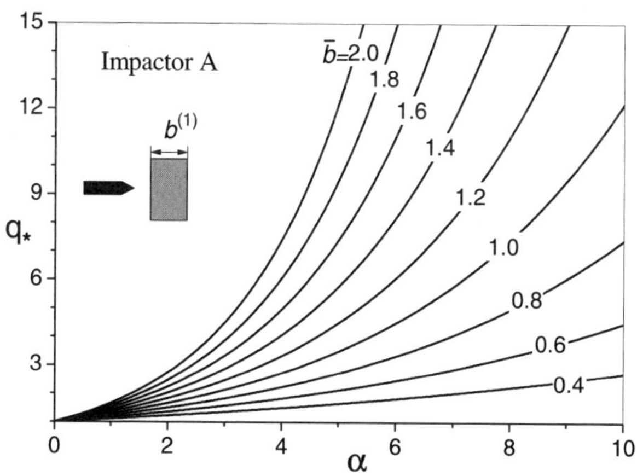

Fig. 24.3.1. Function $q_*(\alpha)$ for different $\bar{b}$.

Taking into account that for $\left|v_{bl}/v_{bl}^0 - 1\right| = \left|\delta - 1\right| << 1$,

$$\left(\frac{v_{bl}}{v_{bl}^0}\right)^2 - 1 \equiv (\delta - 1)^2 + 2(\delta - 1) \approx 2(\delta - 1), \qquad (24.3.6)$$

and Eq. (24.3.5) can be rewritten as follows:

$$e_{abs} - e_{abs}^0 \approx k(1-\delta), \quad k = \frac{2}{q_*}\left(\frac{v_{bl}^0}{v_{imp}}\right)^2 . \qquad (24.3.7)$$

Since $v_{bl}^0 \le v_{imp}$ and $q_* > 1$, then $k < 2$. Moreover, when α increases, the magnitude of q_* increases (see Fig. 24.3.1), that compensates the discussed above increase of δ. Therefore, the model predicts insignificant change of the absorbed energy in the whole practical range

Table 24.3.1. Normalized energy absorption, in percents; numerator–spaced shield, denominator–shield with plates in contact. Based on experimental results by AlmohandesBen-,1996.

Number of shield	Spaced shield structure (sizes in mm)	v_{imp}				
		706.0	754.5	775.4	804.5	826.2
1	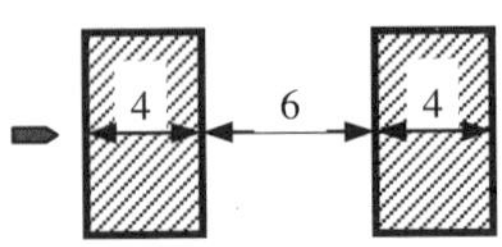	52.2 ----- 52.5	49.6 ----- 50.0	47.4 ----- 46.5	35.9 ----- 36.1	34.5 ----- 34.8
2	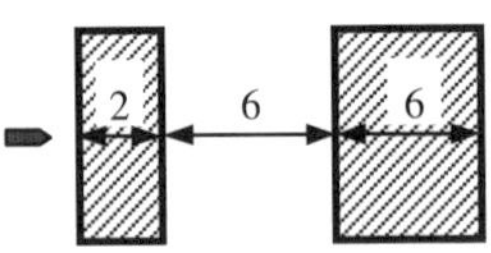	54.9 ----- 54.3	52.2 ----- 53.6	50.7 ----- 49.8	42.1 ----- 42.7	37.4 ----- 36.5
3	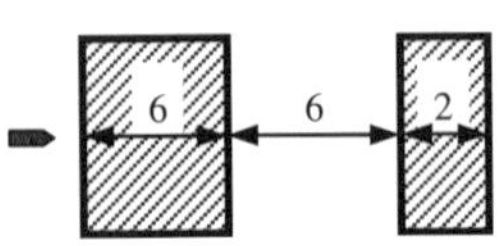	54.4 ----- 55.5	52.0 ----- 52.1	49.7 ----- 51.7	41.5 ----- 41.4	35.1 ----- 37.3

of variation of parameter α. Since the predicted magnitude of the absorbed energy change falls in the range of the experimental errors and model uncertainties, it is conceivable that experimentally observed v_{bl}^{0} и v_{bl} will be different by several percent, and there will be situations when $v_{bl}^{0} < v_{bl}$, and when $v_{bl}^{0} > v_{bl}$.

The latter conclusion is supported by the data in Table 24.3.1 that was obtained using experimental results by Almohandes *et al.*, 1996. In this table for each configuration of the shield and for every value of the impact velocity, we showed a fraction whereby a numerator is a percent of the absorbed impact energy by a corresponding spaced shield, and the dominator is the same characteristic for a shield consisting of the same plates in the same order but without air gaps.

Table 24.3.2. Comparison of normalized energy absorption, in percents, caused by interchanging the order of plates. Based on experimental results by Almohandes *et al.*, 1996.

Number of structure	Spaced shield structure (sizes in mm)	v_{imp}				
		706.0	754.5	775.4	804.5	826.2
1	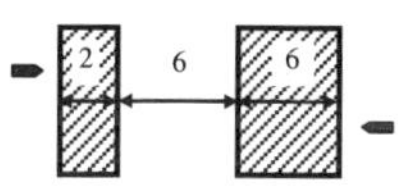	54.9 54.4	52.2 52.0	50.7 49.7	42.1 41.5	37.4 35.1
2	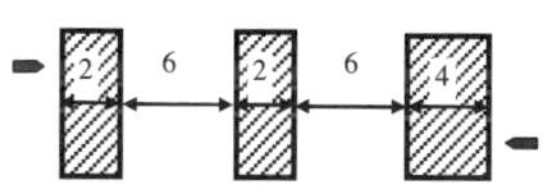	51.3 50.9	49.0 48.4	44.0 44.3	33.8 36.0	32.4 32.7
3	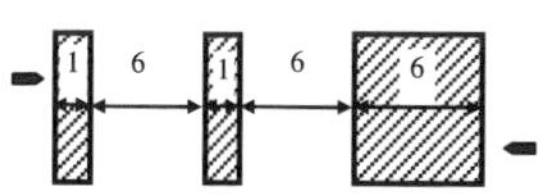	53.0 54.7	49.1 54.3	46.8 52.1	35.9 37.2	34.7 36.4
4	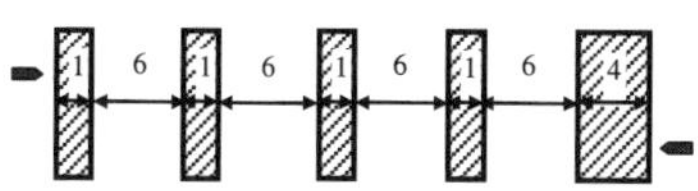	44.2 46.0	42.3 46.3	38.4 37.5	34.0 32.6	31.5 33.2

Table 24.3.2 shows that the change of the order of plates in a spaced shield to the opposite order is not accompanied by significant change of the absorbed energy, and there are also observed situations when $v_{bl}^0 < v_{bl}$ and when $v_{bl}^0 > v_{bl}$.

Slight degradation of protective properties of barriers caused by spacing for projectiles having different nose shapes is reported in a number of experimental and numerical studies (see overview in Chapter 24).

Hence it can be concluded that numerical simulations conducted on the basis of an approximate model as well as available experimental data showed that air gaps cause insignificant change of the BLV and energy absorbed by a shield for non-conical impactors.

Chapter 25

Effect of Order of Plates for Non-Conical Impactors. Numerical Simulation

Nomenclature

$a_0^{(i)}, a_2^{(i)}$	parameters determining the model, Eq. (25.1.1);
a_v	function, Eq. (25.1.2);
$\bar{a}_v$	function, Eq. (25.1.8);
b	total thickness of shield;
$\bar{b}$	$=b/L$;
$b^{(i)}$	thicknesses of plates (layers);
BLV	ballistic limit velocity;
h	instantaneous depth of penetration, Fig. 25.1.1;
$\bar{h}$	$=h/L$;
k	parameter, Eq. (25.1.7);
L	length of the nose of impactor, Fig. 25.1.1;
m	mass of projectile;
M	number of points of interpolation for generatrix of projectile minus 1;
N	number of layers in shield;
Q	function, Eq. (25.1.5);
R	radius of the cylindrical part of impactor, Fig. 25.1.1;
v_{bl}	BLV;
$v_{bl}^{[1-2]}$	BLV for the basic order of plates in shield ;
$v_{bl}^{[2-1]}$	BLV for the inverse order of plates in shield;
x	coordinate associated with impactor, Fig. 25.1.1;

$\overline{x}$	$=x/L$;
$\overline{x}_j$	abscissa of a point of interpolation for generatrix of impactor, Fig. 25.3.1;
$\overline{x}_*$	parameter, Fig. 25.2.1;
α_j, β_j	parameters in the equation of a piecewise linear approximation of impactor generatrix, Eq. (25.3.2);
β	$=1/\overline{\rho}_*$;
δ	$=v_{bl}^{[2-1]}/v_{bl}^{[1-2]}$;
θ, Θ	functions determining impactor-shield interaction surface, Fig.25.1.1, Eq. (24.1.2);
$\overline{\theta}, \overline{\Theta}$	functions in Eq. (25.1.9);
$\overline{\theta}_j^{(i)}, \overline{\Theta}_j^{(i)}$	functions in Eqs. (25.3.4) and (25.3.5)
ξ	coordinate associated with shield, Fig. 25.1.1;
$\xi^{(i)}$	coordinates of the boundaries between plates, Fig. 25.1.1;
ρ	coordinate associated with impactor, Fig. 25.1.1;
$\overline{\rho}$	$=\rho/L$;
$\overline{\rho}_*$	radius of generatrix of the nose of impactor, Fig. 25.2.1;
τ	$=R/L$;
ϕ	coordinate associated with impactor, Fig. 25.1.1;
Φ	function determining the shape of impactor;
$\overline{\Phi}_j$	ordinate of a point of interpolation for generatrix of impactor, Fig. 25.3.1;
$\overline{\Phi}$	$=\Phi/L$;
$\overline{\Phi}_*$	parameter, Fig. 25.2.1;
$\chi^{(i)}$	$=a_0^{(i)}/a_2^{(i)}$;
ψ_v	function, Eq. (25.1.6);
$\omega_j^{(i)}$	function, Eq. (25.3.8);
$\Omega_j^{(i)}$	function, Eq. (25.3.9).

25.1 Mathematical Model and Formulation of Problem

Consider high speed normal penetration of a rigid sharp striker (a body of revolution) into a ductile layered shield having a finite thickness. The main notations are shown in Fig. 25.1.1. The coordinate h, the instantaneous depth of penetration, is defined as the distance between the leading edge of the nose of the impactor and the rear surface of the shield. The coordinate ξ is associated with the shield. In cylindrical coordinates, x, ρ, ϕ, associated with the impactor the surface of the nose with length L is described as $\rho = \Phi(x)$ where $\Phi(x)$ is an increasing, convex upward function.

Assume that the shield consists of N layers (plates in-contact with different mechanical properties) having the thicknesses $b^{(1)}, b^{(2)}, ..., b^{(N)}$. The ith plate is located between the cross-sections $\xi = \xi^{(i-1)}$ and $\xi = \xi^{(i)}$, where $i = 1, 2, ..., N$ and $\xi^{(0)} = 0$. Let b be the total thickness of the shield that equals the sum of the thicknesses of all plates. Hence the part of the lateral surface of the impactor between the cross-sections $x = \theta(h)$ and $x = \Theta(h)$ (see Fig. 25.1.1) interacts with some layers of the shield (see Fig. 25.1.2) where functions $\theta(h)$ and $\Theta(h)$ are determined by Eq. (24.1.2).

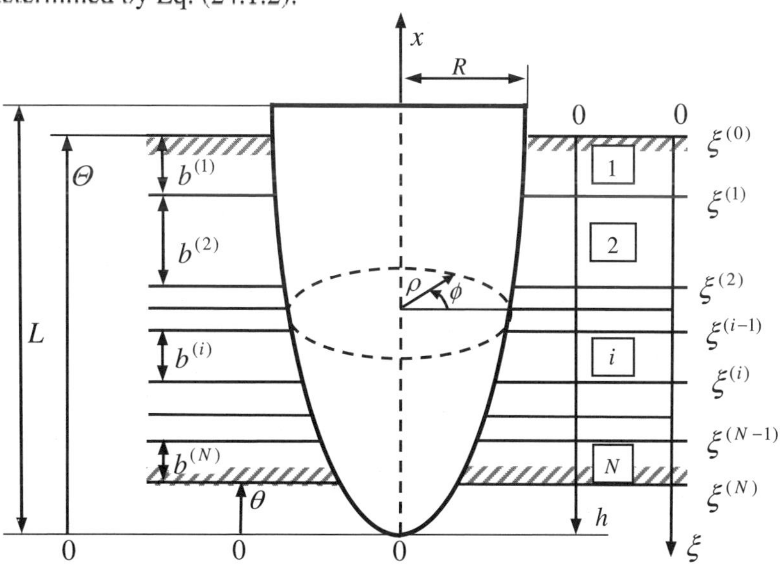

Fig. 25.1.1. Notations.

We assume that the impactor-shield interaction at a given location on the surface of the impactor that is in contact with the i-th plate can be described by the following two-term model:

$$d\vec{F} = [a_2^{(i)}(-\vec{v}^0 \cdot \vec{n}^0)^2 v^2 + a_0^{(i)}]\vec{n}^0 dS\,, \tag{25.1.1}$$

where $d\vec{F}$ is a force acting at the surface element dS of the impactor, $\vec{n}^0$ is the inner normal unit vector at a given location on the impactor surface, $\vec{v}^0$ the unit vector of the impactor velocity, parameters $a_0^{(i)}$ and $a_2^{(i)}$ depend on the properties of the material of the shield, and hereafter the superscript in the parentheses indicates the number of the layer.

In order to adapt Eq. (25.1.1) for a layered shield let us define a step-functions ($\nu = 0,2$):

$$a_\nu(\xi) = \begin{cases} a_\nu^{(1)} & if\ \xi^{(0)} \le \xi < \xi^{(1)} \\ \dots & \\ a_\nu^{(i)} & if\ \xi^{(i-1)} \le \xi < \xi^{(i)} \\ \dots & \\ a_\nu^{(N)} & if\ \xi^{(N-1)} \le \xi \le \xi^{(N)} \end{cases} \tag{25.1.2}$$

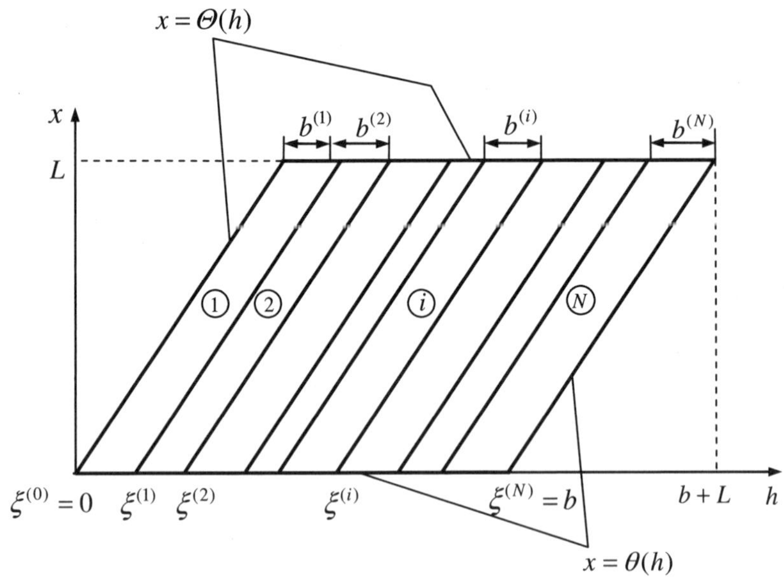

Fig. 25.1.2. Model of layered shield.

Hence Eq. (25.1.1) can be rewritten as follows:

$$d\vec{F} = [a_2(\xi)(-\vec{v}^0 \cdot \vec{n}^0)^2 v^2 + a_0(\xi)]\vec{n}^0 dS\,. \tag{25.1.3}$$

Similar to the previous section we can derive the expression for the ballistic limit velocity (BLV):

$$v_{bl}^2 = k\int_0^{\bar{b}+1} Q(\bar{h})d\bar{h} \int_{\bar{\theta}(\bar{h})}^{\bar{\Theta}(\bar{h})} \tilde{a}_0(\bar{h}-\bar{x})\bar{\Phi}\psi_0(\bar{\Phi}')d\bar{x}\,, \tag{25.1.4}$$

where m is mass of the impactor, and

$$Q(\bar{h}) = \exp\left(k\int_0^{\bar{h}} d\bar{H} \int_{\bar{\theta}(\bar{H})}^{\bar{\Theta}(\bar{H})} \tilde{a}_2(\bar{H}-\bar{x})\bar{\Phi}\psi_2(\bar{\Phi}_x)d\bar{x}\right), \tag{25.1.5}$$

$$\psi_\nu(z) = z\left(\frac{z}{\sqrt{z^2+1}}\right)^\nu, \; \nu = 0{,}2\,, \tag{25.1.6}$$

$$k = \frac{4\pi L^3}{m}, \; \bar{x} = \frac{x}{L}, \; \bar{\Phi} = \frac{\Phi}{L}, \; \bar{\Phi}_x = \frac{d\bar{\Phi}}{dx}, \; \bar{h} = \frac{h}{L}, \; \bar{b} = \frac{b}{L}, \tag{25.1.7}$$

$$\tilde{a}_\nu(\bar{\xi}) = a_\nu(L\bar{\xi}), \quad \nu = 0{,}2\,, \tag{25.1.8}$$

$$\bar{\theta}(\bar{h}) = \begin{cases} 0 & if \quad 0 \le \bar{h} \le \bar{b} \\ \bar{h}-\bar{b} & if \quad \bar{b} \le \bar{h} \le \bar{b}+1 \end{cases}, \quad \bar{\Theta}(\bar{h}) = \begin{cases} \bar{h} & if\ 0 \le \bar{h} \le 1 \\ 1 & if\ \bar{h} \ge 1 \end{cases}. \tag{25.1.9}$$

Ben-Dor *et al.*, 1999b, 1999c showed that the maximum BLV of a layered shield against a conical impactor is attained if the plates are arranged in the increasing order of parameter $\chi = a_0/a_2$. Let us now validate this result for non-conical impactors.

25.2 Ogive-Shaped Generatrix

Since the main goal of this section is to investigate the effect of deviation from a conical shape on the ballistic properties of a multi-layer shield,

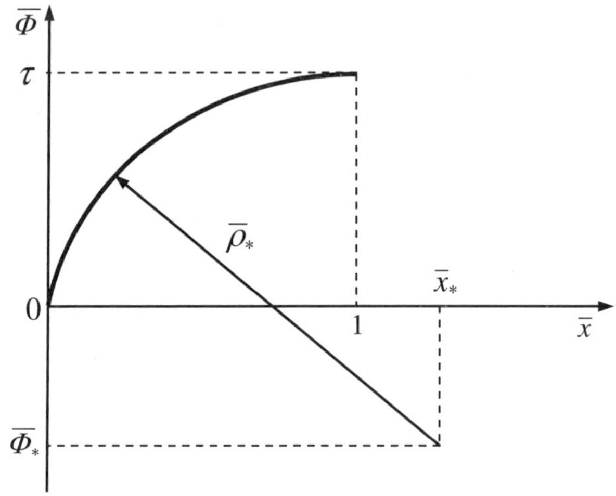

Fig. 25.2.1. Ogive-shaped generatrix.

the most convenient for this purpose are ogive-shaped impactors having a generatrix with a constant curvature. Therefore it is natural to employ curvature as a parameter that characterizes a deviation from a conical shape.

The equation of the circle having radius $\overline{\rho}_*$ (in dimensionless coordinates normalized by the length L) with a center in the point $(\overline{x}_*, \overline{\Phi}_*)$ reads (see Fig. 25.2.1):

$$(\overline{x} - \overline{x}_*)^2 + (\overline{\Phi} - \overline{\Phi}_*)^2 = \overline{\rho}_*^2, \tag{25.2.1}$$

where the arc of the generatrix must pass through the points $(0,0)$ and $(1, \tau)$, and the following constraints must be satisfied:

$$0 \le \overline{x} \le 1, 0 \le \overline{\Phi} \le \tau, \overline{\Phi}_* \le 0, \overline{x}_* \ge 1, \tau = R/L. \tag{25.2.2}$$

After algebraic manipulations we arrive at the following equation for the generatrix:

$$\overline{\Phi} = \frac{\tau}{2} - \frac{2\beta(\overline{x} - 0.5)^2 - 2\tau(\overline{x} - 0.5)\eta - 0.5\beta(\tau^2 + 1)}{\eta + \sqrt{\eta^2 + \beta^2(\tau^2 + 1) - 4\beta^2(\overline{x} - 0.5)^2 + 4\beta\tau\eta(\overline{x} - 0.5)}}, \tag{25.2.3}$$

where

$$\eta = \sqrt{\frac{4}{\tau^2+1} - \beta^2}, \beta = \frac{1}{\overline{\rho}_*}, 0 \le \beta \le \frac{2\min(1,\tau)}{\tau^2+1}. \qquad (25.2.4)$$

Eq. (25.2.3) describes not only a circular arc but also a straight line, $\overline{\Phi} = \tau\overline{x}$, for a conical impactor. This formula for generatrix allows us to avoid computational problems arising for small β.

25.3 Piecewise-Linear Approximation of Generatrix

Eq. (25.1.4) is quite involved and, consequently, it is not convenient for calculations and can introduce uncontrollable numerical error. All these problems can be avoided by using piecewise-linear approximation of the generatrix of a striker.

The equation of a piecewise linear approximation of impactor generatrix can be written as follows (see Fig. 25.3.1):

$$\overline{\Phi}(\overline{x}) = \begin{cases} \alpha_1\overline{x} + \beta_1 & if \quad 0 = \overline{x}_0 \le \overline{x} \le \overline{x}_1 \\ & \quad \dots \\ \alpha_j\overline{x} + \beta_j & if \quad \overline{x}_{j-1} \le \overline{x} \le \overline{x}_j \\ & \quad \dots \\ \alpha_M\overline{x} + \beta_M & if \quad \overline{x}_{M-1} \le \overline{x} \le \overline{x}_M = 1 \end{cases} \qquad (25.3.1)$$

where

$$\alpha_j = \frac{\overline{\Phi}_j - \overline{\Phi}_{j-1}}{\overline{x}_j - \overline{x}_{j-1}}, \beta_j = \frac{\overline{x}_j\overline{\Phi}_{j-1} - \overline{x}_{j-1}\overline{\Phi}_j}{\overline{x}_j - \overline{x}_{j-1}}, \quad j = 1,2,\dots,M. \qquad (25.3.2)$$

The domain determined by Eq. (25.1.9) can be represented as a union of $N \times M$ sub-domains $S_j^{(i)}$ (see Fig. 25.3.2). The parallelogram $A_j^{(i)}B_j^{(i)}C_j^{(i)}E_j^{(i)}$ with the vertices in the points $A_j^{(i)}(\overline{\xi}^{(i-1)} + \overline{x}_{j-1}, \overline{x}_{j-1})$, $B_j^{(i)}(\overline{\xi}^{(i-1)} + \overline{x}_j, \overline{x}_j)$, $C_j^{(i)}(\overline{\xi}^{(i)} + \overline{x}_j, \overline{x}_j)$ and $E_j^{(i)}(\overline{\xi}^{(i)} + \overline{x}_{j-1}, \overline{x}_{j-1})$, encloses the sub-domain $S_j^{(i)}$. The sub-domains $S_j^{(i)}$ are determined by the following formula:

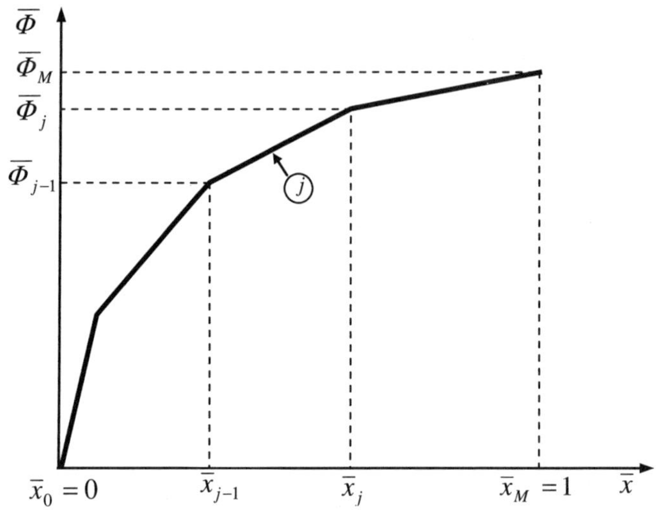

Fig. 25.3.1. Piecewise linear generatrix.

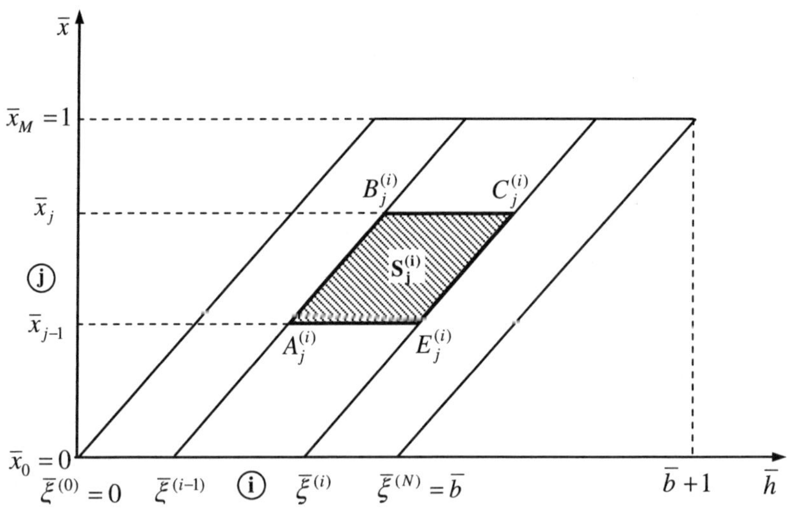

Fig. 25.3.2. Domain of impactor/shield interaction for layered shield and impactor with linear generatrix.

$$\begin{cases} \overline{\xi}^{(i-1)} + \overline{x}_{j-1} & \leq \ \overline{h} \leq \ \overline{\xi}^{(i)} + \overline{x}_j \\ \overline{\theta}_j^{(i)}(h) & \leq \ \overline{x} \leq \ \overline{\Theta}_j^{(i)}(h) \end{cases}, \tag{25.3.3}$$

where

$$\overline{\theta}_j^{(i)}(h) = \begin{cases} 0 & if \ \ \overline{h} \leq \overline{\xi}^{(i-1)} + \overline{x}_{j-1} \\ \overline{x}_{j-1} & if \ \ \overline{\xi}^{(i-1)} + \overline{x}_{j-1} \leq \overline{h} \leq \overline{\xi}^{(i)} + \overline{x}_{j-1} \\ \overline{x} - \xi^{(i)} & if \ \ \overline{\xi}^{(i)} + \overline{x}_{j-1} \leq \overline{h} \leq \overline{\xi}^{(i)} + \overline{x}_j \\ 0 & if \ \ \overline{h} \geq \overline{\xi}^{(i)} + \overline{x}_j \end{cases}, \tag{25.3.4}$$

$$\overline{\Theta}_j^{(i)}(h) = \begin{cases} 0 & if \ \ \overline{h} \leq \overline{\xi}^{(i-1)} + \overline{x}_{j-1} \\ \overline{x}_{j-1} - \overline{\xi}^{(i-1)} & if \ \ \overline{\xi}^{(i-1)} + \overline{x}_{j-1} \leq \overline{h} \leq \overline{\xi}^{(i-1)} + \overline{x}_j \\ \overline{x}_j & if \ \ \overline{\xi}^{(i-1)} + \overline{x}_j \leq \overline{h} \leq \overline{\xi}^{(i)} + \overline{x}_j \\ 0 & if \ \ \overline{h} \geq \overline{\xi}^{(i)} + \overline{x}_j \end{cases}. \tag{25.3.5}$$

Hence the integral in Eq. (25.1.5) can be represented as sum of integrals over sub-domains $S_j^{(i)}$:

$$\begin{aligned} \frac{1}{k}\ln[Q(\overline{h})] &= \sum_{\substack{1 \leq i \leq N \\ 1 \leq j \leq M}} \int_0^{\overline{h}} d\overline{H} \int_{\overline{\theta}_j^{(i)}(\overline{H})}^{\overline{\Theta}_j^{(i)}(\overline{H})} \tilde{a}_2(\overline{\Pi} - \overline{x})\overline{\Phi}\psi_2(\overline{\Phi}')dx \\ &= \sum_{\substack{1 \leq i \leq N \\ 1 \leq j \leq M}} \tilde{a}_2^{(i)}\psi_2(\alpha_j)\int_0^{\overline{h}} d\overline{H} \int_{\overline{\theta}_j^{(i)}(\overline{H})}^{\overline{\Theta}_j^{(i)}(\overline{H})} (\alpha_j \overline{x} + \beta_j)d\overline{x} \\ &= \sum_{\substack{1 \leq i \leq N \\ 1 \leq j \leq M}} \tilde{a}_2^{(i)}\psi_2(\alpha_j)[\Omega_j^{(i)}(\overline{h}) - \omega_j^{(i)}(\overline{h})], \end{aligned} \tag{25.3.6}$$

where

$$\begin{aligned} \omega_j^{(i)}(\overline{h}) &= \int_{\overline{\xi}^{(i-1)} + \overline{x}_{j-1}}^{\overline{h}} \{0.5\alpha_j[\overline{\theta}_j^{(i)}(\overline{H})]^2 + \beta_j \overline{\theta}_j^{(i)}(\overline{H})\}d\overline{H}, \\ \Omega_j^{(i)}(\overline{h}) &= \int_{\overline{\xi}^{(i-1)} + \overline{x}_{j-1}}^{\overline{h}} \{0.5\alpha_j[\overline{\Theta}_j^{(i)}(\overline{H})]^2 + \beta_j \overline{\Theta}_j^{(i)}(\overline{H})\}d\overline{H}. \end{aligned} \tag{25.3.7}$$

The integrals in Eq. (25.3.7) can be calculated taking into account the definition of functions $\overline{\theta}_j^{(i)}(h)$ and $\overline{\Theta}_j^{(i)}(h)$:

$$\omega_j^{(i)}(\overline{h}) = \begin{cases} 0 & if \quad \overline{h} \le \overline{\xi}^{(i-1)} + \overline{x}_{j-1} \\ \lambda_j^{(i)} & if \quad \overline{\xi}^{(i-1)} + \overline{x}_{j-1} \le \overline{h} \le \overline{\xi}^{(i)} + \overline{x}_{j-1} \\ \tilde{\lambda}_j^{(i)} & if \quad \overline{\xi}^{(i)} + \overline{x}_{j-1} \le \overline{h} \le \overline{\xi}^{(i)} + \overline{x}_j \\ \omega_j^{(i)}(\overline{\xi}^{(i)} + \overline{x}_j) & if \quad \overline{h} > \overline{\xi}^{(i)} + \overline{x}_j \end{cases} \tag{25.3.8}$$

$$\Omega_j^{(i)}(\overline{h}) = \begin{cases} 0 & if \quad \overline{h} \le \overline{\xi}^{(i-1)} + \overline{x}_{j-1} \\ \Lambda_j^{(i)} & if \quad \overline{\xi}^{(i-1)} + \overline{x}_{j-1} \le \overline{h} \le \overline{\xi}^{(i-1)} + \overline{x}_j \\ \Lambda_j^{(i)} & if \quad \overline{\xi}^{(i-1)} + \overline{x}_j \le \overline{h} \le \overline{\xi}^{(i)} + \overline{x}_j \\ \Omega_j^{(i)}(\overline{\xi}^{(i)} + \overline{x}_j) & if \quad \overline{h} > \overline{\xi}^{(i)} + \overline{x}_j \end{cases} \tag{25.3.9}$$

where

$$\begin{aligned} \lambda_j^{(i)} &= (0.5\alpha_j \overline{x}_{j-1}^2 + \beta_j \overline{x}_{j-1})(\overline{h} - \overline{\xi}^{(i-1)} - \overline{x}_{j-1}) \\ \tilde{\lambda}_j^{(i)} &= (\alpha_j/6)[(\overline{h} - \overline{\xi}^{(i)})^3 - \overline{x}_{j-1}^3] + 0.5\beta_j[(\overline{h} - \overline{\xi}^{(i)})^2 - \overline{x}_{j-1}^2] \\ \Lambda_j^{(i)} &= (\alpha_j/6)[(\overline{h} - \overline{\xi}^{(i-1)})^3 - \overline{x}_{j-1}^3] + 0.5\beta_j[(\overline{h} - \overline{\xi}^{(i-1)})^2 - \overline{x}_{j-1}^2] \\ \tilde{\Lambda}_j^{(i)} &= (\alpha_j/6)(\overline{x}_j^3 - \overline{x}_{j-1}^3) + 0.5\beta_j(\overline{x}_j^2 - \overline{x}_{j-1}^2) \\ &+ \overline{x}_j(0.5\alpha_j \overline{x}_j + \beta_j)(\overline{h} - \overline{\xi}^{(i-1)}\overline{x}_j) \end{aligned} \tag{25.3.10}$$

Expression for the BLV, Eq. (25.1.4), can be rewritten similar to Eq. (25.3.6) as follows:

$$\begin{aligned} \frac{v_{bl}^2}{k} &= \sum_{\substack{1 \le i \le N \\ 1 \le j \le M}} \int\limits_{\overline{\xi}^{(i-1)} + \overline{x}_{j-1}}^{\overline{\xi}^{(i)} + \overline{x}_j} Q(\overline{H}) d\overline{H} \int\limits_{\overline{\theta}_j^{(i)}(\overline{H})}^{\overline{\Theta}_j^{(i)}(\overline{H})} \tilde{a}_0(\overline{H} - \overline{x}) \overline{\Phi} \psi_0(\overline{\Phi}') d\overline{x} \\ &= \sum_{\substack{1 \le i \le N \\ 1 \le j \le M}} \tilde{a}_0^{(i)} \psi_0(\alpha_j) \int\limits_{\overline{\xi}^{(i-1)} + \overline{x}_{j-1}}^{\overline{\xi}^{(i)} + \overline{x}_j} Q(\overline{h})[\Omega_j^{(i)}(\overline{h}) - \omega_j^{(i)}(\overline{h})] d\overline{h} \end{aligned} \tag{25.3.11}$$

Hence determining the BLV is reduced to calculating one-dimensional integrals.

25.4 Result of Numerical Calculations and Discussion

In numerical calculations we considered a two-layer shield where the first plate and the second plate are manufactured from the soft steel and aluminum, respectively. We used the model given by Eq. (25.1.1) with $a_0^{(1)} = 1850\,MPa$, $a_2^{(1)} = 7830\,kg/m^3$, $a_0^{(2)} = 350\,MPa$ and $a_2^{(2)} = 2765\,kg/m^3$, where the values of the dynamical hardness $a_0^{(i)}$ are adopted from Vitman and Ioffe, 1948. All calculations were performed for the BLVs less than $1000\,m/s$. The latter constraint approximately determines the range of validity of this model.

The following notations are used for describing a structure of a shield. The superscript $[1-2]$ means that the order of the plates in the shield is such that the plate manufactured from material #1 (soft steel) is perforated before the plate manufactured from material #2 (aluminum), the superscript $[2-1]$ indicates the reverse order of the plates. Since the parameter $\chi = a_0/a_2$ decreases with the increase of the ordinal number of the material ($\chi^{(2)} = 0.127 \cdot 10^6\,m^2/s^2$, $\chi^{(1)} = 0.236 \cdot 10^6\,m^2/s^2$), $v_{bl}^{[2-1]} \ge v_{bl}^{[1-2]}$ for conical-nosed impactors (Ben-Dor *et al.*, 1999b, 1999c).

The goal of our calculations is to estimate the effect of the parameter β, that characterizes the effect of deviation from the conical shape on the performance parameter, $\delta = v_{bl}^{[2-1]}/v_{bl}^{[1-2]}$, that quantifies the effect of changing the order of plates on the BLV. Typical results of these calculations are shown in Figs. 25.4.1-25.4.2. It must be noted that $\beta = 0$ for conical impactor.

In all these figures, we showed the dependence δ vs. β, and different curves correspond to different ratios of the thicknesses of the plates in a shield. The curves in Fig. 25.4.1 are plotted for $\tau = 0.5$, while in Fig. 25.4.2 for $\tau = 1$. Clearly, in the second case and for relatively large values of β when the shape of the bluntness is close to spherical, a penetrator is not a sharp-nose body, and the plots have only a formal meaning.

Fig. 25.4.1a corresponds to the dimensionless total thickness of the shield $\bar{b} = 8.0$ and $k = 0.0004\,m^3/kg$. The curves of the dependencies δ vs. β are convex downward, i.e., at the beginning the increase of the

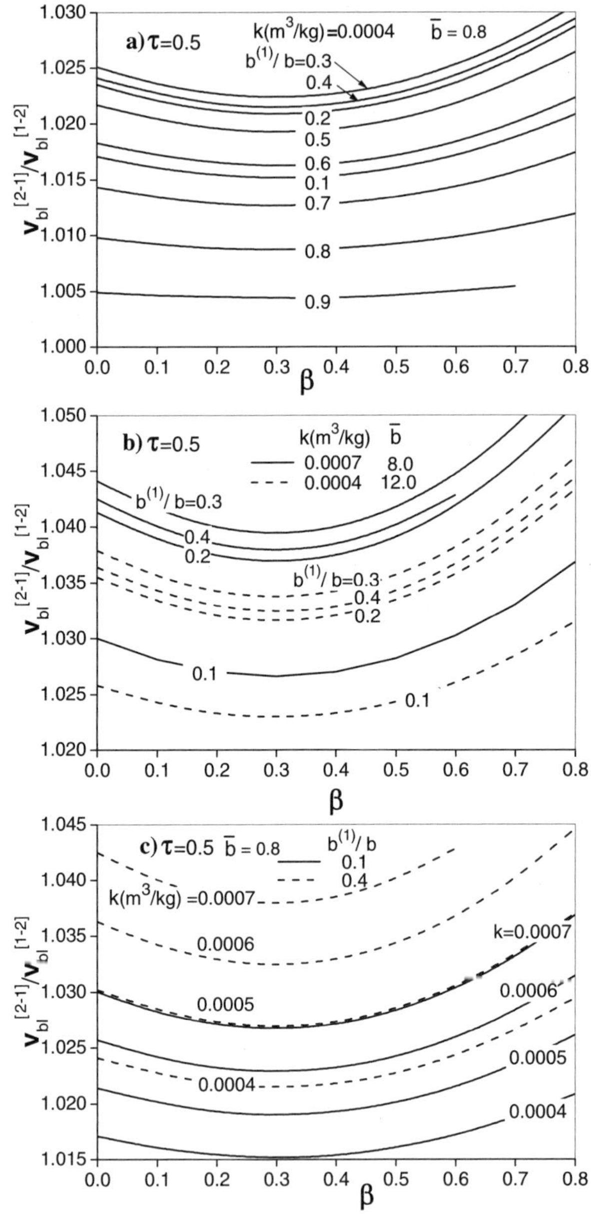

Fig. 25.4.1. Effect of curvature of impactor generatrix on change of ballistic performance due to rearranging the plates in the shield; $\tau = 0.5$.

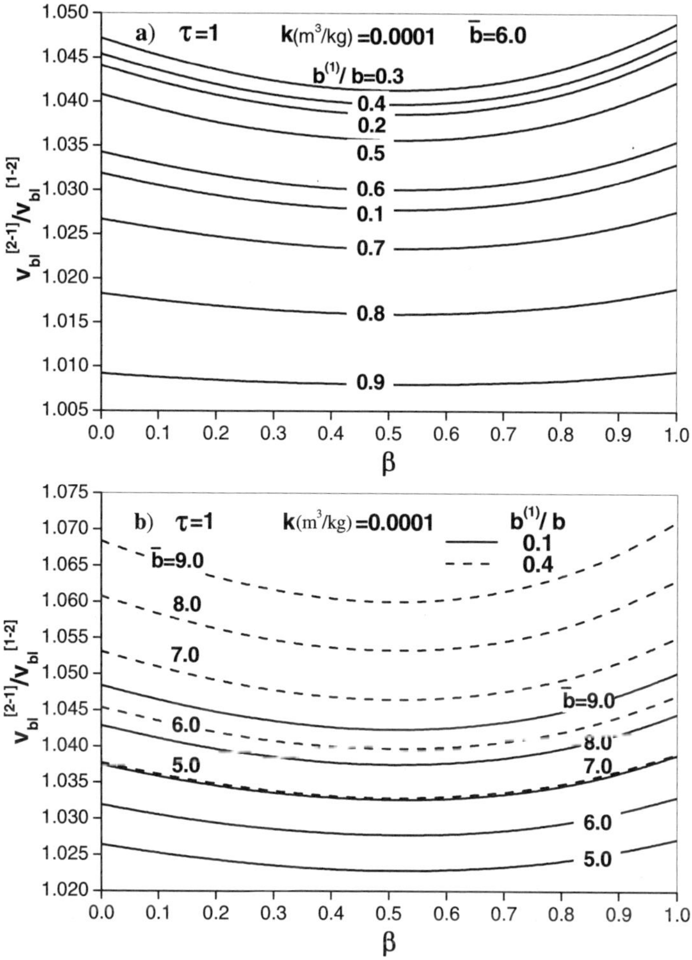

Fig. 25.4.2. Effect of curvature of impactor generatrix on change of ballistic performance due to rearranging the plates in the shield; $\tau = 1$.

curvature of the generatrix of a striker causes reduction of the performance parameter δ, while further increase of β is accompanied by the increase of the parameter δ. Notably, for relatively large β, the magnitude of the parameter δ may become larger than that for a cone-shaped impactor. When the shape of the impactor is specified (β is given) the effect of the change of the order of the plates (magnitude of the parameter δ) depends essentially upon the ratio of the thicknesses of the plates. In the considered case for all ogive-shaped impactors this effect is maximum when $b^{(1)}/b \approx 0.3$.

In Fig. 25.4.1b, two sets of the plots are shown, where the first set is obtained for $k = 0.0007\, m^3/kg$ and $\bar{b} = 8.0$, and the second one is obtained for $k = 0.0004\, m^3/kg$ and $\bar{b} = 12.0$. The first set differs from that in Fig. 25.4.1a by the increased value of k, while in the second set of curves we increased $\bar{b}$. Fig. 25.4.2a for $\tau = 1$ demonstrates the same dependencies as Fig. 25.4.1a. In Fig. 25.4.2b we showed the effect of the total thickness of the shield for two values of the ratio $b^{(1)}/b$ (0.1 and 0.4) on the dependence $\delta = \delta(\beta)$. Inspection of these plots shows that the increase of the total width $\bar{b}$ causes the increase of the ballistic performance parameter δ.

Our calculations demonstrate that replacement of the conical head of impactor by a convex ogive-shaped head can be accompanied either by the increase or decrease of the efficiency of changing the order of the plates in a shield. However we did not encounter situations when the optimal order of the plates in the shield is different for these two different impactor shapes.

Similarly to Section 24.3 we can arrive at the following assertion: the model employed in this study predicts that rearranging the plates in the shield causes the change of the value of the dimensionless absorbed energy that does not exceed twice the ratio of the corresponding BLVs.

On the basis of approximate model it can be concluded that the criterion for the optimal arrangement of the plates in a shield, determined previously for conical impactors by Ben-Dor *et al.*, 1999b, 1999c, is valid also for non-conical impactors. Clearly, these conclusions must be validated by further experimental studies and "exact" numerical simulations.

Chapter 26

Effect of Layering. Theoretical Analysis

Nomenclature

b	shield thickness;
$b_{\min}$	lower bound for argument of function $G(b)$;
b_{sum}	total thickness of layers in shield;
$b^{(j)}$	thickness of jth plate;
BLV	ballistic limit velocity;
g	function, Eq. (26.4.1);
G	function, Eq. (26.1.2);
N	number of plates in layered shield;
v_{bl}	BLV of monolithic shield;
V_{bl}	BLV of layered shield;
$V_{bl}^{(j)}$	BLV of jth plate;
V_{imp}	impact velocity of layered shield;
$V_{imp}^{(j)}$	impact velocity of jth plate;
V_{res}	residual velocity of layered shield;
$V_{res}^{(j)}$	residual velocity of jth plate;
δ	parameter, Eq. (26.2.1);
η_{res}, η_{bl}	parameters, Eq. (26.6.1);
μ_0, μ_1	parameters, Eq. (26.2.4).

26.1 Mathematical Models of the Layered Shield

Consider high speed normal penetration of a rigid sharp-nosed striker into a shield consisting of N metal plates in contact. The plates are numbered in the direction of the impactor motion, and the thickness of the i-th plate is $b^{(i)}$. The common assumption, that the layers do not interact and are perforated independently, implies that the residual velocity of the jth plate, $V_{res}^{(j)}$, is equal to the impact velocity of the $(j+1)$th plate, $V_{imp}^{(j+1)}$, for $j=1\div N-1$. In addition, the impact velocity of the shield, V_{imp}, is equal to $V_{imp}^{(1)}$, and the residual velocity of the shield, V_{res}, is equal to $V_{res}^{(N)}=V_{imp}^{(N+1)}$ (the notation $V_{imp}^{(N+1)}$ is introduced here for convenience). Hereafter, the superscript in the parentheses denotes the ordinal number of the layer.

We use the following relationship between the impact velocity, $V_{imp}^{(j)}$, the residual velocity, $V_{res}^{(j)}$, and the ballistic limit velocity (BLV) of the plates (Recht and Ipson, 1963):

$$V_{imp}^{(j)2}-V_{res}^{(j)2}=V_{bl}^{(j)2}. \tag{26.1.1}$$

Since the plates are manufactured from the same material, the BLV of a plate is a function of the thickness:

$$V_{bl}^{(j)2}=G(b^{(j)}),\quad G(0)=0,\quad j=1,2,...,N, \tag{26.1.2}$$

where G is an increasing, positive function.

Summing Eqs. (26.1.1) from $j=1$ to $j=N$ yields:

$$V_{imp}^{2}-V_{res}^{2}=\sum_{j=1}^{N}G(b^{(j)}), \tag{26.1.3}$$

and, consequently, the expression for the BLV of the layered shield V_{bl} reads:

$$V_{bl}^{2}=\sum_{j=1}^{N}G(b^{(j)}). \tag{26.1.4}$$

Clearly, the BLV of the layered shield does not depend on the arrangement of layers.

In the case of monolithic shield having a thickness

$$b_{sum} = b^{(1)} + b^{(2)} + \ldots + b^{(N)}, \tag{26.1.5}$$

the relationship between the BLV, v_{bl}, and the thickness of the shield can be obtained from Eq. (26.1.4) substituting $V_{bl} = v_{bl}$, $N = 1$ and $b^{(1)} = b_{sum}$:

$$v_{bl}^2 = G(b_{sum}) = G\left(\sum_{j=1}^{N} b^{(j)}\right). \tag{26.1.6}$$

26.2 Comparison of Monolithic and Layered Shields

In this section we compare the values of the BLVs of the layered shield and of the monolithic shield having the same thickness. Clearly, one can consider the difference of squared ballistic velocities:

$$\delta = v_{bl}^2 - V_{bl}^2 = G\left(\sum_{j=1}^{N} b^{(j)}\right) - \sum_{j=1}^{N} G(b^{(j)}), \tag{26.2.1}$$

rather than the difference of the BLVs, $v_{bl} - V_{bl}$.

Hereafter in this section we assume that $G(b)$ is a convex downwards, twice differentiable function which is defined in the interval $b \geq b_{\min} \geq 0$ and satisfies the inequality:

$$b_{\min} G'(b_{\min}) - G(b_{\min}) \geq 0, \tag{26.2.2}$$

where $b_{\min}$ is given. Geometrical interpretation of Eq. (26.2.2) may be found in Section A1 (Appendix A). Property 3 of convex downwards in Appendix A implies that $\delta > 0$ for

$$b^{(j)} \geq b_{\min},\ b^{(j)} > 0,\ j = 1,2,\ldots,N. \tag{26.2.3}$$

Introducing the parameter $b_{\min}$ allows us to separate the range of the considered plate thicknesses. If $b_{\min}=0$, then $G(0)=0$ as can be deduced by simple physically reasoning.

If $G(b)$ is a linear function,

$$G(b)=\mu_1 b+\mu_0\,, \tag{26.2.4}$$

then

$$\delta=\left(\mu_1\sum_{j=1}^{N} b^{(j)}+\mu_0\right)-\sum_{j=1}^{N}(\mu_1 b^{(j)}+\mu_0)=\mu_0(1-N)\,. \tag{26.2.5}$$

Therefore $\delta>0$ if $\mu_0<0$ and $\delta=0$ if $\mu_0=0$.

Consequently, monolithic shield is superior over any layered shield having the same total thickness if $G(b)$ is convex downward function and conditions given by Eqs. (26.2.2) and (26.2.3) are satisfied or if $G(b)$ is a linear function with a negative constant term, μ_0. If $G(b)$ is directly proportional to *b*, then layering does not affect ballistic properties of the shield.

26.3 Worst Layering for a Given Number of Layers

According to the above analysis, ballistic properties of a shield can be impaired by layering. Let us determine the maximum reduction of the ballistic performance of the shield. We will show that the maximum deterioration of ballistic characteristics of a layered shield occurs when the layers have the same thicknesses. It is assumed that the number of the plates in the shield N and the total thickness of the shield b_{sum} are given. The problem is to minimize V_{bl} in Eq. (26.1.4) under constraints given by Eqs. (26.1.5) and (26.2.3). We also assume that $b_{sum}/N\ge b_{\min}$, i.e., the thicknesses of the layers, $b^{(j)}=b_{sum}/N$, $j=1,2,...,N$ are permissible.

This optimization problem for a convex downward function $G(b)$ is reduced to the problem that has been investigated in Section 20.3. The result is that the most unfavorable layering occurs when the monolithic shield is replaced by identical plates having the same total thickness. If $G(b)$ is a linear function given by Eq. (26.2.4) then

$$V_{bl}^2 = \sum_{j=1}^{N} (\mu_1 b^{(j)} + \mu_0) = \mu_1 b_{sum} + \mu_0 N \,, \qquad (26.3.1)$$

i.e., sizes of the layers do not affect ballistic properties of the shield.

26.4 Effect of Number of Layers

Let us determine the dependence of the BLV on the number of plates in the shield assuming that all plates have the same thicknesses. The latter assumption implies the following expression for the BLV of the layered shield:

$$v_{bl}^2 = g(N) = N\, G(b_{sum}/N) \,. \qquad (26.4.1)$$

Assuming that N varies continuously we can calculate the derivative:

$$g'(N) = G(b) - bG'(b), \quad b = b_{sum}/N \,. \qquad (26.4.2)$$

Let us assume that function $G(b)$ satisfies the conditions from Section 26.2 and apply Eq. (A10) from Appendix A. Hence we conclude that Eq. (26.2.2) implies that $g'(N) < 0$ if $b_{sum}/N > b_{\min}$. Therefore $g(N)$ decreases when

$$1 < N \le b_{sum}/b_{\min} \,. \qquad (26.4.3)$$

The conclusion is that the BLV decreases with the increase of the number of layers if all plates have the same thicknesses.

In the case of a linear function $G(b)$, Eq. (26.3.1) implies that the BLV decreases with the increase of the number of layers if $\mu_0 < 0$ ($N < -\mu_1 b_{sum}/\mu_0$) and does not depend on the number of layers if $\mu_0 = 0$.

26.5 Validation of Assumptions

There are relatively few data available for the analysis of the behavior of function $G(b)$ in the case of sharp-nosed impactors.

The model suggested by Ohte *et al.*, 1982 (see Section 5.13) for conical-nosed impactors implies that $G(b) \sim b^{\varepsilon}$ with $\varepsilon \geq 1.5$, i.e., $G(b)$ is a convex downward function for $b \geq 0$ and $G(0) = 0$.

Gupta *et al.*, 2001 describes the following dependence between the BLV of the ogive-nosed impactor and the thickness of the aluminum

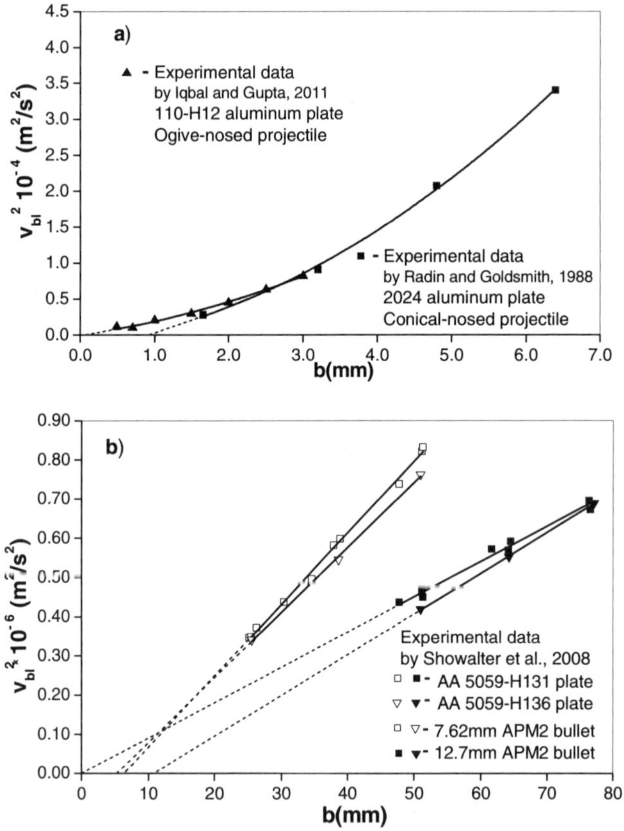

Fig. 26.5.1. Squared BLV vs. thickness of shield. Dotted lines are either tangents to the curves in the point $b = b_{\min}$ or continuations of straight lines to the left of this point.

plate, $v_{bl} = kb^{0.89}$, where k is a constant.. The function $G(b) = kb^{1.78}$ is a convex downward function for $b \geq 0$ and $G(0) = 0$.

Functions $G(b)$ in Fig. 26.5.1a, which we plotted using experimental results by Iqbal and Gupta, 2011 and Radin and Goldsmith, 1988, are convex and satisfy the condition given by Eq. (26.2.2). Processing of the experimental results by Showalter *et al.*, 2008 (Fig. 26.5.1b) yields a linear function, $G(b)$, with $\mu_0 < 0$ in three out of four cases and $\mu_0 = 0$ in one case.

26.6 Comparison with Experiments and Numerical Calculations and Discussion

Results of experimental studies and numerical calculations used in the analysis of our theoretical predictions are summarized in Tables 26.6.1–26.6.8. All experiments/calculations in each table are performed for the impactor having the same shape against the shield manufactured from the same material.

Table 26.6.1. Numerical results of Iqbal *et al.*, 2010 on residual velocities (conical-nosed projectile, Weldox 460E steel shields).

No.	Shield (sizes in mm)	Impact velocity (m/s)					
		600	406	356	318	300	281
1	12.0	555	333	269	217	189	154
2	2 × 6.0	554	337	273	221	195	163

Table 26.6.2. Numerical results of Iqbal *et al.*, 2010 on residual velocities (ogive-nosed projectile, 1100 – H12 aluminum shields).

No.	Shield (sizes in mm)	Impact velocity (m/s)							
		150	113	97	83	82	73	66	57
1	1.0	141	100	81	64	62	49	32	15
2	2 × 0.5	145	101	84	67	65	54	40	20

Table 26.6.3. Experimental results of Almohandes *et al.*, 1996 on residual velocities (7.62mm bullet, mild steel shields).

No.	*Shield (sizes in mm)*	*Impact velocity (m/s)*				
		706	755	775	805	826
1	8.0	446	489	528	611	647
2	2.0+6.0	477	514	549	609	658
3	6.0+2.0	471	522	-	616	654
4	2 × 4.0	486	533	567	643	667

Table 26.6.4. Experimental results of Radin and Goldsmith, 1988 on BLV (conical-nosed projectile, 2024- aluminum shields).

No.	*Shield (sizes in mm)*	*BLV (m/s)*	*No*	*Shield (sizes in mm)*	*BLV (m/s)*	*No*	*Shield (sizes in mm)*	*BLV (m/s)*
1	3.2	95	3	4.8	144	6	6.4	184
2	2 × 1.6	93	4	1.6+3.2	130	7	2×3.2	160
			5	3×1.6	124	8	4×1.6	158

Table 26.6.5. Experimental results of Iqbal *et al.*, 2010 on BLV (ogive-nosed projectile, 1100–H12 aluminum shields).

No.	*Shield (sizes in mm)*	*BLV (m/s)*	*No*	*Shield (sizes in mm)*	*BLV (m/s)*	*No*	*Shield (sizes in mm)*	*BLV (m/s)*	*No*	*Shield (sizes in mm)*	*BLV (m/s)*	*No*	*Shield (sizes in mm)*	*BLV (m/s)*
1	1.0	45.3	3	1.5	54.3	5	2.0	67.1	8	2.5	79.3	10	3.0	90.4
2	2×0.5	39.2	4	3×0.5	51.0	6	2×1.0	62.9	9	5×0.5	66.0	11	2×1.5	78.6
						7	4×0.5	61.7				12	3×1.0	74.2

The structure of the shield is described as follows. In the case of a monolithic shield, the thickness of the shield is indicated. The notation 2×6.0 denotes the shield which consists of two plates having the same thicknesses, 6.0 mm. The notation 6.0+2.0 denotes a two-layer shield having the front plate with the thickness 6.0 mm and the rear plate with the thickness 2.0 mm. In each column, the data concerning monolithic shield are given in the first row, while data on the layered shields are shown below. Tables 26.6.1-26.6.3 present data on pairs: impact velocity – residual velocity. Of two shields, the shield with a smaller residual velocity and the same impact velocity is assumed as having superior protective characteristics.

Table 26.6.6. Numerical results of Teng *et al.*, 2007 on the BLV (conical-nosed projectile, Weldox 460E steel shields).

Conical-nosed projectile			*Light conical-nosed projectile*		
No.	Shield (sizes ir mm)	BLV (m/s)	No	Shield (sizes ir mm)	BLV (m/s)
1	12.0	305.9	3	12.0	525.9
2	2×6.0	282.0	4	2×6.0	524.5

Table 26.6.7. Experimental results of Dey *et al.*, 2007 on the BLV (ogive-nosed projectile, Weldox 460E steel shields).

No.	*Shield (sizes in mm)*	*BLV (m/s)*
1	12.0	318.1
2	2 × 6.0	288.3

Table 26.6.8 . Experimental results of Woodward and Cimpoeru, 1998 (conical-nosed projectile, 2024-T351 aluminum alloy shields).

No.	*Shield (sizes in mm)*	*BLV (m/s)*
1	9.53	462
2	3.18+6.35	421
3	6.35+3.18	474
4	2 × 4.76	484
5	3 ×3.18	473
6	6 × 1.59	454

Table 26.6.9. Validation of theoretical results on residual velocity.

Table	*Compared shields*	v_{imp} (m/s)	$\dfrac{v_{res} - V_{res}}{v_{res}}$
1	1-2	600	+0.002
1	1-2	406	-0.012
1	1-2	356	-0.015
1	1-2	318	-0.018
1	1-2	300	-0.032
1	1-2	281	-0.059
2	1-2	150	-0.028
2	1-2	113	-0.010
2	1-2	97	-0.037
2	1-2	83	-0.047
2	1-2	82	-0.048
2	1-2	73	-0.068
2	1-2	66	-0.250
2	1-2	57	-0.333
3	1-2	706	-0.070
3	1-3	706	-0.056
3	1-4	706	-0.090
3	1-2	755	-0.051
3	1-3	755	-0.067
3	1-4	755	-0.090
3	1-2	775	-0.040
3	1-4	775	-0.074
3	1-2	805	+0.003
3	1-3	805	-0.008
3	1-4	805	-0.052
3	1-2	826	-0.017
3	1-3	826	-0.011
3	1-4	826	-0.031

Tables 26.6.4-26.6.8 present data on the BLVs, and a superior shield has a larger value of the BLV.

Comparison of theoretical results obtained in Section 26.2 with experimental and numerical results is shown in Tables 26.6.9 and 26.6.10. Inequality $\eta_{res} < 0$ or $\eta_{bl} < 0$, where

$$\eta_{res} = \frac{v_{res} - V_{res}}{v_{res}}, \quad \eta_{bl} = \frac{V_{bl} - v_{bl}}{v_{bl}}, \tag{26.6.1}$$

indicates the agreement between theoretical predictions and experimental or numerical results and $\eta_{res} > 0$ or $\eta_{bl} > 0$ denotes a deviation from the theoretical predictions. Inspection of Tables 26.6.9 and 26.6.10 reveals that theoretical predictions are supported by experimental or numerical data in 87% of cases.

Inspection of Table 26.6.3 shows that for all values of impact velocity, the residual velocity for a two-layer shield having the same thicknesses of layers ($2\times 4.0\,mm$) is larger than for any other partitioning of the shield, in compliance with the theoretical predictions in Section 26.3 for the case of convex downward function $G(b)$. However, the BLV in Table 26.6.8 for $2\times 4.76\,mm$ shield is larger than for other two variants of the shield with unequal thicknesses of layers. Effect of thicknesses of layers upon ballistic properties of shields

Table 26.6.10. Validation of theoretical results on the BLV.

Table	*Compared shields*	$\frac{V_{bl} - v_{bl}}{v_{bl}}$
4	1-2	-0.021
4	3-4	-0.097
4	3-5	-0.139
4	6-7	-0.130
4	6-8	-0.165
5	1-2	-0.135
5	3-4	-0.061
5	5-6	-0.063
5	5-7	-0.080
5	8-9	-0.168
5	10-11	-0.131
5	10-12	-0.204
6	1-2	+0.078
6	3-4	+0.003
7	1-2	+0.094
8	1-2	+0.089
8	1-3	-0.026
8	1-4	-0.048
8	1-5	-0.024
8	1-6	+0.017

requires further investigations. Indeed, the maximum adverse effect of layering with equal thicknesses of the layers in our analysis is the consequence of the convexity downward of the function $G(b)$, while the available experimental data afford also a linear function $G(b)$, whereby ballistic efficiency of the shield does not depend upon the thicknesses of layers.

Comparison of different structures of the shield 7-8 in Table 26.6.4, 6-7 and 11-12 in Table 26.6.5 and 4-5-6 in Table 26.6.4 supports theoretical predictions in Section 26.4: increase of the number of layers having the same thickness impairs ballistic properties of the shield.

Therefore, the available experimental and numerical results support our theoretical predictions.

26.7 Concluding Remarks

On the basis of general assumptions that are pertinent to penetration modeling, we investigated the effect of layering of metal shields against sharp-nosed rigid impactors and established/explained a number of general conjectures concerning ballistic properties of shields. Omitting the technical details, the main conclusions of this chapter can be formulated as follows: (i) layering impairs protective properties of a shield against perforation; (ii) the BLV of a multilayer shield does not depend on the order of plates in the shield; (iii) increasing the number of layers having the same thicknesses impairs protective properties of a shield.

Although our theoretical predictions are confirmed by comparing with experimental or numerical results it will be useful to conduct a dedicated experimental study in order to determine the range of validity of our theoretical predictions for various material properties of a shield, shapes of the impactor, thicknesses of the layers and impact velocity.

Chapter 27

Optimization of Multi-Layer Shields

Nomenclature

A	AD of shield;
$\overline{A}$	$= A/(b_* \rho_*)$;
AD	areal density;
b_ν	thickness of ν th layer;
$\overline{b}_\nu$	$= b_\nu / L$;
$b_\nu^{\min}$	lower bound for thickness of ν th plate ;
$b_\nu^{\max}$	upper bound for thickness of ν th plate;
$\overline{b}_\nu^{\min}$	$= b_\nu^{\min} / L$;
$b_\nu^{\max}$	$= b_\nu^{\max} / L$;
BLV	ballistic limit velocity;
$B_{\max}$	upper bound for the total thickness of shield;
$\overline{B}_{\max}$	$= B_{\max} / L$;
c	parameter, Eq. (27.2.20);
k	$= -\eta_1 / \eta_2$;
L	length of impactor nose;
m	mass of impactor;
n	number of materials;
p_ν	characteristics of material of ν th layer, Eq. (27.1.3);
$\overline{p}_\nu$	$= p_\nu / p_*$;
p_*	reference parameter, Eq. (27.1.2);
T_ν	parameter, Eq. (27.1.2);
v_{bl}	BLV of impactor;
v_*	$= 1 m/s$;

$\overline{w}$ $=(v_{bl}/v_*)^2$;
$\overline{W}$ function, Eq. (27.2.3);
x_i variable, Eq. (27.2.5);
x_j^{opt} optimal value of x_j;
α_i parameter, Eq. (27.2.11);
$\beta_i^{\min}$ parameter, Eq. (27.2.11);
$\beta_i^{\max}$ parameter, Eq. (27.2.11);
$\gamma_{\min}$ parameter, Eq. (27.2.11);
γ_{max} parameter, Eq. (27.2.11);
η_i parameter, Eq. (27.2.11);
ϑ half apex angle of conical impactor;
$\lambda_{\max}$ parameter defined in Eq. (27.2.11);
μ parameter of the model, see Eq. (27.1.3);
ξ_ν $=p_\nu/\rho_\nu$;
$\overline{\xi}_\nu$ $=\overline{p}_\nu/\overline{\rho}_\nu$;
Ξ_j notation for j th material;
ρ_ν density of material of ν th layer;
$\overline{\rho}_\nu$ $=\rho_\nu/\rho_*$;
ρ_* reference density, see Eq. (27.1.2);
σ_j parameters in Eq. (27.2.21), $j=1,...,7$.

27.1 Formulation of Problem and Mathematical Model

Consider n materials (metals) $\Xi_1, \Xi_2, ..., \Xi_n$. Hereafter, the subscript indicates the ordinal number of the material. Let us consider different multilayer shields that consist of plates having different thicknesses which are arranged in different order and are manufactured from these materials. Consequently, a shield is determined by the sequences $j_1, j_2, ..., j_n$ and $b_{j_1}, b_{j_2}, ..., b_{j_n}$, where j_i is the number of the material of the i th plate in the direction of penetration and b_{j_i} is the thickness of this plate.

We consider high speed normal penetration of a rigid conical body of revolution into a multilayer shield. The analysis is based on the equation for the BLV derived by Ben-Dor *et al.*, 1999b which is written using the dimensionless variables:

$$\overline{w} = \sum_{i=1}^{n} \overline{\xi}_{j_i} (T_{j_i} - 1) \prod_{\nu=0}^{j_i - 1} T_\nu, \quad T_0 = 1, \tag{27.1.1}$$

where

$$\begin{gathered} T_\nu = \exp(\overline{b}_\nu \overline{\rho}_\nu), \overline{\rho}_\nu = \frac{\rho_\nu}{\rho_*}, \overline{p}_\nu = \frac{p_\nu}{p_*}, \overline{b}_\nu = \frac{b_\nu}{L}, \overline{w} = \left(\frac{v_{bl}}{v_*} \right)^2, \\ \xi_\nu = \frac{p_\nu}{\rho_\nu}, \overline{\xi}_\nu = \frac{\overline{p}_\nu}{\overline{\rho}_\nu}, \rho_* = \frac{m \cos^2 \vartheta}{2\pi\mu L^3 \sin^4 \vartheta}, p_* = \frac{m v_*}{2\pi L^3}, v_* = 1 m/s, \end{gathered} \tag{27.1.2}$$

parameters m, L, ϑ are mass, length of the nose and half apex angle of impactor, correspondingly; ρ_ν is material density; p_ν is a parameter that determines mechanical properties of the material; μ is a constant; v_{bl} is the ballistic limit velocity (BLV) of the shield. Eq. (27.1.1) follows from the assumption that the interaction between the impactor and ν th plate is described by the following two-term model:

$$d\vec{F} = [p_\nu v_n^2 + \mu \rho_\nu] \vec{n}^0 dS, \tag{27.1.3}$$

where $d\vec{F}$ is the force acting on the lateral surface element dS of a projectile that is in contact with the plate, $\vec{n}^0$ is the inner normal unit vector at a given location on the projectile surface, v_n is normal component of the instantaneous velocity of the impactor. This model is determined by Eq. (1.6.14), where $p_\nu = a_2$, $\mu \rho_\nu = a_0$, $v_n = uv$ and superscripts that indicate the ordinal number of the layer are omitted for a_0 and a_2. It is assumed that the layers are perforated subsequently.

Particular models of this kind which can be described by Eq. (27.1.3) are considered in Chapter 6. Note that using the procedure of averaging (see Section 1.6.2.3) one can arrive at the relationship similar to Eq. (27.1.1) for sharp impactors with curvilinear generatrix, i.e., the results of the present study can be applied also to non-conical impactors.

Let us determine the sequence of plates manufactured from different materials and their thicknesses which maximize the BLV of the shield with given areal density (AD) taking into account the constraints on the thicknesses of the plates and on the total thickness of the shield. Mathematically, the problem is to find such sequences $j_1, j_2, ..., j_n$ ($j_r \neq j_s$ for any $r \neq s$) and $b_{j_1}, b_{j_2}, ..., b_{j_n}$, whereby $\overline{w}$ in Eq. (27.1.1) attains the maximum for the impactor with given mass and shape which are determined by parameters m, L, θ.

Maximization is performed under the following constraints:

$$\sum_{i=1}^{n} \overline{b}_{j_i} \overline{\rho}_{j_i} = \overline{A}, \sum_{i=1}^{n} \overline{b}_{j_i} \leq \overline{B}_{\max}, \quad \overline{b}_{j_i}^{\min} \leq \overline{b}_{j_i} \leq \overline{b}_{j_i}^{\max}, \quad i = 1,2,...,n, \quad (27.1.4)$$

where

$$\overline{A} = \frac{A}{b_* \rho_*}, \quad \overline{B}_{\max} = \frac{B_{\max}}{L}, \quad \overline{b}_{\nu}^{\min} = \frac{b_{\nu}^{\min}}{L}, \quad \overline{b}_{\nu}^{\max} = \frac{b_{\nu}^{\max}}{L}, \quad (27.1.5)$$

A is the AD of the shield, $B_{\max}$ is the upper bound on the total thickness of the shield, $b_{\nu}^{\min}$ and $b_{\nu}^{\max}$ are given lower and upper bounds for the thicknesses of the plates manufactured from the ν-th material, correspondingly, and

$$\begin{gathered} \overline{A} > 0, \quad \overline{B}_{\max} > 0, \quad \sum_{\nu=1}^{n} \overline{b}_{\nu}^{\min} < \overline{B}_{\max}, \\ 0 \leq \overline{b}_{\nu}^{\min} < \overline{b}_{\nu}^{\max}, \quad \nu = 1,2,...,n. \end{gathered} \quad (27.1.6)$$

The formulated above problem includes the unknown discrete and continuous parameters. There are no general methods for solving such problems. The attempts to solve these problems numerically are plagued by significant difficulties, and determining an analytical solution is considered as a great success. In the following we demonstrate that in the general case with an arbitrary number of layers, the original problem can be transformed into a problem which can be solved using the methods of geometric programming, and in the cases of three-layer and two-layer shields we even have obtained the exact analytical solutions.

27.2 Analysis of Problem

27.2.1 *General case*

Consider a multi-layer shield and assume that the constraints given by Eq. (27.1.4) are satisfied. If we rearrange the plates in a different order, these constraints are still satisfied for the new shield, i.e., the structure of this shield remains admissible. Ben-Dor *et al.*, 1999b proved that in the case when the thicknesses of the plates are given, the maximum BLV of the shield is attained if the plates are arranged in the order of increasing values of the parameter ξ_{j_i}, i.e., $j_1, j_2, ..., j_n$ must be chosen such that $\xi_{j_1} < \xi_{j_2} < ... < \xi_{j_n}$. Clearly, this is the order of the materials of the layers in the optimal shield. Therefore, the maximization problem is reduced to determining only the thicknesses of the plates. In order to simplify the notations, we assume that the materials $\Xi_1, \Xi_2, ..., \Xi_n$ are originally ordered so that $\xi_1 < \xi_2 < ... < \xi_n$. Then the optimization criterion and the constraints can be rewritten as follows:

$$\overline{w} = \sum_{i=1}^{n} \overline{\xi}_i (T_i - 1) \prod_{\nu=0}^{i-1} T_\nu, \quad T_0 = 1, \tag{27.2.1}$$

$$\sum_{i=1}^{n} \overline{b}_i \overline{\rho}_i = \overline{A}, \quad \sum_{i=1}^{n} \overline{b}_i \le \overline{B}_{\max}, \quad \overline{b}_i^{\min} \le \overline{b}_i \le \overline{b}_i^{\max} \quad i = 1,2,...,n. \tag{27.2.2}$$

After some algebra, the expression for $\overline{w}$ can be rewritten in the following form:

$$\overline{w} = \overline{\xi}_n T_1 T_2 ... T_n - \overline{\xi}_1 - \overline{W}, \quad \overline{W} = \sum_{i=1}^{n-1} (\overline{\xi}_{i+1} - \overline{\xi}_i) \prod_{\nu=1}^{i} T_\nu. \tag{27.2.3}$$

Since the product,

$$T_1 T_2 ... T_n = \exp\left(\sum_{\nu=1}^{n} \overline{b}_\nu \overline{\rho}_\nu \right) = \exp(\overline{A}), \tag{27.2.4}$$

is given, the problem is reduced to the minimization of the function $\overline{W}$.

Let us introduce new variables

$$x_i = \prod_{\nu=1}^{i} T_\nu = \exp\left(\sum_{\nu=1}^{i} \bar{b}_\nu \bar{\rho}_\nu \right), \quad i = 1,2,\ldots,n-1, \tag{27.2.5}$$

and eliminate the variable $\bar{b}_n$ using the constraint on the AD.

Substituting

$$\bar{b}_i = \frac{1}{\bar{\rho}_i} \ln\left(\frac{x_i}{x_{i-1}} \right), i = 1,2,\ldots,n-1, \quad \bar{b}_n = \frac{\bar{A} - \ln x_{n-1}}{\bar{\rho}_n}, \quad x_0 = 1 \tag{27.2.6}$$

into the expression for the criterion $\bar{W}$ in Eqs. (27.2.3) and into the inequalities in Eq. (27.2.2), we obtain the following formulation of the modified optimization problem:

$$\bar{W} = \sum_{i=1}^{n-1} \alpha_i x_i \to \min, \tag{27.2.7}$$

$$\beta_1^{\min} \le x_1 \le \beta_1^{\max}, \quad \gamma_{\min} \le x_{n-1} \le \gamma_{\max}, \tag{27.2.8}$$

$$\beta_i^{\min} \le x_i / x_{i-1} \le \beta_i^{\max}, i = 2,3,\ldots,n-1, \quad n > 2, \tag{27.2.9}$$

$$\prod_{i=1}^{n-1} x_i^{\eta_i} \le \lambda_{\max}, \tag{27.2.10}$$

where

$$\begin{aligned}
&\alpha_i = \bar{\xi}_{i+1} - \bar{\xi}_i > 0, \quad i = 1,2,\ldots n-1, \\
&\beta_i^{\min} = \exp(\bar{\rho}_i \bar{b}_i^{\min}), \ \beta_i^{\max} = \exp(\bar{\rho}_i \bar{b}_i^{\max}), \ i = 1,2,\ldots n-1, \\
&\gamma_{\min} = \exp(\bar{A} - \bar{\rho}_n \bar{b}_n^{\max}), \ \gamma_{\max} = \exp(\bar{A} - \bar{\rho}_n \bar{b}_n^{\min}), \\
&\eta_i = \frac{1}{\bar{\rho}_i} - \frac{1}{\bar{\rho}_{i+1}}, \ i = 1,2,\ldots,n-1, \quad \lambda_{\max} = \exp\left(\bar{B}_{\max} - \frac{\bar{A}}{\bar{\rho}_n} \right).
\end{aligned} \tag{27.2.11}$$

Eqs. (27.2.7)–(27.2.10) constitute a problem of geometric programming (Duffin *et al.*, 1967); there are numerical methods suitable for solving such problems. For most practically interesting cases of two-layer and three-layer shields, the solution can be obtained analytically in the explicit form. After determining the optimal values of $x_1, x_2, ..., x_{n-1}$, the optimal values of $\bar{b}_i$ $(i = 1,2,...,n)$ and the criterion $\bar{w}$ are determined by Eqs. (27.2.3), (27.2.4) and (27.2.6).

27.2.2 *Two-layer shield*

If $n = 2$ Eqs. (27.2.7), (27.2.8) and (27.2.10) yield:

$$\overline{W} = \alpha_i x_i \to \min, \tag{27.2.12}$$

$$\beta_1^{\min} \le x_1 \le \beta_1^{\max}, \quad \gamma_{\min} \le x_1 \le \gamma_{\max}, \quad x_1^{\eta_1} \le \lambda_{\max}. \tag{27.2.13}$$

The problem does not have a solution if both conditions $\eta_1 = 0\,(\rho_1 = \rho_2)$ and $\lambda_{\max} < 1\,(\overline{B}_{\max} < \overline{A}/\overline{\rho}_2)$ are valid. Otherwise, the optimal solution, $x_1^{opt} = x_1^{\min}$, exists if $x_1^{\min} \le x_1^{\max}$, where

$$x_1^{\min} = \begin{cases} \max(\beta_1^{\min}, \gamma_{\min}) & if \quad \eta_1 \ge 0\,(\rho_1 \le \rho_2) \\ \max(\beta_1^{\min}, \gamma_{\min}, \lambda_{\max}^{1/\eta_1}) & if \quad \eta_1 < 0\,(\rho_1 > \rho_2) \end{cases} \tag{27.2.14}$$

$$x_1^{\max} = \begin{cases} \min(\beta_1^{\max}, \gamma_{\max}, \lambda_{\max}^{1/\eta_1}) & if \quad \eta_1 > 0\,(\rho_1 < \rho_2) \\ \min(\beta_1^{\max}, \gamma_{\max},) & if \quad \eta_1 \le 0\,(\rho_1 \ge \rho_2) \end{cases} \tag{27.2.15}$$

27.2.3 *Three-layer shield*

If $n = 3$ Eqs. (27.2.7)–(27.2.10) yield:

$$\overline{W} = \alpha_1 x_1 + \alpha_2 x_2 \to \min, \tag{27.2.16}$$

$$\beta_1^{\min} \le x_1 \le \beta_1^{\max}, \quad \gamma_{\min} \le x_2 \le \gamma_{\max}, \quad \beta_2^{\min} \le x_2/x_1 \le \beta_2^{\max}, \tag{27.2.17}$$

$$x_1^{\eta_1} x_2^{\eta_2} \le \lambda_{\max}, \tag{27.2.18}$$

We do not dwell on the cumbersome secondary details and consider the solution without constraints on the thicknesses of the plates, i.e.

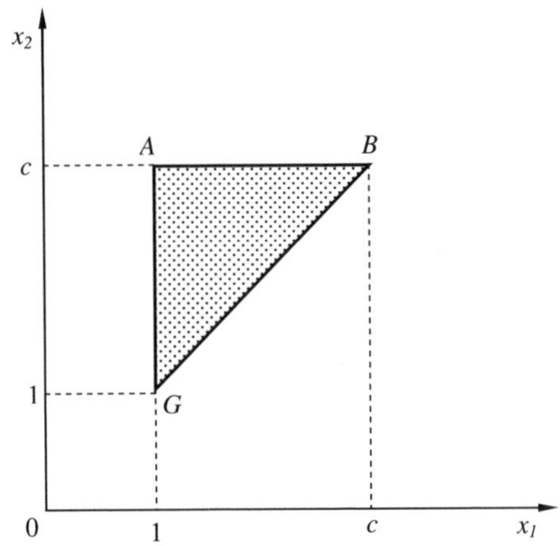

Fig. 27.2.1. Sub-domain of admissible values of x_1 and x_2 determined by Eq. (27.2.17).

$$b_i^{\min} = 0, b_i^{\max} = +\infty, i = 1,2. \tag{27.2.19}$$

In this case

$$\begin{gathered} \beta_1^{\min} = \beta_2^{\min} = 1, \quad \beta_1^{\max} = \beta_2^{\max} = +\infty, \\ \gamma_{\min} = 0, \quad \gamma_{\max} = c, \quad c = \exp(\overline{A}) > 1, \end{gathered} \tag{27.2.20}$$

and the domain determined by Eq. (27.2.17) is shown in Fig. 27.2.1. The domain of the admissible values of x_1 and x_2 is determined by excluding the sub-domain where Eq. (27.2.18) is violated. Since this procedure yields a large number of different variants of the domain of the admissible x_1 and x_2, we skip the detailed analysis and present only the results with illustrations of the typical situations.

Procedure for searching for an optimal solution by sequential examination of various configurations, depending on the values of the initial parameters, is shown in Figs. 27.2.2-27.2.5. In Figs. 27.2.6-27.2.9 we illustrate some particular situations which are incorporated in the procedure. Each of the figures contains a shaded domain of admissible values of x_1 and x_2. The absence of this domain indicates contradictions in the initial data and, consequently, manifests a nonexistence of the solution. The optimal values of x_1 and x_2 are denoted in the figures by bold points.

The following notations are used in the figures:

$$k = -\frac{\eta_1}{\eta_2},\ \sigma_1 = \lambda_{\max}^{1/\eta_1},\ \sigma_2 = \lambda_{\max}^{1/\eta_2},\ \sigma_3 = \sigma_2^{1/(1-k)},\ \sigma_4 = c^{1-k},$$

$$\sigma_5 = \left(-\frac{\alpha_1}{k\alpha_2\sigma_2}\right)^{1/(k-1)},\ \sigma_6 = \left(\frac{c}{\sigma_2}\right)^{1/k},\ \sigma_7 = \sigma_2\sigma_5^k,\ \sigma_8 = \sigma_2 c^k \tag{27.2.21}$$

where σ_3 are coordinates of the interaction point between a curve $x_2 = \sigma_2 x_1^k$ and a straight line $x_2 = x_1$; σ_6 is an abscissa of the point on a curve $x_2 = \sigma_2 x_1^k$, having an ordinate $x_2 = c$; σ_5 and σ_7 are abscissa and ordinate, respectfully, of a point on a hyperbola $x_2 = \sigma_2 x_1^k$ ($k < 0$), where a tangent line is perpendicular to a straight line $\alpha_1 x_1 + \alpha_2 x_2 = const$; σ_8 is an ordinate of a point on a curve $x_2 = \sigma_2 x_1^k$, having an abscissa $x_1 = c$.

Consider in details a case when $\eta_1 < 0, \eta_2 < 0\ (k < 0)$ and hyperbola $x_2 = \sigma_2 / x_1^{|k|}$ intersects with sides AG and BG of a triangle ABG (Figs. 27.2.6b-d). A case when a hyperbola passes through the point (1,1) is included in the situation depicted in Fig. 27.2.6a.

Consider a set of parallel straight lines, $\alpha_1 x_1 + \alpha_2 x_2 = \overline{W}$. Clearly, the smaller is the distance from the origin of the coordinates to the line (along the normal to the line), the smaller is the value of the criterion, $\overline{W}$. Therefore, the problem reduces to finding the straight line having the minimum distance from the origin (hereafter, the minimal line) among all straight lines $\alpha_1 x_1 + \alpha_2 x_2 = \overline{W}$ which pass through the shaded domain.

The point of intersection of this line with the admissible values of x_1 and x_2 provides the optimal values of x_1 and x_2, and the value of $\overline{W}$ that corresponds to this line is the minimal value of the criterion.

Let us select among lines $\alpha_1 x_1 + \alpha_2 x_2 = \overline{W}$ the tangent to the hyperbola $x_2 = \sigma_2 / x_1^{|k|}$ (DEF in Figs. 27.2.6b-d). Hence σ_5 is the abscissa of the point of tangency (a point E). Three cases can occur.

If $\sigma_5 < 1$ (Fig. 27.2.6b) then the minimal line is the line $D^*E^*F^*$ which passes through the point E^*, where the point E^* is the intersection of the vertical boundary of the admissible domain and of the hyperbola; hence, $x_1^{opt} = 1$, $x_2^{opt} = \sigma_2$.

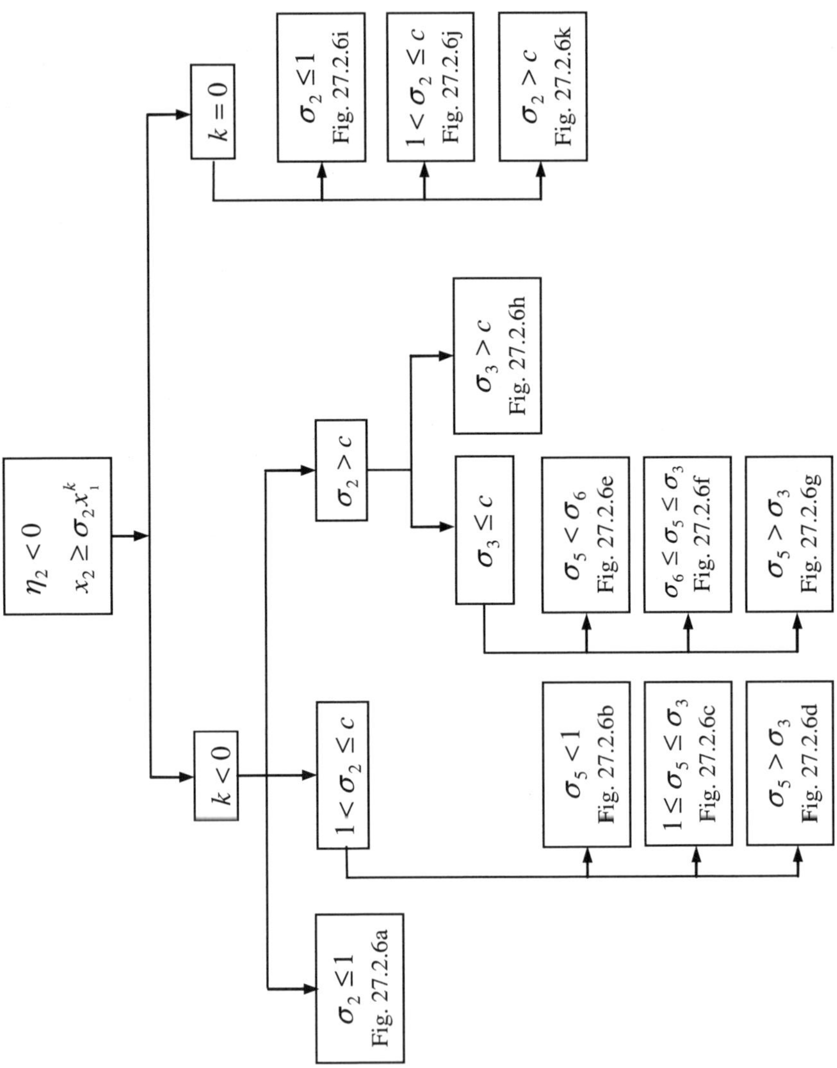

Fig. 27.2.2. Procedure for searching solution in case $\eta_2 < 0, k \leq 0$.

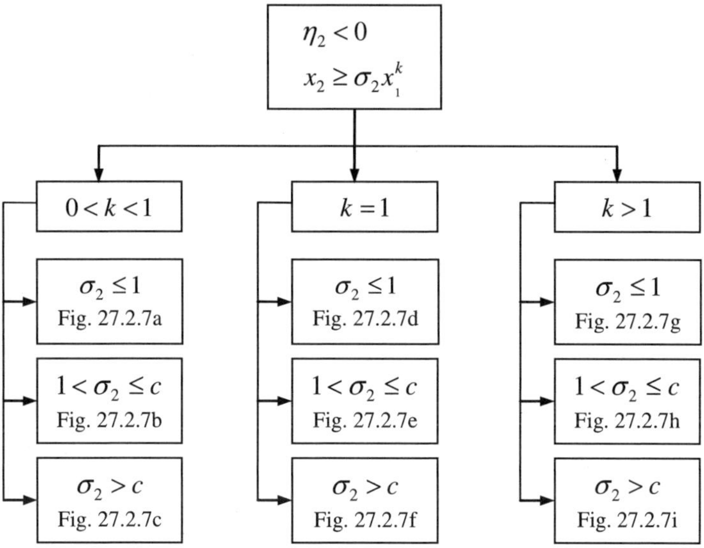

Fig. 27.2.3. Procedure for searching solution in case $\eta_2 < 0, k > 0$.

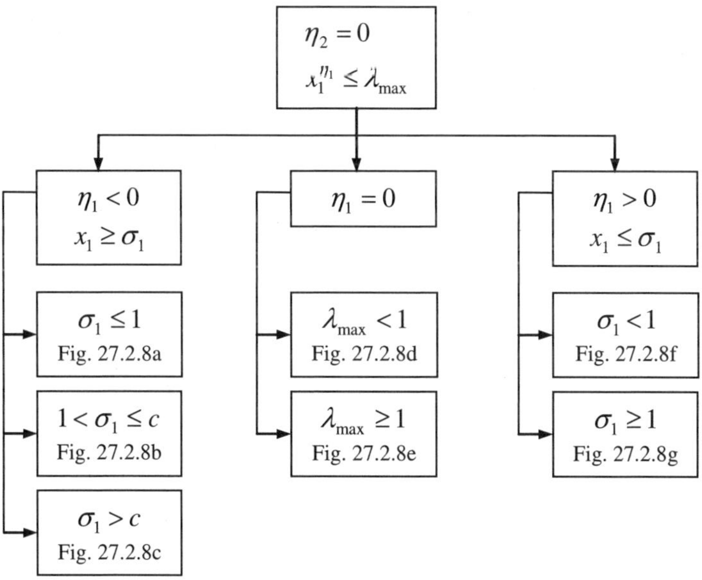

Fig. 27.2.4. Procedure for searching solution in case $\eta_2 = 0$.

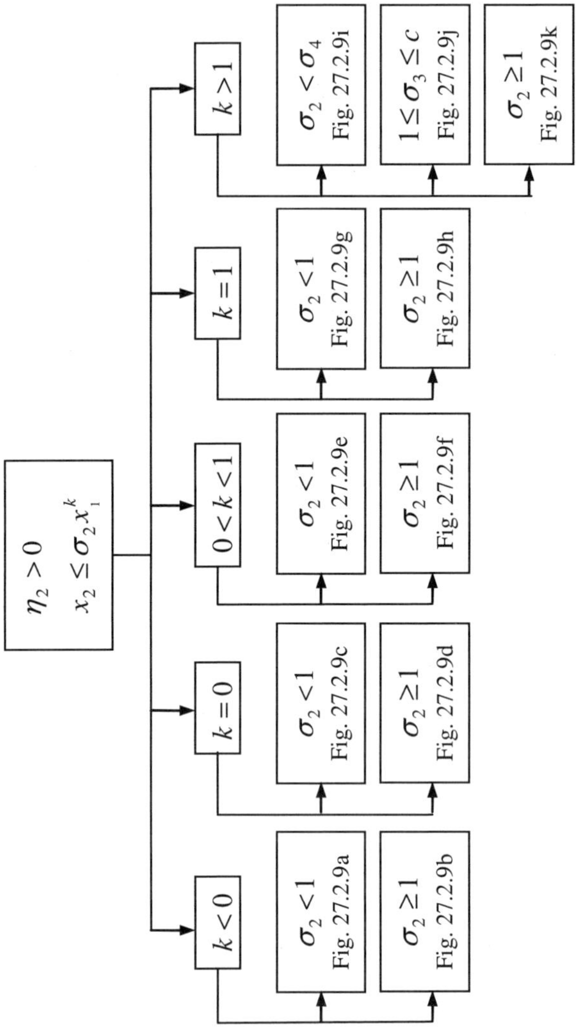

Fig. 27.2.5. Procedure for searching solution in case $\eta_2 > 0$.

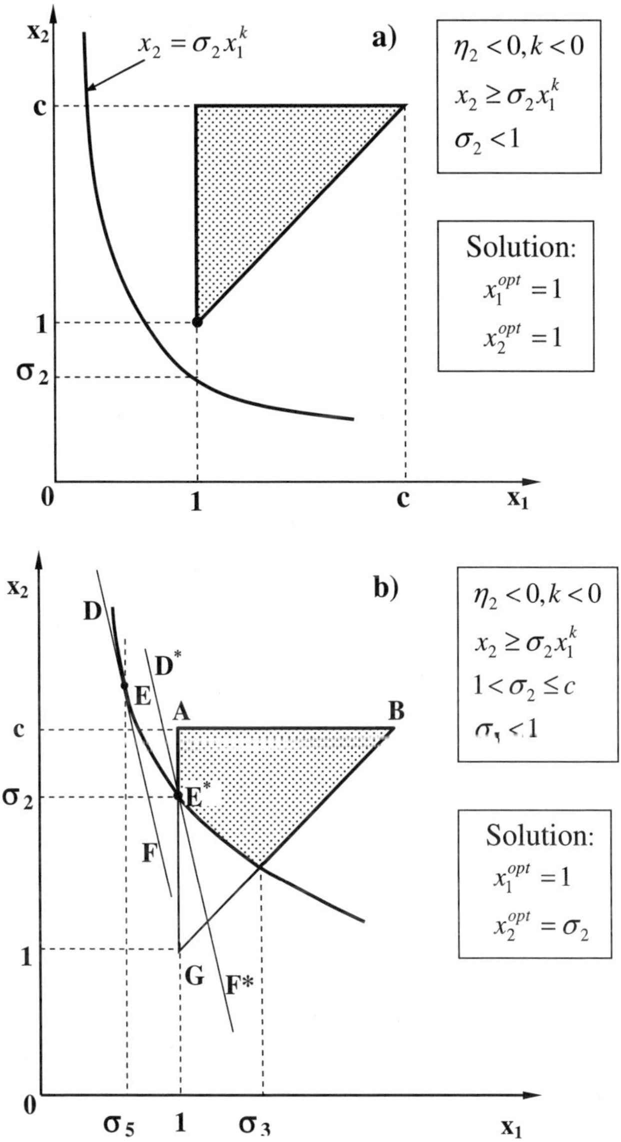

Fig. 27.2.6. Illustration of solution in case $\eta_2 < 0$, $k \leq 0$.

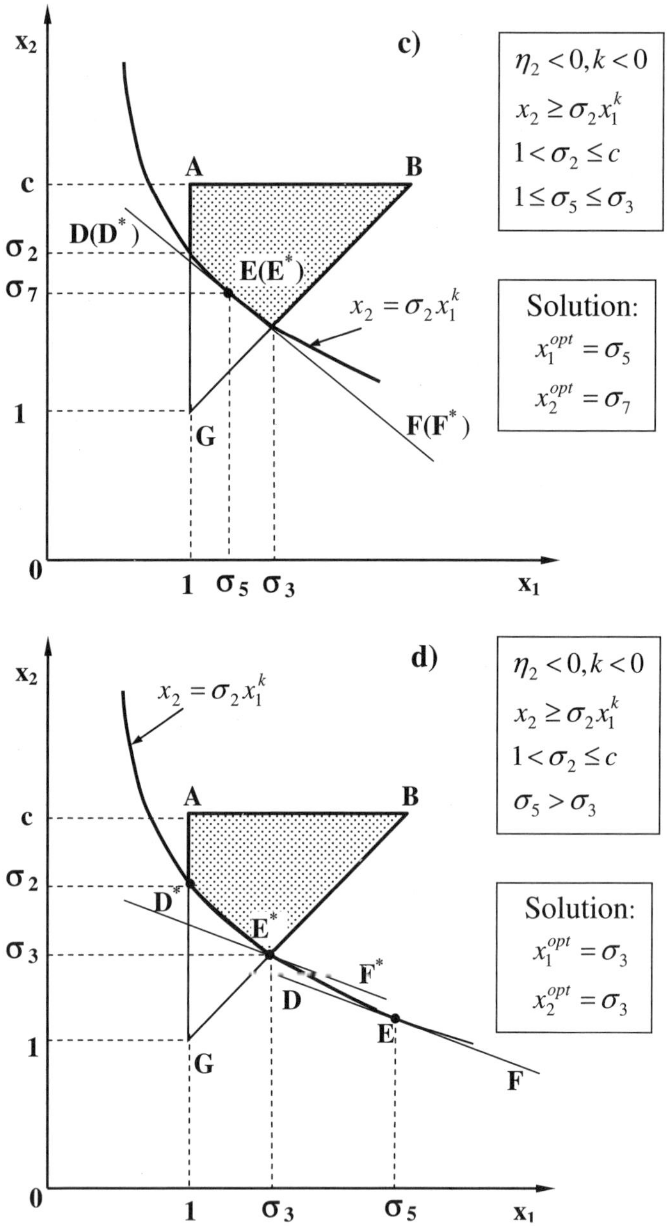

Fig. 27.2.6. Continued.

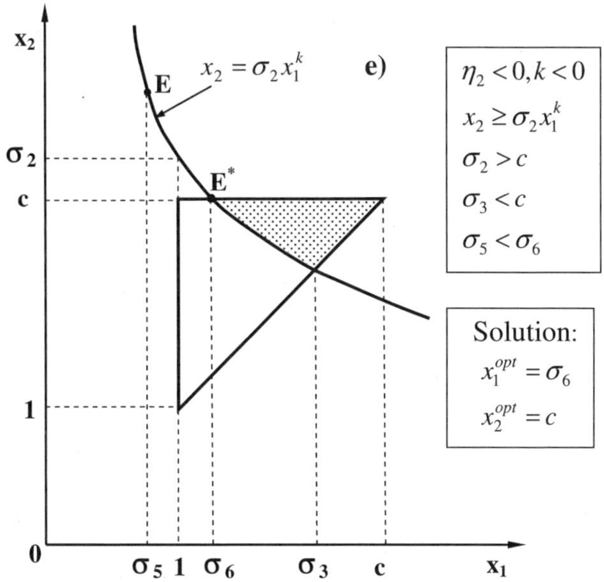

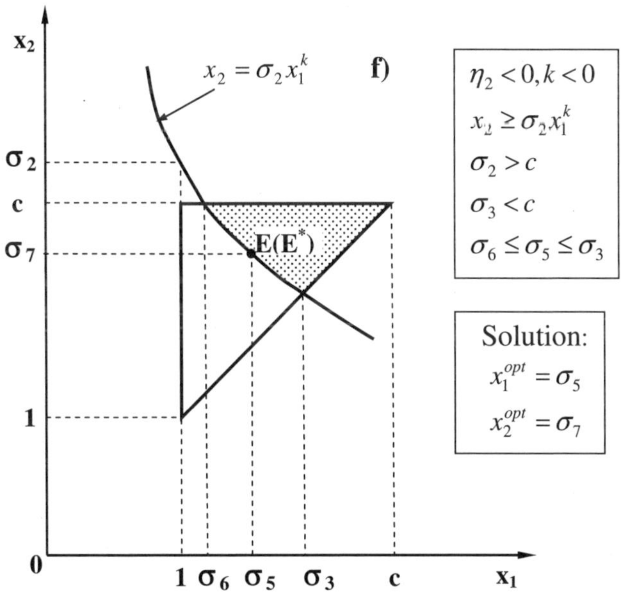

Fig. 27.2.6. Continued.

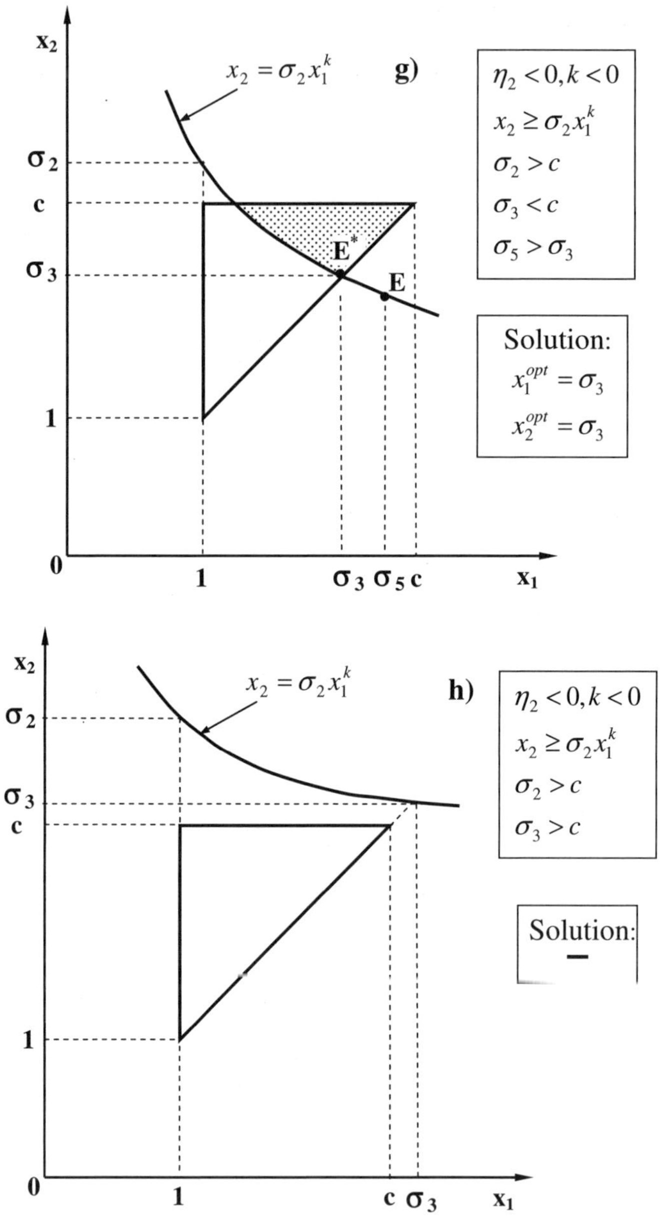

Fig. 27.2.6. Continued.

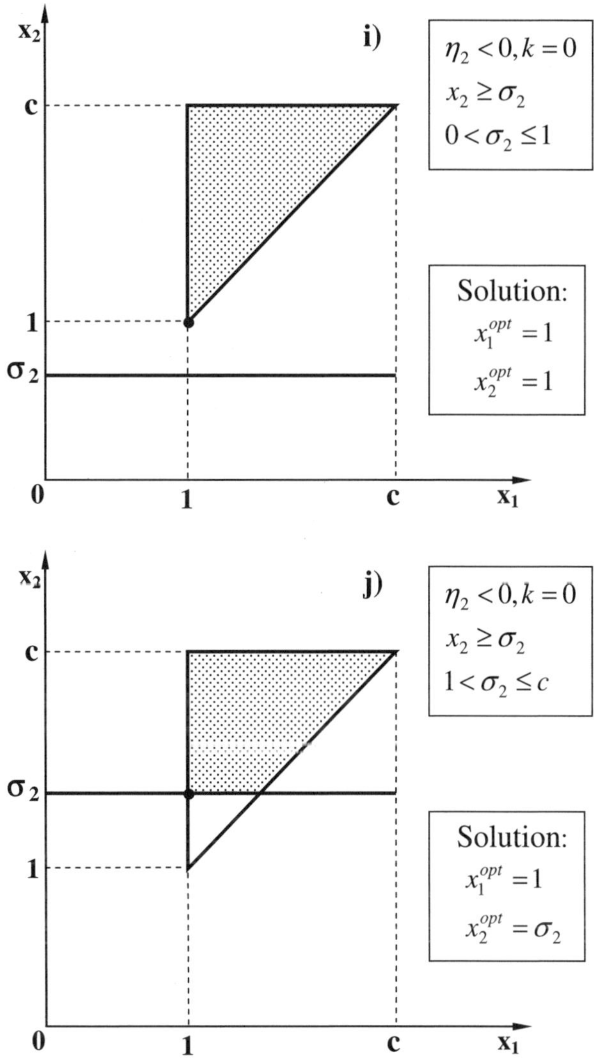

Fig. 27.2.6. Continued.

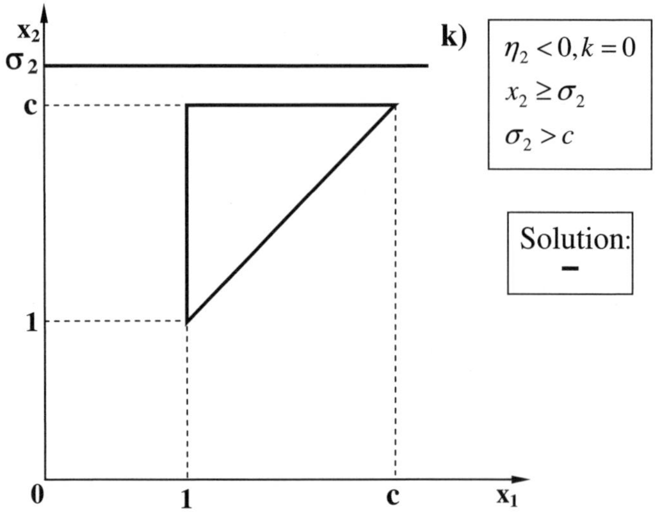

Fig.27.2.6. Continued.

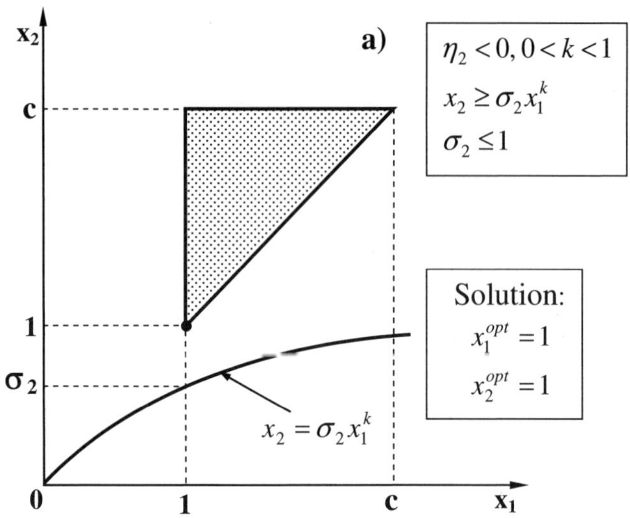

Fig. 27.2.7. Illustration of solution in case $\eta_2 < 0, k > 0$.

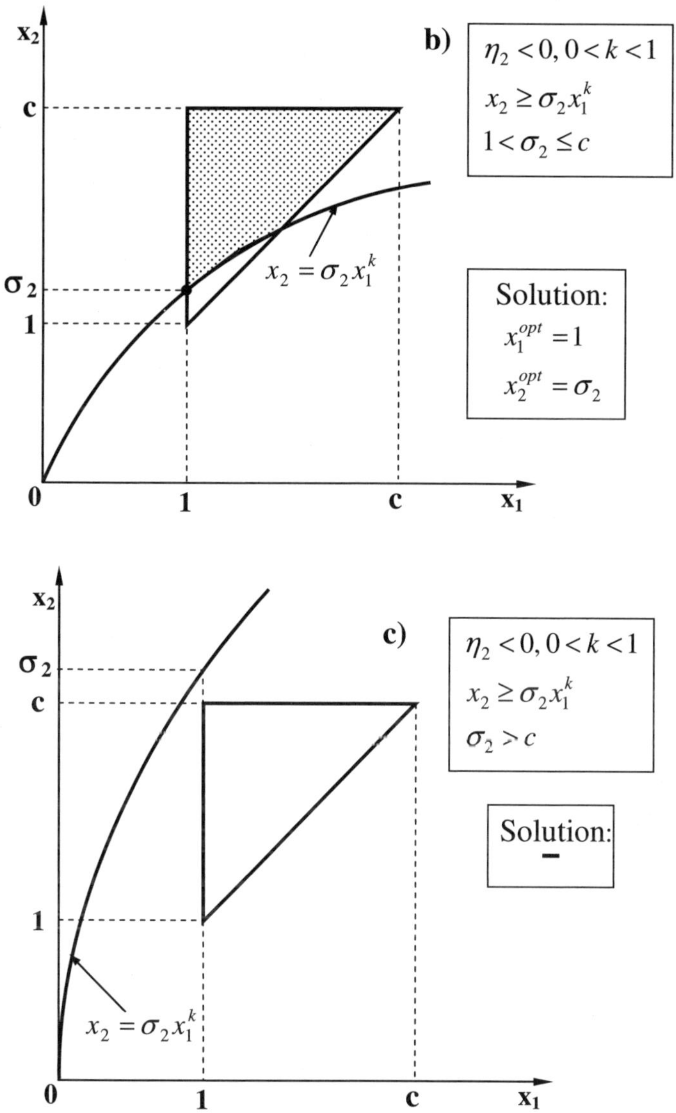

Fig. 27.2.7. Continued.

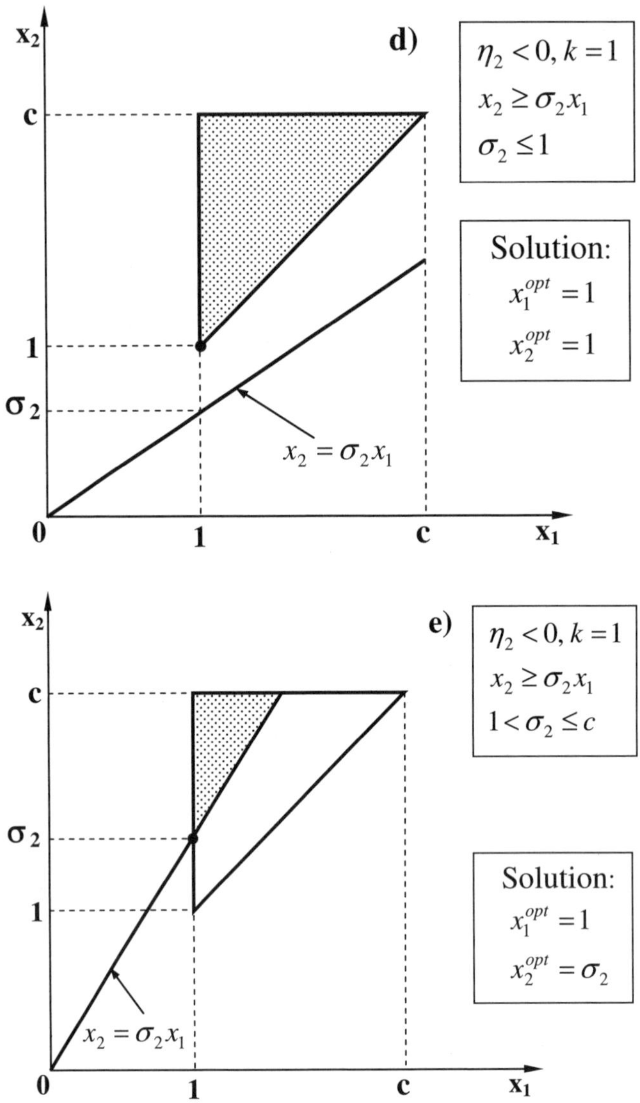

Fig. 27.2.7. Continued.

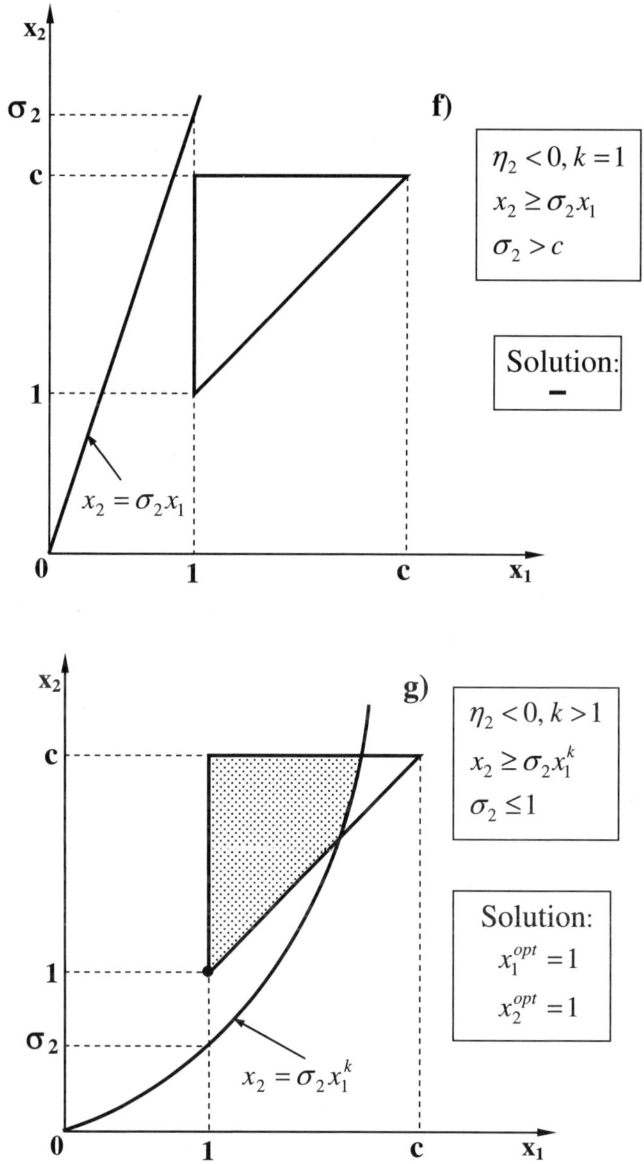

Fig. 27.2.7. Continued.

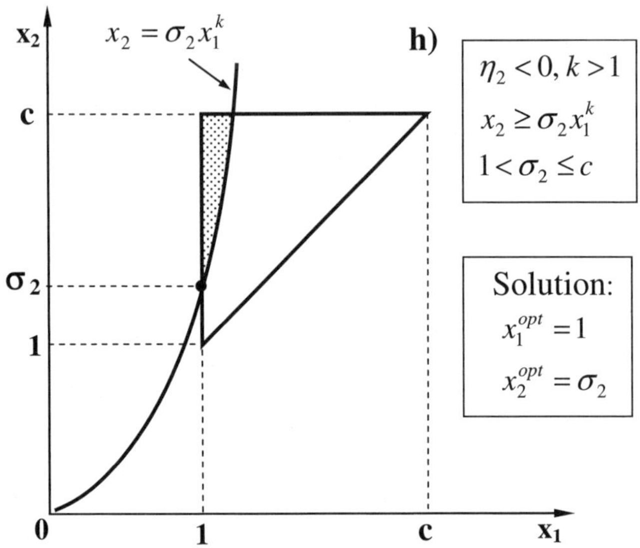

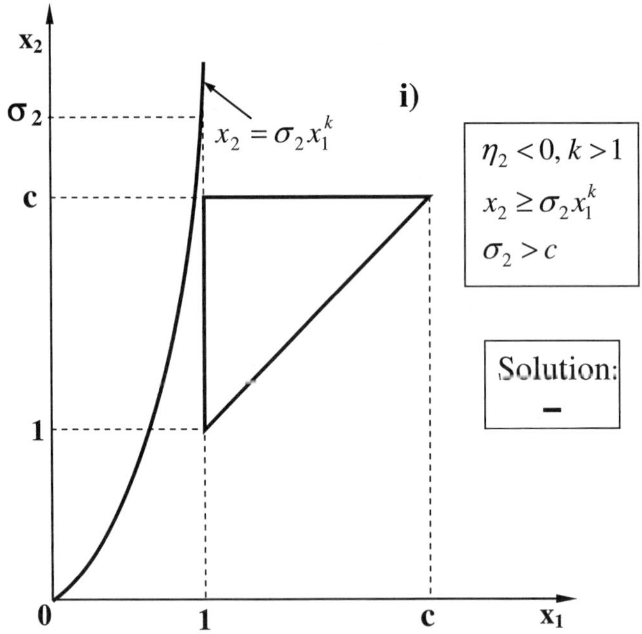

Fig. 27.2.7. Continued.

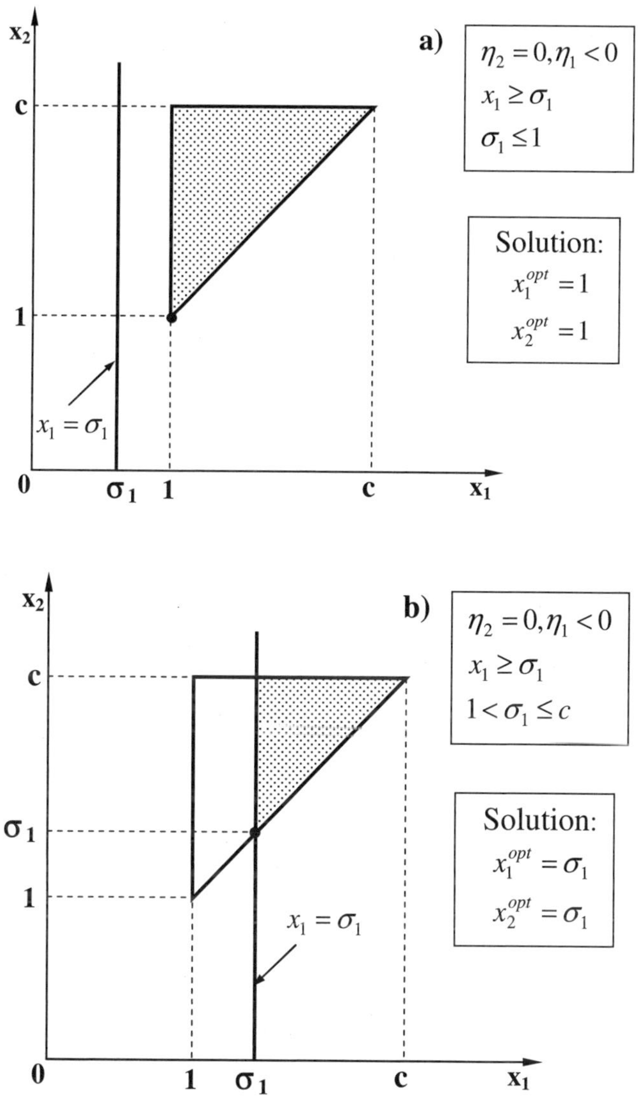

Fig. 27.2.8. Illustration of solution in case $\eta_2 = 0$.

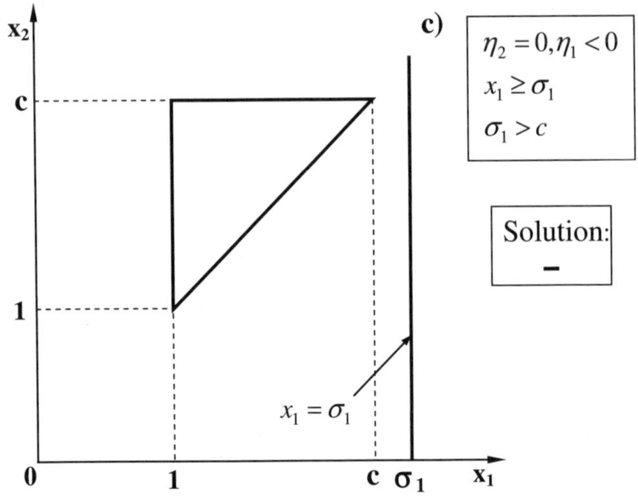

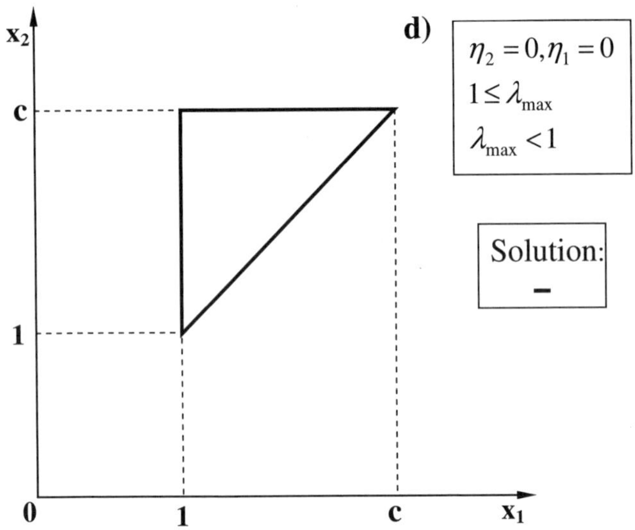

Fig. 27.2.8. Continued.

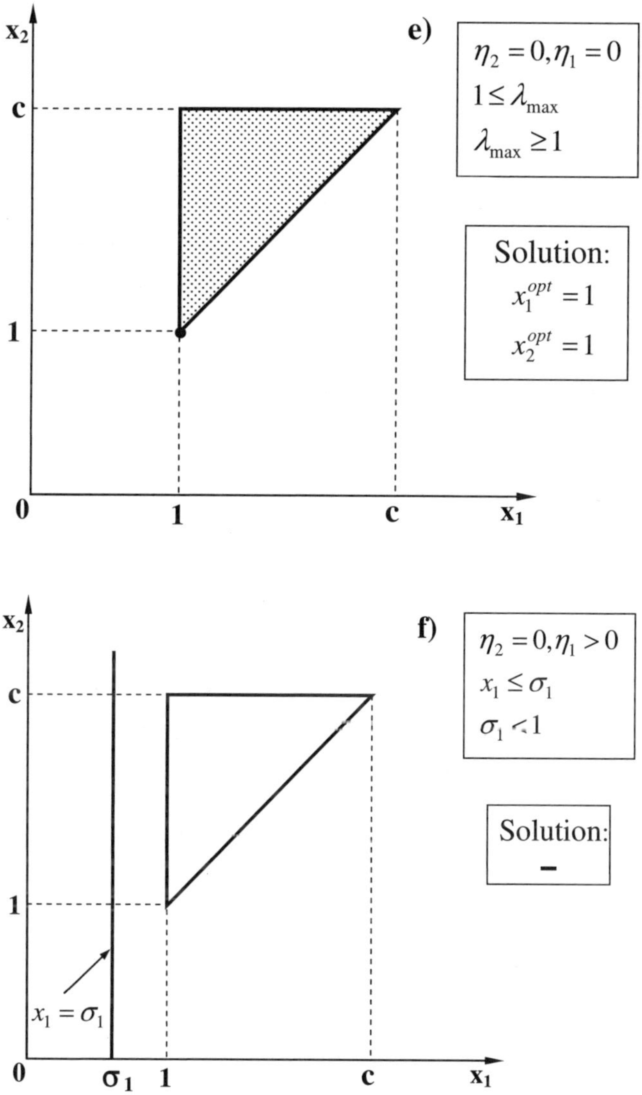

Fig. 27.2.8. Continued.

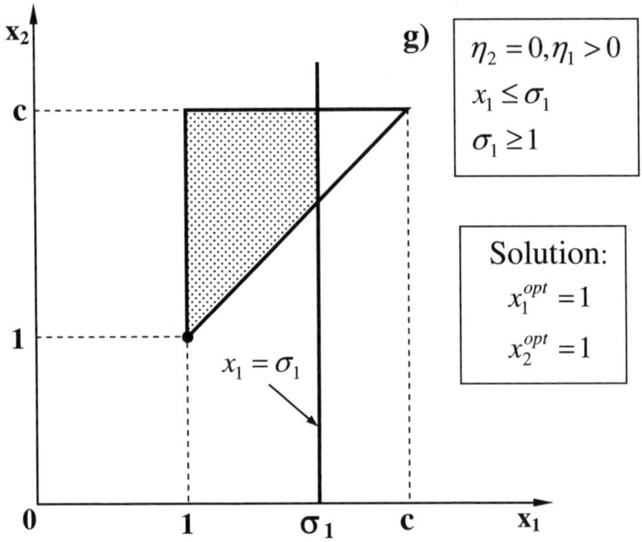

Fig. 27.2.8. Continued.

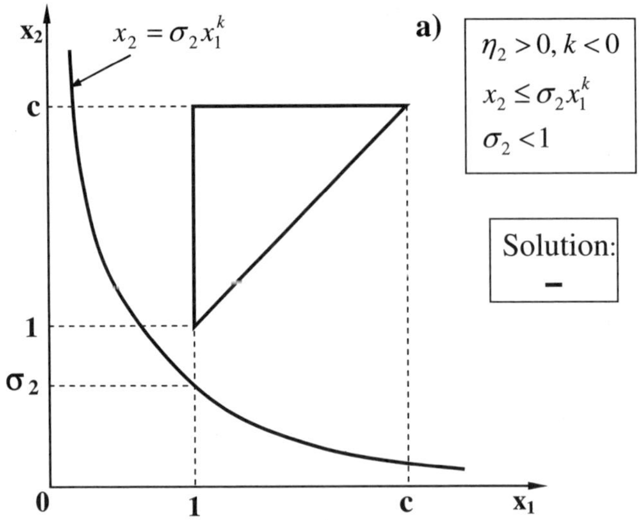

Fig. 27.2.9. Illustration of solution in case $\eta_2 > 0$.

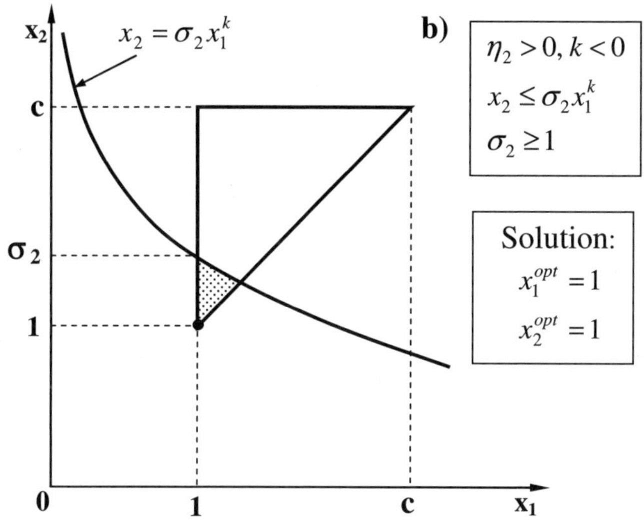

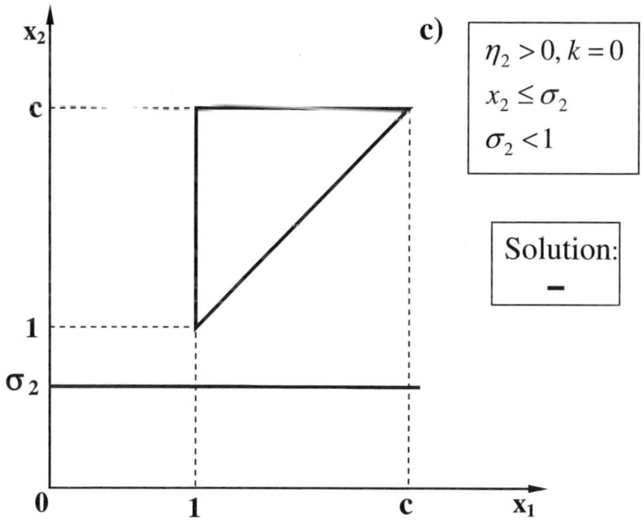

Fig. 27.2.9. Continued.

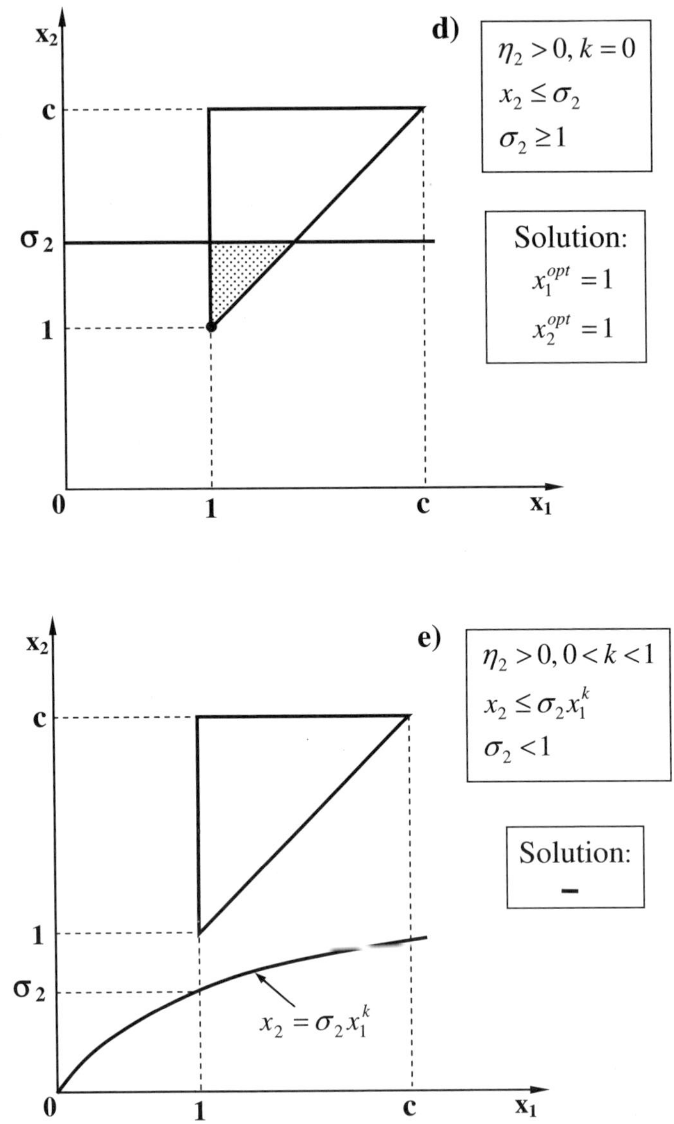

Fig. 27.2.9. Continued.

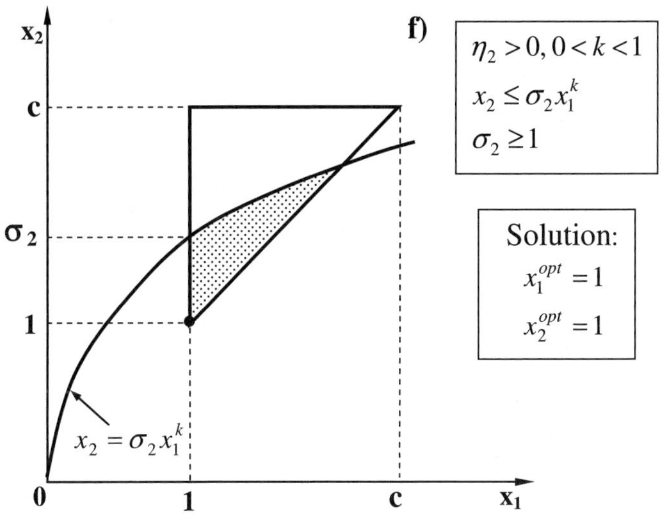

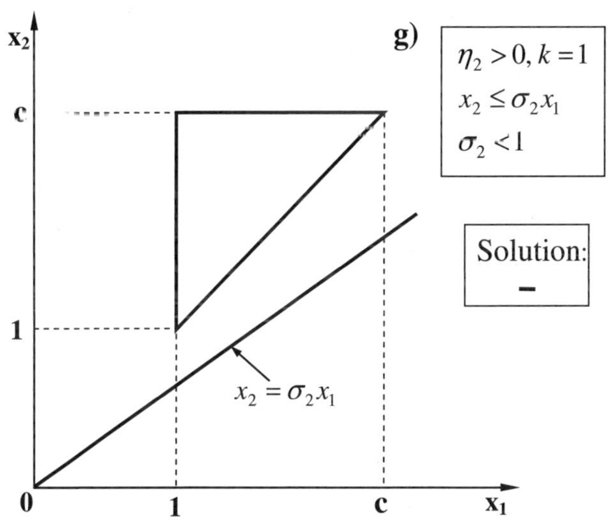

Fig. 27.2.9. Continued.

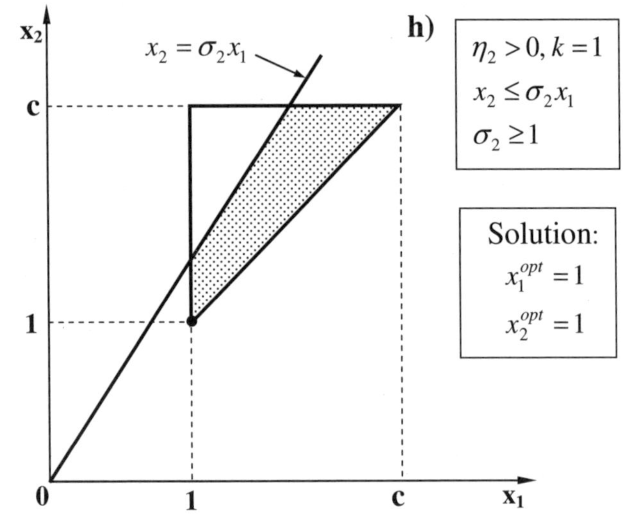

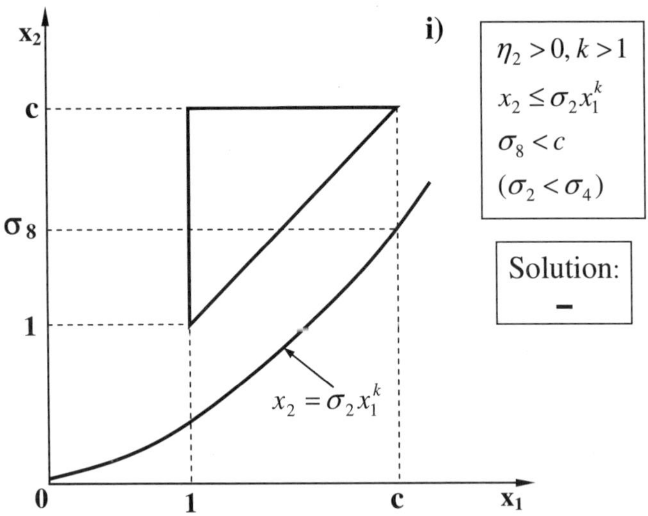

Fig. 27.2.9. Continued.

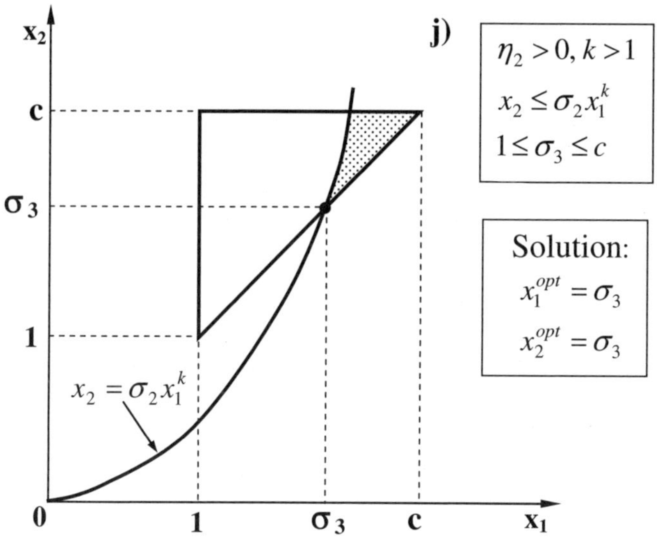

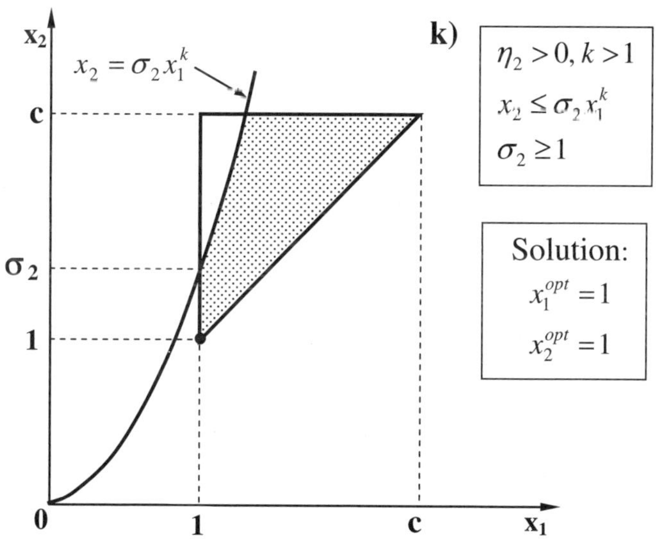

Fig. 27.2.9. Continued.

Table 27.2.1. Summary of solutions in Figs. 27.2.6–27.2.9.

Figure	η_2	η_1, k	σ_1, σ_2	**Others**	x_1^{opt}	x_2^{opt}
27.2.6a	$\eta_2<0$	$k<0$	$\sigma_2\le 1$		1	1
27.2.6b	$\eta_2<0$	$k<0$	$1<\sigma_2\le c$	$\sigma_5<1$	1	σ_2
27.2.6c	$\eta_2<0$	$k<0$	$1<\sigma_2\le c$	$1\le\sigma_5\le\sigma_3$	σ_5	σ_7
27.2.6d	$\eta_2<0$	$k<0$	$1<\sigma_2\le c$	$\sigma_3<\sigma_5$	σ_3	σ_3
27.2.6e	$\eta_2<0$	$k<0$	$c<\sigma_2$	$\sigma_3\le c$, $\sigma_5<\sigma_6$	σ_6	c
27.2.6f	$\eta_2<0$	$k<0$	$c<\sigma_2$	$\sigma_3\le c$, $\sigma_6\le\sigma_5\le\sigma_3$	σ_5	σ_7
27.2.6g	$\eta_2<0$	$k<0$	$c<\sigma_2$	$\sigma_3\le c$, $\sigma_3<\sigma_5$	σ_3	σ_3
27.2.6h	$\eta_2<0$	$k<0$	$c<\sigma_2$	$c<\sigma_3$	-	-
27.2.6i	$\eta_2<0$	$k=0$	$\sigma_2\le 1$		1	1
27.2.6j	$\eta_2<0$	$k=0$	$1<\sigma_2\le c$		1	σ_2
27.2.6k	$\eta_2<0$	$k=0$	$c<\sigma_2$		-	-
27.2.7a	$\eta_2<0$	$0<k<1$	$\sigma_2\le 1$		1	1
27.2.7b	$\eta_2<0$	$0<k<1$	$1<\sigma_2\le c$		1	σ_2
27.2.7c	$\eta_2<0$	$0<k<1$	$c<\sigma_2$		-	-
27.2.7d	$\eta_2<0$	$k=1$	$\sigma_2\le 1$		1	1
27.2.7e	$\eta_2<0$	$k=1$	$1<\sigma_2\le c$		1	σ_2
27.2.7f	$\eta_2<0$	$k=1$	$c<\sigma_2$		-	
27.2.7g	$\eta_2<0$	$k>1$	$\sigma_2\le 1$		1	1
27.2.7h	$\eta_2<0$	$k>1$	$1\le\sigma_2\le c$		1	σ_2
27.2.7i	$\eta_2<0$	$k>1$	$c<\sigma_2$		-	-
27.2.8a	$\eta_2=0$	$\eta_1<0$	$\sigma_1\le 1$		1	1
27.2.8b	$\eta_2=0$	$\eta_1<0$	$1<\sigma_1\le c$		σ_1	σ_1
27.2.8c	$\eta_2=0$	$\eta_1<0$	$c<\sigma_1$		-	-
27.2.8d	$\eta_2=0$	$\eta_1=0$		$\lambda_{\max}<1$	-	-
27.2.8e	$\eta_2=0$	$\eta_1=0$		$\lambda_{\max}\ge 1$	1	1
27.2.8f	$\eta_2=0$	$\eta_1>0$	$\sigma_1<1$		-	-

Figure	η_2	η_1, k	σ_1, σ_2	**Others**	x_1^{opt}	x_2^{opt}
27.2.8g	$\eta_2 = 0$	$\eta_1 > 0$	$\sigma_1 \geq 1$		1	1
27.2.9a	$\eta_2 > 0$	$k < 0$	$\sigma_2 < 1$		-	-
27.2.9b	$\eta_2 > 0$	$k < 0$	$\sigma_2 \geq 1$		1	1
27.2.9c	$\eta_2 > 0$	$k = 0$	$\sigma_2 < 1$		-	-
27.2.9d	$\eta_2 > 0$	$k = 0$	$\sigma_2 \geq 1$		1	1
27.2.9e	$\eta_2 > 0$	$0 < k < 1$	$\sigma_2 < 1$		-	-
27.2.9f	$\eta_2 > 0$	$0 < k < 1$	$\sigma_2 \geq 1$		1	1
27.2.9g	$\eta_2 > 0$	$k = 1$	$\sigma_2 < 1$		-	-
27.2.9h	$\eta_2 > 0$	$k = 1$	$\sigma_2 \geq 1$		1	1
27.2.9i	$\eta_2 > 0$	$k > 1$	$\sigma_2 < \sigma_4$		-	-
27.2.9j	$\eta_2 > 0$	$k > 1$		$1 \leq \sigma_3 \leq c$	σ_3	σ_3
27.2.9k	$\eta_2 > 0$	$k > 1$	$\sigma_2 \geq 1$		1	1

Table 27.2.2. Solution the optimization problem for 3-layer shield.

Conditions	**Solution**
$\eta_2 = 0, \eta_1 = 0, \lambda_{\max} < 1$	Does not exist
$\eta_2 = 0, \eta_1 > 0, \sigma_1 < 1$	Ibid
$\eta_2 = 0, \eta_1 < 0, c < \sigma_1$	Ibid
$\eta_2 > 0, k \leq 1, \sigma_2 < 1$	Ibid
$\eta_2 > 0, k > 1, \sigma_2 < \sigma_4$	Ibid
$\eta_2 < 0, k \geq 0, c < \sigma_2$	Ibid
$\eta_2 < 0, k < 0, c < \sigma_2, c < \sigma_3$	Ibid
$\eta_2 = 0, \eta_1 = 0, \lambda_{\max} \geq 1$	$x_1^{opt} = x_2^{opt} = 1$
$\eta_2 = 0, \eta_1 > 0, \sigma_1 \geq 1$	Ibid
$\eta_2 > 0, \sigma_2 \geq 1$	Ibid
$\eta_2 < 0, \sigma_2 \leq 1$	Ibid
$\eta_2 = 0, \eta_1 < 0, \sigma_1 \leq 1$	Ibid
$\eta_2 = 0, \eta_1 < 0, 1 < \sigma_1 \leq c$	$x_1^{opt} = x_2^{opt} = \sigma_1$

Conditions	Solution
$\eta_2 > 0, k > 1, 1 \le \sigma_3 \le c$	$x_1^{opt} = x_2^{opt} = \sigma_3$
$\eta_2 < 0, k < 0, 1 < \sigma_2 \le c, \sigma_3 < \sigma_5$	Ibid
$\eta_2 < 0, k < 0, c < \sigma_2, \sigma_3 \le c, \sigma_3 < \sigma_5$	Ibid
$\eta_2 < 0, k > 0, 1 < \sigma_2 \le c$	$x_1^{opt} = 1, x_2^{opt} = \sigma_2$
$\eta_2 < 0, k < 0, 1 < \sigma_2 \le c, \sigma_5 \le 1$	Ibid
$\eta_2 < 0, k < 0, 1 < \sigma_2 \le c, 1 \le \sigma_5 \le \sigma_3$	$x_1^{opt} = \sigma_5, x_2^{opt} = \sigma_7$
$\eta_2 < 0, k < 0, c < \sigma_2, \sigma_3 \le c, \sigma_6 \le \sigma_5 \le \sigma_3$	Ibid
$\eta_2 < 0, k < 0, c < \sigma_2, \sigma_3 \le c, \sigma_5 < \sigma_6$	$x_1^{opt} = \sigma_6, x_2^{opt} = c$

If $1 \le \sigma_5 \le \sigma_3$ (Fig. 27.2.6c) then the minimal line $D^* E^* F^*$ is tangent to the hyperbola (a point E^*) at the boundary of the admissible domain; consequently, $x_1^{opt} = \sigma_5, x_2^{opt} = \sigma_7$.

If $\sigma_5 > \sigma_3$ (Fig. 27.2.6d) then the minimal line $D^* E^* F^*$ passes through the point E^* that is the intersection of the intercept BG and the hyperbola; hence, $x_1^{opt} = x_2^{opt} = \sigma_3$.

Situations shown in Figs. 27.2.6–27.2.9 are summarized in Table 27.2.1. In order to present solution in concise and systematic form it is convenient to combine several sets of solutions. For instance, solutions illustrated in Figs. 27.2.9a, 27.2.9c, 27.2.9e, 27.2.9g can be presented in the following concise form: $\eta_2 > 0$, $k \le 1$, $\sigma_2 < 1$. Solutions which are presented in this form are shown in Table 27.2.2.

Part 11

Some Optimization Problems for Non-Homogeneous Non-Ductile Shields

Part 11 includes two chapters.

Chapter 28 is devoted mainly to the investigation of the effect of steel plate lining at the rear face of the reinforced concrete panel on the ballistic properties using a semi-empirical model. We determine the structure of the optimal shield which provides the maximal ballistic limit velocity (BLV) for a given area density, depending on the mechanical properties of concrete and characteristics of the reinforcement. This chapter contains also a brief review of investigations of ballistic properties of multilayer concrete shields.

In Chapter 29, we study bi-component shields consisting of a ceramic front plate and a ductile back plate. A version of the Florence model that takes into account the ballistic resistance of the ceramic plate is used for armor optimization (Ben-Dor *et al.*, 2009a). The direct problems (maximization of the BLV for a given areal density (AD) or a total thickness (TT) of the armor) and the inverse problems (minimizations of the AD or TT for a given impact velocity) are investigated in the case of a single impact. We also consider optimization problems for the case of a repeated impact. Numerical results are presented for the aluminum/alumina armor.

Chapter 28

Optimization of Reinforced Concrete Panels with Steel Liner

Nomenclature

a_1	cross-sectional area of single front rebar, Fig. 28.2.1;
a_2	cross-sectional area of single rear rebar, Fig. 28.2.1;
A	areal density of shield;
b	parameter, Eq. (28.2.10);
$b^{(1)}$	thickness of reinforced concrete plate, Fig. 28.2.1;
$b^{(2)}$	steel plate thickness, Fig. 28.2.1;
$\bar{b}^{(i)}$	$=b^{(i)}/b$;
$\bar{b}_*^{(1)}, \bar{b}_{**}^{(1)}$	parameters, Eq. (28.2.16);
$\bar{b}_\times^{(1)}, \bar{b}_{\times\times}^{(1)}$	parameters, Eq. (28.2.25);
$\bar{b}_{opt}^{(i)}$	optimal value of $\bar{b}^{(i)}$;
B	parameter, Eq. (28.2.2);
BLV	ballistic limit velocity;
c_1	front rebar spacing, Fig. 28.2.1;
c_2	rear rebar spacing, Fig. 28.2.1;
f_{cy}	characteristic compressive strength of concrete;
F	function, Eq. (28.2.13);
k	parameter, Eq. (28.2.2);
l	length of shield, Fig. (28.2.1);
m	mass of projectile;
p	perimeter of impactor;
r	parameter, Eq. (28.2.2);
v_{bl}	BLV of projectile against reinforced concrete plate with liner;

v_{bl}^{0} BLV of projectile against reinforced concrete plate without liner having the same areal density as the optimal plate;

v_{bl}^{opt} optimum BLV of projectile against reinforced concrete plate with liner;

$V_{conc}^{(1)}$ volume of concrete in reinforced concrete plate;

$V_{steel}^{(1)}$ volume of steel in reinforced concrete plate;

w width of the shield, Fig. (28.2.1);

$\overline{W}_{bl}$ dimensionless v_{bl}, Eq. (28.2.8);

α parameter, Eq. (28.2.2);

$\overline{\alpha}$ $=\alpha/b$;

$\overline{\alpha}_{\times}$ parameter, Eq. (28.2.22);

$\overline{\alpha}_{\times\times}$ parameter, Eq. (28.2.29);

$\overline{\beta}$ parameter, Eq. (28.2.16);

δ parameter, Eq. (28.2.31);

$\overline{\theta}$ $=\rho_{steel}/\rho_{conc}$;

ξ parameter in Eq. (28.2.7);

ρ_{conc} density of concrete;

ρ_{steel} density of steel.

28.1 Ballistic Properties of Multi-Layer Concrete Shields

There are only a few publications regarding the effect of layering on the ballistic properties of concrete shields.

In TM 5-855-1, 1986 it is noted that a thick slab manufactured from several layers generally has a lesser perforation resistance than a single slab with the same thickness. Kojima, 1991 performed experiments with spherical-nosed projectile striking the reinforced concrete plates with velocity $200m/s$. He tested a monolithic shield having the thickness $18cm$ and double-layer shields, $9cm+9cm$ and $6cm+12cm$. The conclusion was that the impact resistance of a double-layer shield is inferior to that of a monolithic shield. Shirai *et al.*, 1997 investigated experimentally and numerically the impact resistance of the reinforced concrete plates against the flat-nosed projectile at impact velocity of

$170\,m/s$. They compared the protective effectiveness of the monolithic shield with the thickness of $9\,cm$ and double-layer shields having different structure, $4.5\,cm + 4.5cm$ and $3\,cm + 6cm$. It was found that double-layer shields had a higher impact resistance than monolithic shield. Ben-Dor *et al.*, 2009e investigated theoretically the effect of layering of concrete shields using the model similar to that described in Chapter 26. They found that (i) the ballistic limit velocity (BLV) of the multilayer shield does not depend on the order of the plates in the shield, (ii) monolithic shield is superior to any layered shield having the same thickness, and (iii) the largest deterioration of the BLV occurred when the shield was divided into a number of plates having the same thicknesses.

28.2 Optimization of Reinforced Concrete Panels with Rear Face Steel Liner

28.2.1 *Introduction*

Although experimental and numerical investigations (Barr *et al.*, 1983; Hashimoto *et al.*, 2005; Kojima, 1991, Koshika *et al.*, 1993; Ohno *et al.*, 1994; Tsubota *et al.*, 1993; Walter and Wolde-Tinsae, 1984) demonstrated that the effect of lining can be significant, there are only a few studies which employed analytical models for predicting the effect of steel liner on the ballistic properties of shields. Hashimoto *et al.*, 2005 suggested an empirical relationship between the bulging height of the lining steel plate and the impact velocity. Walter and Wolde-Tinsae, 1984 and Ohno *et al.*, 1994 proposed to assess the effect a liner as an additional concrete slab thickness. The most comprehensive procedure, including the design formulas and recommendations about the range of their applicability, was presented by Barr, 1990. In this section, using these design formulas and the results obtained by Ben-Dor *et al.*, 2009g, we investigate the problem of optimization of a steel liner at the rear face of the concrete plate.

28.2.2 *Mathematical model and formulation of problem*

Consider normal penetration of a rigid flat faced striker (a body of revolution) into a shield consisting of a reinforced concrete plate and a steel plate on its distal face. The notations are shown in Fig. 28.2.1. Following Barr, 1990 (see also Section 3.22), we use the following formula for the BLV, v_{bl}:

$$v_{bl} = k\, b^{(1)4/3}\sqrt{B+0.3}\,, \tag{28.2.1}$$

where

$$B = r + \frac{100b^{(2)}}{b^{(1)}}, \quad r = \frac{100\alpha}{b^{(1)}}, \quad \alpha = \frac{a_1}{3c_1} + \frac{2a_2}{3c_2},$$
$$k = 1.3\rho_{conc}^{1/6} f_{cy}^{1/2}\left(\frac{p}{\pi m}\right)^{2/3}, \tag{28.2.2}$$

a_1 (a_2) is the cross-sectional area of a single front (rear) rebar, c_1 (c_2) is a front (rear) rebar spacing, f_{cy} is a characteristic compressive strength of concrete measured for 150 mm diameter, 300 mm long cylinders, ρ_{conc} is density of concrete, m is mass of impactor, p is

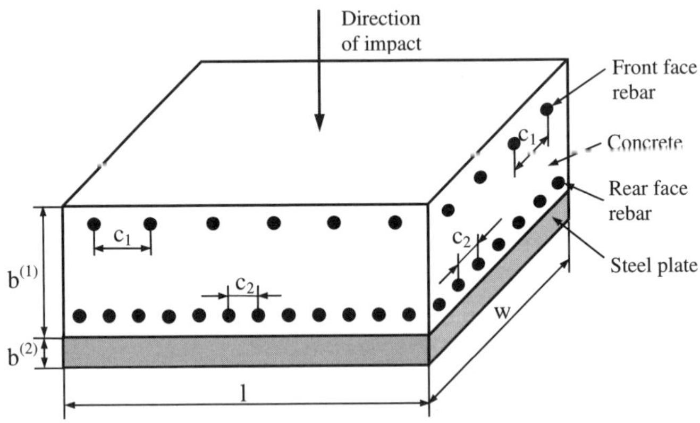

Fig. 28.2.1. Schematic view of the reinforced concrete panel with a rear face steel liner and notations.

perimeter of the impactor, $b^{(1)}$ is thickness of reinforced concrete plate, $b^{(2)}$ is steel plate thickness. Hereafter we use SI units.

Barr, 1990 presents the ranges of parameters where Eq. (28.2.1) is valid. The following constraints are important for the analysis:

$$0 \le r \le 0.75, \quad 1.2 \le B \le 4.3, \tag{28.2.3}$$

where the second constraint is omitted when $b^{(2)} = 0$.

The volume of steel in the reinforced concrete plate, $V^{(1)}_{steel}$, can be calculated as follows:

$$V^{(1)}_{steel} = \sum_{i=1}^{2} \left(\frac{l}{c_i} a_i w + \frac{w}{c_i} a_i l \right) = 2lw \sum_{i=1}^{2} \left(\frac{a_i}{c_i} \right), \tag{28.2.4}$$

Where l and w are the length and the width of the shield, correspondingly. Formulas for the volume of concrete in the reinforced concrete plate, $V^{(1)}_{conc}$, and for the areal density of the shield, A, read:

$$V^{(1)}_{conc} = lwb^{(1)} - V^{(1)}_{steel}, \tag{28.2.5}$$

$$\begin{aligned} A &= \frac{m_{sh}}{lw} = \frac{\rho_{steel}(V^{(1)}_{steel} + V^{(2)}_{steel}) + \rho_{conc} V^{(1)}_{conc}}{lw} \\ &= \rho_{conc} b^{(1)} + \rho_{steel} b^{(2)} + \xi, \end{aligned} \tag{28.2.6}$$

where $V^{(2)}_{steel} = lwb^{(2)}$ is the volume of the steel plate, ρ_{steel} is density of steel, and

$$\xi = 2(\rho_{steel} - \rho_{conc}) \sum_{i=1}^{2} \left(\frac{a_i}{c_i} \right). \tag{28.2.7}$$

The problem is to determine the thicknesses of the reinforced concrete plate and of the steel plates providing the maximum BLV of the shield for given areal density taking into account the constraints in Eq. (28.2.3). It is assumed that $a_1, a_2, c_1, c_2, f_{cy}, \rho_{conc}, \rho_{steel}, m, p$ and A are known.

Using the dimensionless variables the mathematical formulation of the problem is as follows:

$$\overline{W}_{bl} = \frac{v_{bl}}{100k^2 b^{8/3}} = \overline{b}^{(1)5/3}[(\overline{\alpha} + \overline{b}^{(2)}) + 0.003\overline{b}^{(1)}] \to \max \quad (28.2.8)$$

taking into account the following constraints:

$$\begin{aligned} &\overline{b}^{(1)} \geq 0, \quad \overline{b}^{(2)} \geq 0, \quad \overline{b}^{(1)} + \overline{\theta}\overline{b}^{(2)} = 1, \\ &\overline{b}^{(1)} \geq (400/3)\overline{\alpha}, \quad 0.012\overline{b}^{(1)} \leq \overline{\alpha} + \overline{b}^{(2)} \leq 0.043\overline{b}^{(1)}, \end{aligned} \quad (28.2.9)$$

where

$$\overline{b}^{(1)} = \frac{b^{(1)}}{b}, \overline{b}^{(2)} = \frac{b^{(2)}}{b}, \overline{\alpha} = \frac{\alpha}{b}, b = \frac{A-\xi}{\rho_{conc}} > 0, \quad \overline{\theta} = \frac{\rho_{steel}}{\rho_{conc}}. \quad (28.2.10)$$

Hereafter we take into account that

$$3.2 \leq \overline{\theta} \leq 4.1. \quad (28.2.11)$$

28.2.3 *Investigation of problem*

Excluding the variable $\overline{b}^{(2)}$,

$$\overline{b}^{(2)} = (1 - \overline{b}^{(1)})/\overline{\theta}, \quad (28.2.12)$$

we obtain the following optimization problem:

$$F(\overline{b}^{(1)}) = \frac{\overline{\theta}\overline{W}_{bl}}{1 - 0.003\overline{\theta}} = \overline{b}^{(1)5/3}(\overline{\beta} - \overline{b}^{(1)}) \to \max, \quad (28.2.13)$$

subjected to the following constraints:

$$(400/3)\overline{\alpha} \leq \overline{b}^{(1)} \leq 1, \quad (28.2.14)$$

$$\overline{b}_*^{(1)} \leq \overline{b}^{(1)} \leq \overline{b}_{**}^{(1)}, \quad (28.2.15)$$

where

$$\overline{\beta} = \frac{\overline{\alpha}\overline{\theta} + 1}{1 - 0.003\overline{\theta}}, \quad \overline{b}_*^{(1)} = \frac{\overline{\alpha}\overline{\theta} + 1}{1 + 0.043\overline{\theta}}, \quad \overline{b}_{**}^{(1)} = \frac{\overline{\alpha}\overline{\theta} + 1}{1 + 0.012\overline{\theta}}. \quad (28.2.16)$$

Consider the behavior of the function $F(\bar{b}^{(1)})$. Since

$$F'(\bar{b}^{(1)}) = \bar{b}^{(1)2/3}(5\bar{\beta} - 8\bar{b}^{(1)})/3, \tag{28.2.17}$$

the function F increases when $0 < \bar{b}^{(1)} < 0.625\bar{\beta}$ and decreases when $\bar{b}^{(1)} > 0.625\bar{\beta}$, i.e., $\bar{b}^{(1)} = \bar{b}_0^{(1)} = 0.625\bar{\beta}$ is the maximum of F.

If inequalities in Eqs. (28.2.14) and (28.2.15) are mutually exclusive, the solution of the considered problem does not exist. The latter occurs when

$$\bar{b}_*^{(1)} > 1 \text{ or } \bar{b}_{**}^{(1)} < (400/3)\bar{\alpha}. \tag{28.2.18}$$

Since

$$\begin{aligned} \bar{b}_*^{(1)} - 1 &= \frac{(\bar{\alpha} - 0.043)\bar{\theta}}{1 + 0.043\bar{\theta}}, \\ \bar{b}_{**}^{(1)} - (400/3)\bar{\alpha} &= \frac{1 - \bar{\alpha}(0.6\bar{\theta} + 400/3)}{1 + 0.012\bar{\theta}} \end{aligned} \tag{28.2.19}$$

Eq. (28.2.18) implies that

$$\bar{\alpha} > 0.043, \tag{28.2.20}$$

$$\bar{\alpha} > \bar{\alpha}_\times, \tag{28.2.21}$$

where

$$\bar{\alpha}_\times = 1/(0.6\bar{\theta} + 400/3). \tag{28.2.22}$$

Since $1/(0.6\bar{\theta} + 400/3) < 0.043$, the constraint given by Eq. (28.2.20) can be ignored. Therefore the solution of the problem does not exist if Eq. (28.2.21) is valid. In the opposite case, if

$$\bar{\alpha} \le \bar{\alpha}_\times \tag{28.2.23}$$

the solution of the optimization problem can be found.

If inequality (28.2.23) is satisfied, the non-empty interval of the admissible values of $\bar{b}^{(1)}$ for which inequalities (28.2.14) and (28.2.15) are satisfied can be described as follows:

$$\bar{b}_{\times}^{(1)} \le \bar{b}^{(1)} \le \bar{b}_{\times\times}^{(1)}, \tag{28.2.24}$$

where

$$\bar{b}_{\times}^{(1)} = \max\left((400/3)\bar{\alpha}, \bar{b}_{*}^{(1)}\right), \quad \bar{b}_{\times\times}^{(1)} = \min\left(1, \bar{b}_{**}^{(1)}\right). \tag{28.2.25}$$

Since

$$\bar{b}_{*}^{(1)} - \bar{b}_{0}^{(1)} = \frac{(\bar{\alpha}\bar{\theta}+1)(3-0.239\bar{\theta})}{(1-0.003\bar{\theta})(1+0.043\bar{\theta})} > 0, \tag{28.2.26}$$

the function $F(\bar{b}^{(1)})$ decreases in the domain determined by Eq. (28.2.15) and, consequently, it attains the maximum in the point $\bar{b}^{(1)} = \bar{b}_{\times}^{(1)} = \bar{b}_{opt}^{(1)}$:

$$\bar{b}_{opt}^{(1)} = \begin{cases} (400/3)\bar{\alpha} & if \quad (400/3)\bar{\alpha} \ge \bar{b}_{*}^{(1)} \\ \bar{b}_{*}^{(1)} & if \quad (400/3)\bar{\alpha} \le \bar{b}_{*}^{(1)} \end{cases} \tag{28.2.27}$$

After some algebra and taking into account Eq. (28.2.23), the solution given by Eq. (28.2.27) can be written in the following form:

$$\bar{b}_{opt}^{(1)} = \begin{cases} (400/3)\bar{\alpha} & if \quad \bar{\alpha}_{\times\times} \le \bar{\alpha} \le \bar{\alpha}_{\times} \\ \bar{b}_{*}^{(1)} & if \quad 0 < \bar{\alpha} \le \bar{\alpha}_{\times\times} \end{cases} \tag{28.2.28}$$

where

$$\bar{\alpha}_{\times\times} = 3/(14.2\bar{\theta}+400) < \bar{\alpha}_{\times}. \tag{28.2.29}$$

Formula for the thickness of the steel plate corresponding to the optimal solution, $\bar{b}_{opt}^{(2)}$, reads (see Eq. (28.2.12)):

$$\bar{b}_{opt}^{(2)} = (1-\bar{b}_{opt}^{(1)})/\bar{\theta}. \tag{28.2.30}$$

It is convenient to use the ratio

$$\delta = \frac{v_{bl}^{opt}}{v_{bl}^{0}} = \frac{F\left(\bar{b}_{opt}^{(1)}\right)}{F(1)} = \sqrt{\frac{\bar{b}_{opt}^{(1)5/3}(\bar{\beta}-\bar{b}_{opt}^{(1)})}{\bar{\beta}-1}} \tag{28.2.31}$$

for quantifying the advantage of the optimal two-layer shield, where v_{bl}^{0} is the BLV of the reinforced concrete plate with the same areal density.

Results of the investigation of the optimization problem are shown in Figs. 28.2.2 and 28.2.3 and demonstrate the effectiveness of rear lining using a steel plate. Inspection of Fig. 28.2.4 reveals that the effect of the thickness of the steel plate on the BLV of the shield is large, i.e., it is anticipated that optimization results in significant improvement of ballistic characteristics.

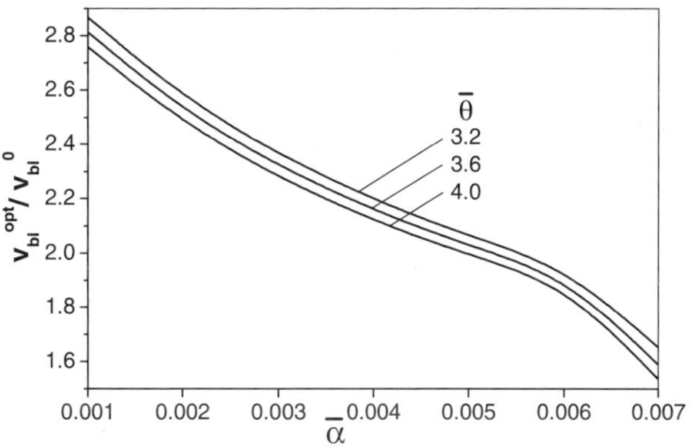

Fig. 28.2.2. Normalized BLV velocity for the optimal shield vs. parameters $\overline{\theta}$ and $\overline{\alpha}$.

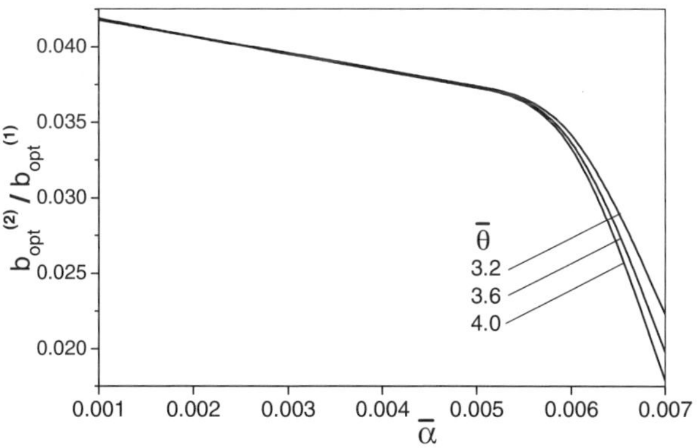

Fig. 28.2.3. Steel liner thickness normalized by the concrete panel thickness of the optimal shield vs. parameters $\overline{\theta}$ and $\overline{\alpha}$.

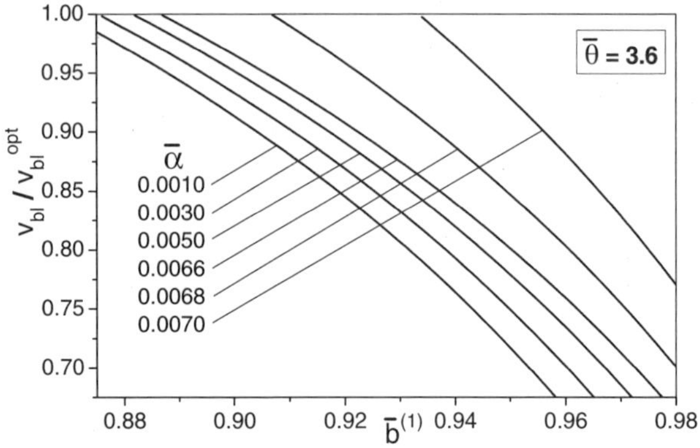

Fig. 28.2.4. BLV for the shield with a given concrete panel thickness normalized by the BLV for the optimal shield, for different parameter $\overline{\alpha}$.

Barr, 1990 claimed that the used above model is valid for $45 < v_{bl} < 300$. At the same time, he recommended using a more accurate formula for the BLV if the BLV is larger than $70\ m/s$:

$$\widetilde{v}_{bl} = v_{bl}(1 + 0.000004v_{bl}^2), \qquad (28.2.32)$$

where v_{bl} is the BLV determined by Eq. (28.2.1). Since $\widetilde{v}_{bl}$ increases as a function of v_{bl}, the obtained optimal solution remains valid when the criterion given by Eq. (28.2.32) is used. However, quantification of the advantage of the optimal solution is less transparent in the this case because the solution depends not only on $\overline{\theta}$ and $\overline{\alpha}$.

Chapter 29

Optimization of Two-Component Armor against Single and Repeated Impacts

Nomenclature

a	parameter, Eq. (29.2.2);
AD	area density of shield;
$\overline{A}_1$	$=\gamma_1 b_1/(\gamma_1 b_1+\gamma_2 b_2)$, dimensionless AD of ceramic plate;
b	total thickness of shield;
b_1	thickness of ceramic plate;
b_1^*	parameter, Eq. (29.5.5);
b_2	thickness of back plate;
BLV	ballistic limit velocity;
$c^{(i)}$	parameters, Eq. (29.3.5);
$e^{(i)}$	parameters, Eq. (29.3.5);
E_1	energy required for formation of conoid;
f_1, f_2	functions, Eq. (29.2.11);
F	criterion, Eq. (29.3.4);
G	function, Eq. (29.5.3);
I	momentum, see Eq. (29.2.1) and the text above;
$\tilde{I}$	momentum transmitted to ceramic plate due to shear strength of plate material;
I_{0-1}	momentum of impactor-conoid system after formation of conoid;
M_0	mass of projectile;
M_1	mass of conoid, Fig. 29.1.1, Eq. (29.2.4);
M_2	mass of back plate under conoid, Eq. (29.2.4);

Q_1, Q_2	functions, Eq. (29.2.11);
R	radius of projectile;
S_1	area of lateral surface of conoid, Eq. (29.2.8);
TT	total thickness of shield;
v_{bl}	BLV of projectile;
v_{0-1}	velocity of impactor-conoid system after formation of conoid;
x	$=\alpha_1 b_1/\mu$;
X	$=Q_2/Q_1$, Eq. (29.3.1);
Y	$=v_{bl}/Q_1$, Eq. (29.3.1);
α_1, α_2	parameters, Eq. (29.4.2);
β_1	parameter, Eq. (29.2.8);
β_2	parameter, Eq. (29.2.2);
γ_1	density of ceramics;
γ_2	density of back plate;
ε_2	breaking strain of ceramics;
ε_{av}	average relative error, Eq. (29.3.4);
θ_1, θ_2	parameters, Eq. (29.2.11);
μ	parameter, Eq. (29.4.2);
ξ_0, ξ_1	parameters, Eq. (29.5.4);
σ_2	ultimate tensile strength of ceramics;
φ	angle of conoid, Fig. 29.1.1;
Superscript	
opt	optimal value of parameter.

29.1 Introduction

Two-component composite armors consisting of a ceramic front plate and a ductile back plate attract considerable interest because of their high ballistic performance and light weight. Since several surveys were recently published on this topic by Zaera, 2011 and Walley, 2009 (see also Abrate, 1998), we only briefly review publications where Florence,

1969 model was applied for optimization of two-component composite armor.

Florence, 1969 modeled impactor as a short cylindrical rod that strikes a ceramic plate (Fig. 29.1.1). The ceramic plate breaks progressively into a cone of fractured material. The impact energy is transferred to the back plate which deforms like a uniform membrane. The simplifying assumptions that allowed deriving an analytical expression for the BLV can be summarized as follows (Hetherington and Rajagopalan, 1991): (i) diameter of the circular area at the back plate over which the momentum is distributed is equal to the base diameter of the fracture conoid in the ceramic facing, and the angle of the conoid is chosen to be equal to 63°; (ii) deformation history of the back plate can be modeled by motion of a membrane clamped around the perimeter of the base of the fractured conoid, and the initial conditions for the motion of membrane are determined by the condition imposed by the projectile impact and by the conservation of momentum in the projectile/shield system; (iii) failure occurs when the maximum tensile strain in the membrane reaches the ultimate breaking strain of the back plate.

Although numerous other analytical models were suggested for analyzing perforation of two-component ceramic shields (Chocron-Benloulo and Sanchez-Galvez, 1998; Den Reijer, 1991;

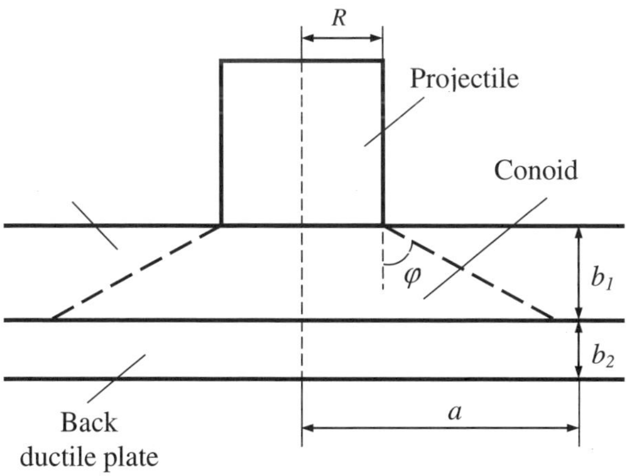

Fig. 29.1.1. Scheme of perforation of two-component armor.

Feli *et al.*, 2010, Feli *et al.*, 2011; Fellows and Barton, 1999; Gonçalves *et al.*, 2004; Kang and Cho, 1991; Li *et al.*, 2002; Navarro *et al.*, 1994; Parameswaran *et al.*, 1999; Ravid *et al.*, 2001; Walker and Anderson, 1996; Woodward, 1990; Zaera and Sanchez-Galvez, 1998; Zaera *et al.*, 2000; Zhang *et al.*, 2002; Zhang *et al.*, 2006a), Florence model was found to be the most suitable for solving problems associated with armor optimization. Some early numerical results obtained with this approach were reported by Florence, 1969 and Hetherington and Rajagopalan, 1991. In the latter study, the original model was slightly reworked, and this modified version was later used for armor optimization against normal impact. Hetherington, 1992a, 1992b investigated analytically the problem of determining the structure of the two-component armor with a given areal density (AD) that provides the maximum ballistic limit velocity (BLV). He suggested an approximate expression for the optimum value of the ratio of the front plate width to the back plate width. Wang and Lu, 1996 investigated a similar problem where the total thickness (TT) of the armor rather than the AD was given. Ben-Dor *et al.*, 2000b, 2005a investigated problems of determining the structure of armor with the maximum BLV for a given areal density and the armor having the minimum AD for a given impact velocity. In these studies, using appropriate dimensionless variables, the solutions of the optimization problems for an arbitrary two-component composite armor was determined in an analytical form. Shi and Grow, 2007 investigated the problem of two-component armor optimization whereby both, the TT and the AD, are used as the constraints. Fawaz *et al.*, 2006 studied the problem numerically using the technique that allows solving the problem using either a traditional formulation or in a generalized formulation when the armor materials are considered as the unknown design parameters.

Hetherington and Lemieux, 1994; Sadanandan and Hetherington, 1997 investigated a modified Florence model for the case of an oblique impact.

Some data regarding the comparison of the results of the armor optimization based on the Florence model with numerical simulations and experiments can be found in Ben-Dor *et al.*, 1999a; Hohler *et al.*,

2001; Ko *et al.*, 1997; Lee and Yoo, 2001; Shokrieh and Javadpour, 2008; Zaera, 2011. Zhang *et al.*, 2011a illustrated the effect of thicknesses of plates on the BLV of the armor using Florence model.

An imperfection of the Florence model which has been repeatedly mentioned in the literature (see, e.g., Zhang *et al.*, 2002; Navarro *et al.*, 1993) is that this model does not take into account mechanical properties of the ceramics and energy dissipation in the course of ceramics fracture. Ben-Dor *et al.*, 2009a proposed a modification of the Florence model which is free of this shortcoming. Using the suggested model, they solved typical problems of two-component armor optimization against a single impact and new problems taking into account a possibility of the repeated impacts which can not be considered on the basis of the original Florence model. These results are used in the subsequent sections. An approach taking into account the dissipation of energy of the ceramic layer was proposed also by Serjouei *et al.*, 2012.

29.2 Mathematical Model

As a basis, we use the intermediate result from Florence, 1969 determining the minimal momentum, I, that must be applied to the system including impactor-conoid-membrane in order to break the membrane at the stage of motion of the system as a whole:

$$I = \sqrt{\pi\beta_2 a^2 b_2 (M_0 + M_1 + M_2)}\,, \qquad (29.2.1)$$

where

$$\beta_2 = \varepsilon_2\sigma_2/0.91, \quad a = R + b_1 \tan\varphi\,, \qquad (29.2.2)$$

M is mass, R is projectile's radius, b is plate thickness, σ is the ultimate tensile strength, ε is the breaking strain, subscripts $0, 1$ and 2 denote projectile, ceramic conoid and portion of the back plate under the conoid, respectively. The model is written in a slightly generalized form: an arbitrary angle φ is introduced instead of 63° as in the original Florence model (Fig. 29.1.1).

Florence, 1969 assumed that I equals to the momentum of the projectile that impacts the shield with the BLV, v_{bl}. Substituting $I = M_0 v_{bl}$ into Eq. (29.2.1) we can obtain the original Florence model in the form of the expression for the BLV:

$$v_{bl} = \sqrt{\pi\beta_2 a^2 b_2 (M_0 + M_1 + M_2)/M_0^2}\,. \qquad (29.2.3)$$

Masses M_1 and M_2 can be expressed through the densities γ_1 and γ_2 and the volumes of the conoid and the cylindrical part of the back plate under the conoid:

$$M_1 = 1/3\,\pi\gamma_1 b_1 (R^2 + aR + a^2), \quad M_2 = \pi\gamma_2 b_2 a^2. \qquad (29.2.4)$$

Substituting the above formula for M_2 and a simplified formula for M_1, $M_1 = \pi\gamma_1 b_1 a$ (the conoid is replaced by the cylinder with the same basis and the same height), into Eq. (29.2.3) we obtain the version of the model suggested by Hetherington and Rajagopalan, 1991 (see also Smith and Hetherington, 1994):

$$v_{bl} = \sqrt{\pi\beta_2 a^2 b_2 [M_0 + \pi a^2 (\gamma_1 b_1 + \gamma_2 b_2)]/M_0^2}\,. \qquad (29.2.5)$$

Eq. (29.2.5) describes the model which is usually used for investigating optimization problems.

In this study we use the expressions for M_1 from Eqs. (29.2.4) and consider the process of penetration in more details.

Eq. (29.2.1) implies a physically incorrect value $v_{bl} = 0$ for $b_2 = 0$ because of neglecting the energy required for fracture of ceramics. This shortcoming can result in the erroneous solution of the optimization problems. Consequently, the modified Florence model must be improved.

To this end, we use momentum conservation at all the stages of penetration and write the momentum balance equation, taking into account formation of conoid, as follows (Recht, 1990):

$$M_0 v_{bl} = I_{0-1} + \tilde{I}, \quad I_{0-1} = (M_0 + M_1) v_{0-1}, \qquad (29.2.6)$$

where $\tilde{I}$ is the momentum transferred to the ceramic plate due to shear strength of plate material, v_{0-1} and I_{0-1} are the velocity and the momentum of the impactor-conoid system after formation of conoid. Taking into account Eqs. (29.2.1) and (29.2.6), the momentum balance equation, and setting $I_{0-1} = I$, we arrive at the following formula:

$$v_{bl} = (\sqrt{\pi\beta_2 a^2 b_2 (M_0 + M_1 + M_2)} + M_0 \tilde{I}) / M_0^2 . \qquad (29.2.7)$$

Using the approach suggested in Cantwell and Morton, 1990 (see also Caprino *et al.*, 2007) for the limiting case of ceramic plate ($b_2 = 0$) we assume that energy required for formation of conoid is determined by the following expression:

$$E_1 = \beta_1 S_1, \quad S_1 = \pi b_1 (R + a) / \cos\varphi \ , \qquad (29.2.8)$$

where β_1 is fracture energy for transverse shearing per unit shear area and S_1 is the area of the lateral surface of the conoid. Here we slightly modified the model proposed in Cantwell and Morton, 1990 by replacing the angle $\varphi = 45^\circ$ by an arbitrary angle φ (Fig. 29.1.1). Substituting $v_{bl} = \tilde{I} / M_0$ and E_1 from Eq. (29.2.8) into the expression $1/2\, M_0 v_{bl}^2 = E_1$ we find that $\tilde{I} = \sqrt{2\beta_1 S_1 M_0}$ and

$$v_{bl} = \sqrt{\pi\beta_2 a^2 b_2 (M_0 + M_1 + M_2)} / M_0 + \sqrt{2\beta_1 S_1 / M_0} \ . \qquad (29.2.9)$$

Substituting M_1 and M_2 from Eq. (29.2.4) we can rewrite Eq. (29.2.9) as follows:

$$v_{bl} = f_1(b_1) + f_2(b_1, b_2), \qquad (29.2.10)$$

where

$$f_i = \theta_i Q_i, \quad \theta_i = \sqrt{\beta_i}, \quad i = 1,2$$

$$Q_1(b_1) = \sqrt{2\pi b_1 (2R + b_1 \tan\varphi) / (M_0 \cos\varphi)} \qquad (29.2.11)$$

$$Q_2(b_1, b_2) = \frac{\sqrt{\pi a^2 b_2 [M_0 + 1/3\, \pi\gamma_1 b_1 (R^2 + aR + a^2) + \pi\gamma_2 b_2 a^2]}}{M_0}$$

29.3 Reduction of Experimental Data for Aluminum/ Alumina Armor

In our calculations we use the experimental data obtained by Wilkins, 1978 in the digital form from Holmquist *et al.*, 1999 for two-component armor consisting of AD85 alumina ceramic front plate ($\gamma_1 = 3430\, kg/m^3$) and 6061-T6 aluminum rear plate ($\gamma_2 = 2765\, kg/m^3$) penetrated by a conical projectile ($M_0 = 8.32\, g$, $R = 3.81\, mm$).

Eqs. (29.2.10)-(29.2.11) imply the following relationship:

$$Y = \theta_2 X + \theta_1, \quad Y = v_{bl} / Q_1, \quad X = Q_2 / Q_1, \tag{29.3.1}$$

which we consider as a linear regression equation with the unknown coefficients θ_1 and θ_2. The set of experimental data obtained by Wilkins, 1978 includes the value of the BLV, v_{bl}^0, for the shield consisting only of the front ceramic plate (without a rear plate) having the thickness b_1^0 which is of particular interest to us. Consequently, we require that the approximating line will pass through the point

$$X = X^{(1)} = 0, Y = Y^{(1)} = v_{bl}^0 / Q_2(b_1^0). \tag{29.3.2}$$

This condition implies that

$$\theta_1 = Y^{(1)} \tag{29.3.3}$$

and only θ_2 must be determined.

A standard criterion of the quality of the approximation, the average relative error, ε_{av}, can be used for determining θ_2. Here we avoid a usual contradiction when the least squares method is used for approximating while the quality of approximation is accessed using another criterion, e.g., the average relative error. Hence the problem is reduced to the following minimization problem:

$$F(\theta_2) = n\varepsilon_{av} = \sum_{i=2}^{n} \left| \frac{\theta_2 X^{(i)} + \theta_1 - Y^{(i)}}{Y^{(i)}} \right| = \sum_{i=2}^{n} e^{(i)} \mid \theta_2 - c^{(i)} \mid \to \min, \tag{29.3.4}$$

where

$$e^{(i)} = \left|\frac{X^{(i)}}{Y^{(i)}}\right|, c^{(i)} = \frac{Y^{(i)} - Y^{(1)}}{X^{(i)}}, X^{(i)} \neq 0, Y^{(i)} \neq 0, \quad i = 2,3,,,n . \quad (29.3.5)$$

In order to avoid rearranging the superscripts we assume they are changed so that $c^{(2)} \leq c^{(3)} \leq ... \leq c^{(n-1)} \leq c^{(n)}$. Hence for some $c^{(j)} \leq \theta_2 \leq c^{(j+1)}$,

$$F(\theta_2) = \left[\sum_{i=2}^{j} e^{(i)} - \sum_{i=j+1}^{n} e^{(i)}\right]\theta_2 + \left[\sum_{i=j+1}^{n} c^{(i)} - \sum_{i=2}^{j} c^{(i)}\right], \quad (29.3.6)$$

i.e., $F(\theta_2)$ is a piecewise linear function. Consequently, taking into account that $F(\theta_2) \to +\infty$ for $\theta_2 \to \pm\infty$, the minimum $F(\theta_2)$ is located in one of the end points of the intervals, $F(c^{(2)}), F(c^{(3)}),..., F(c^{(n)})$, and can be readily found by inspection.

The results of calculations are shown in Fig. 29.3.1 for different values of the angle, φ. Clearly, the scatter of the points around the optimum regression line strongly depends upon the selected value of the angle, φ, and the Florence value $\varphi \approx 63^\circ$ is not the optimal. The latter observation is expressed quantitatively in Fig. 29.3.2 which shows the minimal relative error of the approximation as a function of φ. Consequently, this non-unique choice of φ which has been mentioned previously in the literature (Hetherington and Lemieux, 1994; Wilson and Hetherington, 1995; Zhang *et al.*, 2010) is supported by our calculations. For the further analysis, we select the value $\varphi = 45^\circ$ that is close to the optimum (see Fig. 29.3.2). In this case, $\theta_1 = 865(N/m)^{1/2}$, $\theta_2 = 15797\, m^{-1/2}$ and $\varepsilon_{av} = 7.4\%$.

The goal of the outlined above approach is to afford the best possible description of the experimental data using the suggested model. In many instances, because of the approximate character of the model and experimental errors, the obtained values of β_1 are β_2 differ from the characteristic values for the corresponding materials. Therefore, although in the model these parameters have a clear mechanical meaning, it will be proper to consider them as empirical coefficients which are determined from the experimental data. In order to overcome the latter contradiction the combined criterion for determining the parameters of the model can be used.

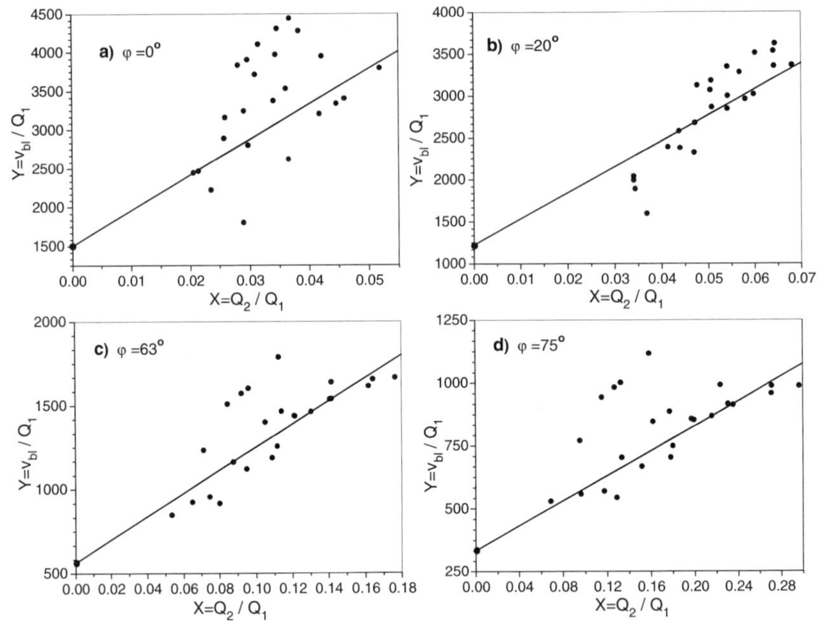

Fig. 29.3.1. Scatter of the experimental data by Wilkins, 1978 around the optimum regression line depending on the choice of angle of fracture conoid.

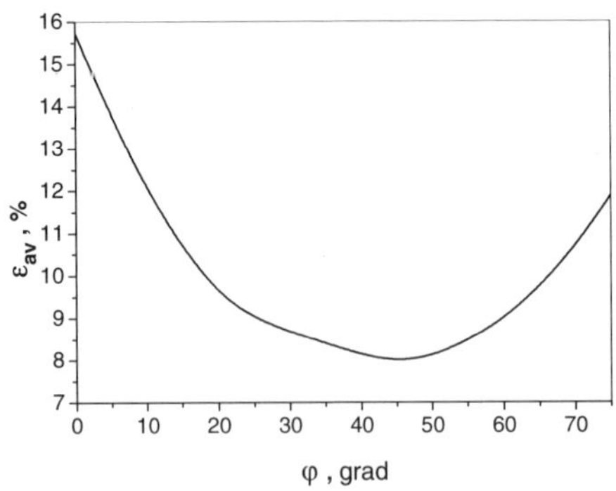

Fig. 29.3.2. Relative error of approximation as a function of angle of fracture conoid.

This combined criterion must take into account both requirements: (i) adequate description of the experimental data and (ii) proximity of the parameters of the model to their values as mechanical properties. It must be emphasized that availability of the mathematical description of the phenomenon is not a necessary condition for deriving a predictive model. Such approaches for modeling bi-component armor which are based (or can be based) exclusively upon the experimental data were employed by Ben-Dor *et al.*, 1999a and Fernández-Fdz and Zaera, 2008.

29.4 Optimal Armor against Single Impact

29.4.1 *Formulation of problem and results of calculations*

Consider a problem of determining b_1 and b_2 that provide the minimum BLV taking into account the following constraints:

$$b_1 \geq 0, \quad b_2 \geq 0, \tag{29.4.1}$$

$$\alpha_1 b_1 + \alpha_2 b_2 = \mu, \tag{29.4.2}$$

where α_1, α_2 and μ are assumed to be known. Two sets of coefficients in Eq. (29.4.2) have a clear physical meaning ($j = 1,2$) in two different optimization problems: (i) $\alpha_j = \gamma_j$, densities of the materials, μ is an areal density (AD) of armor and the problem is to optimize the BLV for an armor with a given AD; and (ii) $\alpha_j = 1$, μ is the total thickness (TT) of the armor; and the problem is to optimize the BLV for the armor with given TT and b.

Let us introduce a new variable, $x = \alpha_1 b_1 / \mu$. After substituting $b_1 = \mu x / \alpha_1$ and $b_2 = \mu(1-x)/\alpha_2$ into the expression for the BLV, the optimization problem is reduced to the maximization of the function of one variable, x, subjected to the linear constraints, $0 \leq x \leq 1$. Because the dependence of the BLV on the parameters is quite involved, the problem cannot be solved analytically while the numerical solution of the problem is straightforward.

Consider the first problem when the AD of the armor is given. The BLV as a function of the normalized AD of the ceramic plate,

$x = \overline{A}_1 = \gamma_1 b_1/(\gamma_1 b_1 + \gamma_2 b_2)$, is shown in Fig. 29.4.1a for different values of AD of the armor. Inspection of Fig. 29.4.1a shows that the maximum BLV is attained for $\overline{A}_1 = \overline{A}_1^{opt} \approx 0.75$, and is practically independent of the given AD of the armor. The optimum ratio of the thicknesses of the plates can be calculated using the following relationship:

$$\frac{b_1}{b_2} = \frac{\gamma_2}{\gamma_1}\frac{\overline{A}_1}{1-\overline{A}_1}. \tag{29.4.3}$$

Substituting $\overline{A}_1 = 0.75$ we find that in the considered case, $b_1^{opt}/b_2^{opt} \approx 2.4$, where b_1^{opt} and b_2^{opt} are the plate thicknesses corresponding to the solution of the optimization problem. The minimization criterion, i.e., the BLV as a function of the given AD, is shown in Fig. 29.4.1b. Since this function is increasing, the same curve is the solution of the inverse problem: minimizing AD for a given BLV. The latter assertion can be easily proved by contradiction.

We solved also several optimization problems similar to those considered above when TT of the armor, b, is considered as a criterion of optimization instead of the AD. The results are shown in Fig. 29.4.2. In this case, $b_1^{opt}/b \approx 0.77$ (Fig. 29.4.2a) and hence, the ratio, $b_1^{opt}/b_2^{opt} \approx 3.3$, is practically independent of the given thickness of the armor.

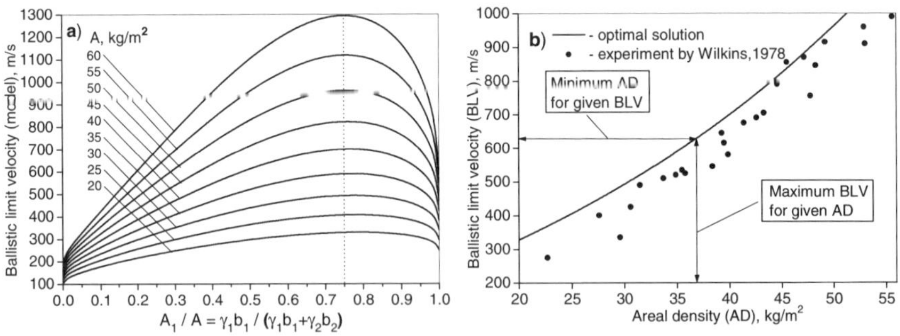

Fig. 29.4.1. Optimization against single impact taking into account total AD of armor; a) BLV as a function of normalized AD of ceramic plate; b) results of solution of the direct and the inverse optimization problems.

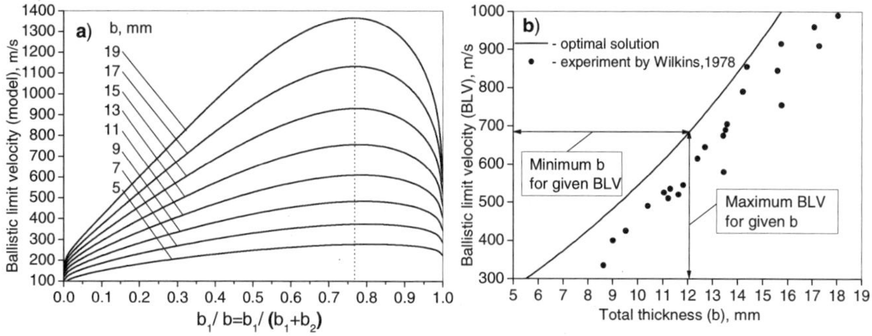

Fig. 29.4.2. Optimization against single impact taking into account TT of armor; a) BLV as a function of normalized thickness of ceramic plate; b) results of solution of the direct and the inverse optimization problems

The circles in Figs. 29.4.1b and 29.4.2b denote the experimental data from Wilkins, 1978. Special attention must be given to the fact that practically every experimental point corresponds to the value of the criterion which is worse than the optimum value obtained as a result of optimization. The latter observation is an important argument in favor of the assertion that the proposed model is not only a good approximation for the experimental data, but it also adequately describes the effect of the plates thicknesses on the BLV of armor. Our results are in good agreement with the numerical and experimental data obtained by Lee and Yoo, 2001 as well as with theoretical analysis by Ben-Dor *et al.*, 2000b, 2005a that is based on the Florence model. A plausible explanation is that in the considered case, the competitive versions of the armor structure do not include shields with small values of the thicknesses of the back plate for which the Florence model is known to provide a crude approximation.

29.4.2 *Characteristic property of optimal solutions*

Calculations show that optimum is attained for the same ratio of plates thicknesses (plates ADs) in both cases, i.e. for the direct and for the inverse problems. In order to demonstrate that this is not a mere coincidence, we will prove the following property of the optimal solutions: if the optimum ratio of the plate thicknesses (plates ADs),

x^{opt}, corresponding to the solution of the direct problem (maximization of the BLV for a given AD (TT), μ) does not depend on AD (TT) of the armor and the BLV increases as a function of μ (for a constant x), then $x = x^{opt}$ is a solution of the inverse problem (minimization of the AD (TT), μ, for a given BLV). We also assume that (i) x^{opt} is the only point of local minimum of the BLV considered as a function x and μ, $v_{bl} = \psi(x,\mu)$, and (ii) optimal solution of the inverse problem is attained in a point of a local maximum. These assumptions are valid for the proposed model. The assumption that x^{opt} equals a constant is no more than an approximation of the solution of the direct problem which yields the useful property of the inverse problem that is valid also with the same accuracy.

Since $x = x^{opt}$ is a point of local minimum of the function, $\psi(x,\mu)$, then

$$\psi_x(x^{opt},\mu) = 0, \quad \psi_{xx}(x^{opt},\mu) < 0 \tag{29.4.4}$$

for any μ. The inverse problem can be formulated as minimization of μ when the BLV, $\tilde{v}_{bl}$, is given, i.e., the equation $\psi(x,v) = \tilde{v}_{bl}$ is valid. Let us differentiate the latter equation considering μ as a function of x:

$$\begin{aligned} &\psi_x(x,v) + \psi_\mu(x,\mu)\,\mu'(x) = 0, \quad \psi_{xx}(x,\mu) + 2\psi_{x\mu}(x,\mu)\,\mu'(x) \\ &+\psi_\mu(x,\mu)[\mu'(x)]^2 + \psi_\mu(x,\mu)\,\mu''(x) = 0. \end{aligned} \tag{29.4.5}$$

Since $\psi_\mu(x,\mu) > 0$ and $\psi_x(x^{opt},\mu) = 0$, the first equation in Eqs. (29.4.5) implies that $\mu'(x^{opt}) = 0$. Taking into account Eq. (29.4.4), the second equation in Eqs. (29.4.5) yields: $\mu''(x^{opt}) = -\psi_{xx}(x^{opt}) / \psi_\mu(x^{opt},\mu) > 0$, where μ is the solution of the equation $\psi(x^{opt},\mu) = \tilde{v}_{bl}$. Consequently, $x = x^{opt}$ is the point of the local minimum that is the point of optimum in the inverse problem. Note that $\psi(x^{opt},\mu) = v_{bl}$ is the equation of the optimum curve shown in Fig. 29.4.1b ($\mu = A$) and in Fig. 29.4.2b ($\mu = b$). The relation between the solutions of the direct and the inverse optimization problems can also be revealed in a more general case when x^{opt} in the direct problem is not a constant but investigating this particular mathematical problem is beyond the scope of this study.

29.5 Optimization of Armor Taking into Account Repeated Impacts

In contrast to the Florence model, the proposed model allows to consider optimization problems for the case when repeated impacts at the same location of the shield are possible. The problem can be formulated as follows. Determine the thicknesses of the plates (b_1 and b_2) in the armor with the minimum AD or TT that shields from a single or repeated impacts by projectiles having velocity v_{imp}. It is assumed that mass and diameter of the impactor and mechanical properties of the materials of the plates are known.

There are two scenarios for perforation of the shield.

The first scenario is realized when armor is perforated as the result of the first impact. In this case, the impact velocity is larger than the BLV, i.e.,

$$f_1(b_1) + f_2(b_1, b_2) \le v_{imp}. \tag{29.5.1}$$

According to the second scenario, only fracture conoid is formed after the first impact whereas perforation occurs after the repeated impact. This scenario is described by the following inequalities:

$$f_1(b_1) + f_2(b_1, b_2) > v_{imp},\ f_1(b_1) \le v_{imp},\ f_2(b_1, b_2) \le v_{imp}. \tag{29.5.2}$$

Domain *ACBOA* (see Fig. 29.5.1a) contains the set of parameters b_1 and b_2 that correspond to the first scenario. It is determined by Eqs. (29.4.1) and (29.5.1) and is located below the curve *ACB* described by the equation $b_2 = G(b_1)$, where

$$G(b_1) = \left(\sqrt{\xi_1^2 + 4\xi_0^2 \xi_2} - \xi_1\right)\Big/(2\xi_2), \quad 0 \le b_1 \le b_1^*, \tag{29.5.3}$$

$$\begin{aligned} &\xi_0 = M_0[v_{imp} - f_1(b_1)]/\theta_2, \\ &\xi_1 = M_0 + 1/3\,\pi\gamma_1 b_1 (R^2 + aR + a^2),\ \xi_2 = 4\pi^2\gamma_2 a^4, \end{aligned} \tag{29.5.4}$$

$$b_1^* = \sqrt{\frac{R^2}{\tan^2\varphi} + \frac{M_0 \cos\varphi\, v_{imp}^2}{2\pi\theta_1^2 \tan\varphi}} - \frac{R}{\tan\varphi}. \tag{29.5.5}$$

Here $G(b_1)$ is the function which is obtained by solving the equation $f_1(b_1) + f_2(b_1, b_2) = v_{imp}$ with respect to b_2, and b_1^* is the solution of the equation $f_1(b_1) + f_2(b_1, 0) = v_{imp}$ that is reduced to the equation $f_1(b_1) = v_{imp}$ because $f_2(b_1, 0) = 0$.

Domain $AEDBCA$ (see Fig. 29.5.1a), which is determined by Eqs. (29.4.1) and (29.5.2), corresponds to the second scenario. It is located between the curves AED and ACB and the line $b_1 = b_1^*$. Equation of the curve AED, $b_2 = g(b_1)$, is obtained by solving the equation $f_2(b_1, b_2) = v_{imp}$ with respect to b_2:

$$g(b_1) = \frac{\sqrt{\xi_1^2 + 4\tilde{\xi}_0^2 \xi_2} - \xi_1}{2\xi_2}, \quad \tilde{\xi}_0 = \frac{M_0 v_{imp}}{\theta_2}, \quad 0 \le b_1 \le b_1^*. \tag{29.5.6}$$

Since $f_1(b_1) > 0$ for $b_1 > 0$ and $f_1(0) = 0$, the curve AED is always located above the curve ACB when $b_1 > 0$ and both curves intersect in the point $b_1 = 0$.

Therefore the domain $AEDBOA$ includes the values of b_1 and b_2 for which the armor is perforated. Accordingly, the domain in the first quadrant outside of the domain $AEDBOA$ represents "not perforation domain" (NPD). The considered optimization problem is to find b_1 and b_2 from the NPD that provide the minimum $\mu = \alpha_1 b_1 + \alpha_2 b_2$ (AD or TT) for a given BLV. Due to the linearity of the optimization criterion, the outline of the solution can be readily illustrated geometrically.

Consider the family of lines, $\alpha_1 b_1 + \alpha_2 b_2 = \mu$, where $\mu > 0$ is a parameter (see Fig. 29.5.1b). The distance from the origin of coordinates to a line $K_1 K_2$, $L_{OH} = \mu / \sqrt{\alpha_1^2 + \alpha_2^2}$, is proportional to μ. Our goal is to find the maximum μ such that at least one point at the corresponding line is located in the NPD. Clearly, this point can belong only to the segment of the border, ADB. Analysis shows that $g(b_1)$ is a concave downward function. Consequently, one of two locations of the optimal line is possible: location $K_1' K_2'$ (the line is tangent to the curve

$b_2 = g(b_1)$ between the points A and D or passes through the point A) or location $K_1''K_2''$ (the line passes through the point B). This property of optimal solution implies the simplest method for determining the best parameter b_1: the choice must provide a minimal value between μ_0 and μ_1, where

$$\mu_0 = \alpha_1 b_1^*, \quad \mu_1 = \min_{0 \le b_1 \le b_1^*} \left(\alpha_1 b_1 + \alpha_2 g(b_1)\right) \tag{29.5.7}$$

and μ_1 is determined by exhaustion.

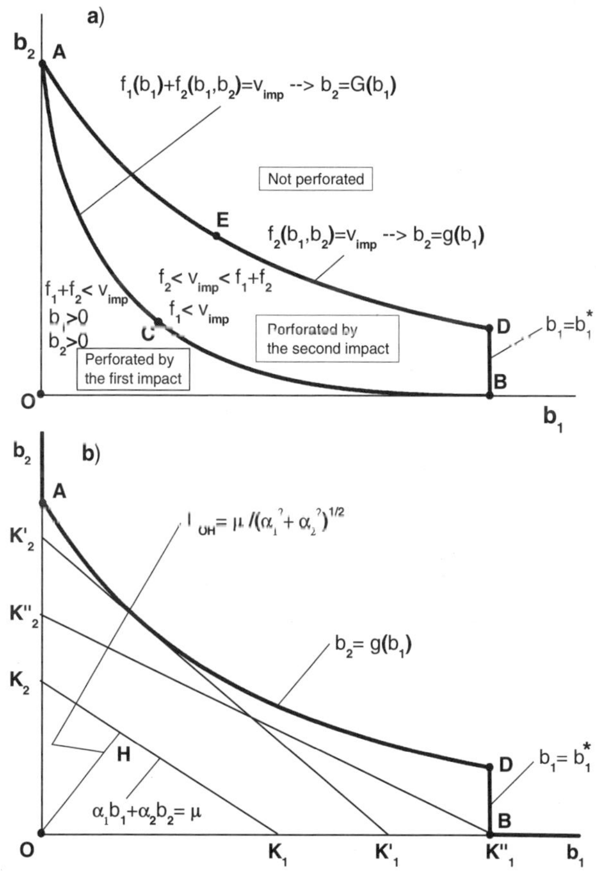

Fig. 29.5.1. Illustration of the method for armor optimization when a repeated impact is possible.

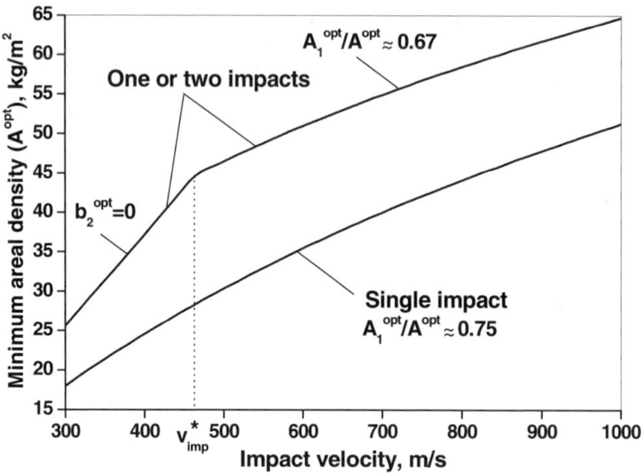

Fig 29.5.2. Minimum AD of armor against single and repeated impacts vs. given impact velocity.

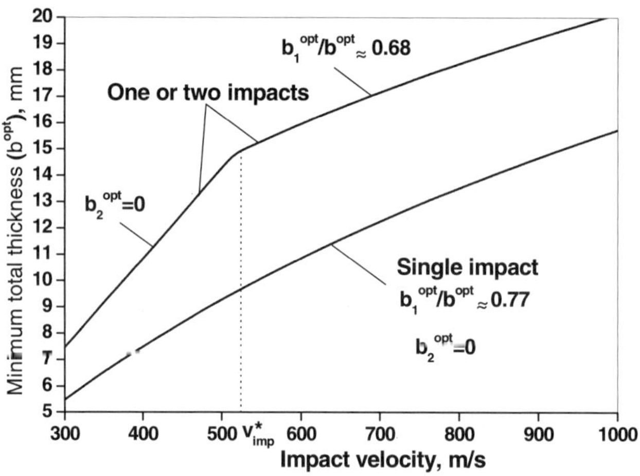

Fig. 29.5.3. Minimum total thickness of armor against single and repeated impacts vs. given impact velocity.

The results of calculations are presented in Figs. 29.5.2 and 29.5.3 for the criterion of minimal AD and of minimal TT, respectively. If the impact velocity is less than some critical value, v^*_{imp}, then the optimal armor consists only of a ceramic plate. If $v_{imp} > v^*_{imp}$ then the optimal armor includes two plates with the approximately constant ratio of the thicknesses. Inspection of Figs. 29.5.2 and 29.5.3 shows that the possibility of the repeated impact considerably increases minimal AD and TT required for defense against perforation.

Appendix A

Properties of Convex/Concave Increasing Positive Functions

In this section we derive some inequalities that are used in different sections of the book.

A.1. Property 1

Let $y = f(x)$ be a convex downward positive increasing function for $0 \le x_{\min} < x < x_{\max}$ that satisfies the condition

$$x_{\min} f'(x_{\min}) - f(x_{\min}) \ge 0\,. \tag{A.1}$$

Let us explain the meaning of inequality given by Eq. (A.1). It implies that the tangent to the curve $y = f(x)$ in the point $x = x_{\min}$ (line AB in Fig. A.1a), intersects with positive x-semi-axis.

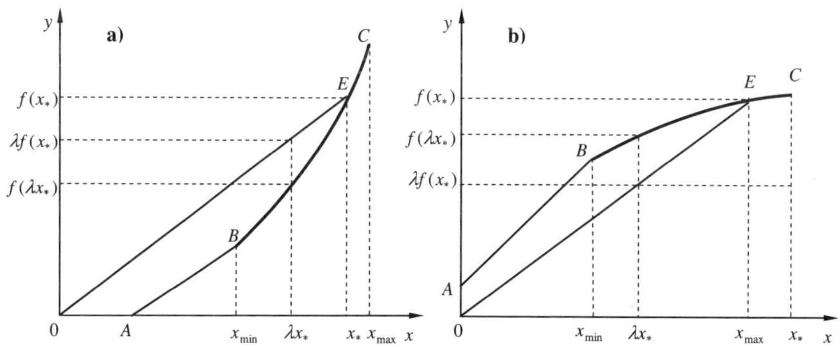

Fig. A.1. Basic properties of function f : a) case of convex downward function, b) case of convex upward function.

Another interpretation of this inequality is that function $f(x)$ can be defined also in the interval $0 \le x < x_{\min}$ where it remains convex downward and $f(0)=0$. If $f(x)$ is defined explicitly as a convex downward function for $0 \le x < x_{\max}$ and $f(0)=0$, then Eq. (A.1) can be omitted.

For some function $y=f(x)$ that is non convex upward for $x_{\min} < x < x_{\max}$, the following property is valid (Rockafellar, 1970): for any $x_{\min} < x_1 < x_{\max}$, $x_{\min} < x_2 < x_{\max}$ and $0 < \lambda < 1$

$$f\big((1-\lambda)x_1 + \lambda x_2\big) \le (1-\lambda) f(x_1) + \lambda f(x_2). \tag{A.2}$$

Consider an arbitrary x_*, such that $x_{\min} < x_* < x_{\max}$ (Fig. A.1a). Since $OABEC$ is a non convex upward curve and O is the origin of coordinates, after substituting $x_1 = 0,\ x_2 = x_*,\ f(x_1)=0$ into Eq. (A.2) we find that

$$f(\lambda x_*) \le \lambda f(x_*) \tag{A.3}$$

if $0 < \lambda < 1$. Clearly (see Fig. A.1a) a strong inequality,

$$f(\lambda x_*) < \lambda f(x_*), \tag{A.4}$$

is valid when $x_{\min}/x_* < \lambda < 1$.

A.2. Property 2

Let $y=f(x)$ be a convex upward positive increasing function for $0 \le x_{\min} < x < x_{\max}$ that satisfies the condition

$$x_{\min} f'(x_{\min}) - f(x_{\min}) \le 0. \tag{A.5}$$

This implies that the tangent to the curve $y=f(x)$ in the point $x = x_{\min}$ (line AB in Fig. A.1b), intersects with the positive y-semi-axis.

Using similar arguments we arrive at the conclusion that

$$f(\lambda x_*) \ge \lambda f(x_*) \tag{A.6}$$

if $0 < \lambda < 1$, and a strong inequality is valid when $x_{\min}/x_* < \lambda < 1$.

If $x_{\min} = 0$ and $f(0) = 0$ Eq. (A.5) can be omitted.

A.3. Property 3

Let $G(T)$ be a convex downwards, twice differentiable function which is defined in the interval $T \geq T_{\min} \geq 0$ and satisfies the inequality:

$$T_{\min} G'(T_{\min}) - G(T_{\min}) \geq 0, \tag{A.7}$$

where $T_{\min}$ is given. Then

$$\delta = G\left(\sum_{j=1}^{N} T^{(j)}\right) - \sum_{j=1}^{N} G(T^{(j)}) > 0, \tag{A.8}$$

for

$$T^{(j)} \geq T_{\min},\ T^{(j)} > 0,\ j = 1,2,\ldots,N. \tag{A.9}$$

In order to prove this claim we use the method of mathematical induction. For convenience, we specify $T^{(j)}$ ($j - 1,2,\ldots,N$) as thicknesses of plates.

As the first step, we prove that the claim is valid for the case of two plates ($N = 2$).

Consider a function $F(T) = TG'(T) - G(T)$. Since $G(T)$ is a convex downward function, $F'(T) = TG''(T) > 0$ and, consequently,

$$TG'(T) - G(T) > T_{\min} G'(T_{\min}) - G(T_{\min}) > 0, \quad T > T_{\min}. \tag{A.10}$$

Assume that $T_1 \geq T_2$ (if $T_1 < T_2$ we need only to replace, $T_1 \leftrightarrow T_2$). Hence using the mean value theorem and Eq. (A.10) we obtain that

$$\begin{aligned} \delta &= G(T_1 + T_2) - G(T_1) - G(T_2) \\ &= [G(T_1 + T_2) - G(T_1)] - G(T_2) \\ &= T_2 G'(\xi) - G(T_2) > T_2 G'(\xi) - T_2 G'(T_2) \\ &= T_2 [G'(\xi) - G'(T_2)] = T_2 G''(\theta), \end{aligned} \tag{A.11}$$

where

$$T_1 < \xi < T_1 + T_2, \quad T_2 < \theta < \xi . \tag{A.12}$$

Since $\theta > T_{\min}$, we proved that $\delta > 0$.

At the second step, we assume that the claim is valid for N plates ($\delta = \delta_N > 0$) and let us prove that the claim is also valid for $N+1$ plates ($\delta_{N+1} > 0$). Using the validity of the statement for two plates with the thicknesses $T^{(1)} + T^{(2)} + ... + T^{(N)}$ and $T^{(N+1)}$ we can write that

$$G\left(\sum_{j=1}^{N+1} T^{(j)}\right) = G\left(\sum_{j=1}^{N} T^{(j)} + T^{(N+1)}\right) > G\left(\sum_{j=1}^{N} T^{(j)}\right) + G\left(T^{(N+1)}\right). \tag{A.13}$$

Now using the validity of the claim for N plates we obtain:

$$\begin{aligned} \delta_{N+1} &= G\left(\sum_{j=1}^{N+1} T^{(j)}\right) - \sum_{j=1}^{N+1} G(T^{(j)}) \\ &> G\left(\sum_{j=1}^{N} T^{(j)}\right) + G\left(T^{(N+1)}\right) - \sum_{j=1}^{N+1} G(T^{(j)}) \\ &= G\left(\sum_{j=1}^{N} T^{(j)}\right) - \sum_{j=1}^{N} G(T^{(j)}) > 0 \end{aligned} \tag{A.14}$$

This completes the proof of Property 3.

Bibliography

[Abrate, 1998] Abrate, S. (1998). *Impact on Composite Structures.* Cambridge University Press, Cambridge.

[ACE, 1946] Fundamentals of protective structures. (1946). Report AT120 AT1207821. Army Corps of Engineers (ACE). Office of the Chief of Engineers.

[ACI, 1978] ACI Standard 318–77 (1978). Building code requirements for reinforced concrete. American Concrete Institute, Detroit, MI.

[Adeli and Amin, 1985] Adeli, H., Amin, A.M. (1985). Local effects of impactors on concrete structures. *Nuclear Engineering and Design,* **88**(3): 301–317.

[Adeli *et al.*, 1986] Adeli, H, Amin, A.M., Sierakowski, R.L. (1986). Earth penetration by solid impactors. *The Shock and Vibration Digest*, **18**(9):14–22

[Alekseeva and Barantsev, 1976] Alekseeva, E.V., Barantsev, R.G. (1976). *Local Method of Aerodynamic Calculations in Rarefied Gas.* Leningrad State University Press, Leningrad (in Russian).

[Al-Hachamee and Azeez, 2010] Al-Hachamee, E.K.S., Azeez, H. (2010). Scabbing and perforation local effect of impactors on concrete structures. *Engineering and Technology J.*, **28**(9): 1757–1770.

[Allen *et al.*, 1957a] Allen, W.A., Mayfield, E.B., Morrison, H.L. (1957). Dynamics of a projectile penetrating sand. *J. of Applied Physics,* **28**(3): 370–376.

[Allen *et al.*, 1957b] Allen, W.A., Mayfield, E.B., Morrison, H.L. (1957). Dynamics of a projectile penetrating sand. Part II. *J. of Applied Physics,* **28**(11): 1331–1335.

[Almohandes *et al.*, 1996] Almohandes, A.A., Abdel-Kader, M.S., Eleiche, A.M. (1996). Experimental investigation of the ballistic resistance of steel-fiberglass reinforced polyester laminated plates. *Composites. Part B,* **27**(5): 447–458.

[Aly and Li, 2008] Aly, S.Y., Li, Q.M. (2008). Critical impact energy for the perforation of metallic plates. *Nuclear Engineering and Design*, **238**(10): 2521–2528

[Ambroso *et al.*, 2005] Ambroso, M.A., Kamien, R.D., Durian, D.J. (2005). Dynamics of shallow impact cratering. *Physical Review E.,* **72**, 041305.

[Amde *et al.*, 1996] Amde, A.M., Mirmiran, A., Walter, T.A. (1996). Local damage assessment of metal barriers under turbine missile impacts. *J. of Structural Engineering*, **122**(1): 99–108.

[Amde *et al.*, 1997] Amde, A.M., Mirmiran, A., Walter, T.A. (1997). Local damage assessment of turbine missile impact on composite and multiple barriers. *Nuclear Engineering and Design,* **178**(1/2): 145–156.

[Amirikian, 1950] Amirikian, A., (1950). Design of protective structures. Report NT-3726, Bureau of Yards and Docks, Department of the Navy, Washington, DC.

[Anderson and Bodner, 1988] Anderson, C.E., Jr., Bodner, S.R. (1988). Ballistic impact: the status of analytical and numerical modeling. *Int. J. of Impact Engineering*, **11**(1): 33–40.

[Anderson *et al.*, 1995] Anderson, C.E., Jr., Walker, J.D., Bless, S.J., Sharron, T.R. (1995). On the velocity dependence of the L/D effect for long-rod penetrators. *Int. J. of Impact Engineering,* **17**(1): 13–24.

[Anderson *et al.*, 1996] Anderson, C.E., Jr., Walker, J.D., Bless, S.J., Partom, Y. (1996). On the L/D effect for long-rod penetrators. *Int. J. of Impact Engineering,* **18**(3): 247–264.

[Anderson *et al.*, 1997] Anderson, C.E., Jr., Subramanian, R., Walker, J.D. Normandia, M.J.,Sharron, T.R. (1997). Penetration mechanics of seg-tel penetrators. *Int. J. of Impact Engineering,* **20**(1–5): 13–26.

[Anderson *et al.*, 1999] Anderson, C.E., Jr., Hohler, V., Walker, J.D., Stilp, A.J. (1999). The influence of projectile hardness on ballistic performance. *Int. J. of Impact Engineering,* **22**(6): 619–632.

[Apshtein and Titow, 1996] Apshtein, E., Titow, O. (1996). Area rule and volume rule for integral quantities. *Acta Mechanica,* **116**(1–4): 45–60.

[Aptukov, 1985] Aptukov, V.N. (1985). Optimal structure of inhomogeneous plate with continuous distribution of properties over the thickness. *Mechanics of Solids,* **20**(3): 148–151.

[Aptukov, 1990] Aptukov, V.N. (1990). Penetration: Mechanical aspects and mathemanical modeling (Review). *Strength of Materials*, **22**(2): 230–240.

[Aptukov, 1991a] Aptukov, V.N. (1991). Optimal interaction of indenter with inhomogeneous plate. In: Dulikracich, G.S., ed., *Proc. of 3rd Int. Conf. on Inverse Design Concepts and Optimization in Engineering Science* (October 23–25, 1991, Washington, DC), pp. 481–488.

[Aptukov, 1991b] Aptukov, V.N. (1991). Expansion of a spherical cavity in a compressible elasto-plastic medium. I. The influence of mechanical characteristics, free surface, and lamination. *Strength of Materials,* **23**(12): 1262–1268.

[Aptukov, 1991c] Aptukov, V.N. (1991). Expansion of a spherical cavity in a compressible elasto-plastic medium. II. Effect of inertial forces. Temperature effects. *Strength of Materials,* **23**(12): 1269–1274.

[Aptukov and Belousov, 1991] Aptukov, V.N., Belousov, V.L. (1991). Analysis of the optimal laminated target made up of discrete set of materials. In: Dulikracich, G.S., ed., *Proc. of 3rd Int. Conf. on Inverse Design Concepts and Optimization in Engineering Science* (October 23–25, 1991, Washington, DC), pp. 489–495.

[Aptukov and Fonarev, 2010] Aptukov, V.N., Fonarev, A.V. (2010). Approximate estimation of penetration depth of a pile into soil under multiple impacts. *Vestnik Permskogo Universiteta. Matematika. Mekhanika. Informatika.* **2**(2): 41–45 (in Russian).

[Aptukov and Khasanov, 2011] Aptukov, V.N., Khasanov, A.R. (2011). Optimal deceleration of a rigid cylinder by an inhomogeneous target for normal impact with friction. *Vestnik Permskogo Universiteta. Matematika. Mekhanika. Informatika.* **3**(7): 19–27 (in Russian).

[Aptukov and Pozdeev, 1982] Aptukov, V.N., Pozdeev, A.A. (1982). Some minimax problems of the technology and strengths of constructions. *Engineering Cybernetics,* **20**(1): 39–46.

[Aptukov *et al.*, 1985] Aptukov, V.N., Petrukhin, G.I., Pozdeev, A.A. (1985). Optimal deceleration of a rigid body by an inhomogeneous plate for the case of normal impact. *Mechanics of Solids,* **20**(1): 155–160.

[Aptukov *et al.*, 1986] Aptukov, V.N., Belousov, V.L., Kanibolotskii, M.A. (1986). Optimization of the structure of a layered slab with the penetration of a rigid striker. *Mechanics of Composite Materials,* **22**(2): 179–183.

[Aptukov *et al.*, 1992] Aptukov, V.N., Murzakaev, A.V., Fonarev, A.V. (1992). *Applied Theory of Penetration*. Nauka, Moscow (in Russian).

[Asatryan *et al.*, 1996] Asatryan, V.L., Bagdoev, A.G., Vantsyan, A.A. (1996). Solution of a dynamic problem of penetration of rigid cone into initially elastic transversely isotropic medium. *Izvestija Akademii Nauk Armjanskoj SSR. Mekhanika.* **49**(4): 77–85 (in Russian).

[Attaway *et al.*, 1998] Attaway, S.W., Mello, F.J., Heinstein, M.W., Swegle, J.W., Ratner, J.A., Zadoks, R.I. (1998). PRONTO3D user's instructions: a transientdynamic code for non-linear structural analysis. Report SAND98–1361. Sandia National Laboratories, Albuquerque, NM.

[Awerbuch, 1970] Awerbuch, J. (1970). A mechanical approach to projectile penetration. *Israel J. of Technology,* **8**(4): 375–383.

[Awerbuch and Bodner, 1974] Awerbuch, J., Bodner, S.R. (1974). Analysis of the mechanics of perforation of projectiles in metallic plates. *Int. J. of Solids and Structures,* **10**(6): 671–684.

[Babaei *et al.*, 2011] Babaei, B., Shokrieh, M.M., Daneshjou, K. (2011). The ballistic resistance of multi-layered targets impacted by rigid projectiles. *Materials Science and Engineering A*, **530** (1): 208–217.

[Babu *et al.*, 2007] Babu, M.G., Velmurugan, R., Gupta, N.K. (2007). Havy mass projectile impact on thin and moderately thick unidirectional fiber/epoxy laminates. *Latin American J. of Solids and Structures*, **4**(3): 247–265.

[Backman and Goldsmith, 1978] Backman, M., Goldsmith, W. (1978). The mechanics of penetration of projectiles into targets. *Int. J. of Engineering Science,* **16**(1): 1–99.

[Bagdoev and Vantsyan, 1981a] Bagdoev, A.G., Vantsyan, A.A. (1981). Penetration of slender bodies into elastic media. *Izvestija Akademii Nauk Armjanskoj SSR. Mekhanika*. **34**(1): 3–14 (in Russian).

[Bagdoev and Vantsyan, 1981b] Bagdoev, A.G., Vantsyan, A.A. (1981). Penetration of slender body into metals and soils. *Izvestija Akademii Nauk Armjanskoj SSR. Mekhanika*. **34**(3): 25–38 (in Russian).

[Bagdoev and Vantsyan, 1983] Bagdoev, A.G., Vantsyan, A.A. (1983). Penetration of slender body into elastic anisotropic media. Izvestija Akademii Nauk Armjanskoj SSR. Mekhanika. **36**(6): 23–30 (in Russian).

[Bagdoev and Vantsyan, 1987] Bagdoev, A.G., Vantsyan, A.A. (1987). Investigation of penetration of slender rigid body into transversely isotropic medium. *Izvestija Akademii Nauk Armjanskoj SSR. Mekhanika*. **40**(4): 3–6 (in Russian).

[Bagdoev and Vantsyan, 1989] Bagdoev, A.G., Vantsyan, A.A. (1989). Penetration of a thin body into a transversely isotropic medium with twisting. *Mechanics of solids*, **24**(2): 182–184.

[Bagdoev *et al.*, 1988] Bagdoev, A.G., Vantsyan, A.A., Grigoryan, M.S. (1988). Influence of anisotropic properties of metal laminated specimen on penetration. *Izvestija Akademii Nauk Armjanskoj SSR. Mekhanika.* **41**(6): 28–34 (in Russian).

[Bagdoev *et al.*, 1989] Bagdoev, A.G., Vantsyan, A.A., Grigoryan, M.S. (1989). Investigation of singularity of stresses in anisotropic plastic medium for cone penetration. *Izvestija Akademii Nauk Armjanskoj SSR. Mekhanika*. **42**(4): 52–57 (in Russian).

[Bagdoev *et al.*, 1997] Bagdoev, A.G., Vantsyan, A.A., Khachatryan, B.K., Khachatryan, L.A. (1997). Investigation of dynamic problem of penetration of rigid indentor into anisotropic medium on the basis of hypothesis of normal cross sections. *Izvestija Akademii Nauk Armjanskoj SSR. Mekhanika*. **50**(1): 44–52 (in Russian).

[Baker *et al.*, 1980] Baker, W.E., Kulesz, J.J., Westine, P.S., Cox, P.A., Wilbeck, J.S. (1980). A Manual for the prediction of blast and fragment loading on structures. Report DOE/TIC-11268. United States Department of Energy, Albuquerque Operation Office, Amarillo Area Office, Amarillo, TX.

[Baker *et al.*, 1991] Baker, W.E., Westine, P.S., Dodge, F.T. (1991). *Similarity Methods in Engineering Dynamics: Theory and Practice of Scale Modeling*. Elsevier, Amsterdam.

[Balagansky and Merzhievsky, 2004] Balagansky, I.A., Merzhievsky, L.A. (2004). *Destructive Effects of Ammunition*. Novosibirsk State Technical University Press, Novosibirsk (in Russian).

[Balakin, 1980] Balakin, V.A. (1980). *Friction and Wear for High Sliding Speeds*. Moscow, Mashinostroenie (in Russian).

[Banach, 1951] Banach, S. (1951). *Mechanics*. Polish Mathematical Society, Warszawa-Wrocław.

[Banerjee, 1968] Banerjee, A.K. (1968). Influence of kinetic friction on the critical velocity of stick-slip motion. *Wear,* **12**(2):107–116.

[Bangash, 2009] Bangash, M.Y.H. (2009).*Shock, Impact and Explosion: Structural Analysis and Design*. Springer, Berlin.

[Bangash and Bangash, 2006] Bangash, M.Y.H., Bangash, T. (2006). *Explosion - Resistant Buildigs: Design, Analysis, and Case Studies.* Springer, Heidelberg, Berlin.

[Banichuk and Ivanova, 2007] Banichuk, N.V., Ivanova, S.Y. (2007). Shape optimization of rigid body penetrating into continuous medium. *Problemy Prochnosti i Plastichnosti*, **69**: 47–58 (in Russian).

[Banichuk and Ivanova, 2008] Banichuk, N.V., Ivanova, S.Y. (2008). Shape optimization of rigid 3-D high-speed impactors penetrating into concrete shields. *Mechanics Based Design of Structures and Machines,* **36**(3): 249–259.

[Banichuk *et al.*, 1969] Banichuk, N.V., Petrov, V.M., Chernous'ko, F.L. (1969). The method of local variations for variational problems involving non-additive functionals. *USSR Computational Mathematics and Mathematical Physics,* **9**(3): 66–76.

[Banichuk *et al.*, 2007] Banichuk, N.V., Ivanova, S.Y., Ragnedda, F. (2007). On shapes of bodies penetrating at maximum depth into solid deformable medium. In: *Proc. of the Conf. "Sovremennye Problemy Mekhaniki Sploshnoy Sredy [Modern Problebs of Continuous Mechanics]"*, (November 26–29, 2007, Rostov on Don, Russia), pp. 44–48. (in Russian)

[Banichuk *et al.*, 2008a] Banichuk, N.V., Ivanova, S.Y., Makeyev, Y.V. (2008). On penetration of rigid non-axisymmetric bodies into elastic-plastic medium. *Problemy Prochnosti i Plastichnosti,* **70**:131–139 (in Russian).

[Banichuk *et al.*, 2008b] Banichuk, N.V., Ivanova, S.Y., Makeev, F.V. (2008b). On penetration of nonaxisymmetric bodies into a deformable solid medium and their shape optimization. *Mechanics of Solids*, **43**(4): 671–677.

[Banichuk *et al.*, 2009] Banichuk, N.V., Ragnedda, F., Serra, M. (2009). On body shapes providing maximum depth of penetration. *Structural and Multidisciplinary Optimization*, **38**(5): 491–498.

[Banichuk *et al.*, 2011] Banichuk, N.V., Ivanova, S.Yu., Ragnedda, F., Serra, M. (2011). On shape of rigid shell with minimum mass and maximum resistance force moving in deformable media. *Problemy Prochnosti i Plastichnosti*, **73**: 36–44 (in Russian).

[Banichuk *et al.*, 2012] Banichuk, N.V., Ivanova, S.Yu., Ragnedda, F., Serra, M. (2012). Multiobjective shape optimization of the rigid shell moving into a condensed media. *Mechanics Based Design of Structures and Machines*, **40**(1): 73–82.

[Baranov and Lopa, 1995] Baranov, V.L., Lopa, I.V. (1995). Variational problem of minimization of drag force of penetrator. *Izvestija Tul'skogo Gosudarsvennogo Universiteta. Serija Matematika, Mekhanika, Informatika*. **1**(2): 18–23 (in Russian).

[Baranov *et al.*, 1991] Baranov, A.V., Konanykhin, Y.P., Pasechnik, L.P., Sugak, S.G. (1991). *Investigation of Dynamic Effect of Impactor on Multilayer Shields.* Universitet Druzhby Narodov, Chernogolovka (in Russian).

[Baranov *et al.*, 1995] Baranov, V.L., Lopa, I.V., Hristov, H.I., Chivikov, Z.Ch. (1995). Variational problem of optimization of the nose shape of planetary penetrator. *Aerospace Research in Bulgaria (Space Research Institute, Sofia)*, **12**: 146–149.

[Baranov *et al.*, 1996] Baranov, V.L., Zubachev, V.I., Lopa, I.V., Schitov, V.N. (1996). *Some Problems of Design of Bullets for Small Arms.* Tula State University Press, Tula (in Russian).

[Baranov *et al.*, 2004] Baranov, V.L., Khromov, I.V., Schitov, V.N. (2004). Determining total drag force of rotating projectile penetrating into soil. *Izvestija Tul'skogo Gosudarsvennogo Universiteta. Serija Problemy Spetsial'nogo Mashinostroenija*, **7**: 28–33 (in Russian).

[Barr, 1990] Barr, P., compiler. (1990) *Guidelines for the Design and Assessment of Concrete Structures Subjected to Impact.* UK Atomic Energy Authority, Safety and Reliability Directorate, UK.

[Barr *et al.*, 1983] Barr, P., Carter, P.G., Howe, W.D., Neilson, A.J., Richards, A.E. (1983). Experimental studies of the impact resistance of steel faced concrete composition. In: *Trans. of 7th Int. Conf. on Structural Mechanics in Reactor Technology (SMiRT-7)* (Aug 22–26, 1983, Chicago, IL), paper j8/4, pp. 395–402.

[Baty *et al.*, 2003] Baty, R.S., Lundgren, R.G., Patterson, W.J. (2003). On pilot-hole assisted penetration. In: *Proc. of 11th Int. Symp. on Interaction of the Effects of Munitions with Structures* (May 5–9, 2003 Mannheim, Germany).

[Bazhenov *et al.*, 2003] Bazhenov, V.G., Bragov, A.M., Kotov, V.L., Kochetkov, A.V. (2003). An investigation of the impact and penetration of solids of revolution into soft earth. *J. of Applied Mathematics and Mechanics*, **67**(4): 611–620.

[Bazhenov and Kotov, 2010] Bazhenov, V.G., Kotov, V.L. (2010). Solution of problems of oblique penetration of axisymmetric projectiles into soft soil based on local interaction models. *J. of Applied Mathematics and Mechanics*, **74**(3): 278–285.

[Bazhenov and Kotov, 2011] Bazhenov, V.G., Kotov, V.L. (2011). Mathematic Modeling of Unsteady Processes of Impact and Penetration of Axisymmetric Bodies and Identification of Properties of Soils. Fizmatlit, Moscow (in Russian).

[Bell, 1987] Bell, L.P. (1987). Evaluation of segmented and solid projectiles during hypervelocity flight and impact. Technical Report AEDC-TRS-87-V6. Arnold Engineering Development Center, Arnold Air Force Base, TN.

[Bellman *et al.*, 1958] Bellman, R.E., Glicksberg, I., Gross, O.A. (1958). *Some Aspects of the Mathematical Theory of Control Processes*. RAND Corporation, Santa Monica, California

[Ben-Dor *et al.*, 1997a] Ben-Dor, G., Dubinsky, A., Elperin, T. (1997). Shape optimization of high velocity impactors using analytical models. *Int. J. of Fracture,* **87**(1): L7–L10.

[Ben-Dor *et al.*, 1997b] Ben-Dor, G., Dubinsky, A., Elperin, T. (1997). Area rules for penetrating bodies. *Theoretical and Applied Fracture Mechanics,* **26**(3): 193–198.

[Ben-Dor *et al.*, 1997c] Ben-Dor, G., Dubinsky, A., Elperin, T. (1997). Optimal 3D impactors penetrating into layered targets. *Theoretical and Applied Fracture Mechanics,* **27**(3): 161–166.

[Ben-Dor *et al.*, 1998a] Ben-Dor, G., Dubinsky, A., Elperin, T. (1998). On the ballistic resistance of multi-layered targets with air gaps. *Int. J. of Solids and Structures,* **35**(23): 3097–3103.

[Ben-Dor *et al.*, 1998b] Ben-Dor, G., Dubinsky, A., Elperin, T. (1998). A model of high speed penetration into ductile targets. *Theoretical and Applied Fracture Mechanics,* **28**(3): 237–239.

[Ben-Dor *et al.*, 1998c] Ben-Dor, G., Dubinsky, A., Elperin, T. (1998). New area rule for penetrating impactors. *Int. J. Impact Engineering,* **21**(1–2): 51–59.

[Ben-Dor *et al.*, 1998d] Ben-Dor, G., Dubinsky, A., Elperin, T. (1998). Effect of air gaps on ballistic resistance of targets for conical impactors. *Theoretical and Applied Fracture Mechanics,* **30**(3): 243–249.

[Ben-Dor *et al.*, 1998e] Ben-Dor, G., Dubinsky, A., Elperin, T. (1998). Analysis of ballistic properties of layered targets using cavity expansion model. *Int. J. of Fracture,* **90**(4): L63–L67.

[Ben-Dor *et al.*, 1998f] Ben-Dor, G., Dubinsky, A., Elperin, T. (1998). Optimization of layered shields with a given areal density. *Int. J. of Fracture,* **91**(1): L9–L14.

[Ben-Dor *et al.*, 1999a] Ben-Dor, G., Dubinsky, A., Elperin, T. (1999). Optimization of light weight armor using experimental data. *Theoretical and Applied Fracture Mechanics,* **100**(4): L29–L33.

[Ben-Dor *et al.*, 1999b] Ben-Dor, G., Dubinsky, A., Elperin, T. (1999) On the order of plates providing the maximum ballistic limit velocity of a layered armor. *Int. J. of Impact Engineering,* **22**(8): 741–755.

[Ben-Dor *et al.*, 1999c] Ben-Dor, G., Dubinsky, A., Elperin, T. (1999). Effect of air gap and order of plates on ballistic resistance of two layered armor. *Theoretical and Applied Fracture Mechanics,* **31**(3): 233–241.

[Ben Dor *et al.*, 1999d] Ben-Dor, G., Dubinsky, A., Elperin, T. (1999). Some ballistic properties of non-homogeneous shields. *Composites. Part A,* **30**(6): 733–736.

[Ben-Dor *et al.*, 2000a] Ben-Dor, G., Dubinsky, A., Elperin, T. (2000). Optimization of the shape of a penetrator taking into account plug formation. *Int. J. of Fracture,* **106**(3): L29–L34.

[Ben-Dor *et al.*, 2000b] Ben-Dor, G., Dubinsky, A., Elperin, T, Frage, N. (2000). Optimization of two component ceramic armor for a given impact velocity. *Theoretical and Applied Fracture Mechanics,* **33**(3): 185–190.

[Ben-Dor *et al.*, 2000c] Ben-Dor, G., Dubinsky, A., Elperin, T. (2000). The optimum arrangement of the plates in a multilayered shield. *Int. J. of Solids and Structures,* **37**(4): 687–696.

[Ben-Dor *et al.*, 2000d] Ben-Dor, G., Dubinsky, A., Elperin, T. (2000). Analytical solution for penetration by rigid conical impactors using cavity expansion models. *Mechanics Research Communications,* **27**(2): 185–189.

[Ben-Dor *et al.*, 2001a] Ben-Dor, G., Dubinsky, A., Elperin, T. (2001). Shape optimization of penetrator nose. *Theoretical and Applied Fracture Mechanics,* **35**(3): 261–270.

[Ben-Dor *et al.*, 2001b] Ben-Dor, G., Dubinsky, A., Elperin, T. (2001). A class of models implying the Lambert-Jonas relation. *Int. J. of Solids and Structures,* **38**(40–41): 7113–7119.

[Ben-Dor *et al.*, 2002a] Ben-Dor, G., Dubinsky, A., Elperin, T. (2002). Optimal nose geometry of the impactor against FRP laminates. *Composite Structures,* **55**(1): 73–80.

[Ben-Dor *et al.*, 2002b] Ben-Dor, G., Dubinsky, A., Elperin, T. (2002). Optimization of the nose shape of an impactor against a semi-infinite FRP laminate. *Composites Science and Technology,* **62**(5): 663–667.

[Ben-Dor *et al.*, 2002c] Ben-Dor, G., Dubinsky, A., Elperin, T. (2002). A model for predicting penetration and perforation of FRP laminates by 3-D impactors. *Composite Structures,* **56**(3): 243–248.

[Ben-Dor *et al.*, 2002d] Ben-Dor, G., Dubinsky, A., Elperin, T. (2002). On the Lambert-Jonas approximation for ballistic impact. *Mechanics Research Communications,* **29**(2–3): 137–139.

[Ben-Dor *et al.*, 2003a] Ben-Dor, G., Dubinsky, A., Elperin, T. (2003). Numerical solution for shape optimization of an impactor penetrating into a semi-infinite target. *Computers and Structures,* **81**(1): 9–14.

[Ben-Dor *et al.*, 2003b] Ben-Dor, G., Dubinsky, A., Elperin, T. (2003). Shape optimization of an impactor penetrating into a concrete or a limestone target. *Int. J. of Solids and Structures,* **40**(17): 4487–4500.

[Ben-Dor *et al.*, 2005a] Ben-Dor, G., Dubinsky, A., Elperin, T. (2005). Optimization of two-component armor against ballistic impact. *Composite Structures,* **69**(1): 89–94.

[Ben-Dor *et al.*, 2005b] Ben-Dor, G., Dubinsky, A., Elperin, T. (2005). Ballistic impact: recent advances in analytical modeling of plate penetration dynamics. A Review. *ASME Applied Mechanics Reviews*, **58**: 355–371.

[Ben-Dor *et al.*, 2006a] Ben-Dor, G., Dubinsky, A., Elperin, T. (2006). *Applied High-Speed Plate Penetration Dynamics*. Springer, Dordrecht.

[Ben-Dor *et al.*, 2006b] Ben-Dor, G., Dubinsky, A., Elperin, T. (2006). Effect of air gaps on the ballistic resistance of ductile shields perforated by nonconical impactors. *J. of Mechanics of Materials and Structures*, **1**(2): 279–299.

[Ben-Dor *et al.*, 2006c] Ben-Dor, G., Dubinsky, A., Elperin, T. (2006). Modeling of high-speed penetration into concrete shields and shape optimization of impactors. *Mechanics Based Design of Structures and Machines*, **34**(2): 139–156.

[Ben-Dor *et al.*, 2006d] Ben-Dor, G., Dubinsky, A., Elperin, T. (2006). Effect of the order of plates on the ballistic resistance of ductile layered shields perforated by nonconical impactors. *J. of Mechanics of Materials and Structures*, **1**(7): 1161–1177.

[Ben-Dor *et al.*, 2007a] Ben-Dor, G., Dubinsky, A., Elperin, T. (2007). Localized interaction models with non-constant friction for rigid penetrating impactors. *Int. J of Solids and Structures*, **44**(7–8): 2593–2607.

[Ben-Dor *et al.*, 2007b] Ben-Dor, G., Dubinsky, A., Elperin, T. (2007). Modification of the method of local variations for shape optimization of penetrating impactors using the localized impactor/shield interaction models. *Mechanics Based Design of Structures and Machines,* **35**(1): 1–14.

[Ben-Dor *et al.*, 2007c] Ben-Dor, G., Dubinsky, A., Elperin, T. (2007). Shape optimization of impactors against a finite width shield using a modified method of local variations. *Mechanics Based Design of Structures and Machines*, **35**(2): 113–125.

[Ben-Dor *et al.*, 2007d] Ben-Dor, G., Dubinsky, A., Elperin, T. (2007). Optimization of High-Speed Penetration by Impactor with Jet Thruster. *Mechanics Based Design of Structures and Machines*, **35**(3): 205–228.

[Ben-Dor *et al.*, 2007e] Ben-Dor, G., Dubinsky, A., Elperin, T. (2007). Penetration by non-monolithic and monolithic high-speed impactors. *Mechanics Based Design of Structures and Machines*, **35**(4): 481–495.

[Ben-Dor *et al.*, 2008a] Ben-Dor, G., Dubinsky, A., Elperin, T. (2008). Optimization of penetration into geological and concrete shields by impactor with jet thruster. *J. of Mechanics of Materials and Structures*, **3**(4): 707–727.

[Ben-Dor *et al.*, 2008b] Ben-Dor, G., Dubinsky, A., Elperin, T. (2008). Optimization of high-speed penetration of segmented impactors using discrete and continuous models. *Mechanics Based Design of Structures and Machines*, **36**(2): 150–168.

[Ben-Dor *et al.*, 2008c] Ben-Dor, G., Dubinsky, A., Elperin, T. (2008). Engineering approach to penetration modeling. *Engineering Fracture Mechanics*, **75**(14): 4279–4282.

[Ben-Dor *et al.*, 2009a] Ben-Dor, G., Dubinsky, A., Elperin, T. (2009). Improved Florence model and optimization of two-component armor against single impact or two impacts. *Composite Structures*, **88**(1): 158–165.

[Ben-Dor *et al.*, 2009b] Ben-Dor, G., Dubinsky, A., Elperin, T. (2009). Method of basic impactors for simplified modeling of penetration. *Engineering Fracture Mechanics*, **76**(4): 614–618.

[Ben-Dor *et al.*, 2009c] Ben-Dor, G., Dubinsky, A., Elperin, T. (2009). An engineering approach to shape optimization of impactors against fiber-reinforced plastic laminates. *Composites Part B*, **40**(3): 181–188.

[Ben-Dor *et al.*, 2009d] Ben-Dor, G., Dubinsky, A., Elperin, T. (2009). Method of Basic Objects and Its Application to Penetration Dynamics. *Mechanics Based Design of Structures and Machines*, **37**(2): 247–258.

[Ben-Dor *et al.*, 2009e] Ben-Dor, G., Dubinsky, A., Elperin, T. (2009). Ballistic properties of multilayered concrete shields. *Nuclear Engineering and Design*, **239**(10): 1789–1794.

[Ben-Dor *et al.*, 2009f] Ben-Dor, G., Dubinsky, A., Elperin, T. (2009). Modeling of penetration by rigid impactors. *Mechanics Research Communications*, **36**(5): 625–629.

[Ben-Dor *et al.*, 2009g] Ben-Dor, G., Dubinsky, A., Elperin, T. (2009). Optimization of reinforced concrete panels with rear face steel liner under impact loading. *Mechanics Based Design of Structures and Machines*, **37**(4): 503–512.

[Ben-Dor *et al.*, 2009h] Ben-Dor, G., Dubinsky, A., Elperin, T. (2009). High-speed penetration modeling and shape optimization of the projectile penetrating into concrete shields. *Mechanics Based Design of Structures and Machines*, **37**(4): 538–549.

[Ben-Dor *et al.*, 2010a] Ben-Dor, G., Dubinsky, A., Elperin, T. (2010). Estimation of perforation thickness for concrete shield against high-speed impact. *Nuclear Engineering and Design*, **240**(5): 1022–1027.

[Ben-Dor *et al.*, 2010b] Ben-Dor, G., Dubinsky, A., Elperin, T. (2010). Effect of layering on ballistic properties of metallic shields against sharp-nosed rigid projectiles. *Engineering Fracture Mechanics,* **77**(14): 2791–2799.

[Ben-Dor *et al.*, 2010c] Ben-Dor, G., Dubinsky, A., Elperin, T. (2010). Some inverse problems in penetration mechanics. *Mechanics Based Design of Structures and Machines,* **38**(4): 468–480.

[Ben-Dor *et al.*, 2010d] Ben-Dor, G., Dubinsky, A., Elperin, T. (2010). Segmentation of high-speed/hypervelocity penetrators: criteria of effectiveness based on approximate analytical models. *Mechanics Based Design of Structures and Machines*, **38**(3): 372–387.

[Ben-Dor *et al.*, 2011] Ben-Dor, G., Dubinsky, A., Elperin, T. (2011). Optimization of multi-layered metallic shield. *Nuclear Engineering and Design,* **241**(6): 2020–2025.

[Berger, 2002] Berger, E.J. (2002). Friction modeling for dynamic system simulation. A review. *Applied Mechanics Reviews,* **55** (6): 535–577.

[Bernard, 1977] Bernard, R.S. (1977). Empirical analysis of projectile penetration in rock. Miscellaneous paper S-77-16. U.S. Army Engineer Waterways Experiment Station, Vicksburg, MS.

[Bernard, 1978] Bernard, R.S. (1978). Depth and motion prediction for earth penetrators. Report WES-TR-S-78-4. U.S. Army Engineer Waterways Experiment Station, Vicksburg, MS.

[Bernard and Creighton, 1979] Bernard, R.S., Creighton, D.C. (1979). Projectile penetration in soil and rock: analysis for non-normal impact. Report WES/TR/SL-79-15. U.S. Army Engineer Waterways Experiment Station, Vicksburg, MS.

[Bernard and Hanagud, 1975] Bernard, R.S., Hanagud S.V. (1975). Development of a projectile penetration theory. Report S-75-9. . Penetration Theory for shallow to moderate depths (Report 1 of Series). U.S. Army Engineer Waterways Experiment Station, Vicksburg, MS.

[Berriaud *et al.*, 1978] Berriaud, C., Sokolovsky, A., Gueraud, R., Dulac, J., Labrot, R. (1978). Local behaviour of reinforced concrete walls undermissile impact. *Nuclear Engineering and Design*; **45**(2): 457–469.

[Beth, 1946] Beth, R.A. (1946). Concrete penetration. Report OSRD-6459. Office of Sci. R&D, Washington, DC.

[Bethe, 1941] Bethe, H.A. (1941). Attempt of a theory of armor penetration. Report. Ordnance Laboratory, Frankford Arsenal, Philadelphia, PA.

[Bishop *et al.*, 1945] Bishop, R.F., Hill, R., Mott, N.F. (1945). The theory of indentation and hardness tests. *Proceedings of the Physical Society,* **57**(321), Part 3: 147–155.

[Bivin, 1999] Bivin, Yu.K. (1999). Comparison of penetration of star-shaped and conic bodies. *Mechanics of Solids*, **34** (4): 94–97.

[Bivin and Simonov, 2010] Bivin, Yu.K., Simonov, I.V. (2010). Mechanics of dynamic penetration into soil medium. *Mechanics of Solids*, **45**(6): 892–920.

[Bivin *et al.*, 1982] Bivin, Yu.K., Kolesnikov, V.A., Flitman, L.M. (1982). Determination of the mechanical properties of a medium by the dynamic penetration method. *Mechanics of Solids*, **17**(5): 180–183.

[BNFL, 2003] BNFL (2003). Reinforced concrete slab local damage assessment, R3 impact assessment procedure, vol. 3, Appendix H. Magnox Electric plc & Nuclear Electric Limited.

[Bo and Pavelescu, 1982] Bo, L.C., Pavelescu, D. (1982). The friction-speed relation and its influence on the critical velocity of stick-slip motion. *Wear,* **82**(3): 277–289.

[Bondarchuk *et al.*, 1982] Bondarchuk, V.S., Vedernikov, Y.A., Dulov, V.G., Minin, V.F. (1982). On optimization of star-shaped impactors. *Izvestija Sibirskogo Otdelenija Akademii Nauk SSSR. Serija Tekhnicheskikh Nauk,* **13**(3): 60–65 (in Russian).

[Børvik *et al.*, 1998] Børvik, T., Langseth, M., Hopperstad, O.S., Malo, K.A. (1998). Empirical equations for ballistic penetration of metal plates. Fortifikatorisk Notat No. 260/98. The Norwegian Defence Construction Service, Central Staff - Technical Division, Oslo, Norway.

[Børvik *et al.*, 1999] Børvik, T, Langseth, M, Hopperstad, O.S, Malo, K.A. (1999). Ballistic penetration of steel plates. *Int. J. of Impact Engineering*, **22**(9–10): 855–886.

[Børvik *et al.*, 2001] Børvik, T, Hopperstad, O.S., Berstad, T., Langseth, M. (2001). Numerical simulation of plugging failure in ballistic penetration. *Int. J. of Solids and Structures*, **38**(34–35):6241–6264.

[Børvik *et al.*, 2002a] Børvik, T., Langseth, M., Hopperstad, O.S., Malo, K.A. (2002). Perforation of 12mm thick steel plates by 20mm diameter projectiles with flat, hemispherical and conical noses. Part I: Experimental study. *Int. J. of Impact Engineering,* **27**(1): 19–35.

[Børvik *et al.*, 2002b] Børvik, T., Langseth, M., Hopperstad, O.S., Malo, K.A. (2002). Perforation of 12mm thick steel plates by 20mm diameter projectiles with flat, hemispherical, and conical noses part II: numerical simulations. *Int. J. of Impact Engineering,* **27**(1): 37–64.

[Børvik *et al.*, 2003] Børvik, T., Hopperstad, O.S., Langseth, M., Malo, K.A. (2003). Effect of target thickness in blunt projectile penetration of Weldox 460 E steel plates. *Int. J. of Impact Engineering*, **28**(4): 413–464.

[Børvik *et al.*, 2004] Børvik, T., Clausen, A. H., Hopperstad, O.S., Langseth, M. (2004). Perforation of AA5083-H116 aluminium plates with conical-nose steel projectiles – experimental study. *Int. J. of Impact Engineering,* **30**(4): 367–384.

[Børvik *et al.*, 2005] Børvik, T., Clausen, A.H., Eriksson, M., Berstad, T., Hopperstad, O.S., Langseth, M. (2005). Experimental and numerical study on the perforation of AA6005-T6 panels. *Int. J. of Impact Engineering*, **32**(1–4): 35–64.

[Børvik *et al.*, 2009a] Børvik, T., Forrestal, M.J., Hopperstad, O.S., Warren, T.L., Langseth, M. (2009). Perforation of AA5083-H116 aluminum plates with conical-nose steel projectiles – calculations. *Int. J. of Impact Engineering*, **36**(3): 426–437.

[Børvik *et al.*, 2009b] Børvik, T., Dey, S., Clausen, A.H. (2009). Perforation resistance of five different high-strength steel plates subjected to small-arms projectiles. *Int. J. of Impact Engineering*; **36**(7): 948–964.

[Børvik *et al.*, 2009c] Børvik, T., Dey, S., Hopperstad, O.S., Langseth, M. (2009). On the main mechanisms in ballistic perforation of steel plates at sub-ordnance impact velocities. In: Hiermaier, A.S., ed. *Predictive Modeling of Dynamic Processes. A Tribute to Professor Klaus Thoma*. Spinger, pp.189–219.

[Børvik *et al.*, 2010] Børvik, T., Forrestal, M.J., Warren, T.L. (2010). Perforation of 5083-H116 aluminum armor plates with ogive-nose rods and 7.62 mm APM2 bullets. *Experimental Mechanics,* **50**(7): 969–978.

[Bowden and Freitag, 1958] Bowden, F.P., Freitag, E.H. (1958). The friction of solids at very high speeds. I: Metal on metal. II: Metal on diamond. *Proceedings of the Royal Society of London. Series A: Mathematical and Physical Sciences*, **248**(1254): 350–367.

[Bowden and Persson, 1961] Bowden, F.P., Persson, P.A. (1961). Deformation, heating and melting on solids at high-speed friction. *Proceedings of the Royal Society of London. Series A: Mathematical and Physical Sciences,* **260**(1303): 433–458.

[Bowden and Tabor, 1964] Bowden, F.P., Tabor, D. (1964). *The Friction and Lubrication of Solids, Part 2*. Clarendon Press, Oxford.

[Brown, 1964] Brown, A. (1964). A quasi-dynamic theory of containment. *Int. J. of Mechanical Science*, **6**(4): 257–262.

[Brown, 1986] Brown, S.J. (1986). Energy release protection for pressurized systems. Part 2. Review of studies into impact/terminal ballistics. *Applied Mechanics Reviews,* **39**(2), Part 1: 177–202.

[Buchar *et al.*, 2001] Buchar, J., Krejčí, J., Forejt, M. (2001). Response of concrete to the impact of fragments simulating projectiles. *Engineering Mechanics*, **8**(1): 63–69.

[Buchar *et al.*, 2002] Buchar, J., Voldřich, J., Rolc, S., Lazar, M. (2002). Response of concrete to the impact of fragments simulating projectiles. *In: Proc. of the 20th Int. Symp. on Ballistics* (23–27 September, 2002, Orlando, FL).

[Bulson, 1997] Bulson, P.S. (1997). Explosive Loading of Engineering Structures. A History of Research and a Review of Recent Developments. E&FN Spon, London.

[Bunimovich and Dubinsky, 1995] Bunimovich, A.I., Dubinsky, A. (1995). *Mathematical Models and Methods of Localized Interaction Theory*. World Scientific, Singapore.

[Bunimovich and Dubinsky, 1996] Bunimovich, A.I., Dubinsky, A. (1996). Development, current state of the art, and applications of local interaction theory. Review. *Fluid Dynamics* **31**(3): 339–349.

[Bunimovich and Dubinsky, 1997] Bunimovich, A., Dubinsky, A., (1997). Development of localized interaction theory. In: *Investigations on History of Physics and Mechanics.* Moscow, Nauka, pp. 198–218 (In Russian)

[Bunimovich and Yakunina, 1987a] Bunimovich A.I., Yakunina, G.Ye. (1987). On the shape of minimum-resistance solids of revolution moving in plastically compressible and elastic-plastic media. *J. of Applied Mathematics and Mechanics,* **51**(3): 386–392.

[Bunimovich and Yakunina, 1987b] Bunimovich, A.I., Yakunina, G.Ye. (1987). The shapes of three-dimensional minimum-resistance bodies moving in compressible plastic and elastic media. *Moscow University Mechanics Bulletin,* **42**(3): 59–62.

[Bunimovich and Yakunina, 1989] Bunimovich, A.I., Yakunina, G.Ye. (1989). On the shape of a minimum resistance solid of rotation penetrating into plastically compressible media without detachment. *J. of Applied Mathematics and Mechanics,* **53**(5): 680 683.

[Buzaud *et al.*, 1999a] Buzaud, E., Don, D., Chapelle, S., Gary, G., Bailly, P. (1999). Perforation studies into Mb 50 concrete slabs In: *Proc. of 9th Int. Symp. on Interaction of the Effects of Munitions with Structures* (3–7 May, 1999, Berlin, Germany), pp. 407–414.

[Buzaud *et al.*, 1999b] Buzaud, E., Laurensou, R., Darrigade, A., Belouet, P., Lissayou, C. (1999). Hard target defeat: An analysis of reinforced concrete perforation process. In: *Proc. of 9th Int. Symp. on Interaction of the Effects of Munitions with Structures* (3–7 May, 1999, Berlin, Germany), pp. 283–290.

[Calder and Goldsmith, 1971] Calder, C.A., Goldsmith, W. (1971). Plastic deformation and perforation of thin plates resulting from projectile impact. *Int. J. of Solids and Structures*, **7**(7): 863–881.

[Cantwell and Morton, 1990] Cantwell, W, Morton, J. (1990). Impact perforation of carbon fibre-reinforced plastic. *Composites Science and Technology,* **38**(2): 119–141.

[Cao *et al.*, 2010] Cao, Z., Deng, Y., Zhang, W. (2010). Numerical investigation of penetration performance of segmented rod projectiles with various connectors. *Key Engineering Materials*, **452–453**: 185–188.

[Caprino *et al.*, 2007] Caprino, G, Lopresto, V, Santoro, D. (2007). Ballistic impact behaviour of stitched graphite/epoxy laminates. Review. *Composites Science and Technology*, **67**(3–4): 325–35.

[Carlucci and Jacobson, 2008] Carlucci, D.E., Jacobson, S.S. (2008). *Ballistics. Theory and Design of Guns and Ammunition*. CRC Press.

[Chakrabarty, 2010] Chakrabarty, J. (2010). *Applied Plasticity*. Springer, New York.

[Chang, 1981] Chang, W.S. (1981). Impact of solid missiles on concrete barriers. *J. of the Structural Division ASCE,* **107**(ST2): 257–271.

[Charters, 1986] Charters, A.C. (1986). The penetration of rolled homogeneous armor by continuous and segmented rods at high velocity: Theory and experiments. Technical Report CR-86-1031. General Research Corporation, Santa Barbara, CA.

[Charters and Locke, 1958] Charters, A.C., Locke, G.S., Jr. (1958). A preliminary investigation of high-speed impact the penetration of small spheres into thick copper targets. NACA Research Memorandum A58B26.

[Charters *et al.*, 1990] Charters, A.C., Menna, T.L., Piekutowski, A.J. (1990). Penetration dynamics of rods from direct ballistic tests of advanced armor components at 2–3 km/s. *Int. J. of Impact Engineering,* **10**(1–4): 93–106.

[Chelapati *et al.*, 1972] Chelapati, C.V., Kennedy, R.P., Wall, I.B. (1972). Probabilistic assessment of hazard for nuclear structures. *Nuclear Engineering and Design,* **19**(2): 333–364.

[Chen, 1989] Chen, E.P. (1989). Penetration into dry porous rock: a numerical study on sliding friction simulation. *Theoretical and Applied Fracture Mechanics,* **11**(2): 135–141.

[Chen and Li, 2002] Chen, X.W., Li, Q.M. (2002). Deep penetration of a non-deformable projectile with different geometrical characteristics. *Int. J. of Impact Engineering,* **27**(6): 619–637.

[Chen and Li, 2003a] Chen, X.W., Li, Q.M. (2003). Perforation of a thick plate by rigid projectiles, *Int. J. of Impact Engineering,* **28**(7): 743–759.

[Chen and Li, 2003b] Chen, X.W., Li, Q.M. (2003). Shear plugging and perforation of ductile circular plates struck by a blunt projectile. *Int. J. of Impact Engineering,* **28**(5): 513–536.

[Chen and Li, 2011] Chen, X.W., Li, Q.M. (2011). Comparative analysis of perforation models of metallic plates by rigid sharp-nosed projectiles. *Int. J. of Impact Engineering*, **38**(7): 613–621.

[Chen *et al.*, 2004] Chen, X.W., Fan, S.C., Li, Q.M. (2004). Oblique and normal perforation of concrete targets by a rigid projectile. *Int. J. of Impact Engineering,* **30**(6): 617–637.

[Chen *et al.*, 2007] Chen, X.W., Li, X.L., Huang, F.L., Wu, H.J., Chen, Y.Z. (2007). Modeling of normal perforation of reinforced concrete slab by rigid projectile. In: *Proc. of the 23rd Int. Symp.on Ballistics* (16–20 April, 2007, Tarragona, Spain), vol. 2, pp. 1235–1242.

[Chen *et al.*, 2008a] Chen, X.W., Li, X.L., Huang, F.L., Wu, H.J., Chen, Y.Z. (2008). Normal perforation of reinforced concrete target by rigid projectile. *Int. J. of Impact Engineering*, **35**(10): 1119–1129.

[Chen *et al.*, 2008b] Chen, X.W., Li, X.L., Huang, F.L., Wu, H.J., Chen, Y.Z. (2008). Damping function in the penetration/perforation struck by rigid projectiles. *Int. J. of Impact Engineering*, **35**(11): 1314–1325.

[Chen *et al.*, 2008c] Chen, X.W., Li, X.L., Deng, K.W. (2008). Damping function in the penetration/perforation dynamics of rigid projectiles. In: Jones, N. and Brebbia, C.A., eds. *Structures Under Shock and Impact X*. WIT Press, Southampton, pp. 263–272.

[Chen *et al.*, 2009] Chen, X., Liang, G., Yao, Y., Wang, R., Tao, J. (2009). Perforation modes of metal plates struck by a blunt rigid projectile. *Chinese J. of Theoretical and Applied Mechanics*, **41**(1): 84–90 (in Chinese).

[Chernousko and Banichuk, 1973] Chernous'ko, F.L., Banichuk, N.V. (1973). *Variational Problems of Mechanics and Control.* Nauka, Moscow (in Russian).

[Chernyi, 1969] Chernyi, G.G. (1969). *Introduction to Hypersonic Flow.* Academic Press, New York.

[Chocron-Benloulo and Sanchez-Galvez, 1998] Chocron-Benloulo, I.S., Sanchez-Galvez, V. (1998). A new analytical model to simulate impact onto ceramic/composite armors. *Int. J. of Impact Engineering,* **21**(6): 461–471.

[Chou and Toland, 1977] Chou, P.C., Toland, R.H. (1977). Experimental study of multiple interior impacts. *Experimental Mechanics*, **17**(6): 201–206.

[Chou *et al.*, 1975] Chou, P.C., Minnich, M., Gause, L. (1975). Experimental study of multiple interior impacts. Technical Report BRL-CR-199. U.S. Army Balltistic Research Laboratory, Aberdeen Proving Ground, MD.

[Choudhury *et al.*, 2002] Choudhury, M.A., Siddiqui, N.A., Abbas, H. (2002). Reliability analysis of a buried concrete target under missile impact. *Int. J. of Impact Engineering,* **27**(8): 791–806.

[Cicala and Miele, 1956] Cicala, P., Miele, A. (1956). Generalized theory of the optimum thrust programming for the level flight of a rocket-powered aircraft. *ARS J.*, **26**(6): 443–455.

[Cohen *et al.*, 2010] Cohen, T., Masri, R., Durban, D. (2010). Ballistic limit predictions with quasi-static cavitation fields. *Int. J. of Protective Structures*, **1**(2): 235–255.

[Corbett and Reid, 1993] Corbett, G.G., Reid, S.R. (1993). Quasi-static and dynamic local loading of monolithic flat-faced long projectiles. *Int. J. of Impact Engineering,* **13**(3): 423–441.

[Corbett *et al.*, 1996] Corbett, G.G., Reid, S.R., Johnson, W. (1996). Impact loading of plates and shells by free-flying projectiles: a review. *Int. J. of Impact Engineering,* **18**(2): 141–230.

[Corran *et al.*, 1983a] Corran, R.S.J., Ruiz, C., Shadbolt, P.J. (1983). On the design of containment shield. *Computers and Structures,* **16**(1–4): 563–572.

[Corran *et al.*, 1983b] Corran, R.S.J., Shadbolt, P.J., Ruiz, C. (1983). Impact loading of plates-an experimental investigation. *Int. J. of Impact Engineering,* **1**(1): 3–22.

[Crull and Swisdak, 2005] Crull, M., Swisdak, M.M., Jr. (2005). Methodologies for calculating primary fragment characteristics. Technical Paper No. 16 (Technical Report DDESB TP 16), Revision 2. Department of Defense Explosives, Safety Board, Alexandria, VA.

[Crull *et al.*, 1999] Crull, M., Taylor, L., Tipton, J. (1999). Estimating ordnance penetration into earth. In: *Proc. of the UXO Forum '99* (25–27 May, 1999, Atlanta, GA).

[Cuadros, 1990] Cuadros, J.H. (1990). Monolithic and segmented projectile penetration experiments in the 2 to 4 km/s impact velocity regime. *Int. J. of Impact Engineering,* **10**(1–4): 147–157.

[Dai *et al.*, 2005] Dai, Q., Yan, L., Jiang, Z. (2005). Analysis on the plugging model of concrete targets struck by rigid projectiles. *J. of Ballistics*, **17**(3): 26–30 (in Chinese)

[Dancygier, 1997] Dancygier, A.N. (1997). Effect of reinforcement ratio on the resistance of reinforced concrete to hard projectile impact. *Nuclear Engineering and Design,* **172**(1–2): 233–245.

[Dancygier, 1998] Dancygier, A.N. (1998). Rear face damage of normal and high-strength concrete elements caused by hard projectile impact. *ACI Structural Journal*, **95**(3): 291–304.

[Dancygier, 2000] Dancygier, A.N. (2000). Scaling of non-proportional non-deforming projectiles impacting reinforced concrete barriers. *Int. J. of Impact Engineering,* **24**(1): 33–55.

[Dancygier and Yankelevsky, 1996] Dancygier, A.N., Yankelevsky, D.Z. (1996). High strength concrete response to hard projectile impact. *Int. J. of Impact Engineering,* **18**(6): 583–599.

[Daudeville and Malécot, 2011] Daudeville, L., Malécot, Y. (2011). Concrete structures under impact. *European J. of Environmental and Civil Engineering,* **15**(SI): 101–140.

[Davis, 2003] Davis, R.N. (2003). Modeling of high-speed friction using multi-step incrementation of the coefficient of sliding friction. In: *Proceedings of the AIAA 54th Annual Southeastern Regional Student Conference* (27-28 March, 2003, Kill Devil Hills, NC), AIAA-RSC2-2003-U-004, pp. 1–10.

[Davis *et al.*, 2003] Davis, R.N., Jones, S.E., Hughes, M.L. (2003). High-speed penetration of concrete using a new analytical model of velocity-dependent friction. In: *Proc. of ASME 2003 Pressure Vessels and Piping Conference* (July 20–24, 2003, Cleveland, OH), Paper No. PVP2003–1823, pp. 111–116.

[De Rosset, 1981] De Rosset, W.S. (1981). Multiple impacts on monolithic steel. In: *Proc. of the 6th Int. Symp. on Ballistics* (October 27–29, 1981, Orlando, FL).

[De Rosset and Kimsey, 1986] De Rosset, W.S., Kimsey, K.D. (1986). Calculation of multiple copper rod impacts on steel targets. In: *Proc. of the 9th Int. Symp. on Ballistics* (29 April - 1 May, 1986, Shrivenham, UK).

[Deere, 1968] Deere, D.U. (1968). Geologic considerations. In: Stagg, K.G. and Zienkiewicz, O.C., eds. *Rock Mechanics in Engineering Practice*. John Wiley & Sons, New York, London, pp. 1–20.

[Degen, 1980] Degen, P.P. (1980). Perforation of reinforced concrete slabs by rigid missiles. *ASCE J. of the Structural Division,* **106** (7): 1623–1642.

[Dehn, 1979] Dehn, J.T. (1979). The particle dynamics of target penetration. Report ARBRL-TR-02188. Army Ballistic Research Laboratory, Aberdeen Proving Ground, MD.

[Dehn, 1986] Dehn, J.T. (1986). A unified theory of penetration. Report BRL-TR-2770. Army Ballistic Research Laboratory, Aberdeen Proving Ground, MD.

[Dehn, 1987] Dehn, J.T. (1987). A unified theory of penetration. *Int. J. of Impact Engineering,* **5**(1–4): 238–248.

[Den Reijer, 1991] Den Reijer, P.S. (1991). Impact on ceramic faced armours. Ph.D. Thesis. Delft University of Technology, Delft.

[Deng *et al.*, 2012] Deng, Y., Zhang, W., Cao, Z. (2012). Experimental investigation on the ballistic resistance of monolithic and multi-layered plates against hemispherical-nosed projectiles impact. Materials and Design, **41**: 266–281.

[Denisov, 1994] Denisov, A.M. (1994). *Introduction to Theory of Inverse Problems.* Moscow State University Press, Moscow (in Russian).

[Dey and Børvik, 2007] Dey, S., Børvik, T. (2007). Ballitic penetration and perforation of layerd steel plates: An experimental and numercal investigation. In: *Proc. of the 23rd Int. Symp.on Ballistics* (16–20 April, 2007, Tarragona, Spain). pp. 1365–1372.

[Dey *et al.*, 2004] Dey, S., Børvik, T., Hopperstad, O.S., Leinum, J.R., Langseth, M. (2004). The effect of target strength on the perforation of steel plates using three different projectile nose shapes. *Int. J. of Impact Engineering,* **30**(8–9): 1005–1038.

[Dey *et al.*, 2006] Dey, S., Børvik, T., Hopperstad, O.S., Langseth, M. (2006). On the influence of fracture criterion in projectile impact of steel plates. *Computational Materials Science,* **38**(1): 176–191.

[Dey *et al.*, 2007] Dey, S., Børvik, T., Teng, X, Wierzbicki, T., Hopperstad, O.S. (2007). On the ballistic resistance of double-layered steel plates: an experimental and numerical investigation. *Int. J. of Solids and Structures,* **44**(20): 6701–6723.

[Dey *et al.*, 2011] Dey, S., Børvik, T, Hopperstad, O.S. (2011). Computer-aided design of protective structures: Numerical simulations and experimental validation. *Applied Mechanics and Materials,* **82**: 686–691.

[Dienes and Miles, 1977] Dienes, J.K., Miles, J.W. (1977). A membrane model for the response of thin plates to ballistic impact. *J. of the Mechanics and Physics of Solids,* **25**(4): 237–256.

[Dikshit and Sundararajan, 1992] Dikshit, S.N., Sundararajan, G. (1992). The penetration of thick steel plates by ogive shaped projectile - experiment and analysis. *Int. J. of Impact Engineering,* **12**(3): 373–408.

[Dikshit and Sundararajan, 1995] Dikshit, S.N., Sundararajan, G. (1995). Author's reply to Liaghat and Malekzadeh. *Int. J. of Impact Engineering*, **16**(4): 693–695.

[DOE, 1981] DOE/TIC-11268. (1981). A manual for the prediction of blast and fragment loadings on structures. Change 1. U.S. Department of Energy, Albuquerque Operations Office, Amarillo, TX.

[DOE, 2006] DOE-STD-3014-2006 (2006). Accident analysis for aircraft crash into hazardous facilities. U.S. Department of Energy, Washington, DC.

[Draper and Smith, 1998] Draper, N.R., Smith, H. (1998). *Applied Regression Analysis.* J. Wiley & Sons, New York.

[Dubinsky and Elperin, 1997] Dubinsky, A., Elperin, T. (1997). Method of basic projectiles for calculating force coefficients. *J. Spacecraft and Rockets,* **34**(4): 558–560.

[Duffin *et al.*, 1967] Duffin, R.J, Peterson, E.L., Zener, C. (1967). *Geometric Programming - Theory and Application*. John Willey & Sons, New-York.

[Durban and Fleck, 1992] Durban, D., Fleck, N.A. (1992). Singular plastic fields in steady penetration of a rigid cone. *J. of Applied Mechanics*, **59**(4): 706–710.

[Durmuş *et al.*, 2011] Durmuş, A., Güden, M., Gülçimen, B., Ülkü, S., Musa, E. (2011). Experimental investigations on the ballistic impact performances of cold rolled sheet metals. *Materials and Design*, **32**(3): 1356–1366.

[Dusenberry, 2010] Dusenberry, D.O. (ed.). (2010). *Handbook for Blast Resistant Design of Buildings*. J. Wiley & Sons, Hoboken, NJ.

[Efimov, 1939] Efimov, M.P. (1939). *Course on Artillery Shells*. OBORONGIZ, Moscow (in Russian).

[Eichelberger, 1956] Eichelberger, R.J. (1956). Experimental test of the theory of penetration by metallic jets. *Journal of Applied Physics*, **27**(1): 63–68.

[Eisler *et al.*, 1998] Eisler, R.D., Chatterjee, A.K., Burghart, G.H., Loan, P. (1998). Simulates the tissue damage from small arms projectiles and fragment penetrating the musculoskeletal system. Final Report. Mission Research Corporation, Fountain Valley, CA.

[Elek *et al.*, 2005] Elek, P., Jaramaz, S., Mickovic, D. (2005). Modeling of perforation of plates and multi-layered metallic targets. *Int. J. of Solids and Structures,* **42**(3–4): 1209–1224.

[Espinosa *et al.*, 2000a] Espinosa, H.D., Patanella, A., Fischer, M. (2000). A novel dynamic friction experiment using a modified Kolsky bar apparatus. *Experimental Mechanics,* **40**(2): 138–153.

[Espinosa *et al.*, 2000b] Espinosa, H.D., Patanella, A., Fischer, M. (2000). Dynamic friction measurements at sliding velocities representative of high-speed machining processes. *J. of Tribology,* **122**(4): 834–848.

[Evstrop'ev-Kudryevatyj, *et al.*, 1992] Evstrop'ev-Kudryevatyj, V.V., Kozhushko, A.A., Sapozhnikov, G.A., Fomin, V.M. (1992). Successive collision of projectiles with barrier. In: *Chislennye Metody Reshenija Zadach Teorii Uprugosti i Plastichnosti [Numerical Methods for Solving Problems of Elasticity and Plasticity]. Proc. of 12th All-Union Conf.* (Novosibirsk, 1992), pp. 264–270 (in Russian).

[Fawaz *et al.*, 2006] Fawaz, Z., Behdinan, K., Xu, Y. (2006). Optimum design of two-component composite armours against high-speed impact. *Composite Structures,* **73**(3): 253–262.

[Fedorov and Veldanov, 2006] Fedorov, S.V., Veldanov, V.A. (2006). Numerical simulation of cavity formation in soil by a flux of high-speed metallic penetrators. *J. of Technical Physics,* **51**(7): 952–955.

[Fedorov *et al.*, 2007] Fedorov, S.V, Veldanov, V.A, Kozlov, V.S. (2007). Numerical analysis of metal projectile penetration into soil in hydrodynamic mode. In: *Proc. of the 23rd Int. Symp.on Ballistics* (16–20 April, 2007, Tarragona, Spain), pp. 1421–1428.

[Feli *et al.*, 2010] Feli, S., Aalami Aaleagha, M.E., Ahmadi, Z. (2010). A new analytical model of normal penetration of projectiles into the light-weight ceramic–metal targets. *Int. J. of Impact Engineering*, 37(5): 561–567.

[Feli *et al.*, 2011] Feli, S., Yas, M.H., Asgari, M.R. (2011). An analytical model for perforation of ceramic/multi-layered planar woven fabric targets by blunt projectiles. *Composite Structures*, **93**(2): 548–556.

[Fellows and Barton, 1999] Fellows, N.A., Barton, P.C. (1999). Development of impact model for ceramic-faced semi-infinite armor. *Int. J. of Impact Engineering,* **22**(8): 793–811.

[Fernández-Fdz and Zaera, 2008] Fernández-Fdz, D., Zaera, R. (2008). A new tool based on artificial neural networks for the design of lightweight ceramic-metal armours against high velocity impact of solids. *Int. J. of Solids and Structures*, **45**(25–26): 6369–6383.

[Florence, 1969] Florence, A.L. (1969). Interaction of projectiles and composite armor. Report AMMRC-CR-69-15, Part 2. Stanford Research Institute, Menlo Park, CA.

[Flores-Johnson *et al.*, 2011] Flores-Johnson, E.A., Saleh, M., Edwards, L. (2011). Ballistic performance of multi-layered metallic plates impacted by a 7.62-mm ARM2 projectile. *Int. J. of Impact Engineering,* **38**(12): 1022–1032.

[Folsom, 1987] Folsom, E.N., Jr., (1987). Projectile penetration into concrete with an inline hole. Master's Thesis. Report UCRL-53786. Lawrence Livermore National Laboratory, University of California, Livermore, CA.

[Fomin, 1999] Fomin, V.M., ed. (1999). *High-Speed Interaction of Bodies*. Nauka, Novosibirsk (in Russian).

[Forrestal, 1986] Forrestal, M.J. (1986). Penetration into dry porous rock. *Int. J. of Solids and Structures,* **22**(12): 1485–1500.

[Forrestal and Luk, 1988] Forrestal, M.J., Luk, V.K. (1988). Dynamic spherical cavity-expansion in a compressible elastic-plastic solid. *J. of Applied Mechanics,* **55**(2): 275–279.

[Forrestal and Luk, 1992] Forrestal, M.J., Luk, V.K. (1992). Penetration into soil targets. *Int. J. of Impact Engineering,* **12**(3): 427–444.

[Forrestal and Romero, 2007] Forrestal, M.J., Romero, L.A. (2007). Comment on "Perforation of aluminum plates with ogive-nose steel rods at normal and oblique impacts" (*Int. J. of Impact Engineering 1996*; **18**: 877–887). *Int. J. of Impact Engineering,* **34**(12): 1962–1964.

[Forrestal and Tzou, 1997] Forrestal, M.J., Tzou, D.Y. (1997). A spherical cavity-expansion penetration model for concrete targets. *Int. J. of Impact Engineering,* **34**(31–32): 4127–4146.

[Forrestal and Warren, 2009] Forrestal, M.J., Warren, T.L. (2009). Perforation equations for conical and ogival nose rigid projectiles into aluminum target plates. *Int. J. of Impact Engineering,* **36**(2): 220–225.

[Forrestal *et al.*, 1981a] Forrestal, M.J., Longcope, D.B., Norwood, F.R. (1981). A model to estimate forces on conical penetrators into dry porous rock. *J. of Applied Mechanics,* **48**(1): 25–29.

[Forrestal *et al.*, 1981b] Forrestal, M.J., Norwood, F.R., Longcope, D.B. (1981). Penetration into targets described by locked hydrostats and shear strength. *Int. J. of Solids and Structures,* **17**(9): 915–924.

[Forrestal *et al.*, 1984] Forrestal, M.J., Lee, L.M., Jenrette, B.D., Setchell, R.E. (1984). Gas-gun experiments determine forces on penetrators into geological targets. *J. of Applied Mechanics,* **51**(3): 602–607.

[Forrestal *et al.*, 1986] Forrestal, M.J., Lee, L.M., Jenrette, B.D. (1986). Laboratory-scale penetration experiments into geological targets to impact velocities of 2.1 km/s. *J. of Applied Mechanics,* **53**(2): 317–320.

[Forrestal *et al.*, 1987] Forrestal, M.J., Rosenberg, Z., Luk, V.K., Bless, S.J. (1987). Perforation of aluminum plates with conical-nosed rods. *J. of Applied Mechanics,* **54**(1): 230–232.

[Forrestal *et al.*, 1988] Forrestal, M.J., Okajima, K., Luk, V.K. (1988). Penetration of 6061-T651 aluminum target with rigid long rods. *J. of Applied Mechanics,* **55**(4): 755–760.

[Forrestal *et al.*, 1990] Forrestal, M.J., Luk, V.K., Brar, N.S. (1990). Perforation of aluminum armor plates with conical-nose projectiles. *Mechanics of Materials,* **10**(1-2): 97–105.

[Forrestal *et al.*, 1991] Forrestal, M.J., Brar, N.S., Luk, V.K. (1991). Perforation of strain-hardening targets with rigid spherical-nose rods. *J. of Applied Mechanics,* **58**(1): 7–10.

[Forrestal *et al.*, 1992] Forrestal, M.J., Luk, V.K., Rosenberg, Z., Brar, N.S. (1992). Penetration of 7075-T651 aluminum targets with ogival-nose rods. *Int. J. of Solids and Structures,* **29**(14/15): 1729–1736.

[Forrestal *et al.*, 1994] Forrestal, M.J., Altman, B.S., Cargile, J.D., Hanchak, S.J. (1994). An empirical equation for penetration depth of ogive-nose projectiles into concrete targets. *Int. J. of Impact Engineering,* **15**(4): 395–405.

[Forrestal *et al.*, 1995] Forrestal, M.J., Tzou, D.Y., Askar, E., Longcope, D.B. (1995). Penetration into ductile metal targets with rigid spherical-nose rods. *Int. J. of Impact Engineering,* **16**(5/6): 699–710.

[Forrestal *et al.*, 1996] Forrestal, M.J., Frew, D.J., Hanchak, S.J., Brar, N.S. (1996). Penetration of grout and concrete targets with ogive-nose steel projectiles. *Int. J. of Impact Engineering,* **18**(5): 465–476.

[Forrestal *et al.*, 2003] Forrestal, M.J., Frew, D.J., Hickerson, J.P., Rohwer, T.A. (2003). Penetration of concrete targets with deceleration-time measurements. *Int. J. of Impact Engineering,* **28**(5): 479–497.

[Forrestal *et al.*, 2010] Forrestal, M.J., Børvik, T., Warren, T.L. (2010). Perforation of 7075-T651 aluminum armor plates with 7.62 mm APM2 bullets. *Experimental Mechanics,* **50**(8): 1245–1251.

[Franzen *et al.*, 1994] Franzen, R.R., Walker, J.D., Orphal, D.L., Anderson, C.E. (1994). An upper limit for the penetration performance of segmented rods with segment - L/D<1. *Int. J. of Impact Engineering,* **15**(5): 661–668.

[Frew *et al.*, 1998] Frew, D.J., Hanchak, S.J., Green, M.L., Forrestal, M.J. (1998). Penetration of concrete targets with ogive-nose steel rods. *Int. J. of Impact Engineering,* **21**(6): 489–497.

[Frew *et al.*, 2000] Frew, D.J., Forrestal, M.J., Hanchak, S.J. (2000). Penetration experiments with limestone targets and ogive-nose steel projectiles. *J. of Applied Mechanics,* **67**(4): 841–845.

[Frew *et al.*, 2006] Frew, D.J., Forrestal, M.J., Cargile, J.D. (2006). The effect of concrete target diameter on projectile deceleration and penetration depth. *Int. J. of Impact Engineering*, **32**(10): 1584–1594.

[Fuchs, 1963] Fuchs, O.P. (1963). Impact phenomena, *AIAA J.,* **1**(9): 2124–2126.

[Fullard and Barr, 1989] Fullard, K., Barr, P. (1989). Development of design guidance for low velocity impacts on concrete floors. *Nuclear Engineering and Design,* **115**(1): 113–120.

[Fullard *et al.*, 1991] Fullard, K, Baum, M.R., Barr, P. (1991). The assessment of impact on nuclear power plant structures in the United Kingdom. *Nuclear Engineering and Design,* **130**(2): 113–120.

[Gagin *et al.*, 1979] Gagin, V.A., Sokolov, S.M., Minaev, V.L. (1979). *Calculation of Depth of Penetration of Short Anchors.* INFORMNEFTEGAZSTROJ, Moscow (in Russian).

[Gama and Gillespie, 2008] Gama, B.A., Gillespie, J.W., Jr. (2008). Punch shear based penetration model of ballistic impact of thick-section composites. *Composite Structures,* **86**(4): 356–369.

[Garcıa-Castillo *et al.*, 2011] Garcıa-Castillo, S.K., Buitrago, B.L., Barbero, E. (2011). Behavior of sandwich structures and spaced plates subjected to high-velocity impacts. *Polymer Composites*, **32**(2): 290–295.

[Gebbeken *et al.*, 2009] Gebbeken, N., Linse, T., Hartmann, T., Teich, M., Pietzsch. (2009). Dimensioning of concrete walls against small calibre impact including models for deformable penetrators and the scattering of experimental results In: Hiermaier, A.S., ed. *Predictive Modeling of Dynamic Processes. A Tribute to Professor Klaus Thoma*. Spinger, pp. 221–238.

[Gendugov *et al.*, 1984] Gendugov, V.M., Romanova, C.V., Romodanova, T.V. (1984). Body of revolution with minimum drag moving in elastic-plastic and plastic compressible media. In: *Problems of Dynamics of Deformable Media*. Armenian Academy of Science, Yerevan, pp. 116–119 (in Russian).

[Gerasimov, 2007] Gerasimov, A.V., Ed. (2007). *Theoretical and Experimental Investigation of High-Speed Interaction of Bodies*. Tomsk State University Press, Tomsk (in Russian).

[Gerasimov *et al.*, 2008] Gerasimov, A.V., Konjaev, A.A., Kul'kov, S.N., Tolkachev, V.M., Khristenko, Yu.F. (2008). Experimental and computational modelling of armor perforation by compact and elongated impactors. *Vestnik Akademii Voennykh Nauk*, **3**(24): 149–154 (in Russian).

[Giere, 1964] Giere, A.C. (1964). Some energy and momentum considerations in the perforation of plates. *AIAA J.*, **2**(8): 1471–1472.

[Glazyrin and Orlov, 2003] Glazyrin, V.P., Orlov, M.Y. (2003). Modeling of perforation of layered shields. *Vychislitel'nyje Technologii*, **8**(4): 143–51 (in Russian).

[Glazyrin *et al.*, 2006] Glazyrin, V.P., Orlov, M.Y., Orlov, Y.N. (2006). Features of perforation of layered barriers by an elongated impactor. In: *Proc of All-Russian Conf. "Fundamental'nye I Prikladnye Problemy Sovremennoj Mehaniki [Fundamental and Applied Problems of Modern Mechanics]"*. Tomsk University Press, Tomsk, pp. 251–253 (in Russian).

[Gogolewski *et al.*, 1996] Gogolewski, R.P., Cunningham, B.J., Riddle, R.O., Lesuer, D. R., Syn, C.K. (1996). On the importance of target material interfaces during low speed impact. In: *Proc. of the 16th Int. Symp. on Ballistics* (23–28 September, 1996, San Francisco, CA), vol.3, pp. 751–760.

[Goldsmith, 1960] Goldsmith, W. (1960). *Impact*. Arnold, London.

[Goldsmith and Finnegan, 1971] Goldsmith, W., Finnegan, S.A. (1971). Penetration and perforation processes in metal targets at and above ballistic velocities. *Int. J. of Mechanical Science*, **13**(10): 843–866.

[Golubev and Medvedkin, 2000] Golubev, V.K., Medvedkin, V.A. (2000). Penetration of a rigid shell into a steel obstacle at moderate impact velocities. *J. of Applied Mechanics and Technical Physics*, **41**(1): 32–34.

[Golubev and Medvedkin, 2001] Golubev, V.K., Medvedkin, V.A. (2001). Penetration of a rigid rod into a thick steel plate at elevated velocities. *Strength of Materials*, **33**(4): 400–405.

[Gomez and Shukla, 2001] Gomez, J.T., Shukla, A. (2001). Multiple impact penetration of semi-infinite concrete. *Int. J. of Impact Engineering,* **25**(10): 965–979.

[Gonçalves *et al.*, 2004] Gonçalves, D.P., de Melo, F.C. L., Klein, A.N., Al-Qureshi, H.A. (2004). Analysis and investigation of ballistic impact on ceramic/metal composite armour. *Int. J. of Machine Tools and Manufacture,* **44**(2–3): 307–316.

[Goodier, 1965] Goodier, J.N. (1965). On the mechanics of indentation and cratering in solid targets of strain-hardening metal by impact of hard and soft spheres. In: *Proc.of 7th Hypervelocity Impact* Symp. (17–19 November, 1964, Tampa, FL), vol. 3, pp. 215–260.

[Gould, 1997] Gould, R.L. (1997). Penetrating vehicle with rocket motor. Patent No. US5596166.

[Grabarek, 1971] Grabarek, C.L. (1971). Penetration of armor by steel and high density penetrators. Memorandum Report BRL MR 2134. Ballistic Research Laboratory, Aberdeen Proving Ground, MD.

[Gradstein and Ryzhik, 1980] Gradstein, I., Ryzhik, I. (1980). *Table of Integrals, Series, and Products*. Academic Press, New York.

[Gran and Frew, 1997] Gran, J.K., Frew, D.J. (1997). In-target radial stress measurements from penetration experiments into concrete by ogive-nose steel projectiles. *Int. J. of Impact Engineering,* **19**(8): 715–726.

[Gran *et al.*, 1999] Gran, J.K., Moxley, R.E., Adley, M.D. (1999). Measurement of Triaxial Stresses during Deep Penetration into Concrete. In: *Proc. of 9th Int. Symp. on Interaction of the Effects of Munitions with Structures* (3–7 May, 1999, Berlin, Germany)

[Greenspon, 1976] Greenspon, J.A. (1976). An approximate non-dimensional representation of the THOR equations. Technical Report No. 173. U.S. Army Materiel System Analysis Activity, Aberdeen Proving Ground, MD.

[Grigoryan, 1979] Grigoryan, S.S. (1979). A new law of friction and mechanism for large-scale avalanches and landslides. *Soviet Physics Doklady,* **24**(2), 110–111.

[Grigoryan, 1993] Grigoryan, S.S. (1993). An approximate solution of the problem of the penetration of a body into soil. *Fluid Dynamics,* **28**(4): 444–449.

[Grigoryan, 2006] Grigoryan, V.A. (2006). *Particular Problems of Terminal Ballistics.* MGTU imeni N.E. Baumana, Moscow (in Russian).

[Guirgis and Guirguis, 2009] Guirgis, S., Guirguis, E. (2009). An energy approach study of the penetration of concrete by rigid missiles. *Nuclear Engineering and Design,* **239**(4): 819–829.

[Gupta and Madhu, 1997] Gupta, N.K., Madhu, V. (1997). An experimental study of normal and oblique impact of hard-core projectile on single and layered plates. *Int. J. of Impact Engineering,* **19**(5–6): 395–414.

[Gupta *et al.*, 2001] Gupta, N.K., Ansari, R., Gupta, S.K. (2001). Normal impact of ogive nosed projectiles on thin plates. *Int. J. of Impact Engineering,* **25**(7): 641–60.

[Gupta *et al.*, 2007] Gupta, N.K., Iqbal, M.A., Sekhon, G.S. (2007). Effect of projectile nose shape, impact velocity, and target thickness on deformation behavior of aluminum plates. *Int. J. of Solids and Structures,* **44**(10): 3411–3439.

[Gupta *et al.*, 2008] Gupta, N.K., Iqbal, M.A., Sekhon, G.S. (2008). Effect of projectile nose shape, impact velocity, and target thickness on the deformation behavior of layered plates. *Int. J. of Impact Engineering,* **35**(1): 37–60.

[Gwaltney, 1968] Gwaltney, R.C. (1968). Missile generation and protection in light water-cooled reactor power plants. Report ORNL-NSIC-22. Oak Ridge National Laboratory, Oak Ridge, TN.

[Hadala, 1975] Hadala, P.F. (1975). Evaluation of empirical and analytical procedures used for predicting the rigid body motion of an earth penetrator. Technical Report S-75-15. U.S. Army Engineer Waterways Experiment Station, Vicksburg, MS.

[Haldar and Hamieh, 1984] Haldar, A., Hamieh, H. (1984) Local effect of solid missiles on concrete structures. *J. of Structural Engineering,* **110**(5): 948–60

[Haldar and Miller, 1982] Haldar, A., Miller, F.J. (1982). Penetration depth in concrete for nondeformable missiles. *Nuclear Engineering and Design,* **71**(1): 79–88.

[Haldar *et al.*, 1983] Haldar, A., Hatami, M., Miller, F.J. (1983). Penetration and spallation depth estimation for concrete structures. In: *Trans. of 7th Int. Conf. on Structural Mechanics in Reactor Technology (SMiRT-7)* (Aug 22-26, 1983, Chicago, IL), paper J7/2.

[Han *et al.*, 2005] Han, Y.-Y., Zhao, G.-Z., Du, Z.-H., Fang, Q. (2005). Performance of telescopic penetrator. *Journal of Nanjing University of Science and Technology,* **29**(3): 270–273 (in Chinese).

[Hankins *et al.*, 2006] Hankins, M.D., Stoltz, B.C., Torres, K.L., Jones, S.E. (2006). Penetration of a coarse sand target by rigid projectiles. In: Jones, N. and Brebbia, C.A., eds. *Structures Under Shock and Impact IX.* WIT Press, Southampton, pp. 187–196.

[Hansson, 2003] Hansson, H. (2003). A note on empirical formulas for the prediction of concrete penetration. Report FOI-R-0968-SE. Swedish Defence Reasearch Agency (FOI), Tumba, Sweden.

[Hansson, 2005] Hansson, H. (2005). Penetration in concrete for projectiles with L/D≈9. Report FOI-R-1659-SE. Swedish Defence Reasearch Agency (FOI), Tumba, Sweden.

[Hashimoto *et al.*, 2005] Hashimoto, J., Takiguchi, K., Nishimura, K., Matsuzawa, K., Tsutsui, M., Ohashi, Y., Kojima, I., Torita, H. (2005). Experimental study on behavior of RC panels covered with steel plates subjected to missile impact. In: *Trans. of 18th Int. Conf. on Structural Mechanics in Reactor Technology (SMiRT-18)* (August 7-12, 2005, Beijing, China), paper J05-4, pp. 2604–2615.

[Hauver and Melani, 1990] Hauver, G.E., Melani, A. (1990). Behavior of segmented rods during penetration. Technical Report BRL-TR-3129. U.S. Army Ballistic Research Laboratory, Aberdeen Proving Ground, MD.

[Hayes and Probstein, 1959] Hayes, W.D., Probstein, R.F. (1959). *Hypersonic Flow Theory*. Academic Press. New York & London.

[He *et al.*, 2007a] He, T., Wen, H.M., Qin, Y. (2007). Penetration and perforation of FRP laminates struck transversely by conical-nosed projectiles. *Composite Structures,* **81**(2): 243–252.

[He *et al.*, 2007b] He, T., Wen, H.M., Qin, Y. (2007). Ballistic penetration and perforation of thick FRP laminates by ogival-nosed projectiles. *J. of Composite Materials,* **41**(23): 2829–2842.

[He *et al.*, 2008] He, T., Wen, H.M., Qin, Y. (2008). Finite element analysis to predict penetration and perforation of thick FRP laminates struck by projectiles. *Int. J. of Impact Engineering,* **35**(1): 27–36.

[He *et al.*, 2011] He, T., Wen, H.M., Guo, X.-J. (2011). A spherical cavity expansion model for penetration of ogival-nosed projectiles into concrete targets with shear-dilatancy. *Acta Mechanica Sinica*, **27**(6): 1001–1012.

[Healey and Weissman, 1974] Healey, J.J., Weissman, S. (1974). Primary fragment characteristics and impact effects in protective design. In: *Proc. of 16th Explosives Safety Seminar* (September, 1974, Hollywood, FL).

[Healey *et al.*, 1975] Healey, J., Weissman, S., Werner, H., Dobbs, N., Price, P. (1975). Primary fragment characteristics and impact effects on protective barriers. Technical Report No. 4903. Picatinny Arsenal, Dover, NJ.

[Heimdahl and Schulz, 1986] Heimdahl, O.E.R., Schulz, J.C. (1986). A note on the obtainment of instantaneous penetration information from final penetration data. *J. of Applied Mechanics,* **53**(1): 226–227.

[Herbette, 1989] Herbette, G. (1989). The influence of projectile shape on penetration power. In: *Proc. of the 11th Int. Symp. on Ballistics* (May 9–11, 1989, Brussels, Belgium), pp. 561–567.

[Herrmann and Jones, 1961] Herrmann, W., Jones, A.H. (1961). Survey of hypervelocity impact information. Report No. 99–1. Massachusetts Institute of Technology (MIT), Aeroelastic and Structures Research Laboratory, Cambridge, MA.

[Hetherington, 1992a] Hetherington, J.G. (1992). Optimization of two component composite armours. *Int. J. of Impact Engineering,* **12**(3): 409–414.

[Hetherington, 1992b] Hetherington, J.G. (1992). Correspondence on "An investigation into the energy absorbed during ballistic perforation of composite armors". *Int. J. of Impact Engineering,* **12**(2): 325–327.

[Hetherington and Lemieux, 1994] Hetherington, J. G., Lemieux, P. F. (1994). Effect of obliquity on the ballistic performance of two component composite armors. *Int. J. of Impact Engineering,* **15**(2): 131–137.

[Hetherington and Rajagopalan, 1991] Hetherington, J.G., Rajagopalan, B.P. (1991). An investigation into the energy absorbed during ballistic perforation of composite armors. *Int. J. of Impact Engineering,* **11**(1): 33–40.

[Heuzé, 1990] Heuzé, F.E. (1990). An overview of projectile penetration into geological materials, with emphasis on rocks. *Int. J. of Rock Mechanics and Mining Science & Geomechanics Abstracts,* **27**(1): 1–14.

[Hibbs, 1952] Hibbs, A.R. (1952). Optimum burning program for horizontal flight. *ARS J.,* **22**(4): 204–212.

[Hiermaier and Thoma, 1997] Hiermaier, S., Thoma, K. (1997). Numerical simulation of penetration in concrete using different types of code. In: *Proc. of the 8th Int. Symp. on Interaction of the Effects of Munitions with Structures* (April 22–25, 1997, McLean, VA).

[Hill, 1950a] Hill, R. (1950). *The Mathematical Theory of Plasticity.* Clarendon Press, Oxford.

[Hill, 1950b] Hill, R. (1950). A theory of the plastic bulging of a metal diaphragm by lateral pressure. *Philosophical Magazine. Series 7,* **41**(322): 1133–1142.

[Hill, 1980] Hill, R. (1980). Cavitation and the influence of headshape in attack of thick targets by non-deforming projectiles. *J. of Mechanics and Physics of Solids,* **28**(5–6): 249–263.

[Hohler and Stilp, 1990] Hohler, V., Stilp, A.J. (1990). Penetration performance of segmented rods at different spacing-comparison with homogeneous rods at 2.5–3.5 km/s. In: *Proc. of the 12th Int. Symp. on Ballistics (30 October - 1 November, 1990, San Antonio, TX),* vol. 3, pp. 178–187.

[Hohler *et al.*, 2001] Hohler, V., Weber, K., Tham, R., James, B., Barker, A., Pickup, I. (2001). Comparative analysis of oblique impact on ceramic composite systems. *Int. J. of Impact Engineering,* **26**(1–10): 333–334.

[Holland *et al.*, 1990] Holland, P.H., Gordon, J.T., Menna, T.L., Charters, A.C. (1990). Hydrocode results for the penetration of continuous, segmented and hybrid rods compared with ballistic experiments. *Int. J. of Impact Engineering,* **10**(1–4): 241–250.

[Holmquist *et al.*, 1999] Holmquist, T.J., Rajendran, A.M., Templeton, D.W., Bishnoi, K.D. (1999). A ceramic armor material database. Report TARDEC-TR-13754. TACOM Research, Development and Engineering Center, Warren, MI.

[Holt *et al.*, 1993] Holt, W.H., Mock, W., Jr., Soper, W.G., Coffey, C.S., Ramachandran, V., Armstrong, R.W. (1993). Reverse-ballistic impact study of shear plug formation and displacement in Ti_6Al_4V alloy. *J. of Applied Physics,* **73**(8): 3753–3759.

[Honda *et al.*, 1930] Honda, K., Takamae, G., Watanabe, T. (1930). On the measurement of the resistance of shield plates to penetration by a rifle bullet. *Tohoku Imperial University. 1st Series,* **19**: 703–725.

[Hopkins, 1960] Hopkins, H.G. (1960). Dynamic expansion of spherical cavities in metals. In: Hill, R. and Sneddon, I.N., eds. *Progress in Solid Mechanics,* **1**. Pergamon Press, Oxford, pp. 84–164.

[Hsueh, 1988] Hsueh, D.-W. (1988). Normal perforation of a thin infinite plate by a flat-headed cylindrical projectile. *Applied Mathematics and Mechanics*, **9**(12): 1113–1119.

[Hsueh, 1989] Hsueh, D.-W. (1989). Normal perforation of a thin infinite plate by a flat-headed cylindrical projectile. *Int. J. of Impact Engineering*, **8**(2): 133–139.

[Huang *et al.*, 2011a] Huang, G.-Y., Wu, G., Feng, S.-S., Zhou, T. (2011). Theory and experiment study of normal penetration of slender nose projectile into steel target. *Applied Mechanics and Materials*, **80–81**: 137–142.

[Huang *et al.*, 2011b] Huang, G.-Y., Feng, S.-S., Guang, W., Li, S.-P. (2011). Performance of different projectile nose shapes in normal penetrating armor targets. *Advanced Materials Research*, **308–310**: 1420–1425.

[Hughes, 1984] Hughes, G. (1984). Hard missile impact on reinforced concrete. *Nuclear Engineering and Design,* **77**(1): 23–35.

[Hunkler, 1990] Hunkler, R. (1990). Numerical simulation of segmented rods. In: *Proc. of the 12th Int. Symp. on Ballistics* (30 October - 1 November, 1990, San Antonio, TX).

[Hurlich, 1950] Hurlich, A. (1950). Spaced armor. Report WAL-710/930-1. Watertown Arsenal Laboratory, Watertown, MA.

[Hyde, 1991] Hyde, D.W. (1991). CONWEP: Conventional Weapons Effects Program. U.S. Army Engineer Waterways Experiment Station, Vicksburg, MS.

[Iqbal and Gupta, 2008] Iqbal, M.A, Gupta, N.K. (2008). Energy absorption characteristics of aluminum plates subjected to projectile impact. *Latin American J. of Solids and Structures,* **5**(4): 259–287.

[Iqbal and Gupta, 2011] Iqbal, M.A, Gupta, N.K. (2011). Ballistic limit of single and layered aluminium plates. Strain. **47**: e205–e219.

[Iqbal *et al.*, 2010] Iqbal, M.A, Chakrabarti, A., Beniwal, S., Gupta, N.K. (2010). 3D numerical simulations of sharp nosed projectile impact on ductile targets. *Int. J. of Impact Engineering,* **37**(2): 185–95.

[Iqbal *et al.*, 2012] Iqbal, M.A, Gupta, P.K., Deore, V.S., Tak, S.K., Tiwari, G., Gupta, N.K. (2012). Effect of target span and configuration on the ballistic limit. *Int. J. of Impact Engineering,* **42**: 11–24.

[Ilyushin, 1935] Ilyushin, A.A. (1935). Application of multilayer armours and methods of calculation of their perforation. In: Ilyushina, E.A. and Korotkina, N.R., compilers. (2003). Ilyushin A.A. *Works. Vol.1 (1935–1945).* Fizmatlit, Moscow, pp. 285–294 (in Russian).

[Irfan and Prakash, 2000] Irfan, M.A., Prakash, V. (2000). Time resolved friction dry sliding of metal on metal. *Int. J. of Solids and Structures,* **37**(20): 2859–2882.

[Isbell *et al.*, 1992] Isbell, W.M., Anderson, C.E, Jr., Asay, J.R., Bless, S.J., Grady, D.E., Sternberg, J. (1992). Penetration mechanics research in the former Soviet Union. Technical Assessment Report. Science Applications International Corporation, San Diego, CA. Foreign Applied Sciences Assessment Center.

[Jaeger, 1979] Jaeger, C. (1979). *Rock Mechanics and Engineering*. Cambridge University Press, Cambridge.

[Jankov *et al.*, 1976] Jankov, Z.D., Shanahan, J.A., White, M.P. (1976). Missile tests of quarter-scale reinforced concrete barriers. In: *Proc. of a Symp. on Tornadoes, Assessment of Knowledge and Implications for Man* (June 1976, Lubbock, TX).

[Jenq *et al.*, 1988] Jenq, S.T., Goldsmith, W.,. Kelly, J.M. (1988). Effect of target bending in normal impact of a flat-ended cylindrical projectile near the ballistic limit. *Int. J. of Solids and Structures,* **24**(12): 1243–1266.

[Jiang *et al.*, 2004] Jiang, Z.-G., Zeng, S.-Y., Zhou, J.-P. (2004). Analyses of critical impact energy for plugging failure of thin metallic plates. *Engineering Mechanics*, **21**(5): 203–208 (in Chinese)

[Jiang *et al.*, 2005a] Jiang, Z.G., Zeng, S.Y., Zhou, J.P. (2005). A two-stage model for analysis of ballistic limit of metallic targets. *Engineering Mechanics*, **22**(4): 229–234 (in Chinese).

[Jiang *et al.*, 2005b] Jiang, Z., Zeng, S., Zhou, J. (2005). Two analytical models for ballistic limit of metallic targets struck by rigid sharp-nosed projectiles. *Chinese J. of Applied Mechanics*, **22**(1): 21–25. (in Chinese)

[Jiang *et al.*, 2007a] Jiang, Z.-G., Zeng, S.-Y., Zhou, J.-P. (2007). A three-stage model for the perforation of moderately thick metallic plates. *Acta Armamentarii*, **28** (9): 1046–1052. (in Chinese)

[Jiang *et al.*, 2007b] Jiang, Z.-G., Zeng, S.-Y., Zhou, J.-P. (2007). An analytical model for penetration into finite thickness metallic target struck by rigid sharp-nosed projectiles. *Acta Armamentarii*, **28**(8): 923–929 (in Chinese)

[Jonas and Zukas, 1978] Jonas, G.H., Zukas, J.A. (1978). Mechanics of penetration: analysis and experiment. *Int. J. of Engineering Science,* **16**(1): 879–903.

[Jones and Kim, 1997a] Jones, N., Kim, S.-B. (1997a). A Study on the large ductile deformations and perforation of mild steel plates struck by a mass-Part I: Experimental results. J. of Pressure Vessel Technology, 1997, **119**(2): 178–184.

[Jones and Kim, 1997b] Jones, N., Kim, S.-B. (1997b). A Study on the large ductile deformations and perforation of mild steel plates struck by a mass- Part II: Discussion. J. of Pressure Vessel Technology, 1997, **119**(2): 185–191

[Jones and Rule, 2000] Jones, S.E., Rule, W.K. (2000). On the optimal nose geometry for a rigid penetrator, including the effects of pressure-dependent friction. *Int. J. of Impact Engineering,* **24**(4): 403–415.

[Jones *et al.*, 1998] Jones, S.E., Rule, W.K., Jerome, D.M., Klug, R.T. (1998). On the optimal nose geometry for a rigid penetrator. *Computational Mechanics,* **22**(5): 413–417.

[Jones *et al.*, 2003] Jones, S.E., Hughes, M.L., Tonnes, O.A., Davis, R.N. (2003). A one-dimensional analysis of rigid-body penetration with high-speed friction., *J. of Mechanical Engineering Science (Proc. of the Institution of Mechanical Engineers. Part C),* **217**(C4): 411–422.

[Jowett, 1986] Jowett, J. (1986). The effects of missile impact on thin metal structures. Report S.R.D.R 378. United Kingdom Atomic Energy Authority (UKAEA), Safety and Reliability Directorate, Culcheth, UK.

[Kane *et al.*, 2008] Kane, A., Børvik, T., Hopperstad, O.S., Langseth, M. (2008). Finite element analysis of plugging failure in steel plates struck by blunt projectiles. *J. of Applied Mechanics*, **76**(5) (051302).

[Kang and Cho, 1991] Kang, K.J., Cho, K.Z. (1991). An estimation of ballistic limit for ceramic-FRP composite armor. *Korean Society of Mechanical Engineers (KSME) Int. J.*, **5**(2): 140–148.

[Kar, 1977] Kar, A.K. (1977). Projectile penetration of earth media. In: *Trans. of 4th Int. Conf. on Structural Mechanics in Reactor Technology (SMiRT-4)* (August 15–19, 1977, San-Francisko, CA).

[Kar, 1978] Kar, A.K. (1978). Local effects of tornado generated missiles. *J. of the Structural Division,* **104**(ST5): 809–816.

[Kar, 1978a] Kar, A.K. (1978). Projectile penetration into buried structures. *J. of the Structural Division*, **104**(ST1): 125–139.

[Kar, 1979] Kar, A.K. (1979). Residual velocity for projectiles. *Nuclear Engineering and Design*, **53**(1): 87–95.

[Kasano and Abe, 1997] Kasano, H., Abe, K. (1997). Perforation characteristics prediction of multi-layered composite plates subjected to high velocity impact. *In: Proc. of 11th Int. Conf. on Composite Materials* (14-18 July, 1997, Gold Coast, Australia), vol. 2, pp. 522–531.

[Katsuragi and Durian, 2007] Katsuragi, H., Durian, D.J. (2007). Unified force law for granular impact cratering. *Nature Physics,* **3**(6): 420–423.

[Kennedy, 1976] Kennedy, R.P. (1976). A review of procedures for the analysis and design of concrete structures to resist missile impact effects. *Nuclear Engineering and Design,* **37**(2): 183–203.

[Khromov, 2003] Khromov, I.V. (2003). Optimization of penetrator nose. *Izvestija Tul'skogo Gosudarsvennogo Universiteta. Serija: Problemy Spetsial'nogo Mashinostroenija,* **6**(1): 252–255 (in Russian).

[Khromov, 2004] Khromov, I.V. (2004). Dynamics of penetration of rigid rotating penetrators into soil. Ph.D. Thesis. Tula State University, Tula (in Russian).

[Kivity *et al.*, 1989] Kivity, Y., Yitzhak, E., Hirsch, E. (1989). Penetration of segmented rods into homogeneous targets. In: *Proc. of the 11th Int. Symp. on Ballistics* (May 9-11, 1989, Brussels, Belgium).

[Klepaczko, 2001] Klepaczko, J.R. (2001). Surface layer thermodynamics of steel penetrators at high and very high sliding velocities. Report AFRL-MN-EG-TR-2001-7076. Air Force Research Laboratory, Eglin Air Force Base, FL.

[Klepaczko, 2002] Klepaczko, J.R. (2002). Surface layer thermodynamics of steel penetrators at high and very high velocities. Report AFRL-MN-EG-TR-2002-7076. Air Force Research Laboratory, Eglin Air Force Base, FL.

[Ko *et al.*, 1997] Ko, K.F., Geshury, A.J., Song, J.W. (1997). Behavior of gradient designed composite under ballistic impact. *In: Proc. of 11th Int. Conf. on Composite Materials* (14-18 July, 1997, Gold Coast, Australia), vol. 2, pp. 463–472.

[Kojima, 1991] Kojima, I. (1991). An experimental study on local behavior of reinforced concrete slabs to missile impact. *Nuclear Engineering and Design,* **130**(2): 121–132.

[Korn and Korn, 1968] Korn, G.A., Korn, T.M. (1968). *Mathematical Handbook for Scientists and Engineers.* McGraw-Hill Book Comp., New York.

[Koshika *et al.*, 1993] Koshika, N., Tsubota, H., Kasai, Y., Morikawa, H., Sawamoto, Y., Kobayashi, N. (1993). Quantitative studies on impact resistance of a reinforced concrete panels with a steel lines under impact loading. Part 2: Analytical study for small-scale and full-scale tests. In: *Trans. of 12th Int. Conf. on Structural Mechanics in Reactor Technology* (SMiRT-12) (August 15-20, 1993, Stuttgart, Germany), paper J7/2, pp. 175–180.

[Kosmodemiansky, 1966] Kosmodemiansky, A.A. (1966). *Course on Theoretical Mechanics. Part 2.* Prosveschenije, Moscow (in Russian).

[Kotov, 2008] Kotov, V.L. (2008). Investigation of applicability of self-similarity analysis to problem of expansion of a spherical cavity in a compressible medium for determining pressure on striker-soil contact surface. *Problemy Prochnosti I Plastichnosti,* **70**: 123–130 (in Russian).

[Kotov *et al.*, 2002] Kotov, V.L., Krylov, S.V., Balandin, V.V. (2002). Penetration of finite angle wedge into sand. *Vestnik of the Nizhnii Novgorod State University*, **1**(4): 133–141 (in Russian).

[Krafft, 1955] Krafft, J.M. (1955). Surface friction in ballistic penetration. *J. of Applied Physics,* **26**(10): 1248–1253.

[Kraghelsky, 1965] Kraghelskii, I.V. (1965). *The Friction and Wear.* Butterworths, London.

[Kraghelsky and Vinogradova 1962] Kraghelsky, I.V., Vinogradova, I.E. (1962). *Coefficients of Friction.* MASHGIZ, Moscow (in Russian).

[Kraghelsky *et al.*, 1982] Kraghelsky, I.V., Dobychin, M..N., Kombalov, V.S. (1982). *Friction and Wear Calculation Methods.* Pergamon Press, Oxford.

[Kravchenko *et al.*, 1994] Kravchenko, V.P., Skorkin, N.A., Sapozhnikov, A.A. (1994). Penetration of solid body of revolution into rock and soils. Preprint No. 18. Research Institute of Technical Physics (VNIITF), Cheliabinsk-70 (in Russian).

[Krishna Teja Palleti *et al.*, 2012] Krishna Teja Palleti, H.N., Gurusamy, S., Kumar, S., Soni, R., John, B., Vaidya, R., Bhoge, A., Naik, N.K. (2012). Ballistic impact performance of metallic targets. *Materials and Design*, **39**: 253–263.

[Krotov, 1995] Krotov, V.F. (1995). *Global Methods in Optimal Control Theory.* Marcel Dekker, New York.

[Kucher, 1967] Kucher, V. (1967). Penetration with optimal work. Report BRL-R-1384. Army Ballistic Research Laboratory, Aberdeen Proving Ground, MD.

[Kucher, 1982] Kucher, V. (1982). Multiple impacts on monolithic steel. Technical Report BRL-TR-02406. U.S. Army Ballistic Research Laboratory, Aberdeen Proving Ground, MD.

[Lambert, 1978] Lambert, J.P. (1978). A residual velocity predictive model for long rod penetrators. Report ARBRL-MR-02828. Army Ballistic Research Laboratory, Aberdeen Proving Ground, MD.

[Lambert and Jonas, 1976] Lambert, J.P., Jonas, G.H. (1976). Towards standardization of in terminal ballistic testing: velocity representation. Report BRL-R-1852. Army Ballistic Research Laboratory, Aberdeen Proving Ground, MD.

[Landgrov and Sarkisyan, 1984] Landgrov, I.F., Sarkisyan, O.A. (1984). Piercing plastic-material barriers with a rigid punch. *J. of Applied Mechanics and Technical Physics,* **25**(5): 771–773.

[Landkof and Goldsmith, 1993] Landkof, B., Goldsmith, W. (1993). Petalling of thin, metallic plates during penetration by cylindro-conical projectiles. *Int. J. of Solids and Structures*, **21**(3): 245–266.

[Lang *et al.*, 2011] Lang, L., Chen, X.-W., Lei, J.-S. (2011). Numerical simulations on long rods and segmented rods penetrating into steel targets. *Explosion and Shock Waves*, **31**(2): 127–134 (in Chinese).

[Latif *et al.*, 2011] Latif, Q.B.A.I., Rahman, I.A., Zaidi, A.M.A. (2011). Development of empirical formula prediction on critical impact energy for perforation phenomena on concrete structures. *J. of Mathematics Research*, **3**(1): 83–87.

[Lee, 2001] Lee, M. (2001). A numerical comparison of the ballistic performance of unitary rod and segmented rods against stationary and moving oblique plates. *Int. J. of Impact Engineering,* **26**(1–10): 399–407.

[Lee and Normandia, 1999] Lee, M., Normandia, M.J. (1999). Successive impact of segmented rods at high-velocity. *Korean Society of Mechanical Engineers (KSME) Int. J.*, **13**(4): 312–320.

[Lee and Yoo, 2001] Lee, M., Yoo, Y.H. (2001). Analysis of ceramic/metal armour systems. *Int. J. of Impact Engineering,* **25**(9): 819–829.

[Leitmann, 1962] Leitmann, G., ed. (1962). *Optimization Techniques with Applications to Aerospace Systems*. Academic Press, NewYork, London.

[Levy and Goldsmith, 1984] Levy, N., Goldsmith, W. (1984). Normal impact and perforation of thin plates by hemispherically tipped projectiles - 1. *Int. J. of Impact Engnineerng*, **2**(3): 209–229.

[Li and Chen, 2002] Li, Q.M., Chen, X.W. (2002). Penetration into concrete targets by a hard projectile. In: Jones, N. *et al.*, eds., *Proc. of 7th Int. Conf. on Structures under Shock and Impact* (27-29, May, 2002, Montreal, Canada). WIT Press, Southampton, pp. 91–100.

[Li and Chen, 2003] Li, Q.M., Chen, X.W. (2003). Dimensionless formulae for penetration depth of concrete target impacted by a non-deformable projectile. *Int. J. of Impact Engineering,* **28**(1): 93–116.

[Li and Tong, 2003] Li, Q.M., Tong, D.J. (2003). Perforation thickness and ballistic limit of concrete target subjected to rigid projectile impact. *J. of Engineering Mechanics,* **129**(9): 1083–1091.

[Li *et al.*, 2002] Li, M., Jiang, C., Wang, Z. (2002). Ballistic limit analytical model of tungsten ball perforating ceramic/aluminum composite target. J. of Beijing Institute of Technology, 4: 389–392.

[Li *et al.*, 2004] Li, Q.M., Weng, H.J., Chen, X.W. (2004). A modified model for the penetration into moderately thick plates by a rigid, sharp-nosed projectile. *Int. J. of Impact Engineering,* **30**(2): 193–204.

[Li *et al.*, 2005] Li, Q.M, Reid, S.R., Wen, H.M., Telford, A.R. (2005). Local impact effects of hard missiles on concrete targets. *Int. J. of Impact Engineering,* **32**(1–4):224–284

[Li *et al.*, 2006] Li, Q.M., Reid, S.R., Ahmad-Zaidi, A.M. (2006). Critical impact energies for scabbing and perforation of concrete target. *Nuclear Engineering and Design,* **236**(11): 1140–1148.

[Li *et al.*, 2007] Li, X.L., Liu, C.L., Huang, F.L., Wu, H.J., Chen, Y.Z., Chen, X.W. (2007). On the normal perforation of thick metallic plates by a sharp nose rigid projectile. In: *Proc. of the 23rd Int. Symp.on Ballistics* (16-20 April, 2007, Tarragona, Spain), vol. 2, pp. 1173–1180.

[Liaghat and Malekzadeh, 1995] Liaghat, G.H, Malekzadeh, A. (1995). A comment on the penetration theory by Dikshit and Sundararajan. *Int. J. of Impact Engineering*, **16**(4): 691–693.

[Liaghat and Malekzadeh, 1999] Liaghat, G.H., Malekzadeh, A. (1999). A modification to the mathematical model of perforation by Dikshit and Sundararajan. *Int. J. of Impact Engineering*, **22**(5): 543–550.

[Liaghat *et al.*, 2005] Liaghat, G.H., Moslemi Naeini, H, Feli, S. (2005). The mechanics of normal and oblique penetration of conical projectiles into multi layer metallic targets. *Iranian J. of Science and Technology. Transaction B. Engineering,* **29**(b2): 241–251.

[Liang *et al.*, 2003] Liang, C.-C., Jen, C.-Y., Yang, M.-F., Wu, P.-W. (2003). Resistant performance of perforation in protective structures using a semi-empirical method with marine applications. *Ocean Engineering*, **30**(9): 1137–1162.

[Liang *et al.*, 2005] Liang, C.C., Yang, M.-F., Wu, P.W., Teng, T.L. (2005). Resistant performance of perforation of multi-layered targets using an estimation procedure with marine application, *Ocean Engineering,* **32**(3–4): 441–468.

[Lim and Ashby., 1987] Lim, S.C., Ashby, M.F. (1987). Wear-mechanism maps. *Acta Metallurgica,* **35**(1): 1–24.

[Lim *et al.*, 1989] Lim, S.C., Ashby, M.F., Brunton, J.H. (1989). The effect of sliding conditions on the dry friction of metals. *Acta Metallurgica,* **37**(3): 767–772.

[Linderman *et al.*, 1974] Linderman, R.B., Rotz, J.V., Yeh, G.C.K. (2004). Design of structures for missile impact. Topical Report BC-TOP-9, Revision 2. Bechtel Power Corporation, San Francisco, California.

[Liss *et al.*, 1983] Liss, J., Goldsmith, W., Kelly, J.M. (1983). A phenomenological penetration model of plates. *Int. J. of Impact Engineering,* **1**(4): 321–341.

[Littlefield, 2001] Littlefield, D.L. (2001). Effect of alignment on the penetration of segmented rods. *Int. J. of Impact Engineering,* **26**(1–10): 421–431.

[Littlefield *et al.*, 1997] Littlefield, D.L., Anderson, C.E., Jr., Partom, Y., Bless, S.J. (1997). The penetration of steel targets finite in radial extent. *Int. J. of Impact Engineering,* **19**(1): 49–62.

[Littlefield *et al.*, 1999] Littlefield, D.L., Garcia, R.M., Bless, S.J. (1999). The effect of offset on the performance of segmented penetrators. *Int. J. of Impact Engineering,* **23**(1): 547–560.

[Lixin *et al.*, 2000] Lixin, Q., Yunbin, Y., Tong, L. (2000). A semi-analytical model for truncated-ogive-nose projectiles penetration into semi-infinite concrete targets. *Int. J. of Impact Engineering,* **24**(9): 947–955.

[Longcope *et al.*, 1999] Longcope, D.B., Jr., Tabbara, M.R., Jung, J. (1999). Modeling of oblique penetration into geologic targets using cavity expansion penetrator loading with target free-surface effects. Report SAND99-1104. Sandia National Laboratories, Albuquerque, NM.

[Luk and Amos, 1991] Luk, V.K., Amos, D.E. (1991). Dynamic cylindrical cavity expansion of compressible strain-hardering materials. *J. of Applied Mechanics,* **58**(2): 334–340.

[Luk and Forrestal, 1987] Luk, V.K., Forrestal, M.J. (1987). Penetration into semi-infinite reinforced-concrete targets with spherical and ogival nose projectiles. *Int. J. of Impact Engineering,* **6**(4): 291–301.

[Luk and Forrestal, 1989] Luk, V.K., Forrestal, M.J. (1989). Comment on "Penetration into semi-infinite reinforced-concrete targets with spherical and ogival nose projectiles (*Int. J. of Impact Engineering,* **6**: 291–301, 1987)". *Int. J. of Impact Engineering,* **8**(1): 83–84.

[Luk *et al.*, 1991] Luk, V.K., Forrestal, M.J., Amos, D.E. (1991). Dynamic spherical cavity expansion of strain-hardening materials. *J. of Applied Mechanics,* **58**(1): 1–6.

[Lynch *et al.*, 2001] Lynch, N.J., Subramanian, R., Brown, S., Alston, J. (2001). The influence of penetrator geometry and impact velocity on the formation of crater volume in semi-infinite targets. In: *Proc. of the 19th Int. Symp. on Ballistics* (7-11 May, 2001, Interlaken, Switzerland), pp. 1265–1271.

[Macek and Duffey, 2000] Macek, R.W., Duffey, T.A. (2000). Finite cavity expansion method for near-surface effects and layering during earth penetration. *Int. J. of Impact Engineering,* **24**(3): 239–258.

[Madhu *et al.*, 2003] Madhu, V., Bhat, T.B., Gupta, N.K. (2003). Normal and oblique impacts of hard projectiles on single and layered plates – an experimental study. *Defence Science J.*, **53**(2): 147–156.

[Magier, 2008] Magier, M. (2008). Numerical analysis of influence of modification of two-segment penetrator of a subcaliber projectile on the depth of RHA penetration. *Problemy Techniki Uzbrojenia*, **107**(5): 43–60 (in Polish).

[Magier, 2010] Magier, M. (2010). The conception of the segmented kinetic energy penetrators for tank guns. *Journal of Applied Mechanics*, **77**(5): 051802.

[Magnusson *et al.*, 2001] Magnusson, J., Unosson, M., Carlberg, A. (2001). High performance concrete "HPC". Field experiments and production. Technical Report FOI-R-0256-SE. Swedish Defence Reasearch Agency (FOI), Tumba, Sweden.

[Malanin and Pensky, 2007] Malanin, V.V., Pensky, O.G. (2007). Conjugate Models of Dynamics of Heat Impulse Machines and Penetration of Rigid Bodies into Continuous Medium. Perm State University Press, Perm (in Russian).

[Marom and Bodner, 1979] Marom, I., Bodner, S.R. (1979). Projectile perforation of multi-layered beams. *Int. J. of Mechanical Science,* **21**(8): 489–504.

[Martins *et al.*, 1990] Martins, J.A.C., Oden, J.T., Simoes, M.F. (1990). A study of static and kinetic friction. *Int. J. of Engineering Science,* **28**(1): 29–92.

[Masri, 2010] Masri, R. (2010). Estimation of deep penetration resistance with spherical specific cavitation energy. *Int. J. of Protective Structures*, **1**(3): 391–408.

[Masri and Durban, 2005] Masri, R., Durban, D. (2005). Dynamic spherical cavity expansion in an elastoplastic compressible mises solid. *J. of Applied Mechanics*, **72**(6): 887–898.

[Masri and Durban, 2006] Masri, R., Durban, D. (2006). Dynamic cylindrical cavity expansion in an incompressible elastoplastic medium. *Acta Mechanica*, **181**(1–2): 105–123.

[Masri and Durban, 2007] Masri, R., Durban, D. (2007). Cylindrical cavity expansion in compressible Mises and Tresca solids. *European J. of Mechanics. A/Solids*, **26**(4): 712–727.

[Masri and Durban, 2009] Masri, R., Durban, D. (2009). Deep penetration analysis with dynamic cylindrical cavitation fields. *Int. J. of Impact Engineering*, **36**(6): 830–841.

[Masri *et al.*, 2010] Masri, R., Cohen, T., Durban, D. (2010). Enlargement of a circular hole in a thin plastic sheet: taylor–bethe controversy in retrospect. *Quarterly J. of Mechanics and Applied Mathematics*, **63**(4): 589–616.

[Mayersak, 2004] Mayersak, J.R. (2004). Kinetic energy cavity penetrator weapon. Patent application number: 10/443,621. Publication number: US 2004/0231552 A1.

[Me-Bar, 1997] Me-Bar, Y. (1997). A method for scaling ballistic penetration phenomena. *Int. J. of Impact Engineering,* **19**(9–10): 821–829.

[Miele, 1957] Miele, A. (1957). An extension of the theory of the optimum burning program for the level flight of a rocket-powered aircraft. *J. of Aerospace Engineering,* **24**(12): 874–884

[Miele, 1962a] Miele, A. (1962). A study of the slender body of revolution of minimum drag using the Newton-Busemann pressure coefficient law. Technical Report No.

62. Boeing Scientific Research Laboratories, Flight Sciences Laboratory, Seattle, WA.

[Miele, 1962b] Miele, A. (1962). The calculus of variations in applied aerodynamics and flight mechanics. In: Leitmann, G., ed. *Optimization Techniques with Applications to Aerospace Systems.* Academic Press, New York, London, pp. 258–301.

[Miele, 1962c] Miele, A. (1962). Extremization of linear integrals by Green's theorem. In: Leitmann, G. (ed). *Optimization Techniques with Applications to Aerospace Systems.* Academic Press, New York, London, pp. 69–98.

[Miele, 1965] Miele, A., ed. (1965). *Theory of Optimum Aerodynamic Shapes.* Academic Press, New York.

[Mileiko and Sarkisyan, 1981] Mileiko, S.T., Sarkisyan, O.A. (1981). Phenomenological model of punch-through. *J. of Applied Mechanics and Technical Physics,* **22**(5): 711–713.

[Mileiko *et al.*, 1994] Mileiko, S.T., Sarkisyan, O.A., Kondakov, S.F. (1994). Ballistic limits of Al-6% Mg allow laminated by diffusion bonding. *Theoretical and Applied Fracture Mechanics,* **21**(1): 9–16.

[Mines *et al.*, 1999] Mines, R.A.W., Roach, A.M., Jones, N. (1999). High velocity perforation behaviour of polymer composite laminates. *Int. J. of Impact Engineering,* **22**(6): 561–588.

[Miroshin and Khalidov, 1991] Miroshin, R.N., Khalidov, U.A. (1991). *Locaized Interaction Theory.* Leningrad State University Press, Leningrad (in Russian).

[Miroshin and Khalidov, 2002] Miroshin, R.N., Khalidov, U.A. (2002). *Local Methods in Continuum Mechanics.* St.-Petersburg University Press, St. Petersburg (in Russian).

[Molinari *et al.*, 1999] Molinari, A., Estrin, Y., Mercier, S. (1999). Dependence of the coefficient of friction on the sliding conditions in the high velocity range. *J. of Tribology,* **121**(1): 35–41.

[Montgomery, 1976] Montgomery, R.S. (1976). Friction and wear at high sliding speeds. *Wear,* **36**(3): 275–298.

[Moore, 1979] Moore, R.E. (1979). *Methods and Applications of Interval Analysis.* Society for Industrial and Applied Mathematics (SIAM), Philadelphia, PA.

[Murphy, 1984] Murphy, M.J., (1984). Performance analysis of two-stage munitions. In: *Proc. of the 8th Int. Symp. on Ballistics* (October 23–25, 1984, Orlando, FL), paper TB 23–29.

[Murthy *et al.*, 2010] Murthy, A.R.C., Palani, G.S., Iyer, N.R. (2010). Impact analysis of concrete structural components. *Defence Science J.*, **60**(3): 307–319.

[Navarro *et al.*, 1993] Navarro, C., Martinez, M. A., Cortes, R., Sanchez-Galvez, V. (1993). Some observations on the normal impact on ceramic faced armours backed by composite plates. *Int. J. of Impact Engineering,* **13**(1): 145–156.

[Navarro *et al.*, 1994] Navarro, C., Zaera, R., Cortes, R., Martinez, M.A., Sanchez-Galvez, V. (1994). The response of ceramic faced lightweight armours under projectile impact. In: Bulson, P.S., ed., *Proc. of the* 3th *Int. Conf. on Structures under Shock and Impact* (June 1-3, 1994, Madrid), WIT Press, Southampton, pp. 323–330.

[Naz and Lehr, 1990] Naz, P., Lehr, H.F. (1990). The crater formation due to segmented rod penetrators. *Int. J. of Impact Engineering,* **10**(1–4): 413–425.

[NDRC, 1946] Effects of impact and explosion (1946). Effects of impact and explosion. Summary Technical Report of Division 2, vol. 1. National. Defence Research Committee (NDRC), Washington, DC.

[Neilson, 1985] Neilson, A.J. (1985). Empirical equations for the perforation of mikd steel plates. *Int. J. of Impact Engineering,* **3**(2): 137–142.

[Nelson, 2002] Nelson, R.W. (2002). Low-yield earth-penetration nuclear weapons. *Science and Global Security,* **10:** 1–20.

[Newton, 1687] Newton, I. (1687). *Philosophiae Naturalis Principia Mathematica.* (*Mathematical Principles of Natural Philosophy*. Translated into English by A. Motte, revised by F. Cajori. University of California Press, 1966).

[Nia and Hoseini, 2011] Nia, A.A., Hoseini, G.R. (2011). Experimental study of perforation of multi-layered targets by hemispherical-nosed projectiles. *Materials and Design*, **32**(2): 1057–1065.

[Ning *et al.*, 2005] Ning, J., Song, W., Wang, J. (2005). Study of the perforation of stiffened plates by rigid projectiles. *Acta Mechanica Sinica.* **21**(6): 582–591.

[Nishiwaki, 1951] Nishiwaki, J. (1951). Resistance to the penetration of a bullet through an aluminium plate. *J. of Physics Society of Japan,* **6**(5): 374–378.

[Nixdorff, 1983] Nixdorff, K. (1983). Some remarks on the penetration theory of J. Awerbuch and S.R. Bodner. Transactions of the Canadian Society for Mechanical Engineering, **7**(3): 148–153.

[Nixdorff, 1984a] Nixdorff, K. (1984). Application of the penetration theory of J. Awerbuch and S. R. Bodner on multilayered targets. *J. of Applied Mathematics and Mechanics / Zeitschrift für Angewandte Mathematik und Mechanik,* **64**(4): T147–T149.

[Nixdorff, 1984b] Nixdorff, K. (1984). Some applications of the impact theory of J. Awerbuch and S. R. Bodner, Transactions of the Canadian Society for Mechanical Engineering, **8**(1): 16–20.

[Nixdorff, 1987a] Nixdorff, K. (1987). Discussion of two theories on the penetration of multilayer metallic targets. Transactions of the Canadian Society for Mechanical Engineering, **11**(3): 161–178.

[Nixdorff, 1987b] Nixdorff, K. (1987). On the efficiency of different head shapes to perforate thin targets. Transactions of the Canadian Society for Mechanical Engineering, **11**(2): 109–112.

[Normandia and Lee, 1999] Normandia, M.J., Lee, M. (1999). Penetration performance of multiple segmented-rods at 2.6 km/s. *Int. J. of Impact Engineering,* **23**(1): 675–686.

[Nurick and Walters, 1990] Nurick, G.N., Walters, C.E. (1990). The ballistic penetration of multiple thin plates separated by an air gap. In: *Proc. of the SEM Spring Conference on Experimental Mechanics* (June 3–6, 1990, Albuquerque, NM), pp. 631–637.

[Oden and Martins, 1985] Oden, J.T., Martins, J.A.C. (1985). Models and computational methods for dynamic friction phenomena. *Computer Methods in Applied Engineering*, **52**(1–3): 527–634.

[Ogawa, 1997] Ogawa, K. (1997). Impact friction test method by applying stress wave. *Experimental Mechanics,* **37**(4): 398–402.

[Ohno *et al.*, 1994] Ohno, T., Uchida, N., Ishikawa, Y., Kasai, H., Tsubota, M., Ueda, A., Kambayashi, A., Shirai, T. (1994). Improvement on impact resistance of reinforced concrete panels against projectile impact. In: Bulson, P.S., ed., *Proc. of the 3th Int. Conf. on Structures under Shock and Impact* (June 1–3, 1994, Madrid), WIT Press, Southampton, pp. 254–260.

[Ohnuma *et al.*, 1985] Ohnuma, H., Ito, C., Nomachi, S.G. (1985). Dynamic response and local rupture of reinforced concrete beam and slab under impact loading. In: *Trans. of 8th Int. Conf. on Structural Mechanics in Reactor Technology (SMiRT-8)* (August 19–23, 1985, Brussels, Belgium), paper J5/3, pp.179–184.

[Ohte *et al.*, 1982] Ohte, S., Yoshizawa, H., Chiba, N., Shida, S. (1982). Impact strength of steel plates struck by projectiles. *Bulletin of the Japan Society of Mechanical Engineering,* **25**(206): 1226–1231.

[Orlov, 2008] Orlov, M.Y. (2008). Features of perforation of steel layered barriers. In: Proc. of the 6th All-Russian Conf. "Fundamental'nye I Prikladnye Problemy Sovremennoj Mehaniki [Fundamental and Applied Problems of Modern Mechanics]" (September 30 - October 2, 2008, Tomsk, Russia), Tomsk University Press, Tomsk, pp. 277–278 (in Russian).

[Orphal, 2006] Orphal, D.L. (2006). Explosions and impacts. *Int. J. of Impact Engineering,* **33**(1–12): 496–545.

[Orphal and Franzen, 1990] Orphal, D.L., Franzen, R.R. (1990). Penetration mechanics and performance of segmented rods against metal targets. *Int. J. of Impact Engineering,* **10**(1–4): 427–438.

[Orphal and Miller, 1991] Orphal, D.L., Miller, C.W. (1991). Penetration performance of nonideal segmented rods. *Int. J. of Impact Engineering,* **11**(4): 457–461.

[Osipenko and Simonov, 2002] Osipenko, K.Yu., Simonov, I.V. (2002). A model of 3D dynamics of a body of revolution interacting with low-strength media and nonsymmetric cavitation. *Mechanics of Solids*, **37**(1): 119–128.

[Ostapenko, 1997] Ostapenko, N.A. (1997). *Optimum Shapes of Bodies Moving in Dense Media.* Vladar, Moscow (in Russian).

[Ostapenko and Yakunina, 1992] Ostapenko, N.A., Yakunina, G.Ye. (1992). Least-drag bodies moving in media subject to locality hypothesis. *Fluid Dynamics,* **27**(1): 71–80.

[Ostapenko and Yakunina, 1999] Ostapenko, N.A., Yakunina, G.Ye. (1999). The shape of slender three-dimensional bodies with maximum depth of penetration into dense media. *J. of Applied Mathematics and Mechanics,* **63**(6): 953–967.

[Ostapenko *et al.*, 1994] Ostapenko, N.A., Romanchenko, V.I., Yakunina, G. Ye. (1994). Optimum forms of three-dimensional bodies for penetration of dense media. *J. of Applied Mechanics and Technical Physics,* **35**(4): 515–521.

[Parameswaran *et al.*, 1999] Parameswaran, V., Bentley, W., Shukla, A., Prosser, R.A. (1999). A new approach for improving ballistic performance of composite armor. *Experimental Mechanics*, **39**(2): 103–110.

[Park *et al.*, 2005] Park, M., Yoo, J., Chung, D.-T. (2005). An optimization of a multi-layered plate under ballistic impact. *Int. J. of Solids and Structures,* **42**(1): 123–137.

[Partom, 1996] Partom, Y. (1996). Static cavity expansion model for partially confined targets. Report IAT.R-0092. The University of Texas at Austin, Institute for Advanced Technology, Austin, TX.

[Partom *et al.*, 2001] Partom, Y., Anderson, C.E., Jr., Yaziv, D. (2001). Penetration of AP projectiles into spaced ceramic targets. In: *Proc. of the 19th International Symposium on Ballistics* (7–11 May, 2001, Interlaken, Switzerland), pp. 1175–1181.

[Paul and Zaid, 1958] Paul, B., Zaid, M. (1958). Normal perforation of a thin plate by truncated projectiles. *J. of the Franklin Institute*, **265**(4): 317–335.

[Pedregal, 2003] Pedregal, P. (2003). *Introduction to Optimization.* Springer, New York.

[Pelak *et al.*, 2000] Pelak, R.A., Rightleya, P., Hammerberg, J. E. (2000). Friction in high-speed impact experiments. *AIP Conf. Proc.*, **505**(1): 1221–1224.

[Pelak *et al.*, 2001] Pelak, R.A., Rightleya, P., Hammerberg, J. E. (2001). Erratum: Friction in high-speed impact experiments. *AIP Conf. Proc.*, **620**(1): 1212–1214.

[Pensky, 2007] Pensky, O.G. (2007). Mathematical modeling of dynamics of driving rigid bodies into continuous medium using heat impulse machines. Sci.D. Thesis. Perm State University, Perm (in Russian).

[Petry, 1910] Petry, L. (1910). *Monographies de Systemes d'Artillerie*. Brussels.

[Philippon *et al.*, 2004] Philippon, S., Sutter, G., Molinari, A. (2004). An experimental study of friction at high sliding velocities. *Wear*, **257**(7–8): 777–784.

[Pilyugin, 2004] Pilyugin, N.N. (2004). A simulation of the shape of a crater in an organic-glass target under high-velocity impact. *High Temperature,* **42**(3): 481–488.

[Pol *et al.*, 2009] Pol, M.H., Bidi, A., Hoseini, A.V., Liaghat, G.H. (2009). Analysis of normal penetration of ogive-nose projectiles into thin metallic plates. *World Academy of Science, Engineering and Technology,* **38**: 239–242.

[Prakash, 1995] Prakash, V. (1995). A pressure-shear plate impact experiment for investigating transient friction. *Experimental Mechanics,* **35**(4): 329–336.

[Prakash, 1998] Prakash, V. (1998). Friction response of sliding interfaces subjected to time varying normal pressures. *Journal of Tribology,* **120**(1): 97–102.

[Prakash and Clifton, 1993] Prakash, V., Clifton, R.J. (1993). Time resolved dynamic friction measurements in pressure shear. In: Ramesh, K.T., ed. *Experimental Techniques in the Dynamics of Deformable Bodies*. American Society of Mechanical Engineers, Applied Mechanics Division, **165**, pp. 33–48.

[Presnell and Rajendran, 2010] Presnell, M.B., Rajendran, A.M. (2010). A Comparative study of segmented tungsten rod penetration into a thick steel target plate. *ASME Early Career Technical Journal*, **9**: 153–157.

[Raatschen *et al.*, 1987] Raatschen, H.J., Pavel, W., Fuchs, S., Senf, H., Rothenhausler, H. (1987). Penetration efficiency of segmented rods. In: Proc. of the 11th Int. Symp. on Ballistics (May 9–11, 1989, Brussels, Belgium), pp. 4–11.

[Rabinowicz, 1995] Rabinowicz, E. (1995). *Friction and Wear of Materials*. John Wiley and Sons, New York.

[Radin and Goldsmith, 1988] Radin, J., Goldsmith, W. (1988). Normal projectile penetration and perforation of layered targets. *Int. J. of Impact Engineering,* **7**(2): 229–259.

[Ragnedda and Serra, 2010] Ragnedda, F., Serra, M. (2010). Optimum shape of high speed impactor for concrete targets using PSOA heuristic. *Engineering*, **2**(4): 257–262.

[Raguraman *et al.*, 2009] Raguraman, M., Deb, A., Gupta, N.K. (2009). CAE-based prediction of projectile residual velocity for impact on single and multi-layered metallic armour plates. *Latin American J. of Solids and Structures*, **6**(3): 247–263.

[Raguraman *et al.*, 2010] Raguraman, M., Deb, A., Gupta, N..K. (2010). Semi-empirical procedures for estimation of residual velocity and ballistic limit for impact on mild steel plates by projectiles. *Latin American J. of Solids and Structures*, **7**(1): 63–76.

[Rahman *et al.*, 2010a] Rahman, I.A., Zaidi, A.M.A., Latif, Q.B.A.I., Ismail, M.Y. (2010). Empirical formula prediction on critical impact energy for scabbing phenomena on concrete structures. *Applied Physics Research*, **2**(2): 192–197.

[Rahman *et al.*, 2010b] Rahman, I.A., Zaidi, A.M.A., Latif, Q.B.A.I. (2010). Review on empirical studies of local impact effects of hard missile on concrete structures. *Int. J. of Sustainable Construction Engineering and Technology*, **1**(1): 73–97.

[Rahmatulin and Bagdoev, 1982] Rahmatulin, K.A., Bagdoev, A.G. (1982). The penetration of thin rigid body in initially elastic media. *Int. J. of Engineering Science*, **20**(1): 101–111.

[Rahmatulin *et al.*, 1964] Rahmatulin, K.A., Sagomonyan, A.Ya., Alekseev, N.A. (1964). *Soils Dynamics*. Moscow State University Press, Moscow (in Russian).

[Rajagopalan and Prakash, 1999] Rajagopalan, S., Prakash, V. (1999). A modified torsional Kolsky bar for investigating dynamic friction. *Experimental Mechanics,* **39**(4), 295–303.

[Rajagopalan *et al.*, 1999] Rajagopalan, S., Irfan, M.A., Prakash, V. (1999). Novel experimental techniques for investigating time resolved high speed friction. *Wear,* **225–229**(2): 1222–1237.

[Ramm, 2005] Ramm, A.G. (2005). *Inverse Problems: Mathematical and Analytical Techniques with Applications to Engineering*. Springer, New York.

[Ravid and Bodner, 1983] Ravid, M., Bodner, S.R. (1983). Dynamic perforation of viscoplastic plates by rigid projectiles. *Int. J. of Engineering Science*, **21**(6): 577–591.

[Ravid and Bodner, 1994] Ravid, M., Bodner, S.R. (1994). Penetration into thick targets – refinement of a 2D dynamic plasticy approach. *Int. J. of Impact Engineering*, **15**(4): 491–499.

[Ravid *et al.*, 2001] Ravid, M., Bodner, S.R., Chocron, I.S. (2001). Penetration analysis of ceramic armor with composite material backing. In: *Proc. of the 19th International Symposium on Ballistics* (7–11 May, 2001, Interlaken, Switzerland), paper TB 45, pp. 1401–1407.

[Recht, 1990] Recht, R.F. (1990). High velocity impact dynamics: Analytical modeling of plate penetration dynamics. In: Zukas, J.A., ed. *High Velocity Impact Dynamics.* John Wiley and Sons, New York, pp. 443–513.

[Recht and Ipson, 1963] Recht, R.F., Ipson, T.W. (1963). Ballistic perforation dynamics. *J. of Applied Mechanics,* **30**(3): 384–390.

[Reid and Wen, 2000] Reid, S.R. Wen, H.M. (2000). Perforation of FRP laminates and sandwich panels subjected to missile impact. In: Reid, S.R. and Zhou, G., eds. *Impact Behaviour of Fibre-Reinforced Composite Materials and Structures.* Woodhead Publishing Limited, Cambridge, pp. 239–279.

[Reid and Wen, 2001] Reid, S.R., Wen, H.M. (2001). Predicting penetration, cone cracking, scabbing and perforation of reinforced concrete targets struck by flat-faced projectiles. Report ME/AM/02.01/TE/G/018507/Z. University of Manchester Institute of Science and Technology (UMIST), Manchester, UK.

[Reid *et al.*, 1999] Reid, S.R., Wen, H.M., Soden, P.D., Reddy, T.Y. (1999). Response of single skin laminates and sandwich panels to projectile impact. In: Wang, S.S., Williams, J.J., Lo, K.H., eds. *Composite Materials for Offshore Operation-2.* Amer. Bur. Shipp., pp. 593–617.

[Reyes-Villanueva and Cantwell, 2004] Reyes-Villanueva, G., Cantwell, W.J. (2004). The high velocity impact response of composite and FML-reinforced sandwich structures. *Composites Science and Technology,* **64**(1): 35–54.

[Richmond, 1980] Richmond, P.W. (1980). Influence of nose shape and L/D ratio on projectile penetration in frozen soil. Special Report No. 80–17. U.S. Army Cold Regions Research and Engineering Laboratory, Hanover, NH.

[Riedel *et al.*, 1999] Riedel, W., Thoma, K., Hiermaier, S., Schmolinske, E. (1999). Penetration of reinforced concrete by BETA-B-500. Numerical analysis using a new macroscopic concrete model for hydrocodes. *In: Proc. of 9th Int. Symp. on Interaction of the Effects of Munitions with Structures* (3–7 May, 1999, Berlin, Germany), paper TP054.

[Riedel *et al.*, 2010] Riedel, W., Nöldgen, M., Straßburger, E., Thoma, K., Fehling, F. (2010). Local damage to ultra high performance concrete structures caused by an impact of aircraft engine missiles. *Nuclear Engineering and Design,* **240**(10): 2633–2642.

[Riera, 1989] Riera, J.D. (1989). Penetration, scabbing and perforation of concrete structures hit by solid missiles. *Nuclear Engineering and Design,* **115**(1), 121–131.

[Roberts, 1964] Roberts, S.M. (1964). *Dynamic Programming in Chemical Engineering and Process Control*. Academic Press, New York, London.

[Rockafellar, 1970] Rockafellar, R.T. (1970). *Convex Analysis*. Princeton University Press, New Jersey.

[Rosenberg and Dekel, 1994] Rosenberg, Z., Dekel, E. (1994). The relation between the penetration capability of long rods and their length to diameter ratio. *Int. J. of Impact Engineering,* **15**(2): 125–129.

[Rosenberg and Dekel, 2008] Rosenberg, Z., Dekel, E. (2008). A numerical study of the cavity expansion process and its application to long-rod penetration mechanics. *Int. J. of Impact Engineering*, **35**(3): 147–154.

[Rosenberg and Dekel, 2009a] Rosenberg, Z., Dekel, E. (2009). On the deep penetration and plate perforation by rigid projectiles. *Int. J. of Solids and Structures*, **46**(24): 4169–4180.

[Rosenberg and Dekel, 2009b] Rosenberg, Z., Dekel, E. (2009). The penetration of rigid long rods - revisited. *Int. J. of Impact Engineering*, **36**(4): 551–564.

[Rosenberg and Dekel, 2009c] Rosenberg, Z., Dekel, E. (2009). Analytical solution of the spherical cavity expansion process. *Int. J. of Impact Engineering*, **36**(2): 193–198.

[Rosenberg and Dekel, 2010a] Rosenberg, Z., Dekel, E. (2010). The deep penetration of concrete target by rigid rods - Revisited. *Int. J. of Protective Structures,* **1**(1): 125–144.

[Rosenberg and Dekel, 2010b] Rosenberg, Z., Dekel, E. (2010). Revisiting the perforation of ductile plates by sharp-nosed rigid projectiles. *Int. J. of Solids and Structures,* **47**(22–23): 3022–3033.

[Rosenberg and Forrestal, 1988] Rosenberg, Z., Forrestal, M.J. (1988). Perforation of aluminum plates with conical-nosed rods-additional data and discussion. *J. of Applied Mechanics,* **55**(1): 236–238.

[Ross and Hanagud, 1969] Ross, B., Hanagud, S. (1969). Penetration studies of ice with application to arctic and subarctic warfare. Final Report for Submarine Arctic Warfare and Scientific Program. Contract NOOO14–68-A0243. Stanford Research Institute, Menlo Park, CA.

[Ross *et al.*, 1971] Ross, B., Hanagud, S., Sidhu, G. (1971). Penetration studies of ice with application to arctic and subarctic warfare. Final Report (Phase 4) for Submarine Arctic Warfare and Scientific Program. Contract NOOO14–68-A0243. Stanford Research Institute, Menlo Park, CA.

[Rotz, 1975] Rotz, J.V. (1975). Results of missile impact tests on reinforced concrete panels. In: *Proc. of the 2nd ASCE Specialty Conf. on Structural Design of Nuclear Plant Facilities* ((December 1975, New Orleans, LA), vol. 1-A, pp. 720–738.

[Rotz, 1976] Rotz, J.V. (1976). Evaluation of tornado missile impact effects on structures. In: *Proc. of a Symp. on Tornadoes, Assessment of Knowledge and Implications for Man* (June 1976, Lubbock, TX).

[Roy *et al.*, 1987] Roy, P.K., Rao, K.R., Patkar, M.R. (1987). Penetration dynamics of earth penetration warhead into composite target media. *Defence Science J.*, **37**(3): 347–360.

[Rule and Jones, 1999] Rule, W.K., Jones, S.E. (1999). Some remarks on the optimal nose geometry of a rigid penetrator in the presence of friction. In: Jerome, D.M., ed., *Structures under extreme loading conditions. ASME Pressure Vessels and Piping Conf.* (August 1–5, 1999, Boston, MA). ASME Pressure Vessels Piping Division Publication (PVP), **394**, pp. 11–17.

[Sabet *et al.*, 2011] Sabet, A., Fagih, N., Beheshty, M.H. (2011). Effect of reinforcement type on high velocity impact response of GRP plates using a sharp tip projectile. *Int. J. of Impact Engineering,* **38**(8–9): 715–722.

[Sadanandan and Hetherington, 1997] Sadanandan, S., Hetherington, J.G. (1997). Characterization of ceramic/steel and ceramic/aluminum armours subjected to oblique impact. *Int. J. of Impact Engineering,* **19**(9–10): 811–819.

[Sagomonyan, 1960] Sagomonyan, A.Ya. (1960). Penetration of sharp bodies of revolution into soils. *Doklady Akademii Nauk SSSR*, **134**(6): 1320–1323 (in Russian).

[Sagomonyan, 1974] Sagomonyan, A.Ya. (1974). *Penetration: Penetration of Solids into Compressible Continuous Media.* Moscow State University Press, Moscow (In Russian).

[Sagomonyan, 1975] Sagomonyan, A.Ya. (1975). Plate piercing by slender solid projectile. *Moscow University Mechanics Bulletin*, **30**(5–6): 26–31.

[Sagomonyan, 1977] Sagomonyan, A.Ya. (1977). On problem of barrier perforation by cylinder. *Vestnik Moskovskogo Universiteta. Matematika Mekhanika*, 5: 111–118 (in Russian)

[Sagomonyan, 1988] Sagomonyan, A.Ya. (1988). *Dynamics of Barriers Perforation.* Moscow State University Press, Moscow (In Russian).

[Sagomonyan, 1997] Sagomonyan, A.Ya. (1997). High-speed penetration of a solid sphere into the soil. *Moscow University Mechanics Bulletin*, **52**(3): 9–13.

[Sagomonyan *et al.*, 1982] Sagomonyan, A.Ya., Gendugov, V.M., Sagomonyan, E.A. (1982). On the problem of perforation of a barrier. *Moscow University Mechanics Bulletin*, **37**(1–2): 14–18.

[Samuely and Hamann, 1939] Samuely, F.J, Hamann, C.W. (1939). Civil protection: the Application of the Civil Defence Act and Other Government Requirements for Air Raid Shelters etc. Architectural Press, London.

[Satapathy, 1997] Satapathy, S., (1997). Application of cavity expansion analysis to penetration problems. Report IAT.R-0136. The University of Texas at Austin, Institute for Advanced Technology, Austin, TX.

[Scheffler, 1989] Scheffler, D.R. (1989). Two-Dimensional computer simulations of segmented penetrators. Technical Report BRL-TR-3013. U.S. Army Ballistic Research Laboratory, Aberdeen Proving Ground, MD.

[Scheffler, 1990] Scheffler, D.R. (1990). 2-D computer simulation of segmented penetrator impacting semi-infinite steel target. *Int. J. of Impact Engineering,* **9**(1): 35–43.

[Scheffler and Zukas, 1990] Scheffler, D.R., Zukas, J.A. (1990). Numerical simulation of segmented penetrator impact. *Int. J. of Impact Engineering,* **10**(1–4): 487–497.

[Scott, 1980] Scott, C.R. (1980). *An Introduction to Soil Mechanics and Foundations.* Applies Science Publishers, London.

[Seah *et al.*, 2011] Seah, C.C., Børvik, T., Remseth, S., Pan, T-C (2011). Penetration and perforation of rock targets by hard projectiles. In: Zhou, Y., Zhao, J., eds., *Advances in Rock Dynamics and Applications.* CRC Press, Boca Raton, FL, Chapter 7, pp.143–162.

[Sedgwick *et al.*, 1987] Sedgwick, R.T., Waddell, J.L., Wilkinson, G.M. (1987). High velocity long rod impact: Theory and experiment. In: *Proc. of the 10th Internation Symposium on Ballistics* (October 27–29, 1987, San Diego, CA).

[Seguin *et al.*, 2008] Seguin, A., Bertho, Y., Gondret, P. (2008). Influence of confinement on granular penetration by impact. *Physical Review E,* **78**(1), (010301).

[Seifoori and Liaghat, 2011] Seifoori, S., Liaghat, G.H. (2011). A semianalytical and numerical study of penetration and perforation of an ogive-nose projectile into concrete targets under normal impact. *J. of Mechanical Engineering Science (Proc. of the Institution of Mechanical Engineers, Part C),* **225**(8): 1782–1797.

[Seliverstov, 2001] Seliverstov, V.S. (2001). Estimation and optimization of resistibility of layered barriers to perforation. In: *Proc. of the Conf. "VI Zababakhinskie Nauchnye Chtenija [6th Zababakhin Scientific Talks*]" (24–28 September, 2001, Snezhinsk, Russia) (in Russian).

[Serjouei *et al.*, 2012] Serjouei, A., Sridhar, I., Hua, W.E. (2012). On the design of bi-layer armor materials. *Solid State Phenomena,* **185**: 48–50.

[Shaw, 2012] Shaw, A. (2012). Penetration of rigid objects into semi-infinite compressible solids. *Mechanics of Materials*, **50**: 22–35.

[Shi and Grow, 2007] Shi, J., Grow, D. (2007). Effect of double constraints on the optimization of two-component armor systems. *Composite Structures,* **79**(3): 445–53.

[Shiqiao *et al.*, 2004] Shiqiao, G., Lei, J., Haipeng, L. (2004). Dynamic response of a projectile perforating multi-plate concrete targets. *Int. J. of Solids and Structures,* **41**(18–19): 4927–4938.

[Shiqiao *et al.*, 2006] Shiqiao, G., Haipeng, L., Kejie, L., Fenglei, H., Lei, J. (2006). Normal expansion theory for penetration of a projectile against concrete target. *Applied Mathematics and Mechanics,* **27**(4): 485–492.

[Shiqiao *et al.*, 2009] Shiqiao, G., Haipeng, L., Lei, J. (2009). A fuzzy model of the penetration resistance of concrete targets. *Int. J. of Impact Engineering*, **36**(4): 644–649.

[Shirai *et al.*, 1997] Shirai, T., Kambayashi, A., Ohno, T., Taniguchi, H., Ueda, M., Ishikawa, N. (1997). Experiment and numerical simulation of double-layered RC plates under impact loading. *Nuclear Engineering and Design*, **176**(3): 195–205.

[Shiu *et al.*, 2008] Shiu, W., Donzé, F.V., Daudeville, L. (2008). Penetration prediction of missiles with different nose shapes by the discrete element numerical approach. *Computers and Structures,* **86**(21–22): 2079–2086.

[Shokrieh and Javadpour, 2008] Shokrieh, M.M., Javadpour, G.H. (2008). Penetration analysis of a projectile in ceramic composite armor. *Composite Structures,* **82**(2): 269–76.

[Shoukry *et al.*, 1991] Shoukry, M., Nair, S., Kalpakjian, S. (1991). Effect of shear deformation on the dynamic plastic bending of metallic plates during normal penetration. *Int. J. of Engineering Science,* **29**(9): 1035–1052.

[Showalter *et al.*, 2008] Showalter, D.D., Placzankis, B.E., Burkins, M.S. (2008). Ballistic performance testing of aluminum alloy 5059-H131 and 5059-H136 for armor applications. Report ARL-TR-4427. US Army Research Laboratory, Adelphi, MD.

[Siddiqui and Abbas, 2002] Siddiqui, N.A., Abbas, H. (2002). Mechanics of missile penetration into geo-materials. *Structural Engineering and Mechanics*, **13**(6): 639–652.

[Siddiqui *et al.*, 2002] Siddiqui, N.A., Choudhury, M.A., Abbas, H. (2002). Reliability analysis of projectile penetration into geological targets. *Reliability Engineering and System Safety,* **78**(1): 13–19.

[Siddiqui *et al.*, 2009] Siddiqui, N.A., Khan, F.H., Umar, A. (2009). Reliability of underground concrete barriers against normal missile impact. *Computers and Concrete*, **6**(1): 79–93.

[Sierakowski *et al.*, 1977] Sierakowski, R.L., Malvern, L.E., Milton, J.E., Ross, C.A., Collins, J.A. (1977). Penetrator impact studies of soil/concrete. Final Report AFOSR TR 78–0164. University of Florida, Department of Engineering Science, Gainesville, FL.

[Simonov, 1993] Simonov, I.V. (1993). Cavitating penetration of a plastic medium by bodies of minimum resistance. *J. of Applied Mathematics and Mechanics*, **57**(6): 1067–1075.

[Sjøl and Teland, 2000] Sjøl, H., Teland, J.A. (2000). Prediction of concrete penetration using Forrestal's formula. Report FFI/RAPPORT-99/04415. Norwegian Defence Research Establishment (FFI), Kjeller, Norway.

[Sjøl and Teland, 2001] Sjøl, H., Teland J.A. (2001). Perforation of concrete targets. Report FFI/RAPPORT-2001/05786. Norwegian Defence Research Establishment (FFI), Kjeller, Norway.

[Sjøl and Teland, 2003] Sjøl, H., Teland, J.A. (2003) Analytical model to determine the ballistic limit of concrete targets. In: *Proc. of 11th Int. Symp. on Interaction of the Effects of Munitions with Structures* (May 5–9, 2003 Mannheim, Germany).

[Sjøl *et al.*, 2002] Sjøl, H., Teland, J.A., Kaldheim, Ø. (2002). Penetration into concrete – Analysis of small scale experiments with 12 mm projectiles. Report FFI/RAPPORT-2002/04867. Norwegian Defence Research Establishment (FFI), Kjeller, Norway.

[Sliter, 1980] Sliter, G.E. (1980). Assessment of empirical concrete impact formulas. *J. of the Structural Division (ASCE),* **106** (ST5): 1023–1045.

[Smith and Hetherington, 1994] Smith, P.D., Hetherington, J.G. (1994). *Blast and Ballistic Loading of Structures.* Butterworth Heinemann, Oxford.

[Sodha and Jain, 1958] Sodha, M.S., Jain, V.K. (1958). On physics of armor penetration. *J. of Applied Physics,* **29**(12): 1769–1770.

[Sommerfeld, 1952] Sommerfeld, A. (1952). *Mechanics*. Academic Press, New York.

[Song *et al.*, 1985] Song, Y, Wang, D., Yang, G. (1985). Analysis of penetration of axisymmetric projectiles into semi-infinite targets. *Acta Mechanica Sinica*, **1**(2): 147–157.

[Song *et al.*, 2005] Song, D., Jiang, Z., Zeng, S. (2005). Analysis on energy absorption of metallic targets struck by rigid sharp-nosed projectiles. *J. of Ballistics*, **17**(2): 28–32 (in Chinese).

[Sorensen *et al.*, 1990] Sorensen, B.R., Kimsey, K.D., Silsby, G.F., Scheffler, D.R., Sherrick, T.M., de Rosset, W.S. (1990). High velocity penetration of steel targets. *Int. J. of Impact Engineering,* **11**(1): 107–119.

[Srivathsa and Ramakrishnan, 1997] Srivathsa, B., Ramakrishnan, N. (1997). On the ballistic performance of metallic materials. *Bulletin of Material Science*, **20**(1): 111–123.

[Srivathsa and Ramakrishnan, 1998] Srivathsa, B., Ramakrishnan, N. (1998). A ballistic performance index for thick metallic armour. *Computer Simulation Modelling in Engineering*, **3**(1): 33–39.

[Srivathsa and Ramakrishnan, 1999] Srivathsa, B., Ramakrishnan, N. (1999). Ballistic performance maps for thick metallic armour. *J. of Materials Processing Technology*, **96**(1–3): 81–91.

[Stone, 1994] Stone, G.W. (1994). Projectile penetration into representative targets. Report SAND-94-1490. Sandia National Laboratories, Albuquerque, NM.

[Sun and Yuan, 2012] Sun, W., Yuan, J. (2012). Penetration deep of non-deformable projectile into concrete targets. *Advanced Materials Research*, **446–449**: 3604–3608.

[Szuladziński, 2010] Szuladziński, G. (2010). *Formulas Mechanical and Structural Shock and Impact*. CRC Press, Bosa Roca.

[Tate, 1990] Tate, A. (1990). Engineering modeling of some aspects of segmented rod penetrators. *Int. J. of Impact Engineering,* **9**(3): 327–341.

[Taylor, 1948] Taylor, G.I. (1948). The formation and enlargement of a circular hole in a thin plastic sheet. *Quarterly J. of Mechanics and Applied Mathematics*, **1**(1): 103–124.

[Taylor and Flanagan, 1987] Taylor, L.M., Flanagan, D.P. (1987). PRONTO2D - a two dimensional transient solid dynamics program. Report SAND86–0594. Sandia National Laboratories, Albuquerque, NM.

[Teland, 1998] Teland, J.A. (1998). A review of empirical equations for missile impact effects on concrete. Report FFI/RAPPORT-97/05856. Norwegian Defence Research Establishment (FFI), Kjeller, Norway.

[Teland, 1999] Teland, J.A. (1999). A review of analytical penetration mechanics. Report FFI/RAPPORT-99/01264. Norwegian Defence Research Establishment (FFI), Kjeller, Norway.

[Teland, 2001] Teland, J.A. (2001). Cavity expansion theory applied to penetration of targets with pre-drilled cavities. In: *Proc. of the 19th Int. Symp.on Ballistics* (7–11 May, 2001, Interlaken, Switzerland), Paper TB 36, pp. 1329–1335.

[Teland and Moxnes, 2003] Teland, J.A., Moxnes, J.F. (2003). Analytical cavity expansion penetration models compared with numerical simulations. Report FFI/RAPPORT-2003/00934. Norwegian Defence Research Establishment (FFI), Kjeller, Norway.

[Teland and Sjøl, 2000] Teland, J.A., Sjøl, H. (2000). Boundary effects in penetration into concrete. Report FFI/RAPPORT-2000/05414. Norwegian Defence Research Establishment (FFI), Kjeller, Norway.

[Teland and Sjøl, 2004] Teland, J.A., Sjøl, H. (2004). Penetration into concrete by truncated projectiles. *Int. J. of Impact Engineering,* **30**(4): 447–464.

[Teng and Wierzbicki, 2005] Teng, X., Wierzbicki, T. (2005). Dynamic shear plugging of beams and plates with an advancing crack. *Int. J. of Impact Engineering*, **31**(6): 667–698.

[Teng *et al.*, 2007] Teng, X., Dey, S., Børvik, T., Wierzbicki, T. (2007). Protection performance of double-layered metal shield against projectile impact. *J. of Mechanics of Materials and Structures,* **2**(7): 1309–1331.

[Teng *et al.*, 2008] Teng, X., Wierzbicki, T., Huang, M. (2008). Ballistic resistance of double-layered armor plates. *Int. J. of Impact Engineering,* **35**(8): 870–884.

[Tertychny-Dauri, 2004] Tertychny-Dauri, V.Y. (2004). *Hyperreactive Mechanics.* FIZMATLIT, Moscow (in Russian).

[Tham, 2006] Tham, C.Y. (2006). Numerical and empirical approach in predicting the penetration of a concrete target by an ogive-nosed projectile. *Finite Elements in Analysis and Design*, **42**(14–15): 1258–1268.

[Thompson and Mitchell, 1971] Thompson, J.B., Mitchell, J.K. (1971). Determination of in situ soil properties utilizing an impact penetrometer. In: *Lunar Surface Engineering Properties Experiment Definition.* NASA Research Contract NAS 8–21432, Final Report, Vol. I, Chapter 2, University of California, Berkeley, CA.

[Thomson, 1955] Thomson, W. T. (1955). An approximate theory of armor penetration. *J. of Applied Physics,* **26**(1): 80–82.

[THOR, 1961] The resistance of various metallic materials to perforation by steel fragments; empirical relationships for fragment residual velocity and residual weigh. Project THOR Technical Report No. 47, The Johns Hopkins University, Institute for Cooperative Research, Ballistic Analysis Laboratory, Baltimore, MD.

[THOR, 1963] Project THOR Technical Report No. 51 (1963).The resistance of various non-metallic materials to perforation by steel fragments; Empirical relationships for fragment residual velocity and residual weight. Ballistics Research Laboratory, Aberdeen Proving Ground, MD.

[Tirosh *et al.*, 2009] Tirosh, J., Tylis, A., Davidi, G. (2009). Foreign object damage: Penetration of rigid projectiles into solids. *Mechanics of Materials*, **41**(5): 535–544.

[TM 5–1300, 1990] Army TM 5–1300, Navy NAVFAC P-397, Air Force AFR 88–22 (1990). Structures to resist the effect of accidental explosions.

[TM 5–855–1, 1986] TM 5–855–1. (1986). Fundamentals of protective design for conventional weapons. Technical Manual. US Department of Army, Washington, DC.

[Trushkov and Kramshonkov, 2008a] Trushkov, V.G., Kramshonkov, E.N. (2008). Penetration of thick shields by rods and segmented impactors. *Izvestija Vuzov. Fizika,* **9**(2): 1–9 (in Russian).

[Trushkov and Kramshonkov, 2008b] Trushkov, V.G., Kramshonkov, E.N. (2008). Analysis of effectiveness of segmented impactors in wide range of impact velocities. In: *Proc. of the 5th All-Russian Scientific Conf. "Matematicheskoe Modelirovanie i Kraevye Zadachi [Mathematical Modeling and Boundary-Value Problems]"* (29–31 May, 2008, Samara, Russia), part 1. Samara State Technical University, Samara, pp. 327–329. (in Russian)

[Trushkov *et al.*, 2007] Trushkov, V.G., Kramshonkov, E.N., Yagnitsina, E.A. (2007). Destruction of dangerous cosmic objects by means of high-speed impact. *Izvestija Vuzov. Fizika,* 9: 1–6 (in Russian).

[Tsubota *et al.*, 1993] Tsubota, H., Kasai, Y., Koshika, N., Morikawa, H., Uchida, T., Ohno, T., Kogure, K. (1993). Quantitative studies on impact resistance of reinforced concrete panels with a steel liner under impact loading. Part I: Scaled model impact test. In: *Trans. of 12th Int. Conf. on Structural Mechanics in Reactor Technology (SMiRT-12)* (August 15–20, 1993, Stuttgart, Germany), paper J07/1, pp. 169–174.

[Uehara *et al.*, 2003a] Uehara, J., Ambroso, M., Ojha, R., Durian, D. (2003). Low-speed impact craters in loose granular media. *Physical Review Letters,* **90**(19), (194301).

[Uehara *et al.*, 2003b] Uehara, J., Ambroso, M., Ojha, R., Durian, D. (2003). Erratum: Low-speed impact craters in loose granular media [*Physical Review Letters* **90**, 194301, 2003]. *Physical Review Letters,* **91**(14), (149902).

[Ulven *et al.*, 2003] Ulven, C., Vaidya, U.K., Hosur, M.V. (2003). Effect of projectile shape during ballistic perforation of VARTM carbon/epoxy composite panels. *Composite Structures,* **61**(1–2): 143–150.

[Vahedi *et al.*, 2008] Vahedi, K., Latefi, M., Khosravi, F. (2008). Investigation and analysis of ogive-shape nose steel projectile into concrete target. *Turkish J. of Engineering and Environmental Sciences*, **32**(5): 295–302.

[Van Brunt, 2003] Van Brunt, B. (2003). *The Calculus of Variations*. Springer, NewYork.

[Vantsyan, 1981] Vantsyan, A.A. (1981). Determining depth of penetration of slender body into metals. *Doklady Akademii Nauk Armjanskoj SSR*, **72**(2): 95–101 (in Russian).

[Vedernikov and Shchepanovsky, 1995] Vedernikov, Y.A., Shchepanovsky, V.A. (1995). *Optimization of Reactive Gas Dynamic Systems.* Nauka, Novosibirsk (in Russian).

[Vedernikov *et al.*, 1995] Vedernikov, Y.A., Khudiakov, Y.S., Omelaev, A.I., (1995). *Ballistics: From Arrows to Rockets*. Nauka, Novosibirsk (in Russian).

[Veldanov, 1989] Veldanov, V.A. (1989). Penetration of impactors having nontraditional shapes into soil. *Trudy MVTU imeni Baumana*, 530: 85–97.

[Vitman and Ioffe, 1948] Vitman, F.F., Ioffe, B.S. (1948). A simple method of determining dynamic hardness of metals using a double cone. *Zavodskaja Laboratorija* **14**(6): 727–732 (in Russian).

[Vitman and Stepanov, 1959] Vitman, F.F., Stepanov, V.A. (1959). Effect of strain rate on resistance of metals to deformation at impact velocities of 100-1000 m/s. In: *Nekotoryje Problemy Prochnosti Tvjordogo Tela.* USSR Academy of Science, Moscow-Leningrad, pp. 207–221 (in Russian).

[Vossoughi *et al.*, 2007] Vossoughi, F., Ostertag, C.P., Monteiro, P.J.M., Johnson, G.C. (2007). Resistance of concrete protected by fabric to projectile impact. *Cement and Concrete Research*, **37**(1): 96–106.

[Vretblad, 1988] Vretblad, B. (1988). Penetration of projectiles in concrete according to Fort II.1. In: *Proc. from Workshop on Weapon Penetration into Hard Targets (30–31 May, 1988,* Norwegian Defence Research Establishment (FFI), Kjeller, Norway).

[Vukolov and Zharov, 2005] Vukolov, L.A., Zharov, V.A.(2005). Comparative characteristics of brake shoes from different suppliers. *Vestnik VNIIZhT*, 2: 16–20 (in Russian).

[Walker and Anderson, 1996] Walker, J.D., Anderson, C.E., Jr. (1996). An analytical model for ceramic-faced light armors. In: *Proc. of the 16th Int. Symp. on Ballistics* (23–28 September, 1996, San Francisco, CA), vol.3, pp. 289–298.

[Walley, 2009] Walley, S.M. (2009). Historical review of high strain rate and shoch properties of ceramics relevant to their application in armour. *Advances in Applied Ceramics*, **109**: 446–466.

[Walter and Wolde-Tinsae, 1984] Walter, T.A., Wolde-Tinsae, A.M. (1984). Turbine missile perforation of reinforced concrete. *J. of Structural Engineering,* **110**(10): 2439–2455.

[Wang and Lu, 1996] Wang, B., Lu, G., (1996). On the optimisation of two-component plates against ballistic impact. *J. of Materials Processing Technology,* **57**(1–2): 141–145.

[Wang and Zhu, 2010] Wang, X.-Q., Zhu, X. (2010). Study on high-velocity blunt-nosed projectiles penetrating moderate thickness metallic targets. *Engineering Mechanics*, **27**(12): 213–218, 223 (in Chinese)

[Wang *et al.*, 1995] Wang, X., Zhao, G., Shen, P., Zha, H. (1995). High velocity impact of segmented rods with an aluminum carrier tube. *Int. J. of Impact Engineering,* **17**(4–6): 915–923.

[Wang *et al.*, 2005a] Wang, Y.B., Wang, X.J., Hu, X.Z., Wang, F. (2005). Experimental study of ballistic resistance of Kevlar laminates. *Engineering Mechanics,* **22**(3): 76–81 (in Chinese).

[Wang *et al.*, 2005b] Wang, H., Jiang, Z., Zeng, S. (2005). Analysis of the critical cone angle of the projectile nose for plugging failure of metallic plates. *J. of Ballistics*, **17**(4): 44–49 (in Chinese)

[Wang *et al.*, 2010] Wang, X.-Q., Zhu, X., Wang, Y.-H. (2010). An analytical model for ballistic limit of sharp-nosed projectiles penetrating a metallic target. *J. of Vibration and Shock*, **29**(5): 96–100, 140 (in Chinese)

[Warren, 1999] Warren, T.L. (1999). The effect of strain rate on the dynamic expansion of cylindrical cavities. *J. of Applied Mechanics,* **66**(3): 818–821.

[Warren and Forrestal, 1998] Warren, T.L., Forrestal, M.J. (1998). Effect of strain hardening and strain rate sensitivity on penetration of aluminum targets with spherical-nosed rods. *Int. J. of Solids and Structures,* **35**(28–29): 3737–3753.

[Warren *et al.*, 2004] Warren, T.L., Hanchak, S.J., Poormon, K.L. (2004). Penetration of limestone targets by ogive-nosed VAR 4340 steel projectiles at oblique angles: experiments and simulations. *Int. J. of Impact Engineering.* **30**(10): 1307–1331

[Wei *et al.*, 2012] Wei, Z., Yunfei, D., Sheng, C.Z., Gang, W. (2012). Experimental investigation on the ballistic performance of monolithic and layered metal plates subjected to impact by blunt rigid projectiles. Int. J. of Impact Engineering **49**: 115–129.

[Weidemaier *et al.*, 1993] Weidemaier, P., Senf, H., Rothenhäusler, H., Filbey, G.L., Gooch, W.A. (1993). On the ballistic resistance of laminated steel targets: experiments and numerical calculations. In: *Proc. of the 14th Int. Symp. On Ballistics* (26–29 September, 1993, Quebec, Canada), pp. 681–690.

[Weihrauch and Wollmann, 1993] Weihrauch, G., Wollmann, E. (1993). Segmented penetrators. Propellants, explosives. *Pyrotechnics,* **18**(5): 270–274.

[Wen, 2000] Wen, H.M. (2000). Predicting the penetration and perforation of FRP laminates struck normally by projectiles with different nose shapes. *Composite Structures,* **49**(3): 321–329.

[Wen, 2001] Wen, H.M. (2001). Penetration and perforation of thick FRP laminates. *Composites Science and Technology,* **61**(8): 1163–1172.

[Wen, 2002a] Wen, H.M. (2002). Predicting the penetration and perforation of targets struck by projectiles at normal incidence. *Mechanics of Structures and Machines,* **30**(4): 543–577.

[Wen, 2002b] Wen, H.M. (2002). Penetration and perforation of thick metallic targets under impact by missiles. *Chinese J. of High Pressure Physics,* **16**(2): 94–104 (in Chinese).

[Wen and Jones, 1992] Wen, H.M., Jones, N. (1992). Semi-empirical equations for the perforation of plates struck by the mass. In: Bulson, P.S., ed. *Strucktures Under Shock and Impact II.* WIT Press, Southampton, pp. 369–380.

[Wen and Xu, 2004] Wen, H.-M., Xu, H.-B. (2004). Theoretical study on the penetration of rocks struck by rigid projectiles. *Chinese J. of High Pressure Physics.* **18**(4): 302–308 (in Chinese).

[Wen *et al.*, 1998] Wen, H.M., Reddy, T.Y., Reid, S.R., Soden, P.D. (1998). Indentation penetration and perforation of composite laminates and sandwich panels under quasi-static and projectile loading. *Key Engineering Materials,* **141–143**: 501–552.

[Westerling *et al.*, 1997] Westerling, L., Lundberg, P., Holmberg, L., Lundberg, B. (1997). High velocity penetration of homogeneous, segmented and telescopic projectiles into alumina targets. *Int. J. of Impact Engineering,* **20**(6–10): 817–827.

[Whiffen, 1943] Whiffen, P. (1943). UK Road Research Laboratory Note No. MOS/311.

[Wierzbicki, 1999] Wierzbicki, T. (1999). Petalling of plates under explosive and impact loading. *Int. J. of Impact Engineering,* **22**(9–10): 935–954.

[Wijk, 2005] Wijk, G. (2005). Initially increasing penetration resistance. Friction and target size effects in connection with rigid projectile penetration and perforation of steel and metallic targets. Report FOI-R-1631-SE. Swedish Defence Reasearch Agency (FOI), Tumba, Sweden.

[Wijk *et al.*, 2005] Wijk, G., Hartmann, M., Tyrberg, A. (2005). A model for rigid projectile penetration and perforation of hard steel and metallic targets. Report FOI-R-1617-SE. Swedish Defence Reasearch Agency (FOI), Tumba, Sweden.

[Wilkins, 1978] Wilkins, M.L. (1978). Mechanics of penetration and perforation. *Int. J. of Engineering Science,* **16**(11): 793–807.

[Williams, 1994] Williams, M.S. (1994). Modeling of local impact effects on plain and reinforced concrete. *ACI Structural Journal,* **91**(2): 178–187

[Wilson and Hetherington, 1995] Wilson, D., Hetherington, J.G. (1995). Analysis of ballistic impact on ceramic faced armour using high-speed photography. In: *Proc. of Lightweight Armour System Symposium (1995).* Royal Military Colleage of Science, Cranfield, UK.

[Woodward, 1978a] Woodward, R.L. (1978). The penetration of metal targets by conical projectiles. *Int. J. of Mechanical Science,* **20**(6): 349–359.

[Woodward, 1978b] Woodward, R.L. (1978). The penetration of metal targets which fail by adiabatic shear plugging. *Int. J. of Mechanical Science,* **20**(9): 599–607.

[Woodward, 1990] Woodward, R.L. (1990). A simple one-dimensional approach to modeling ceramic composite armor defeat. *Int. J. of Impact Engineering,* **9**(4): 455–474.

[Woodward and Cimpoeru, 1998] Woodward, R.L., Cimpoeru, S.J. (1998). A study of the perforation of aluminium laminate targets. *Int. J. of Impact Engineering*, **21**(3): 117–131.

[Wu and Batra, 1996] Wu, Y., Batra, R.C. (1996). An engineering penetration/perforation model of hemispherical nosed rigid cylindrical rods into strain-hardening targets. *Computers and Structures*, **58**(1): 51–58.

[Xu *et al.*, 1997] Xu, Y., Keer, L.M., Luk, V.K. (1997). Elastic-cracked model for penetration into unreinforced concrete targets with ogival nose projectiles. *Int. J. of Solids and Structures,* **34**(12): 1479–1491.

[Xue *et al.*, 2010] Xue, L., Mock, W., Jr., Belytschko, T. (2010). Penetration of DH-36 steel plates with and without polyurea coating. *Mechanics of Materials,* **42**(11): 981–1003.

[Yadav *et al.*, 1995] Yadav, S., Chichilli, D.R., Ramesh, K.T. (1995). The mechanical response of a 6061-T6 Al/Al203 metal matrix composite at high rates of deformation. *Acta Metallurgica et Materialia*, **43**(12): 4453–4464.

[Yaghoubi and Liaw, 2012] Yaghoubi, A.S., Liaw, B. (2012). Thickness influence on ballistic impact behaviors of GLARE 5 fiber-metal laminated beams: Experimental and numerical studies. *Composite Structures*, **94**(8): 2585–2598.

[Yakunina, 2001] Yakunina, G.Ye. (2001). On body shapes providing maximum penetration depth in dense media. *Doklady Physics*, **46**(2): 140–143.

[Yakunina, 2005a] Yakunina, G.Ye. (2005). Effects of sliding friction on the optimal 3D nose geometry of rigid rods penetrating media. *Optimization and Engineering*, **6**(3): 315–338.

[Yakunina, 2005b] Yakunina, G.Ye. (2005). On the optimal shapes of bodies in dense media. *Doklady Physics*, **50**(12): 650–654.

[Yakunina, 2005c] Yakunina, G.Ye. (2005). The optimum shapes of bodies moving in a medium taking friction into account. *J. of Applied Mathematics and Mechanics*, **69**(5): 680–692.

[Yan *et al.*, 2008] Yan, L.-H., Zeng, S.-Y., Zhou, J.-P. (2008). Experiment and theoretical model of high-strength armor normally penetrated by armor-piercing bullet. *J. of National University of Defense Technology*, **30**(4): 33–38 (in Chinese).

[Yankelevsky, 1983a] Yankelevsky, D.Z. (1983). Projectile penetration through a narrow drill in soil. *Int. J. of Impact Engineering,* **1**(4): 377–391.

[Yankelevsky, 1983b] Yankelevsky, D.Z. (1983). Optimal shape of an earth penetrating projectile. *Int. J. of Solids and Structures*, **19**(1): 25–31.

[Yankelevsky, 1985] Yankelevsky, D.Z. (1985). Cavitation phenomena in soil-projectile interaction. *Int. J. of Impact Engineering*, **3**(3): 167–178.

[Yankelevsky, 1997] Yankelevsky, D.Z. (1997). Local response of concrete slabs to low velocity missile impact. *Int. J. of Impact Engineering,* **19**(4): 331–343.

[Yankelevsky and Adin, 1980] Yankelevsky, D.Z., Adin, M.A. (1980). A simplified analytical method for soil penetration analysis. *Int. J. for Numerical and Analytical Methods in Geomechanics,* **4**(3): 233–254.

[Yankelevsky and Gluck, 1980] Yankelevsky, D.Z., Gluck, J. (1980). Nose shape effect on high velocity soil penetration. *Int. J. of Mechanical Science,* **22**(5): 297–311.

[Yankelevsky and Leibowitz, 1999] Yankelevsky, D.Z., Leibowitz, O. (1999). Punching shear in concrete slabs. *Int. J. of Mechanical Science*, **41**(1): 1–15.

[Yarin *et al.*, 1995] Yarin, A.L., Rubin, M.B., Roisman, I.V. (1995). Penetration of a rigid projectile into an elastic-plastic target of finite thickness. *Int. J. of Impact Engineering*, **16**(5/6): 801–831.

[Yossifon *et al.*, 2002] Yossifon, G., Yarin, A.L., Rubin, M.B. (2002). Penetration of a rigid projectile into a multi-layered target: theory and numerical computations. *Int. J. of Engineering Science,* **40**(12): 1381–1401.

[Young, 1967] Young, C.W. (1967). The development of empirical equations for predicting depth of an earth–penetrating projectile. Report SC-DR-67-70 Sandia National Laboratories, Albuquerque, NM.

[Young, 1972] Young, C.W. (1972). Empirical equations for predicting penetration performation in layered earth materials for complex penetrator configurations Report SC-DR-72-0523. Sandia National Laboratories, Albuquerque, NM.

[Young, 1997] Young, C.W. (1997). Penetration equations. Report SAND97-2426. Sandia National Laboratories, Albuquerque, NM.

[Yu, 2000] Yu, H.S. (2000). *Cavity Expansion Methods in Geomechanics.* Kluwer, Dordrecht.

[Yuan *et al.*, 2009] Yuan, F., Liou, N-S., Prakash, V. (2009). High-speed frictional slip at metal-on-metal interfaces. *Int. J. of Plasticity*, **25**(4): 612–634.

[Zaera, 2011] Zaera, R. (2011). Ballistic impacts on polymer matrix composites, composite armor, personal armor. In: Abrate, S., Ed. *Impact Engineering of Composite Structures*. Springer, Wien, New-York, pp. 305–403.

[Zaera and Sanchez-Galvez, 1998] Zaera, R., Sanchez-Galvez, V. (1998). Analytical modeling of normal and oblique ballistic impact on ceramic/metal lightweight armors. *Int. J. of Impact Engineering,* **21**(3): 133–148.

[Zaera *et al.*, 2000] Zaera, R., Sanchez-Saez, S., Perez-Castellanos, J.L. Navarro, C. (2000). Modelling of the adhesive layer in mixed ceramic/metal armours subjected to impact. *Composites. Part A,* **31**(8): 823–833.

[Zaid and Paul, 1957] Zaid, M., Paul, B. (1957). Mechanics of high speed projectile perforation. *J. of the Franklin Institute,* **264**(2): 117–126.

[Zaid and Travis, 1974] Zaid, A.I.O., Travis, F.W. (1974). A comparison of single and multi-plate shields subjected to impact by a high speed projectile. In: Harding, J., ed., *Proc. of the Conference of Mechanical Properties at High Rates of Strain.* Institute of Physics (UK) Conf. Series, No. 21, pp. 417–428

[Zaid *et al.*, 1973] Zaid, A.I.O., El-Kalai, A., Travis, F.W. (1973). An examination of the perforation of mild steel plate by a flat-ended cylindrical projectile. *Int. J. of Mechanical Science,* **15**(2): 129–143.

[Zaidi *et al.*, 2010a] Zaidi, A.M.A., Latif, Q.B.A.I., Rahman, I.A., Ismail, M.Y. (2010). Development of empirical prediction formula for penetration of ogive nose hard missile into concrete targets. *American J. of Applied Sciences,* **7**(5): 711–716.

[Zaidi *et al.*, 2010b] Zaidi, A.M.A., Rahman, I.A., Latif, Q.B.A.I. (2010). Study on the analytical behaviour of concrete structure against local impact of hard missile. *Int. J. of Sustainable Construction Engineering and Technology*, **1**(2): 11–39.

[Zhang and Mu, 2000] Zhang, T.G., Mu, J.C. (2000). Discing around perforations of thin plates and tubes by hemispherical-nosed missiles or spheres. *Key Engineering Materials*, **177–180**: 75–80.

[Zhang *et al.*, 2002] Zhang, Z., Shen, J., Zhong, W., Sun, Z. (2002). A dynamic model of ceramic/fibre-reinforced plastic hybrid composites under projectile striking. *J. of*

Aerospace Engineering (Proc. of the Institute of Mechanical Engineering, Part G): **216**(6): 325–331.

[Zhang *et al.*, 2006a] Zhang, X.-Q., Yang, G.-T., Huang, X.-Q. (2006). Analytical model of ceramic/metal armor impacted by deformable projectile. *J. of Applied Mathematics and Mechanics,* **27**(3): 287–294.

[Zhang *et al.*, 2006b] Zhang, D.-Z., Lin, J.-D., Tang, R.-D., Zhang, X.-R. (2006). Empirical equation for penetration depth of projectiles into high-strength rock targets. *Acta Armamentarii*, **27**(1): 15–18 (in Chinese).

[Zhang *et al.*, 2010] Zhang, Z., Wang, M., Song, S., Li, M., Sun, Z. (2010). Influence of panel/back thickness on impact damage behavior of alumina/aluminum armors. *J. of the European Ceramic Society*, **30**(4): 875–887.

[Zhang *et al.*, 2011a] Zhang, J., Zhu, Z., Ma, D., Wang, R., He, X. (2011). Optimum design of composite armor against transverse impact. *Advanced Material Reseach,* **335–336**: 101–104.

[Zhang *et al.*, 2011b] Zhang, W., Deng, Y.-F., Cao, Z.-S., Wei, G., Zhang, B.-X. (2011). Effect of air gap on penetration of double-layered A3 steel shields against hemispherical-nosed projectiles. *Acta Armamentarii*, **32**(SUPPL. 2): 325–329 (in Chinese)

[Zhao *et al.*, 1995] Zhao, G., Shen, P., Wang, X. (1995). An analysis on penetration efficiency of the segmented rods. In: *Proc. of the 15th Int. Symp. on Ballistics* (21–24 May, 1995, Jerusalem, Israel), pp. 419–426.

[Zhou and Stronge, 2008] Zhou, D.W., Stronge, W.J. (2008). Ballistic limit for oblique impact of thin sandwich panels and spaced plates. *Int. J. of Impact Engineering,* **35**(11): 1339–1354.

[Zhou and Wen, 2006] Zhou, H., Wen, H.M. (2006). Dynamic cylindrical cavity expansion model and its application to penetration problems. *Chinese Journal of High Pressure Physics*, **20**(1): 67–78 (in Chinese)

[Zhu *et al.*, 1992a] Zhu, G., Goldsmith, W., Dharan, C.K.H. (1992). Penetration of laminated Kevlar by projectiles - I. Experimental investigation. *Int. J. of Solids and Structures,* **29**(4): 399–420.

[Zhu *et al.*, 1992b] Zhu, G., Goldsmith, W., Dharan, C.K.H. (1992). Penetration of laminated Kevlar by projectiles – II. Analytical model. *Int. J. of Solids and Structures,* **29**(4): 421–36.

[Zook, 1977] Zook, J. (1977). An analytical model of kinetic energy projectile/fragment penetration. Report BRL-MR-2797. Army Ballistic Research Laboratory, Aberdeen Proving Ground, MD.

[Zukas, 1982] Zukas, J.A. (1982). Penetration and perforation of solids. In: Zukas J.A., Nicholas, T., Swift, H.F., Greszczuk, L.B., Curran, D.R., eds. *Impact Dynamics.* J. Wiley and Sons, New York, pp. 155–214.

[Zukas, 1990] Zukas, J.A. (1990) Numerical simulation of semi-infinite target penetration by continuous and segmented rods. Technical Report BRL-TR-3081. U.S. Army Ballistic Research Laboratory, Aberdeen Proving Ground, MD.

[Zukas, 1996] Zukas, J.A. (1996). Effect of lamination and spacing on finite thickness plate perforation. In: Jones, N., Brebbia, C.A., Watson, J.A., eds. *Structures Under Shock and Impact IV*. WIT Press, Southampton, pp. 103–115.

[Zukas and Scheffler, 2001] Zukas, J.A., Scheffler, D.R. (2001). Impact effects in multilayered plates. *Int. J. of Solids and Structures.* **38**(19): 3321–3328.

[Zukas and Walters, 1990] Zukas, J.A., Walters, W.P. (1990). Analytical models for kinetic energy penetration. In: Zukas J.A., ed. *High Velocity Impact Dynamics*. John Wiley and Sons, New York, pp. 405–441.

Author's Index